Industrial Biotechnology

Industrial Biotechnology

Edited by **Wendell Carter**

SYRAWOOD
PUBLISHING HOUSE
New York

Published by Syrawood Publishing House,
750 Third Avenue, 9th Floor,
New York, NY 10017, USA
www.syrawoodpublishinghouse.com

Industrial Biotechnology
Edited by Wendell Carter

International Standard Book Number: 978-1-68286-233-9(Hardback)

Printed in China.

Contents

Preface

Industrial biotechnology also referred to as white biotechnology, uses biological components or organisms to produce relevant products and raw materials for different industrial sectors like chemicals, paper, bioenergy, etc. Medical diagnostics, enzyme formation, food processing, industrial fermentation, biofuel production, etc. are some of the diverse topics covered in this book which aim to address the varied applications of biotechnology. For all those who are interested in the field of industrial biotechnology, this book can prove to be an essential guide that will provide comprehensive knowledge to the readers.

All of the data presented henceforth, was collaborated in the wake of recent advancements in the field. The aim of this book is to present the diversified developments from across the globe in a comprehensible manner. The opinions expressed in each chapter belong solely to the contributing authors. Their interpretations of the topics are the integral part of this book, which I have carefully compiled for a better understanding of the readers.

At the end, I would like to thank all those who dedicated their time and efforts for the successful completion of this book. I also wish to convey my gratitude towards my friends and family who supported me at every step.

Editor

1

Hydrolysis of *Agave fourcroydes* Lemaire (henequen) leaf juice and fermentation with *Kluyveromyces marxianus* for ethanol production

Pablo A Villegas-Silva, Tanit Toledano-Thompson, Blondy B Canto-Canché, Alfonso Larqué-Saavedra and Luis F Barahona-Pérez*

Abstract

Background: Carbon sources for biofuel production are wide-ranging and their availability depends on the climate and soil conditions of the land where the production chain is located. Henequen (*Agave fourcroydes* Lem.) is cultivated in Yucatán, Mexico to produce natural fibers from the leaves, and a juice containing fructans is produced during this process. Fructans can be hydrolyzed to fructose and glucose and metabolized into ethanol by appropriate yeasts. In Mexico, different *Agave* species provide the carbon source for (distilled and non-distilled) alcoholic beverage production using the stem of the plant, whilst the leaves are discarded. In this work, we investigated the effect of thermal acid and enzymatic hydrolysis of the juice on the amount of reducing sugars released. Growth curves were generated with the yeasts *Saccharomyces cerevisiae* and *Kluyveromyces marxianus* and fermentations were then carried out with *Kluyveromyces marxianus* to determine alcohol yields.

Results: With thermal acid hydrolysis, the greatest increase in reducing sugars (82.6%) was obtained using 5% H_2SO_4 at 100°C with a 30 min reaction time. Statistically similar results can be obtained using the same acid concentration at a lower temperature and with a shorter reaction time (60°C, 15 min), or by using 1% H_2SO_4 at 100°C with a 30 min reaction time. In the case of enzymatic hydrolysis, the use of 5.75, 11.47 and 22.82 U of enzyme did not produce significant differences in the increase in reducing sugars. Although both hydrolysis processes obtained similar results, the difference was observed after fermentation. Ethanol yields were 50.3 ± 4 and $80.04 \pm 5.29\%$ of the theoretical yield respectively.

Conclusions: Final reducing sugars concentrations obtained with both thermal acid and enzymatic hydrolysis were similar. *Saccharomyces cerevisiae*, a good ethanol producer, did not grow in the hydrolysates. Only *Kluyveromyces marxianus* was able to grow in them, giving a higher ethanol yield with the enzymatic hydrolysate. The leaves account for a non-negligible weight of the total agave plant biomass, so this work complements the knowledge already developed on agave fermentations by making it possible to produce ethanol from almost the entire plant (stem and leaves).

Keywords: Biofuel, Sugars, Oligofructans, Hydrolysis, Pretreatments

* Correspondence: barahona@cicy.mx
Centro de Investigación Científica de Yucatán AC, Calle 43 No. 130 Col. Chuburná de Hidalgo, Mérida, Yucatán 97200, Mexico

Background

The use of biofuels such as bioethanol, biodiesel and biogas in the transport sector offers a valuable alternative in order to minimize greenhouse gas emissions and fossil-fuel dependence, and is being developed in many countries. Bioethanol is currently produced at competitive prices from two main raw materials: sugar and starch [1]. However, in order to avoid the conflict between using land for food or energy production, the use of forestry and/or agricultural residues for bioethanol production is attracting the interest of scientists with a view to establishing economical and sustainable processes. The availability of local raw materials is therefore an important issue that has to be addressed when attempting to assess their viability. Carbon sources for bioethanol production are wide-ranging [2] and their availability depends on the climate and soil conditions of the land where the production chain is located. The north of the Yucatán peninsula, Mexico, consists mainly of semi-arid lands with poor soils that do not support the cultivation of sugar cane. Corn is cultivated on a small scale, but cannot be used for fuel purposes because this crop forms the basis of the traditional Mexican diet. Nevertheless, agaves can grow well in these soils due to the fact that their metabolism is adapted to arid conditions [3]. Although biomass productivity of henequen (*Agave fourcroydes* Lem.) is not as high as sugar cane or sweet sorghum, it is similar to a variety of other energy crops, as shown in Table 1 [4,5]. Henequen is cultivated in this region to produce natural fibers from the leaves, and a juice containing fructans is produced during this process [6]. These fructans can be hydrolyzed to fructose and glucose and metabolized into ethanol by appropriate yeasts [7]. In Mexico, different *Agave* species provide the carbon source for (distilled and non-distilled) alcoholic beverage production and both traditional and modern processes have been discussed at length in the literature [8]. They all use the stem of the plant as the source of sugars, whilst the leaves are discarded. The stems are cooked to hydrolyze the fructan chains into their monomers, and pressed to extract a fermentable sugar-rich juice. In the case of henequen, the older leaves at the base of the plant are harvested every 3-4 months and decorticated to obtain the fibers. During processing, water is used to wash the fibers, resulting in the dilution of the components present in the juice.

Table 1 Comparison of *Agave fourcroydes* L. biomass productivity with other energy crops

Crop	Dry biomass productivity ($t \cdot ha^{-1} \cdot year^{-1}$)
Agave fourcroydes	15
Agave tequilana	25
Sugarcane	50-67
Corn	17-18
Sugar beet	11-17
Sweet sorghum	24-47
Switchgrass	10-15
Miscanthus	10-13

Although producing ethanol from the juice (stem and/or leaves) of agaves is not a direct fermentation of simple sugars (as with sugar cane or sweet sorghum), it is not as difficult as producing it from lignocellulosic materials. There is no need for solid material pretreatments, but a hydrolysis step is needed because fermentable sugars are not directly available. Unlike fructans from chicory, which are mainly composed of fructose molecules joined by β(2-1) linkages, agave fructans contain an important amount of β(2-6) linkages that result in branched molecules [7,9,10]. This can affect, for example, the choice of enzyme for the hydrolysis of the fructan chains. Commercial enzyme preparations that contain endo- and exo-inulinases can be used to achieve good yields in agave fructans hydrolysis. Inulinases can mainly be produced by microorganisms such as yeast strains (*Candida* sp., *Sporotrichum* sp., *Pichia* sp., and *Kluyveromyces* sp.) or fungi (*Aspergillus* and *Penicillium* species) [11]. Thermal acid hydrolysis can also be used, but this process is associated with the production of unwanted by-products (hydroxymethyl furfural and fructose dianhydride) that can affect further biological steps or the purity of final products [12].

The juice from the leaves of the henequen plant has previously been studied for ethanol production [6]. In that work, the juice was diluted to simulate the concentrations obtained from the mills. The low sugar content was compensated for by the addition of molasses, another industrial residue. However, technical proposals exist to obtain pure juice from the mills, meaning that sugar concentrations could be higher and there would be no need to add other carbon sources. The use of a mixture of yeast strains, *Saccharomyces cerevisiae* and *Kluyveromyces marxianus*, helped to improve ethanol yields, although they were still low. The aim of this work was to investigate the effect of thermal acid and enzymatic hydrolysis of the juice on the amount of reducing sugars released from these processes. Temperature, time and sulfuric acid concentrations were varied for thermal acid hydrolysis. Temperature and enzyme concentration were varied for enzymatic hydrolysis. The treatment that gave the highest reducing sugars yields was used to continue the study. Growth curves were generated with the yeasts *Saccharomyces cerevisiae* and *Kluyveromyces marxianus* and fermentations were then carried out with *Kluyveromyces marxianus* to determine alcohol yields.

Results and discussion

Thermal acid hydrolysis

The juice obtained from henequen leaves had a soluble solids concentration of 10°Bx, which was similar to that obtained previously [6], and an initial reducing sugars concentration of 40.2 ± 0.64 g·L^{-1}. Time had little effect on thermal acid hydrolysis of the juice, as can be seen in the surface response graphs in Figure 1. The reducing sugars concentrations remained almost the same when reaction temperature (Figure 1a) or acid concentration (Figure 1b) were kept the same and only time was changed. Acid concentration showed a greater effect on reducing sugars release than reaction temperature within the ranges employed in this work. As expected, a lower release of reducing sugars was obtained when lower values were used for acid concentration and reaction temperature, as shown in Figure 2. Good yields were also obtained at low acid concentrations and high temperatures. Reducing sugars values after the thermo-acid treatments are shown in Table 2. The greatest increases in reducing sugars concentrations were obtained when 5% H_2SO_4 was added to the juice. The only experiment that produced a statistically similar increase in reducing sugars using 1% H_2SO_4 was treatment 7, carried out at 100°C for 30 min. Although it requires more energy and a longer reaction time, the amount of H_2SO_4 is 5 times lower and neutralization requires less NaOH. Treatment 7 conditions were used in further experiments in this work. The selection of parameters and their ranges depends not only on better results, but also on economic, environmental and practical factors [13]. For larger-scale production, procedures requiring smaller energy inputs are preferred, and even Treatment 4 conditions could be used without a detrimental effect on reducing sugars production. To what extent the use of less H_2SO_4 or lower temperature and reaction times produce greater savings is beyond the scope of this work, but should be studied. There are few reports on agave fructan hydrolysis, and most of them are on the stems of *Agave tequilana* W. because this is used for tequila production [9,10]. Waleckx et al. recently reported that 98% hydrolysis efficiency was obtained after 25.5 h of stem cooking [7]. However, it was possible to use shorter heating times in this work due to the liquid nature of the juice. Avila et al. [14] reported a hydrolysis efficiency of greater than

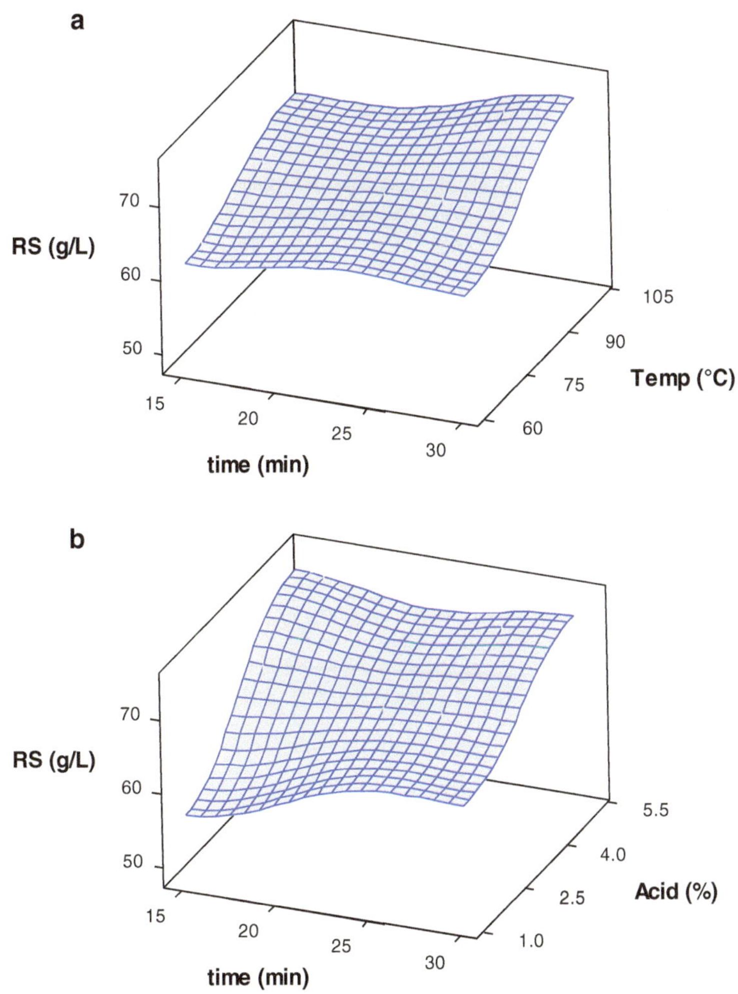

Figure 1 Response surfaces for the effect of a) reaction temperature and b) H_2SO_4 concentration on reducing sugars concentration.

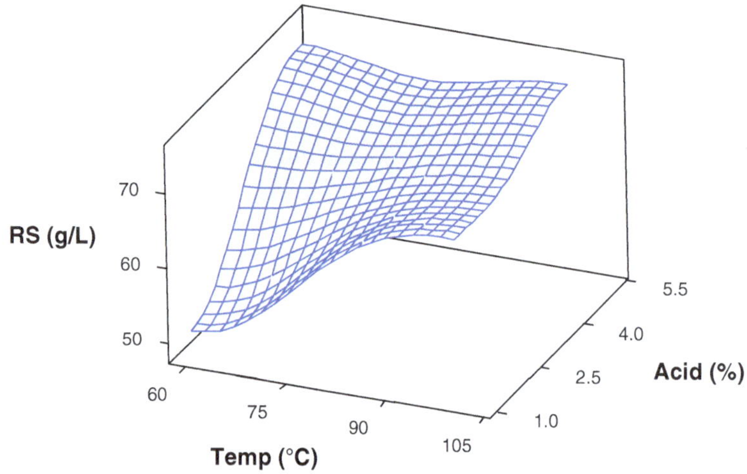

Figure 2 Response surface for the combined effect of reaction temperature and H_2SO_4 concentration on reducing sugars concentration.

90% at 60°C with a 10 minute reaction time after fructans from *A. tequilana* stems were extracted and then thermally hydrolyzed using 0.54 N HCl.

Enzymatic hydrolysis

The effect of temperature on reducing sugars concentration during enzymatic hydrolysis of henequen juice is shown in Figure 3. As can be seen, the maximum release of reducing sugars is obtained after a 30 min reaction at both temperatures (50 and 60°C) using 22.82 U of enzyme as indicated by the manufacturer. A slightly higher concentration of reducing sugars was obtained at 60°C. The reference (henequen juice without enzyme) was also conducted at 50 and 60°C and the results were the same for both temperatures, showing that no increase in reducing sugars concentrations occurred without enzyme. It has been reported [11] that inulinase from the marine yeast *Pichia guilliermondii* is most active at 60°C. However, Ricca et al. [15] reported that higher inulinase activity is obtained at 60°C, but 30% activity loss occurred after 5 h. This must be considered when scaling-up the process, as procedures are longer than at laboratory level, meaning that lower temperatures may give better results. In order to obtain the maximum concentration of reducing sugars, experiments on enzymatic hydrolysis of henequen juice were conducted at 60°C in this study.

To determine the effect of enzyme concentration on reducing sugars release, lower concentrations than 22.82 U were tested. As can be seen in Figure 4, hydrolysis of the fructans present in the juice occurs during the first 10 min of the reaction. This experiment also showed that the use of 4 times less enzyme (5.75 U) is enough to attain the same levels of reducing sugars release as using 22.82 U. This suggests that there is no saturation of the enzyme at lower levels. There were no significant differences in the increase in reducing sugars using 5.75, 11.47 and 22.82 U of enzyme. The reference was henequen juice incubated under the same conditions without enzyme; in

Table 2 Increase in reducing sugars after thermal acid hydrolysis of henequen juice

Treatment	Tempeature (°C)	Heating time (min)	H_2SO_4 conc. (%v/v)	Reducing sugars[2] (g L^{-1})	Reducing sugars increase (%)
Reference[1]	-	-	-	40.25 ± 0.64[a]	-
1	60	15	1	48.84 ± 2.58[b]	21.3
2	60	15	5	74.59 ± 1.05[c]	85.3
3	60	30	1	53.55 ± 1.83[d]	33
4	60	30	5	73.36 ± 1.7[c]	82.3
5	100	15	1	64.38 ± 2.3[e]	60
6	100	15	5	74.01 ± 1.49[c]	83.9
7	100	30	1	74.4 ± 3.29[c]	84.8
8	100	30	5	74.95 ± 1.41[c]	86.2

[1]Raw henequen leaf juice was used as a reference.
[2]Different letters indicate statistically significant differences.

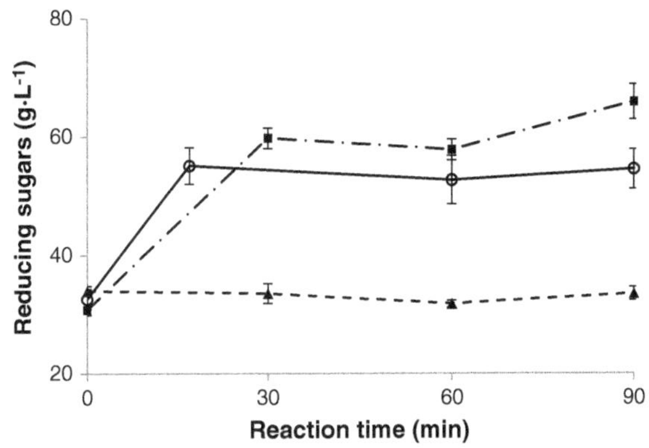

Figure 3 Effect of temperature on enzymatic hydrolysis of henequen juice. (■) 60°C; (○) 50°C; (▲) without enzyme at both 50°C and 60°C.

this case, no change in reducing sugars concentrations was observed. The concentrations of reducing sugars obtained by enzymatic hydrolysis are similar to those obtained by thermal acid hydrolysis.

Fermentation

A growth curve was first generated for the yeast *Saccharomyces cerevisiae* with both thermo-acid and enzymatic henequen juice hydrolysates. The yeast did not grow in either of the hydrolysates. This behavior could be due to the presence of inhibitory compounds in the henequen juice. It is well documented that agave plants produce saponins with inhibitory activity [16,17]. To corroborate this, Cira et al. [18] produced a modified strain of *S. cerevisiae* which resists the saponins present in the musts of *Agave tequilana* Weber var. azul and *Agave salmiana*, and succeeded in fermenting them into ethanol. Saponins can be hydrolyzed to their corresponding sapogenins, but higher acid concentrations and longer reaction times than the ones used in this work are needed [19]. This lack of performance has been reported in the literature [20], where lower ethanol yields were obtained with *Saccaromyces cerevisiae* (70-76%) compared to a *Kluyveromyces marxianus* strain (92-96%) when fermenting *Agave tequilana* musts. On the contrary, the yeast *Kluyveromyces marxianus* was able to grow in the henequen juice hydrolysates, as shown in Figure 5. The yeast could also grow in raw juice (the juice was only heated to 80°C and cooled immediately to lower microbial contamination). This can be explained by the fact that the strain used in this work was isolated from the henequen plant and is adapted to the metabolites present in the juice or has the enzymes to hydrolyze them. In a spontaneous fermentation of *A. fourcroydes* stem must, a great diversity of yeasts was present and *K. marxianus* and *S. cerevisiae* were the dominant species by the end of the fermentation [Dr. Patricia Lappe, personal communications]. *K. marxianus* presented a lag phase of approximately 6 hours. In Figure 5, it can be seen that the yeast performed better in the enzymatic hydrolysate and in the raw juice than in the thermal acid hydrolysate at the beginning of the growth curves. It has been reported that materials subjected to acid hydrolysis are harder to ferment [21]. After 18 hours of growth, cell concentrations were the same in both hydrolysates.

Alcoholic fermentations were carried out with enzyme and thermal acid hydrolyzed henequen juice. Reducing sugars were consumed during the first 24 hours of fermentation as shown in Figure 6. After 48 hours of fermentation, ethanol concentrations were 13.74 ± 0.99 $g \cdot L^{-1}$ for the thermal acid hydrolyzed juice and 16.5 ± 0.56 $g \cdot L^{-1}$ for the enzyme-hydrolyzed juice. As shown in Figure 7a, the difference between the two results was found to be statistically significant. Based on the amount of reducing sugars consumed during the fermentations, ethanol yields of 50.3 ± 4 and $80.04 \pm 5.29\%$ were obtained respectively. As show in Figure 7b, the difference between the two results was

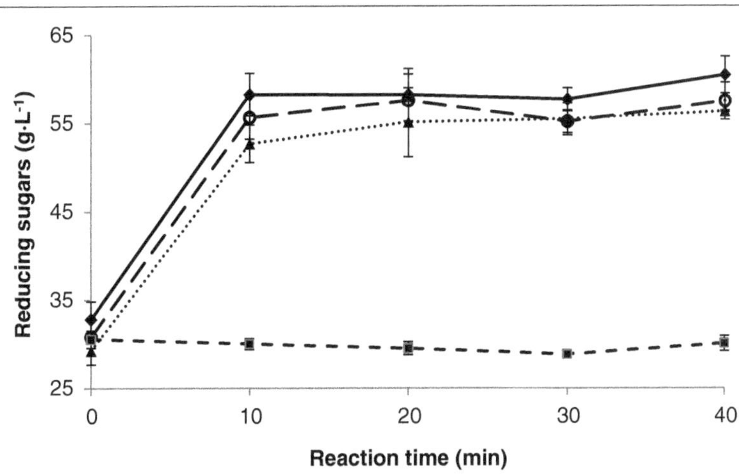

Figure 4 Effect of enzyme concentration on enzymatic hydrolysis of henequen juice. (♦) 22.82 U; (○) 11.47 U; (▲) 5.75 U; (■) without enzyme.

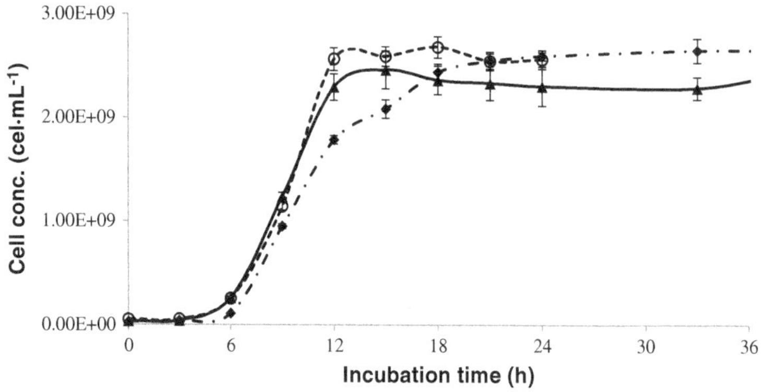

Figure 5 Growth curves of *Kluyveromyces marxianus*. Enzymatic (▲), thermal acid (♦) and non-hydrolyzed (○) henequen juice hydrolysates.

statistically significant. The yeast was able to consume the reducing sugars, but did not efficiently transform them into ethanol in the case of the thermal acid hydrolysate. Ethanol yields depend on fermentation conditions. When inulinase-hydrolyzed artichoke tubers were fermented using a strain of *S. cerevisiae*, yields of 80-84% of the theoretical value were obtained [22] which are similar to those obtained in this work. This indicates that *S. cerevisiae* can assimilate the sugars released by an enzymatic hydrolysis of oligofructans, so henequen leaves contain compounds that inhibit growth of this yeast. Ethanol yields of up to 90% of the theoretical value were obtained when *S. cerevisiae* was used to ferment YM broth with nitrogen supplementation [23]. In another study carried out in our laboratories, the same *S. cerevisiae* strain used in this work gave an ethanol yield of 98% of the theoretical value when a thermal acid hydrolysate of jatropha kernel fruit was fermented. The yeast *S. cerevisiae* is known for its higher tolerance to ethanol concentrations [24], whilst *K. marxianus* is known for its thermotolerance [25,26].

The selection of one yeast over the other depends on the nature of the must and fermentation conditions. The use of a mixture of both yeasts can even be envisaged [6].

An ethanol production curve was generated using an enzymatic hydrolysate. This juice was obtained during the dry season and was more concentrated. It has been documented that the juice of henequen stems is more concentrated in the dry season [27] and we found the same effect with the leaf juice. It was decided not to dilute the juice to find out if there was any effect on the hydrolysis and fermentation processes. A higher concentration of reducing sugars was obtained after enzymatic hydrolysis, indicating that the amount of enzyme chosen previously can perform well with any juice concentration obtained during the year. Reducing sugars uptake and ethanol production are shown in Figure 8. It can be seen that after 30 hours of fermentation, sugar consumption is already complete with no statistically significant differences in ethanol concentrations until 48 h. An ethanol yield of 77 ± 5.86% of the theoretical yield was obtained.

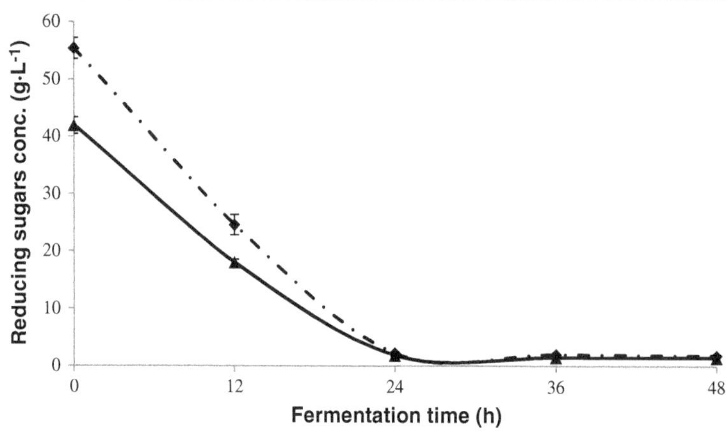

Figure 6 Reducing sugars consumption during fermentation of henequen juice hydrolysates. Enzymatic (▲) and thermal acid (♦) hydrolysates.

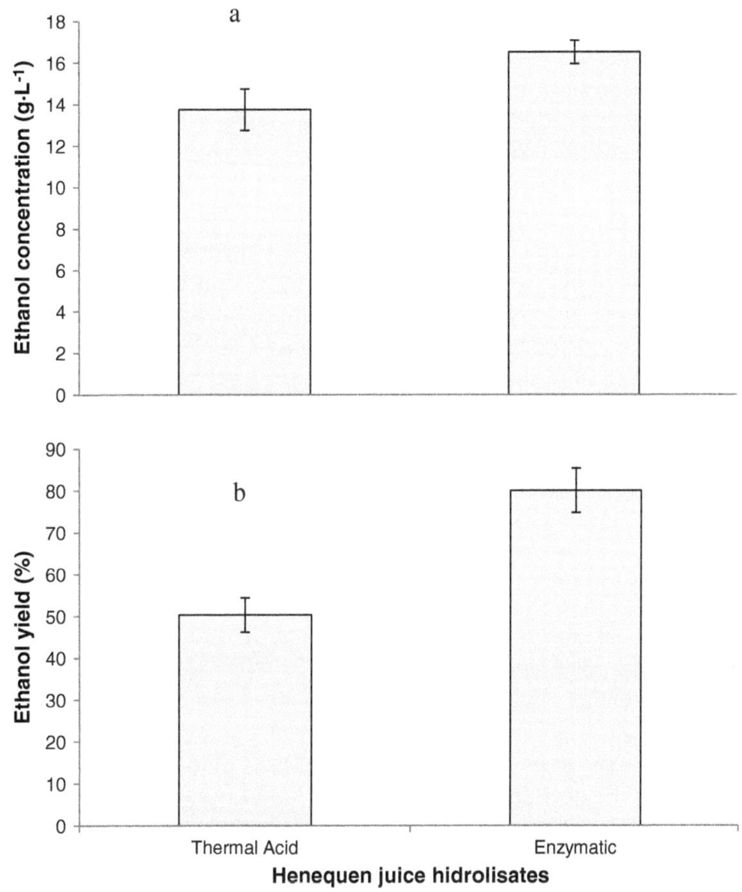

Figure 7 Comparison of a) ethanol concentrations and b) ethanol yields obtained after fermentation of thermal acid and enzymatic hydrolysates of henequen juice.

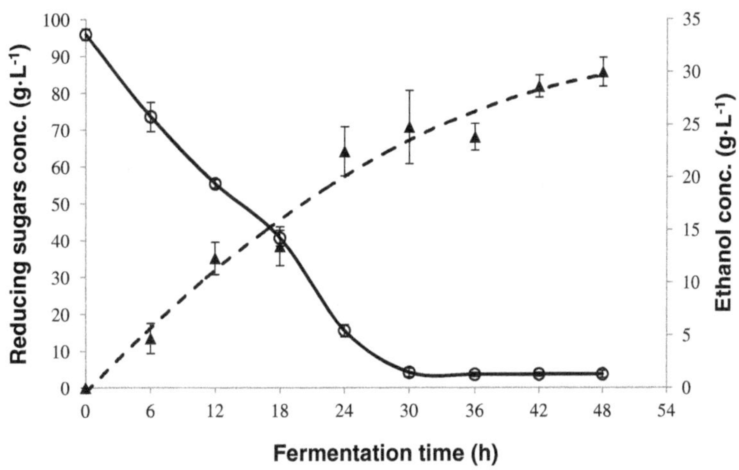

Figure 8 Reducing sugars consumption (○) and ethanol production (▲) during fermentation of an enzymatic hydrolysate of henequen juice.

With the higher yields obtained in this work, it is possible to produce 0.038 L of ethanol per liter of leaf juice. Taking into account that a henequen leaf contains 0.78-0.89 L of juice and approximately 250 million leaves are processed per year, 7.125 million liters of ethanol could therefore be produced with this residue. As for production costs, they will be higher than for processes employing direct fermentation of simple sugars, such as sugar cane or sweet sorghum juice, due to the inclusion of the enzymatic hydrolysis step. The amount of enzyme used, however, is minimal. Production costs will be lower than for solid lignocellulosic residues, because there is no pre-treatment step before saccharification. It is easier and cheaper to process liquid materials.

To our knowledge, no data has previously been reported on hydrolyzed agave leaf juice fermentations to produce ethanol.

Conclusions

Final reducing sugars concentrations obtained with both thermal acid and enzymatic hydrolysis were similar, but enzymatic hydrolysis gave better ethanol yields after fermentation. *Saccharomyces cerevisiae*, a good ethanol producer and one used in agave stem must fermentations, could not grow in either hydrolysate. The development of strains resistant to inhibitory compounds present in agave leaves is important for commercial exploitation of this part of the plant. Fermentation of thermal acid and enzymatic hydrolysates with *Kluyveromyces marxianus* resulted in theoretical ethanol yields of 50.3 ± 4 and 80.04 ± 5.29% respectively. This yeast is capable of producing inulinase, but attempts to ferment the leaf juice without hydrolysis failed, meaning that preliminary hydrolysis of the fructans present in the juice appears to be necessary. *K. marxianus* can also be manipulated in order to attain higher ethanol yields. These studies, together with the isolation of new strains from henequen leaf juice, are being carried out by our group. This work complements the knowledge already developed on agave juice fermentation. Agave leaves (*fourcroydes* or *tequilana*) account for a non-negligible weight of total plant biomass and comprise an important carbohydrate source that can be used to produce ethanol as biofuel.

Methods

Reagents

All reagents were analytical grade. Inulinase enzyme (Sigma-Aldrich) was obtained from *Aspergillus niger* with a declared activity of 2000 $U \cdot mL^{-1}$ and a density of 1.13 $g \cdot cm^{-3}$. One enzymatic unit is defined as the amount of enzyme necessary to release one g of reducing sugars per minute at 50°C.

Plant material

Henequen leaves were obtained from healthy plants grown in the gardens of Centro de Investigación Científica de Yucatán AC. They were rinsed with tap water and passed through a three-roller mill to extract the juice, which was filtered to eliminate solid leaf residues and stored at -20°C until use.

Thermal acid hydrolysis

The variables considered important in this process were temperature, heating time and H_2SO_4 concentration. A 3^2 full factorial design leading to eight sets of experiments was used to determine the effect of the above mentioned variables on the release of reducing sugars. The levels for each variable were determined on the basis of previous experience with henequen juice hydrolysis and were decided as follows: temperature = 60 and 100°C, heating time = 15 and 30 min, H_2SO_4 concentration = 1 and 5% (v/v). These levels were decided not only based on reaction system characteristics, but also based on economic and practical factors. Thermal acid hydrolysis experiments were carried out in 250 mL Erlenmeyer flasks with 80 mL of henequen leaf juice. Fresh juice was used as a reference for the initial reducing sugars concentration.

Enzymatic hydrolysis

To study enzymatic hydrolysis, a more classical (one variable at a time) approach was used due to the lack of previous data. Conditions of preliminary assays were based on enzyme data provided by the manufacturer. They were carried out in 25 mL Erlenmeyer flasks with 5 mL of raw henequen juice at pH 4.5. Two temperatures, 55 and 60°C, were tested using 22.82 U of enzyme and 90 min incubation time. Subsequently, three enzyme concentrations were tested at 60°C: 5.75, 11.47 and 22.82 U of enzyme and 40 min incubation time.

For fermentation assays, enzymatic hydrolysis was carried out in 250 mL Erlenmeyer flasks with 180 mL of raw henequen juice. An enzyme concentration of 5.75 U was used. Incubation was carried out at 60°C for 30 min and pH 4.5. All enzymatic hydrolysis incubations were performed at 150 rpm.

Yeast strains

Kluyveromyces marxianus was isolated from the base of henequen leaves following classical isolation procedures. It was characterized by phenotypical and molecular tests [28]. Commercial *Saccharomyces cerevisiae* was obtained from Safmex S.A. de C.V. (Mexico). Both yeasts were maintained in Petri dishes on YPGA medium containing glucose (20 $g \cdot L^{-1}$), yeast extract (5 $g \cdot L^{-1}$), peptone (10 $g \cdot L^{-1}$) and agar (20 $g \cdot L^{-1}$), and incubated at 30 ± 2°C in darkness. The cultures were collected in sterile water and kept at 4°C until use. Viability of the cell suspensions was measured by staining with methylene blue. Cell suspension viabilities ≥ 95% of both strains were used in all experiments.

Growth curves

Growth curves were generated in 250 mL Erlenmeyer flasks with 100 mL of thermal acid or enzyme hydrolyzed henequen juice adjusted to pH 4.5. Prior to hydrolysis, soluble solids of the raw henequen juice were adjusted to 6°Bx. Ammonium sulfate (1.5 $g{\cdot}L^{-1}$) was used as the nitrogen source [23] and a cell concentration of $3{\times}10^7$ $cell{\cdot}mL^{-1}$ was employed in all cases. The flasks were incubated at 30 ± 2°C with 150 rpm agitation for 33 h. Samples were harvested every 3 h. Non-hydrolyzed henequen juice was used as a reference. In this case, to avoid contamination, the juice was heated to 80°C and immediately cooled to room temperature after adjusting soluble solids to 6°Bx.

Fermentations

Inocula for fermentations were prepared with thermal acid or enzymatically hydrolyzed henequen juice with a soluble solids concentration of 6°Bx and an initial cell concentration of $3{\times}10^7$ $cell{\cdot}mL^{-1}$. pH was adjusted to 4.5 and 1.5 $g{\cdot}L^{-1}$ ammonium sulfate was added. Growth was maintained at 30 ± 2°C for 18 hours. The amount of the inoculum was 10% of total fermentation volume. Fermentations were carried out in 250 mL Erlenmeyer flasks with 180 mL of non-diluted thermal acid or enzymatically hydrolyzed henequen juice. pH was adjusted to 4.5 and 1.5 $g{\cdot}L^{-1}$ ammonium sulfate was added. After inoculation, the flasks were kept at 30 ± 2°C for 48 hours without agitation.

Analytical methods

Soluble solids concentration was measured with a portable refractometer (Cole-Parmer FG103/113) and expressed as °Bx. Reducing sugars concentrations were determined in the hydrolysates and during fermentations by the 3, 5-dinitrosalicilic acid method [29]. The absorbance of the samples was read at 550 nm. For ethanol quantification, 25 mL of fermented sample was diluted with 25 mL of distilled water and distilled at 100°C until 25 mL of distillate was recovered. Ethanol concentration was determined by the dichromate method [30]. The absorbance of the samples was read at 585 nm.

Statistical analysis

All experiments were performed in triplicate (n = 3). For statistical analysis, the Statgraphics Centurion XV (Statpoint technologies Inc., Warrenton, VA, USA) software was employed. Significant differences among treatments were determined using the Tukey HSD test with $p \leq 0.05$. For response surface analysis, the MINITAB 16 (Minitab Inc., State College, PA, USA) software was used.

Competing interests

The authors declare that they have no competing interests.

Authors' contributions

PAVS carried out the experimental work and participated in the design of the study. TTT participated in the reducing sugar and statistical analysis. BBCC participated in the design and analysis of the enzymatic hydrolysis experiments. ALS participated in the design of the study and commented on the manuscript. LFBP conceived the study, participated in its design and drafted the manuscript. All authors read and approved the final manuscript.

Authors' information

PAVS received his M.Sc. in Renewable Energy working on this project. TTT is a laboratory technician at CICY working in the Renewable Energy Department. BBCC is head of the Biotechnology Department at CICY. Her research interests are on enzymatic hydrolysis. ALS is a researcher in the Natural Resources Department at CICY. His main area of expertise is Biomass and Productivity. LFBP is head of the Renewable Energy Department. His research area involves biofuel production.

Acknowledgements

The authors would like to thank Gerardo Rivera for his advice and María Guadalupe Serrano for her support during the experimental work.

References

1. Hahn-Hägerdal B, Galbe M, Gorwa-Grauslund MF, Lidén G, Zacchi G: **Bio-ethanol – the fuel of tomorrow from the residues of today.** *Trends Biotechnol* 2006, **24**(12):549–556.
2. Lin Y, Tanaka S: **Ethanol fermentation from biomass resources: current state and prospects.** *Appl Microbiol Biotechnol* 2006, **69**:627–642.
3. García-Moya E, Romero-Manzanares A, Nobel PS: **Highlights for *Agave* productivity.** *Glob Change Biol Bioenergy* 2011, **3**:4–14.
4. Escamilla-Treviño LL: **Potential of plants from the genus *Agave* as bioenergy crops.** *Bioenerg Res* 2012, **5**:1–9.
5. Günnur K, Nilgün C: **An overview of biofuels from energy crops: current status and future prospects.** *Renew Sust Energ Rev* 2013, **28**:900–916.
6. Cáceres-Farfán M, Lappe P, Larqué-Saavedra A, Magdub-Méndez A, Barahona-Pérez L: **Ethanol production from henequen (*Agave fourcroydes* Lem.) juice and molasses by a mixture of two yeasts.** *Bioresour Technol* 2008, **99**:9036–9039.
7. Waleckx E, Gschaedler A, Colonna-Ceccaldi B, Monsan P: **Hydrolysis of fructans from *Agave tequilana* Weber var. azul during the cooking step in a traditional tequila elaboration process.** *Food Chem* 2008, **108**:40–48.
8. Lappe-Oliveras P, Moreno-Terrazas R, Arrizón-Gaviño J, Herrera-Suárez T, García-Mendoza A, Gschaedler-Mathis A: **Yeasts associated with the production of Mexican alcoholic nondistilled and distilled Agave beverages.** *FEMS Yeast Res* 2008, **6**:1037–1052.
9. López MG, Mancilla-Margalli NA, Mendoza-Díaz G: **Molecular structures of fructans from *Agave tequilana* Weber var. azul.** *J Agric Food Chem* 2003, **51**(27):7835–7840.
10. Mancilla-Margalli NA, López MG: **Generation of Maillard compounds from inulin during the thermal processing of *Agave tequilana* Weber Var. azul.** *J Agric Food Chem* 2002, **50**(4):806–812.
11. Gong F, Zhang T, Chi Z, Sheng J, Li J, Wang X: **Purification and characterization of extracellular inulinase from a marine yeast *Pichia guilliermondii* and inulin hydrolysis by the purified inulinase.** *Biotechnol Bioprocess Eng* 2008, **13**:533–539.
12. Singh RS, Dhaliwal R, Puri M: **Production of high fructose syrup from asparagus inulin using immobilized exoinulinase from *Kluyveromyces marxianus* YS-1.** *J Ind Microbiol Biotechnol* 2007, **34**:649–655.
13. Dequan Zhou D, Xu X, Mu H, Høy CE, Adler-Nissen J: **Synthesis of structured triacylglycerols containing caproic acid by lipase-catalyzed acidolysis: optimization by response surface methodology.** *J Agric Food Chem* 2001, **49**:5771–5777.

14. Avila-Fernández A, Galicia-Lagunas N, Rodríguez-Alegría ME, Olvera C, López-Munguía A: **Production of functional oligosaccharides through limited acid hydrolysis of agave fructans.** *Food Chem* 2011, **129**:380–386.
15. Ricca E, Calabrò V, Curcio S, Iorio G: **Optimization of inulin hydrolysis by inulinase accounting for enzyme time- and temperature-dependent deactivation.** *Biochem Eng J* 2009, **48**:81–86.
16. Ohtsuki T, Koyano T, Kowithayakorn T, Sakai S, Kawahara N, Goda Y, Yamaguchi N, Ishibashi M: **New chlorogenin hexasaccharide isolated from *Agave fourcroydes* with cytotoxic and cell cycle inhibitory activities.** *Bioorg Med Chem* 2004, **12**(14):3841–3845.
17. Yokosuka A, Mimaki Y: **Steroidal saponins from the whole plants of *Agave utahensis* and their cytotoxic activity.** *Phytochemistry* 2009, **70**:807–815.
18. Cira L, González G, Torres J, Pelayo C, Gutiérrez M, Ramírez J: **Heterologous expression of *Fusarium oxysporum* tomatinase in *Saccharomyces cerevisiae* increases its resistance to saponins and improves ethanol production during the fermentation of *Agave tequilana* Weber var. azul and *Agave salmiana* must.** *Antonie Van Leeuwenhoek* 2008, **93**:259–266.
19. Eskander J, Lavaud C: **Steroidal saponins from the leaves of *Agave macroacantha*.** *Fitoterapia* 2010, **81**:371–374.
20. López-Alvarez A, Díaz-Pérez AL, Sosa-Aguirre C, Macías-Rodríguez L, Campos-García J: **Ethanol yield and volatile compound content in fermentation of agave must by *Kluyveromyces marxianus* UMPe-1 comparing with *Saccharomyces cerevisiae* baker's yeast used in tequila production.** *J Biosci Bioeng* 2012, **113**(5):614–618.
21. Galbe M, Zacchi G: **Pretreatment of lignocellulosic materials for efficient bioethanol production.** *Adv Biochem Eng Biotechnol* 2007, **108**:41–65.
22. Szambelan K, Nowak J, Czarnecki Z: **Use of *Zymomonas mobilis* and *Saccharomyces cerevisiae* mixed with *Kluyveromyces fragilis* for improved ethanol production from Jerusalem artichoke tubers.** *Biotechnol Lett* 2004, **26**:845–848.
23. Breisha GZ: **Production of 16% ethanol from 35% sucrose.** *Biomass Bioenergy* 2010, **34**:1243–1249.
24. Bai FW, Anderson WA, Moo-Young M: **Ethanol fermentation technologies from sugar and starch feedstocks.** *Biotechnol Adv* 2008, **26**:89–105.
25. Hacking AJ, Taylor IWF, Hanas CM: **Selection of yeast able to produce ethanol from glucose at 40°C.** *Appl Microbiol Biotechnol* 1984, **19**:361–363.
26. Banat IM, Nigam P, Singh D, Marchant R, McHale AP: **Review: ethanol production at elevated temperatures and alcohol concentrations: part I - yeasts in general.** *World J Microbiol Biotechnol* 1998, **14**:809–821.
27. Rendón-Salcido LA, Colunga-García P, Barahona-Pérez LF, Pimienta-Barrios E, Magdub-Méndez A, Larqué-Saavedra A: **Sugars and alcoholic byproducts from henequen (Agave fourcroydes) as influenced by plant age and climate.** *Rev Fitotec Mex* 2009, **32**(1):39–44.
28. Pérez-Brito D, Tapia-Tussell R, Quijano-Ramayo A, Larqué-Saavedra A, Lappe P: **Molecular characterization of *Kluyveromyces marxianus* strains isolated from *Agave fourcroydes* Lem. in Yucatan, Mexico.** *Mol Biotechnol* 2007, **37**:181–186.
29. Miller GL: **Use of dinitrosalicylic acid reagent for determination of reducing sugar.** *Anal Chem* 1959, **31**:426–428.
30. Williams MB, Reese HD: **Colorimetric determination of ethyl alcohol.** *Anal Chem* 1950, **22**:1556–1561.

2

Simultaneous multi-analyte urinary protein assay for bladder cancer detection

Charles J Rosser[1,2,3*], Yunfeng Dai[4], Makito Miyake[1], Ge Zhang[1] and Steve Goodison[1,2,5]

Abstract

Background: The ability to accurately measure multiple proteins simultaneously in a single assay has the potential to markedly improve the efficiency of a myriad of clinical assays. Here, we tested the performance of a new, multiplex protein array platform to quantitate three bladder cancer-associated proteins in urine samples. The following analytes, interleukin 8 (IL8), matrix metallopeptidase 9 (MMP9), and vascular endothelial growth factor A (VEGFA) were monitored using Q-plex, a customized multiplex ELISA system from Quansys Biosciences, and individual target commercial ELISA kits. The performance of the two approaches was compared by evaluating the diagnostic accuracy of the biomarker assays in samples from a cohort of 73 subjects of known bladder cancer status.

Results: The combination biomarker panel analyses revealed an AUROC value of 0.9476 for the Q-plex assay, and 0.9119 for the combination of the single-target ELISA assays. The Q-plex assay achieved an overall diagnostic sensitivity of 0.93 and specificity of 0.81, and the individual target ELISA assays achieved an overall sensitivity of 0.77 and specificity of 0.91.

Conclusion: Based on these encouraging preliminary data, we believe that the Q-Plex technology is a viable platform that can be exploited as an efficient, highly accurate tool to quantitate multiplex panels of diagnostic proteins in biologic specimens.

Keywords: Biomarkers, Bladder cancer, Q-plex, Protein, Urine

Practical applications

While there continues to be an increasing number of biomarker discovery studies published, very few new cancer biomarkers have entered the clinic over the past 30 years. One reason for this has been the reliance on assays of single biomarkers for the evaluation of cancers that have a broad spectrum of molecular changes. Coupled with the observed variation between individuals and the heterogeneity within tumors, there has been a necessary shift to molecular signatures comprised of multiple biomarkers. Technologies that can efficiently, simultaneously monitor molecular signatures would be pivotal in moving promising biomarker panels towards clinical utility. Quansys Biosciences have developed the Q-Plex, a multiplex ELISA system that prints capture antibodies as an array in a single well. In this study, we showed that this system could simultaneously and accurately monitor three target proteins in urine samples. Multiplexed ELISAs such as the Q-plex system can maximize data generation from each sample, and provide a low cost per data point. The further optimization and development of robust Q-plex assays in complex biological fluids will allow this technology to become more widely adopted.

Background

The advent of advanced molecular profiling techniques has enabled the derivation of molecular signatures that hold promise for more accurate, even individualized patient evaluation [1]. A number of molecular signature assays are now being incorporated into clinical practice [2,3], but the assays employed to monitor multiple targets per sample are, to date, rather complex and thus, expensive, and often require centralized processing and analysis. We have previously employed a range of proteomic [4,5] and genomic [6,7] technologies to profile

* Correspondence: crosser@cc.hawaii.edu
[1]MD Anderson Cancer Center Orlando, Orlando, FL, USA
[2]Nonagen BioScience Corp, Jacksonville, FL, USA
Full list of author information is available at the end of the article

naturally voided urine samples with the aim of identifying disease-associated biomarkers that could be developed for the non-invasive detection of bladder cancer.

The objective of the current feasibility study was to test the ability of a customized multiplex assay system to accurately and simultaneously monitor three urinary protein biomarkers from one of our validated bladder cancer signatures [8]. The levels of the three proteins, interleukin 8 (IL8), matrix metallopeptidase 9 (MMP9), and vascular endothelial growth factor A (VEGFA), were quantitated in real-world urine samples and the diagnostic performance of the multiplex system was compared with analyses performed using three individual target commercial ELISA kits.

The multiplex assay used Q-Plex™ technology (Quansys BioScience, Logan, UT), which involves the micro-spotting of individual groups of capture antibody in either a cartesian or polar coordinate system on the bottom of a 96 well plate with each spot being its own 'micro ELISA' assay [9]. Micro-spotted systems have the advantage of higher assay sensitivities and faster reaction kinetics due to minimizing diffusion constraints for analyte/antibody binding [10]. Standard ELISA incubation steps such as initial sample incubation, washing, secondary antibody incubation, washing, incubation are employed, but the labeling and reporting system used in the Q-Plex Array™ is chemiluminescent. Chemiluminescent ELISAs have been shown to be more sensitive than colorimetric detection systems [11,12].

In our study, the multiplex ELISA assay was shown to be highly sensitive and to be consistently accurate across wells and plates. Most importantly, the assay achieved comparable diagnostic performance with respect to bladder cancer detection as the individual target commercial ELISA assays [13-17]. The system can be applied to a range of biological biomarker monitoring applications, and we are working to expand the multiplex platform to provide a high-throughput approach to quantitatively detect more comprehensive bladder cancer-associated diagnostic protein signatures in voided urine samples.

Results and discussion

Q-plex assay characterization

The physical components, a library of capture antibodies, and the secondary reagents of the Q-Plex™ system have undergone extensive optimization for consistent implementation [18,19]. Ranges for each analyte assay were evaluated by dilution of standards to determine upper ranges where high-end hook effect and apparent antibody saturation are avoided and lower ranges that are above detection limits. Lower limits of detection (LLD) were calculated based on 2× the standard deviation of the background of 20 negative wells. LLD for IL8, MMP9 and VEGFA were 0.4 pg/ml, 297 pg/ml and 0.5 pg/ml, respectively, demonstrating sufficient sensitivity to detect proteins present in small amounts in voided urine samples. The LLD data for the individual ELISA kits (1.02 pg/ml for IL8, 66.1 pg/ml for MMP9, and 23.7 pg/ml for VEGFA) were comparable in range for the most part. Any differences are presumably due to the antigen binding characteristics of the proprietary antibodies. Intra assay precision was measured with acceptance criteria of a coefficient of variation (%CV) of less than 15.0. Median inter assay variability across all plates was also determined to be less than 15% CV for each analyte. As the technology is an array, all components were checked for cross reactivity with other components in the antigen and antibody cocktails and confirmed to have less than 0.5% cross reactivity. Lower limits of quantification (LLOQ) were determined to be the lowest point of the 10-point positive standard curve where the back-fit regression values were within 20% of the known value. The minimal LLOQ, maximum and average LLOQ for IL8, MMP9, and VEGFA were 0.3 pg/ml, 0.7 pg/ml, and 0.7 pg/ml; 35373 pg/ml, 52433 pg/ml, and 40782 pg/ml; and 19 pg/ml, 613 pg/ml, and 419 pg/ml, respectively. Only three of the 73 urine samples subsequently tested (two for IL8 and one for VEGFA) fell outside the standard curve and required extrapolation.

Urine sample analysis

After confirming the robustness of the multiplex assay for these specific biomarkers, it was used to test a voided urine sample set. Seventy-three clinical urine samples from MD Anderson Cancer Center Orlando were made available for analysis. Demographics and disease characteristics of the entire cohort are summarized in Table 1. The ability of each of the test biomarkers within the multiplex array to predict the presence of bladder cancer

Table 1 Demographic and clinicopathologic characteristics of 73 subjects comprising study cohort

	Bladder cancer	Benign controls
	n = 31	n = 42
Median Age (range, y)	71 (54–93)	62 (24–81)
Male : Female ratio	26:5	35:7
Suspicious/positive cytology	32%	0%
Clinical stage		
Tis	3	N/A
Ta	12	N/A
T1	2	N/A
≥T2	14	N/A
Tumor grade		
Low	4	N/A
High	27	N/A

was analyzed using nonparametric ROC analyses, according to National Cancer Institute guidelines [20]. Urinary IL8 was the most accurate single biomarker for bladder cancer detection with an AUROC of 0.907 (95% CI: 0.830 - 0.985), a sensitivity of 90%, specificity of 86%, PPV of 82% and NPV of 92%. Urinary MMP9 data generated an AUROC of 0.533 (95% CI: 0.392 - 0.674), sensitivity of 45%, specificity of 76%, PPV of 58% and NPV of 65%, and VEGFA as an individual analyte was noted to have an AUROC of 0.524 (95% CI: 0.386 - 0.661), sensitivity of 17%, specificity of 95%, PPV of 71% and NPV of 61%. Through combinatorial analysis of all 3 biomarkers using optimal cutoff values defined by Youden index calculations, the AUROC (Figure 1) for the diagnostic panel using the multiplex array was 0.9476 [95% CI: 0.903 - 0.992]. The combination assay achieved an overall sensitivity of 0.93, specificity of 0.81, PPV of 0.78 and NPV of 0.94 for bladder cancer classification.

Spearman Correlation coefficient was not lower than 0.417 for any biomarker, thus comparison of the results of the Q-Plex™ assay to those of the commercial ELISA kits directed towards IL8, MMP9 and VEGFA were encouraging (Tables 2 & 3). The ability of each of the test biomarkers of the individual commercial ELISA assays to predict the presence of bladder cancer was analyzed using nonparametric ROC analyses. As with the multiplex assay, urinary IL8 was the most accurate single biomarker when monitored by individual ELISA with an AUROC of 0.878 (95% CI: 0.783 - 0.972), a sensitivity of 87%, specificity of 86%, PPV of 82% and NPV of 90%. Urinary MMP9 data generated an AUROC of 0.548 (95% CI: 0.409 - 0.688), sensitivity of 45%, specificity of 76%, PPV of 58% and NPV of 65%, while VEGFA was noted to have an AUROC of 0.493 (95% CI: 0.356 - 0.630), sensitivity of 16%, specificity of 95%, PPV of 71% and NPV of 61%. Through combination of data from all 3 biomarkers, the AUROC for the diagnostic panel from the ELISA assays was 0.9119 [95% CI: 0.796 - 0.968], with an overall sensitivity of 0.77, specificity of 0.91, PPV of 0.86 and NPV of 0.84 for bladder cancer classification. Thus, the comparison of multiplex results with standard ELISA of these three diagnostic biomarkers showed similar results and trends.

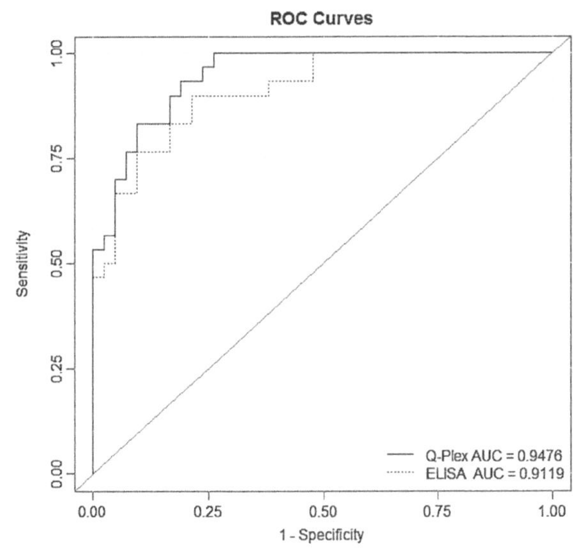

Figure 1 Diagnostic performance of bladder cancer-associated molecular panel comprised of three biomarkers. ROC curves were plotted to illustrate the performance characteristics of the 3-biomarker signature for the detection of bladder cancer in urine samples using the Q-Plex multiplex assay (solid line) and commercial ELISA assays (dotted line). Based on the area under the ROC curve (AUROC), Youden Index cutoff values that maximized the sum of sensitivity and specificity were determined for the combination of biomarkers. The Q-Plex multiplex assay achieved an overall sensitivity of 0.93 and specificity of 0.81 (AUROC 0.9476). The combination of data from the individual target ELISA assays achieved an overall sensitivity of 0.77 and specificity of 0.91 (AUROC 0.9119). Thus, the comparison of multiplex results with standard ELISA of these three diagnostic biomarkers showed similar results and trends.

The Quansys Multiplex ELISA is now being optimized to measure the remaining proteins of our highly accurate bladder cancer-associated diagnostic panel comprised of 10 biomarkers [8]. We are currently producing specific antibodies (Nonagen BioScience) for the 10 analytes in our diagnostic panel, which can potentially be incorporated into a Quansys BioScience custom array system for an integrated bladder cancer detection assay. The ability to analyze multiple proteins per well, in addition to dozens of samples per microplate, results in marked improvements in efficiency and considerable cost savings compared to single analyte ELISA.

Our study has several limitations. Notably, the study cohort of 73 subjects was relatively small, and only three targets from our validated 10-biomarker panel were tested in this study. However this is a proof-of-principle study in which we investigated the potential for multiplexing analyte detection in real-world urine samples. Lastly, as part of a biomarker discovery and validation program, archived urines were retrieved from tissue banks for analysis. Prolonged storage can result in protein degradation and mute the performance of diagnostic protein tests on such samples. Thus, validation of the performance of the multiplexed detection of a bladder cancer biomarker panel in freshly voided urines will be required.

There is growing demand to integrate multiplex signatures to obtain favorable assay properties, such as reduced sample volume, decreased processing time, low cost analysis and low reagent consumption. Several multiplex protein services are available (*e.g.*, Sample Testing Services of Quansys Biosciences Inc., microplate-based; Aushon Biosystems, SearchLight Assays Services, microplate-based; Milliplex MAP, bead-based; and RayBiotech,

Table 2 Comparison of Q-plex™ technology to traditional ELISA assays

	Q-plex™ array		Commercial ELISA assay	
	Cancer	Benign	Cancer	Benign
IL8 (pg/ml)				
Median	127.7	2.2	100.8	144.3
(Min, Max)	(0.4, 5,087.5)	(0.5, 185.3)	(0, 4,355.3)	(0, 506.3)
Mean ± SD*	423.6 ± 931.2	14.6 ± 40.9	276.8 ± 768.9	180.0 ± 138.2
MMP9 (pg/ml)				
Median	268,688	248,730.5	2,833.2	545.3
(Min, Max)	(176,103.9, 3,783,990.1)	(162,655.6, 867,918.5)	(0, 12,988.9)	(0, 7,146)
Mean + SD	469,075.4 ± 655,960.4	320,746.6 ± 171,092.6	5,236.0 ± 4,144.5	876.5 ± 1,353.0
VEGFA (pg/ml)				
Median	12,599.4	12,677.5	143.6	19.5
(Min, Max)	(2,093.3, 498,114.9)	(810.1, 262,356.7)	(1.1, 9,874.1)	(0, 184.2)
Mean + SD	33,483.5 ± 88,403.7	20,690.5 ± 39,702.7	1061.6 ± 2132.9	36.2 ± 47.4

*, $p < 0.05$ comparing cancer cohort to benign cohort for both Q-Plex™ Array and Commercial ELISA Assay.

Inc., slide-based). Several studies have reported that the multiplex ELISA procedures appear suitable and reliable for tissue lysate and serum [21,22]. While there are inherent, usually subtle differences between the various multiplex technologies, the overall technique is rapid, cost effective, and reliable. The reliability may be due in part to specific capture and detection antibody pairs that are optimized for the specific protein in question without cross-reactivity. The Q-Plex system combines powerful technology that allows the simultaneous determination of the expression levels of many proteins in biologic samples (*e.g.,* urine, blood, sputum, cerebrospinal fluid). The continued identification of biomarker proteins, which are associated with diagnosis or prognosis, will result in this technology being more widely adopted.

Conclusion

Based on these encouraging preliminary data, we believe that the Q-Plex technology is a viable new platform that can be exploited to be a simple, yet accurate tool to quantitate a panel of diagnostic proteins in biologic specimens, in this case the detection of diagnostic bladder cancer biomarkers in voided urine samples. Importantly, this novel platform has acceptable inter assay variability, as well as very sensitive levels of detection. The platform can be readily implemented in a clinical laboratory and may be used as the basis for further work by incorporation of our entire bladder cancer-associated diagnostic signature.

Methods

Patients and specimen processing

Under MD Anderson Cancer Center Orlando Institutional Review Board approval and informed consent, prospectively collected voided urine samples (50 mL) were assigned a unique identifying number before immediate laboratory processing. Each urine sample was centrifuged at 600 × *g* 4°C for 5 min. The supernatant was decanted and aliquoted, and stored at -80°C in our

Table 3 Performance comparison of Q-plex™ and individual-target ELISA assays for the detection of bladder cancer in urine samples

Urinary analytes	Assay platform	AUROC [95% CI]	Sensitivity (%)	Specificity (%)	PPV (%)	NPV (%)
IL8	Q-Plex™ Array	0.907 [0.830 - 0.985]	90	86	82	92
	Commercial ELISA	0.878 [0.783 - 0.972]	87	86	82	90
MMP9	Q-Plex™ Array	0.533 [0.392 - 0.674]	45	76	58	65
	Commercial ELISA	0.548 [0.409 - 0.688]	45	76	58	65
VEGFA	Q-Plex™ Array	0.524 [0.386 - 0.661]	17	95	71	61
	Commercial ELISA	0.493 [0.356 - 0.630]	16	95	71	61
Combination	Q-Plex™ Array	0.948 [0.903 - 0.992]	93	81	78	94
	Commercial ELISA	0.912 [0.796 - 0.968]	77	91	86	84

genitourinary tissue bank prior to analysis. The genitourinary tissue bank was queried for suitable samples to analyze in the current study. Patients with known renal disease or documented renal insufficiency were not selected for inclusion in the current study. The study cohort consisted of 42 subjects with no previous history of urothelia carcinoma, gross hematuria, active urinary tract infection or urolithiasis (controls), and 31 subjects with newly diagnosed urothelial carcinoma. All 73 subjects a) had their urines analyzed by individual commercial ELISA kits for IL8, MMP9 and VEGFA [8] and b) had at least 5 mL of urine in the tissue bank for the current analysis. Frozen supernatant aliquots from the 73 subjects were shipped on dry ice to Quansys Bioscience, biotechnology company geared towards standardize or customize singleplex and multiplex ELISA assays, for analysis. Clinical information associated with these urine samples was collected. Data is reported using the STARD criteria [23]. Aliquots of urine supernatants were thawed and analyzed. The total protein concentration for each sample was determined via absorbance at 280 nm using a Nanodrop ND-1000 spectrophotometer (Thermo Scientific, Wilmington DE), and creatinine assays were performed on each sample for normalization with respect to urine volume [17].

Q-plex multiplex assay

The Q-Plex technology is based on placement of immobilized capture antibody in 350–500 μm spots at the bottom of polypropylene 96-well plates to capture target proteins (for further details see http://www.quansysbio.com/assay-development/). Each spot is printed with a different analyte capture antibody, in this case against IL8, MMP9 and VEGFA. The sandwich immunoassay complex that forms is labeled via biotin-streptavidin with Horse Radish Peroxidase that generates light from a chemiluminescent substrate. Two internal control assays are designed within the assay to ensure integrity of results from each well through monitoring for the presence of interfering factors.

Thawed urine samples were handled on ice and diluted with Human Sample Dilution Buffer (Quansys), designed to reduce effects from heterophilic antibodies and other interferents [24]. Samples and standards (50 μl) were loaded onto the Q-Plex™ plate with a multichannel pipettor in order to reduce pipetting error and allowed to incubate for two hours. Laboratory personnel were blinded to final diagnosis. Subsequently, the sandwich immunoassay complex that forms was incubated with biotin-streptavidin/Horse Radish Peroxidase reagents for an additional two hours. Antigen standard curves were performed in duplicate diluting the antigen standard 1:3 for 11 points with a single negative point. Samples were diluted at ratios of 1 to 2 (sample to buffer) (50%), 1 to 20 (5%) and 1 to 200 (0.5%). Each dilution was loaded into three wells and measured in triplicate, a total of 9 wells per sample. Standard curves were constructed using Q-View Software™, which allows for the selection of multiple non-linear and linear equations to fit the standard curve. Optimal curve fits were determined by visual graph evaluation and comparison of Aikake's information criteria (AIC) values [25]. A composite or stacked image composed of individual exposures of 30, 60, and 180 seconds with camera noise background subtraction was performed using Q-View Imager™. Levels of chemiluminescent units, or pixel intensity units were measured by Q-View Software™. The range of pixel intensity units is from 0 to 65536 (216 for a 16 bit image). Lower limits of quantification (LLOQ) were determined to be the lowest point of the 10-point positive standard curve where the back-fit regression values were within 20% of the known value.

Commercial Enzyme-Linked Immunosorbent Assays (ELISA)

The levels of human Interleukin 8 (IL8, Cat # ab46032 Abcam, Cambridge, MA, USA), Matrix Metalloproteinase 9 (MMP9, Cat # DMP900 R&D Systems Inc., Minneapolis, MN, USA) and Vascular Endothelial Growth Factor A (VEGFA, Cat # 100663 Abcam, Cambridge, MA, USA) were monitored in voided urine samples, as previously reported [8].

Data analysis

Nonparametric receiver operating characteristic (ROC) curves in which the value for sensitivity is plotted against false-positive rate (1-specificity) were generated for association of biomarkers with disease status and assay performance comparison. We defined a diagnostic test as positive or negative for bladder cancer detection by defining an optimal analyte concentration cutoff value (Youden index), selected to maximize the sum of the sensitivity and specificity [26,27]. Threshold cut-offs used were: IL8, 83 pg/ml; MMP9, 365,525 pg/ml; and VEGFA, 21,286 pg/ml. The relative ability of the combination of biomarkers to indicate bladder cancer was estimated by calculating the area under the ROC curves (AUC), and the sensitivity and specificity was defined by calculation of the Youden index. Statistical significance in this study was set at $p < 0.05$ and all reported p values were 2-sided. All analyses were performed using SAS software version 9.3 (SAS Institute Inc., Cary, NC).

Abbreviations

IL8: Interleukin 8; MMP9: Matrix metallopeptidase 9; VEGFA: Vascular endothelial growth factor A; CV: Coefficient of variation; AUROC: Area under the receiver operating characteristic; PPV: Positive predictive value; NPV: Negative predictive value; ELISA: Enzyme-linked immunosorbent assay; STARD: STAndards for the Reporting of Diagnostic; AIC: Aikake's information criteria; ROC: Receiver operating characteristic; AUC: Area under the concentration curve; LLD: Lower limits of detection; LLOQ: Lower limits of quantification; CI: Confidence interval; SD: Standard deviation; CLIA: Clinical laboratory improvement amendments.

Competing interests

CJR and SG are officers at Nonagen BioScience Corp. YD, MM and GZ have no interests to disclose.

Authors' contributions

CJR – Study concept and design, drafting of manuscript, administrative support and funding. YD - Statistical analysis and drafting of manuscript. MM - Study design, statistical analysis and drafting of manuscript. GZ Sample collection, sample preparation and creation of database. SG Study concept and design, drafting of manuscript. All authors read and approved the final manuscript.

Acknowledgements

This work was supported by Florida Department of Health James and Esther King Team Science Award 10KT-01 (CJR) and NIH/NCI R44CA173921 (SG). The funding bodies did not have a role in the collection, analysis, and interpretation of data; in the writing of the manuscript; and in the decision to submit the manuscript for publication.

Author details

[1]MD Anderson Cancer Center Orlando, Orlando, FL, USA. [2]Nonagen BioScience Corp, Jacksonville, FL, USA. [3]University of Hawaii Cancer Center, Honolulu, HI, USA. [4]Department of Biostatistics, The University of Florida, Gainesville, FL, USA. [5]Department of Health Sciences Research, Mayo Clinic, Jacksonville, FL, USA.

References

1. Diamandis EP: **The failure of protein cancer biomarkers to reach the clinic: why, and what can be done to address the problem?** *BMC Med* 2012, **10**:87.
2. Nguyen B, Cusumano PG, Deck K, Kerlin D, Garcia AA, Barone JL, Rivera E, Yao K, de Snoo FA, van den Akker J, Stork-Sloots L, Generali D: **Comparison of molecular subtyping with BluePrint, MammaPrint, and TargetPrint to local clinical subtyping in breast cancer patients.** *Ann SurgOncol* 2012, **19**:3257–3263.
3. Malo TL, Lipkus I, Wilson T, Han HS, Acs G, Vadaparampil ST: **Treatment choices based on oncotypeDx in the breast oncology care setting.** *J Cancer Epidemiol* 2012, **2012**:941495.
4. Yang N, Feng S, Shedden K, Xie X, Liu Y, Rosser CJ, Lubman DM, Goodison S: **Urinary glycoprotein biomarker discovery for bladder cancer detection using LC/MS-MS and label-free quantification.** *Clin Cancer Res* 2011, **17**:3349–3359.
5. Kreunin P, Zhao J, Rosser C, Urquidi V, Lubman DM, Goodison S: **Bladder cancer associated glycoprotein signatures revealed by urinary proteomic profiling.** *J Proteome Res* 2007, **6**:2631–2639.
6. Urquidi V, Goodison S, Cai Y, Sun Y, Rosser CJ: **A candidate molecular biomarker panel for the detection of bladder cancer.** *Cancer Epidemiol Biomarkers Prev* 2012, **12**:2149–2158.
7. Rosser CJ, Liu L, Sun Y, Villicana P, McCullers M, Porvasnik S, Young PR, Parker AS, Goodison S: **Bladder cancer-associated gene expression signatures identified by profiling of exfoliated urothelia.** *Cancer Epidemiol Biomarkers Prev* 2009, **2**:444–453.
8. Rosser CJ, Ross S, Chang M, Dai Y, Mengual L, Zhang G, Kim J, Urquidi V, Alcaraz A, Goodison S: **Multiplex protein signature for the detection of bladder cancer in voided urine samples.** *J Urol* 2013, **190**:2257–2262.
9. Quansys Corp: http://www.quansysbio.com.
10. Ekins RP: **Ligand assays: from electrophoresis to miniaturized microarrays.** *Clin Chem* 1998, **44**:2015–2030.
11. Hatch W, Scalarone G: **Comparison of colorimetric and chemiluminescent ELISAs for the detection of antibodies to blastomyces dermatitidis.** *J Med Biol Sci* 2009, **3**:19–25.
12. Samineni S, Parvataneni S, Kelly C, Gangur V, Karmaus W, Brooks K: **Optimization, comparison, and application of colorimetric vs. chemiluminescence based indirect sandwich ELISA for measurement of human IL-23.** *J Immunoassay Immunochem* 2006, **27**:183–193.
13. Jantzie L, Tanay V, Todd K: **Methods in immunochemistry.** In *Handbook of neurochemistry and molecular neurobiology: Practical neurochemistry methods.* Edited by Lajtha A. New York, NY: Springer Science and Business Media, LLC; 2007:193–218.
14. Urquidi V, Goodison S, Ross S, Chang M, Dai Y, Rosser CJ: **Diagnostic potential of urinary alpha1-antitrypsin and apolipoprotein E in the detection of bladder cancer.** *J Urol* 2012, **188**(6):2377-2383.
15. Urquidi V, Kim J, Chang M, Dai Y, Rosser CJ, Goodison S: **CCL18 in a multiplex urine-based assay for the detection of bladder cancer.** *PLoS One* 2012, **7**:e37797.
16. Urquidi V, Chang M, Dai Y, Kim J, Wolfson ED, Goodison S, Rosser CJ: **IL-8 as a urinary biomarker for the detection of bladder cancer.** *BMC Urol* 2012, **12**:12.
17. Urquidi V, Goodison S, Kim J, Chang M, Dai Y, Rosser CJ: **Vascular endothelial growth factor, carbonic anhydrase 9, and angiogenin as urinary biomarkers for bladder cancer detection.** *Urol* 2012, **79**:1185.e1-6.
18. Goodison S, Chang M, Dai Y, Urquidi V, Rosser CJ: **A multi-analyte assay for the non-invasive detection of bladder cancer.** *PLoS One* 2012, **7**(10):e47469. doi:10.1371/journal.pone.0047469. Epub 2012 Oct 19.
19. Kelley M, DeSilva B: **Key elements of bioanalytical method validation for macromolecules.** *Aaps J* 2007, **9**:E156–E163.
20. Viswanathan CT, Bansal S, Booth D, DeStafano AJ, Rose MJ, Sailstad J, Shah VP, Skelly JP, Swann PG, Weiner R: **Quantitative bioanalytical methods validation and implementation: best practices for chromatographic and ligand binding assays.** *Pharm Res* 2007, **24**:1962–1973.
21. Edgar R, Domrachev M, Lash AE: **Gene expression omnibus: NCBI gene expression and hybridization array data repository.** *Nucleic Acids Res* 2002, **30**:207–210.
22. Trune DR, Larrain BE, Hausman FA, Kempton JB, MacArthur CJ: **Simultaneous measurement of multiple ear proteins with multiplex ELISA assays.** *Hear Res* 2011, **275**:1–7.
23. Zeng L, Liu J, Wang Y, Wang L, Weng S, Chen S, Yang GY: **Cocktail blood biomarkers: prediction of clinical outcomes in patients with acute ischemic stroke.** *Eur Neurol* 2013, **69**:68–75.
24. Bossuyt PM, Reitsma JB, Bruns DE, Gatsonis CA, Glaszious PP, Irwig LM, Lijmer JG, Moher D, Rennie D, de Vet HC, STARD Group: **Towards complete and accurate reporting of studies of diagnostic accuracy: the STARD initiative.** *Fam Pract* 2004, **21**:4–10.
25. Bjerner J, Bormer OP, Nustad K: **The war on heterophilic antibody interference.** *Clin Chem* 2005, **51**:9–11.
26. Motulsky H, Christopoulos H: *Fitting models to biological data using linear and nonlinear regression: a practical guide to curve fitting, 351.* Oxford; New York: Oxford University Press; 2004:17.
27. Fluss R, Faraggi D, Reiser B: **Estimation of the Youden index and its associated cutoff point.** *Biom J* 2005, **47**:458–472.

Solid fermentation of wheat bran for hydrolytic enzymes production and saccharification content by a local isolate *Bacillus megatherium*

Reda M El-Shishtawy[1*], Saleh A Mohamed[2*], Abdullah M Asiri[1,3], Abu-bakr M Gomaa[4], Ibrahim H Ibrahim[2] and Hasan A Al-Talhi[2]

Abstract

Back ground: For enzyme production, the costs of solid state fermentation (SSF) techniques were lower and the production higher than submerged cultures. A large number of fungal species was known to grow well on moist substrates, whereas many bacteria were unable to grow under this condition. Therefore, the aim of this study was to isolate a highly efficient strain of *Bacillus* sp utilizing wheat bran in SSF and optimizing the enzyme production and soluble carbohydrates.

Results: A local strain *Bacillus megatherium* was isolated from dung sheep. The maximum production of pectinase, xylanase and α-amylase, and saccharification content (total soluble carbohydrates and reducing sugars) were obtained by application of the *B. megatherium* in SSF using wheat bran as compared to grasses, palm leaves and date seeds. All enzymes and saccharification content exhibited their maximum production during 12–24 h, at the range of 40–80% moisture content of wheat bran, temperature 37-45°C and pH 5–8. An ascending repression of pectinase production was observed by carbon supplements of lactose, glucose, maltose, sucrose and starch, respectively. All carbon supplements improved the production of xylanase and α-amylase, except of lactose decreased α-amylase production. A little increase in the yield of total reducing sugars was detected for all carbon supplements. Among the nitrogen sources, yeast extract induced a significant repression to all enzyme productivity. Sodium nitrate, urea and ammonium chloride enhanced the production of xylanase, α-amylase and pectinase, respectively. Yeast extract, urea, ammonium sulphate and ammonium chloride enhanced the productivity of reducing sugars.

Conclusions: The optimization of enzyme production and sccharification content by *B. megatherium* in SSF required only adjustment of incubation period and temperature, moisture content and initial pH. Wheat bran supplied enough nutrients without any need for addition of supplements of carbon and nitrogen sources.

Keywords: *Bacillus megatherium*, Enzymes, Saccharification, Solid fermentation

Background

Agricultural residues have an enormous potential as renewable carbon and energy sources. The main potential applications of agricultural residues are in food, animal feed, biofuel and pharmaceutical industries. Saccharification of agricultural residues by microbial hydrolytic enzymes (cellulases, xylanases, amylases and pectinases) is the first step of bioconversion of organic material into reducing sugars, like glucose and xylose [1]. In the saccharification of agricultural residues, a potential effect was detected in presence of two or more enzymes [2]. Cellulases for cellulose hydrolysis [2], xylanases for hemicelluloses hydrolysis [3], amylase for amylose hydrolysis [4] and pectinase for pectin hydrolysis [5] are cooperatively needed in the saccharification of agricultural residues. The reducing sugars obtained from these hydrolyzing actions could be utilized as carbon and energy sources in the fermentation industry, such as lactic acid [6], hydrogen [7] and ethanol [8]. In addition, microbial hydrolytic enzymes utilized in

* Correspondence: elshishtawy@hotmail.com; saleh38@hotmail.com
[1]Chemistry Department, Faculty of Science, King Abdulaziz University, P.O. Box 80203, Jeddah 21589, Saudi Arabia
[2]Biochemistry Department, Faculty of Science, King Abdulaziz University, Jeddah, Kingdom of Saudi Arabia
Full list of author information is available at the end of the article

several applications, such as food, textile, paper, pulp and detergent industries [9-12].

Solid state fermentation (SSF) is the growth of organisms on moist substrates in the absence of free-flowing water. The use of SSF for production of enzymes and other products has many advantages over submerged fermentation [13]. These advantages included: easier recovery of products, the absence of foam formation and smaller reactor volumes. Moreover, contamination risks are significantly reduced due to the low water contents and, consequently, the volume of effluents decreases. Another very important advantage is that, it permits the use of agricultural and agro-industrial residues as substrates which are converted into products with high commercial value like secondary metabolites [13,14]. Furthermore, the utilization of these compounds helps in solving pollution problems, which otherwise cause their disposal [15]. For enzyme production, the costs of these techniques are lower and the production is higher than submerged cultures [16,17]. A large number of fungal species was known to grow well on moist substrates in the absence of free-flowing water, whereas many bacteria are unable to grow under this condition [18-21]. As a result, most studies involving SSF have been conducted by using fungi. However, there are little reports of bacterial strains being used successfully for the production of enzymes by using SSF [4,5,22,23]. Therefore, the aim of this study is to isolate strain of *Bacillus* sp. capable of using wheat bran in SSF to produce α-amylase, xylanase and pectinase. The saccharification content, total soluble carbohydrates and reducing sugars, of wheat bran was studied. Studies on optimizing production of enzymes and saccharification content were also carried out.

Methods

Isolation, identification and efficiency of the cellulose decomposing bacilli

Five isolates of *Bacillus* spp. were isolated from different samples i.e., sheep dung, horses waste, manure compost and rhizosphere soil. The efficiency of the five strains in cellulose decomposion was estimated using caboxymethyl cellulose (CMC) agar medium containing g/l: CMC, 5; peptone, 5; NaCl, 5; beef extract, 3; agar, 18 and pH was adjusted to 7 [24]. The most efficient strain in cellulose decomposion was identified according to Bergey's Manual of Systematic Bacteriology [25]. The highest efficient strain in cellulose decomposion was isolated from sheep dung and identified as *B. megatherium*.

Agricultural residues

Four dried agricultural residues, i.e. wheat bran, date seeds, grass and palm leaves were used as substrates for solid state fermentation (SSF).

Physicochemical parameters of SSF

Physicochemical parameters of SSF were studied for optimization production conditions of soluble carbohydrates, reducing sugars, α-amylase, pectinase and xylanase by *B. megatherium*. The agricultural residues were sperately sterilized in an autoclave for 20 min at 121°C. *B. megatherium* was grown in 50 ml Erlenmeyer flask included 5 g of the respective sterilized agricultural residue and appropriate amount of water needed to adjust the moisture of dried substrate, which contained 10% moisture after dring. Optimized physicochemical parameters including: incubation period, incubation temperature, and moisture content of the substrate and incubation pH. The pH was adjusted using 0.1 M NaOH or HCl. The influence of supplementation of carbon sources (glucose, maltose, starch, sucrose, and lactose at 1% w/v) and nitrogen sources (yeast extract, urea, sodium nitrate, ammonium sulphate, and ammonium chloride at 1% w/v) has been studied. Each experiment was done in triplicate.

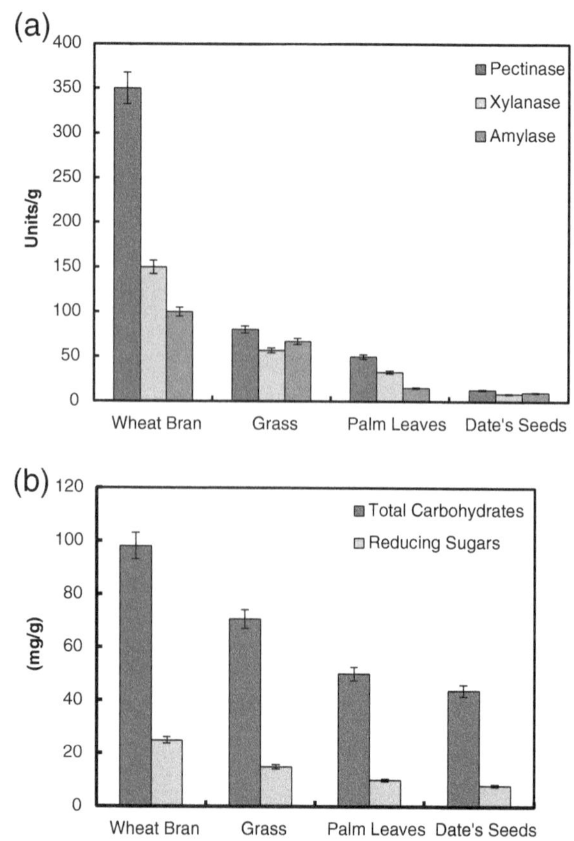

Figure 1 Effect of different agriculture wastes on the production of pectinase, xylanase and α-amylase (a) and total carbohydrate and reducing sugars (b) by *B. megatherium* in SSF. Process conditions: incubation time 48 h., initial moisture content 50% (by volume per mass) and temperature at 37°C.

Soluble carbohydrate and enzyme extraction

Soluble carbohydrate and enzyme were extracted by mixing the fermented substrate with 50 ml distilled water and shaked on a rotary shaker at 180 rpm overnight. The suspension was then centrifuged at 12000 rpm for 10 min and the supernatant was designated as a crude extract.

Determination of total reducing sugars

Total reducing sugars were determined by the method of Miller [26]. The reaction mixture contained 0.5 ml of crude extract and 0.5 ml dinitrosalicylic acid reagent. The tubes were heated in a boiling water bath for 10 min. After cooling to room temperature, the absorbance was measured at 560 nm. Glucose served as the calibration standard for total reducing sugar determination.

Determination of total soluble carbohydrates

Total soluble carbohydrates were determined by the method of Dubois et al. [27]. The reaction mixture contained 25 μl of a 4:1 mixture of phenol and water, 0.8 ml of crude extract and 2 ml of concentrated sulfuric acid. Then mixed well, and heated in a boiling water bath for 30 min. The absorbance was determined at 480 nm. Glucose served as the calibration standard for total carbohydrate determination.

Enzymes assays

α-Amylase, pectinase and xylanase activities were assayed by determining the liberated reducing end products using maltose, galacturonic acid and xylose as standards, respectively [26]. Substrates used were starch, polygalacturonic acid and birchwood xylan for α-amylase, pectinase and xylanase, respectively. The reaction mixture (0.5 ml) contained 1% substrate, 0.05 M sodium acetate buffer pH 5.5 and 0.1 ml crude extract. Assays were carried out at 37°C for 1 h. Then 0.5 ml dinitrosalicylic acid reagent was added to each tube. Then the reaction mixture was mixed well, and heated in a boiling water bath for 10 min. After cooling to room temperature, the absorbance was measured at 560 nm. One unit of enzyme activity is defined as the amount of enzyme which liberated one μmol of reducing sugar per min under standard assay conditions.

All the experimental work was run in triplicates.

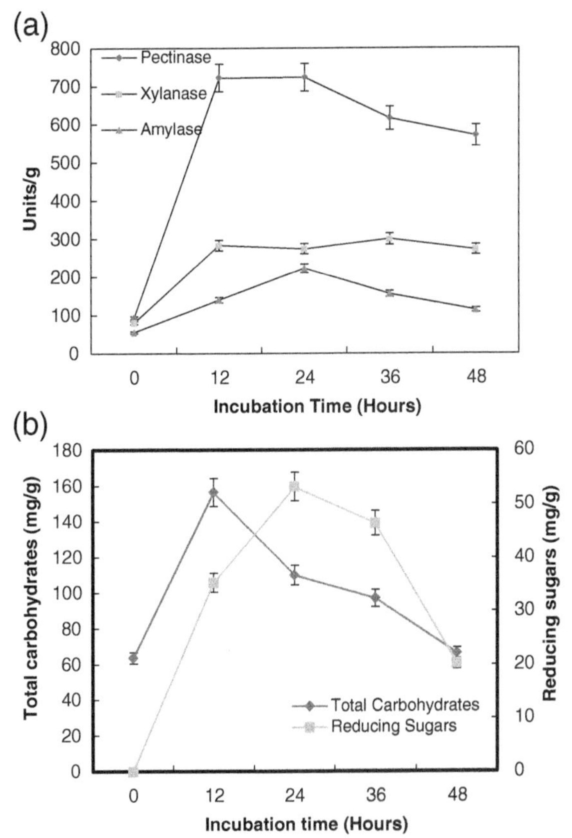

Figure 2 Effect of incubation time on the production of pectinase, xylanase and α-amylase (a), and total carbohydrate and reducing sugars (b) by *B. megatherium* in SSF using wheat bran as substrate. Process conditions: initial moisture content 50% (by volume per mass) and temperature at 37°C.

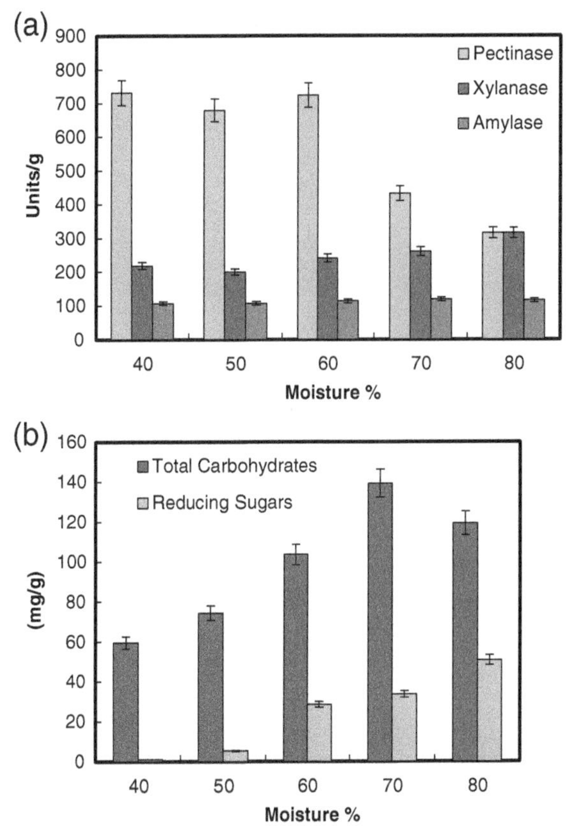

Figure 3 Effect of moisture % on the production of pectinase, xylanase and α-amylase (a), and total carbohydrate and reducing sugars (b) by *B. megatherium* in SSF using wheat bran as substrate. Process conditions: incubation times 48 h and temperature at 37°C.

Statistical analysis

The obtained data were statistically analyzed as a randomized complete block design with three replicates by analysis of variance (ANOVA) using the statistical package software SAS (SAS Institute Inc., 2000, Cary, NC., USA). Comparisons between means were made by *F*-test and the least significant differences (LSD) at level $P = 0.05$. Correlations coefficient among the different parameters were also calculated by SAS.

Results and discussion

The effect of agricultural residues

Production of pectinase, xylanase and α-amylase, and saccharification content (total soluble carbohydrates and reducing sugars) by *B. megatherium* were tested in different SSF sources using wheat bran, grasses, palm leaves and date seeds. Figure 1a showed the maximum production of pectinase, xylanase and α-amylase by *B. megatherium* (350, 150 and 100 units/g solid, respectively) in SSF using wheat bran as compared to other agriculture residues. The saccharification content of these agricultural residues improved these results, whereas the productivity of total soluble carbohydrates and reducing sugars exhibited maximum production by using wheat bran (Figure 1b). Similarly, among the lignocellulosic substrates tested, wheat bran gave maximum yield of xylanase by *Cellulosimicrobium* sp. as compared to other substrates viz. gram bran, rice husk, rice bran, wood dust and apple pomace [28]. The best substrate for xylanase production was wheat bran due to its nutritional content and large surface area [29], where wheat bran contained xylan and protein, which were served as carbon and nitrogen sources for microorganisms, respectively [30]. α-Amylase was also exhibited the maximum production by *Clostridium thermosulforegenes* using wheat bran [31]. Wheat bran characterized by its better air circulation, loose particle binding and efficient penetration by mycelia and cheaper, therefore it showed a better prospect economically in fermentation processes [32]. Therefore, physicochemical parameters of SSF using wheat bran for optimization production of enzymes and saccharification content of *B. megatherium* was performed in the following study.

The effect of incubation period

Figure 2a showed the effect of incubation period on production of pectinase, α-amylase and xylanase by *B. megatherium* grown on wheat bran in SSF. Pectinase, α-amylase and xylanase exhibited their maximum production after

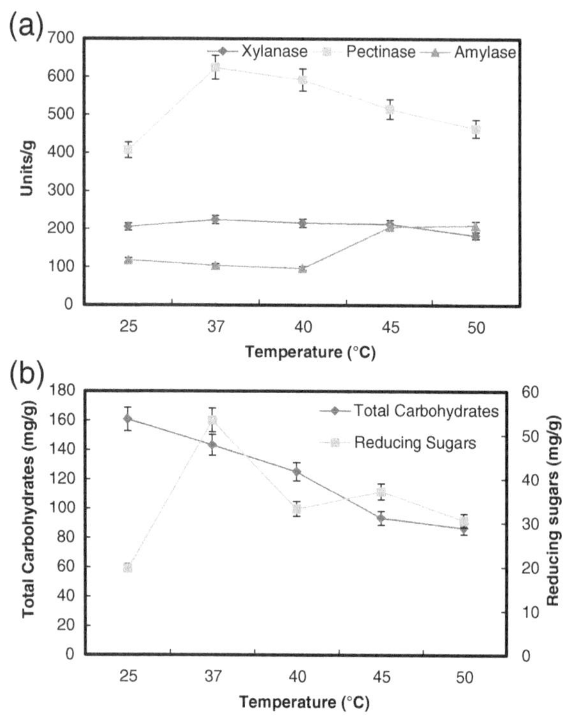

Figure 4 Effect of incubation temperature on the production of pectinase, xylanase and α-amylase (a), and total carbohydrate and reducing sugars (b) by *B. megatherium* in SSF using wheat bran as substrate. Process conditions: incubation times 48 h and initial moisture content 50% (by volume per mass).

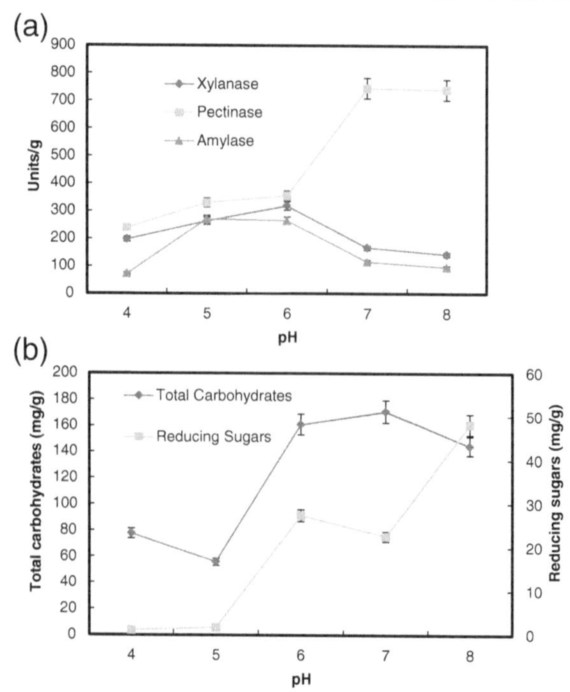

Figure 5 Effect of pH on the production of pectinase, xylanase and α-amylase (a), and total carbohydrate and reducing sugars (b) by *B. megatherium* in SSF using wheat bran as substrates. Process conditions: incubation times 48 h, initial moisture content 50% (by volume per mass) and temperature at 37°C.

24 h (720, 220 and 280 units/g solid, respectively). The maximum production of total soluble carbohydrates (160 mg/g solid) and reducing sugars (55 mg/g solid) was detected after 12 and 24 h, respectively (Figure 2b). The maximum production of reducing sugars was inagreement with the maximum production of enzymes. Similarly, the maximum activity of α-amylase was noted in enzyme extracts harvested after 24 h of SSF *B. subtilis* [33]. The formation of xylanase started from 18 h and reached its maximum after 72 h by *Bacillus* sp. [34]. *B. licheniformis* produced maximum of pectinase after 48 h [5].

The effect of initial moisture content

In SSF, the moisture content is an important factor that influences the growth and product yield of microbes [35]. Moisture is reported to cause swelling of the substrates, thereby facilitating better utilization of the substrate by microorganisms [36,37]. The data presented in t Figure 3a, clearly indicated a maximum production of the enzymes pectinase (720 units/g solid), xylanase (310 units/g solid) and α-amylase (115 units/g solid) by *B. megatherium* at 40, 80 and 70% moisture, respectively. In addition, the maximum saccharification content of total soluble carbohydrates (140 mg/g solid) and reducing sugars (55 mg/g solid) was observed at 70 and 80% moisture, respectively (Figure 3b). Other results showed that the yield of α-amylase and amylopullulanase by *Clostridium thermosulfurogenes* increased with an increase in solid to moistening agent ratio from 1:0.5 to 1:2.5 (30 to 75%) with a maximum at 1:2.25 (73%) [31]. Optimum xylanase production from *Bacillus* sp. PKD-9 was observed when wheat bran was used at 1:4 ratio of solid substrate-to-moisture [23]. Any further increase in the ratio resulted in the decrease of enzyme yields may be due to clumping of solid particles which results in the decrease of interparticle space and diffusion of nutrients [29,37]. In contrast, the low moisture content leads to the decreased solubility of nutrients present in the wheat bran thereby decreased enzyme yields [38].

The effect of incubation temperature

The usual temperature maintained in SSF systems is in the range of 25-32°C, depending on the growth kinetics

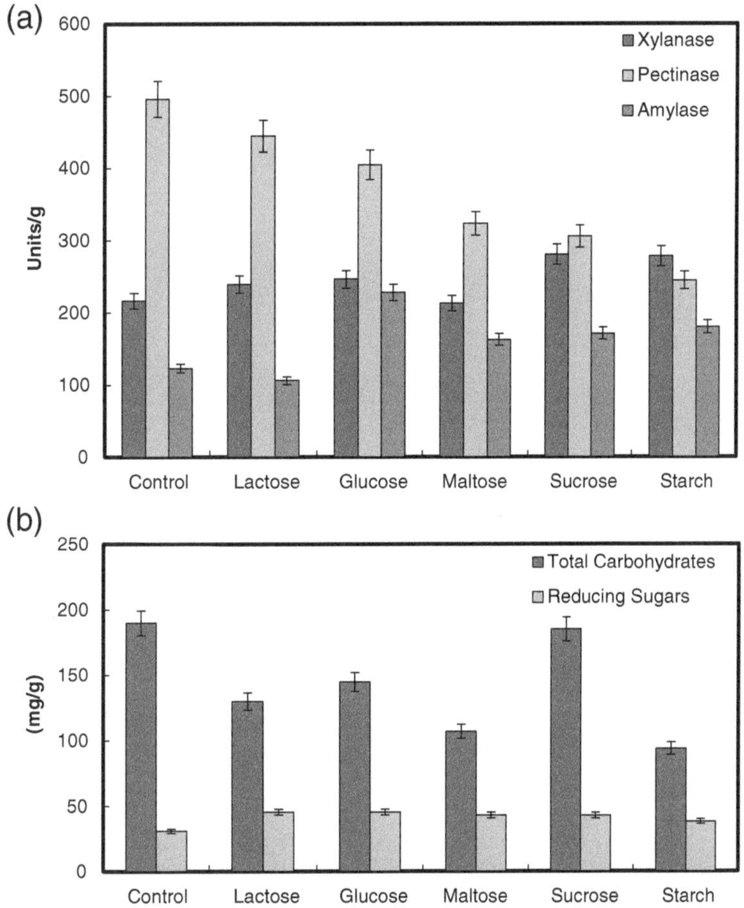

Figure 6 Effect of carbon source (1%) supplementation on the production of pectinase, xylanase and α-amylase (a), and total carbohydrate and reducing sugars (b) by *B. megatherium* in SSF using wheat bran as substrates.

of microorganism employed for fermentation purposes [39]. In the present study, the maximum production of both pectinase (620 units/g solid) and xylanase (220 units/g solid) by *B. megatherium* was achieved at 37°C, and 45°C for α-amylase (220 units/g solid) (Figure 4a). These results were confirmed by maximum saccharification content of reducing sugars (54 mg/g solid) at 37°C, where the maximum of total soluble carbohydrates (160 mg/g solid) was detected at 25°C (Figure 4b). In other study, the optimum temperature recorded for maximum growth and α-amylase production by *B. subtilis* was 35°C [33]. Maximum production of pectinase and xylanase were obtained by *B. licheniformis* [5] and *Bacillus* sp. PKD-9 [23] at 37°C in solid wheat bran.

The effect of pH

The initial pH of fermentation medium has a significant effect on bacterial growth and enzyme production [40]. In order to study the effect of pH on enzyme production and saccharification content of wheat bran by *B. megatherium*, the production medium having pH's ranged from 4 to 8. Figure 5a shows a maximum production of pectinase (750 units/g solid), xylanase (280 units/g solid) and α-amylase (250 units/g solid) by *B. megatherium* at pH 7, 6 and 5, respectively. The high activity of pectinase results was reinforced by maximum saccharification content of total soluble carbohydrates (170 mg/g solid) and reducing sugars (50 mg/g solid) and at pH 7–8 (Figure 5b). However, most of *Bacillus* species were found to be produce maximum pectinase at different pH's ranging from 7 to 9 [41]. Further, the type of buffer used in nutrient solution is a key factor in governing α-amylase production by the *B. subtilis*.

The effect of supplementation carbon and nitrogen sources

The influence of supplementary carbon sources such as starch, sucrose, maltose, lactose or glucose at 1% (by

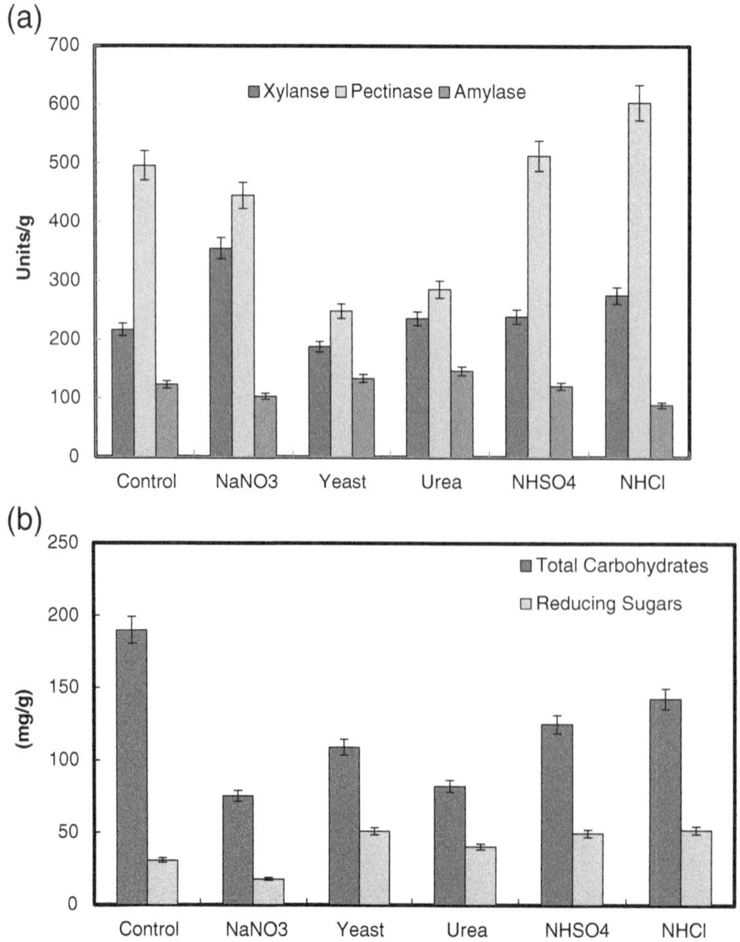

Figure 7 Effect of nitrogen source (1%) supplementation on the production of pectinase, xylanase and α-amylase (a), and total carbohydrate and reducing sugars (b) by *B. megatherium* in SSF using wheat bran as substrates. Process conditions: incubation times 48 h, initial moisture content 50% (by volume per mass) at temperatures 37°C.

mass) to wheat bran on production of pectinase, xylanase and α-amylase by *B. megatherium* was studied (Figure 6a). An ascending inhibition of pectinase production by lactose, glucose, maltose, sucrose and starch was detected, respectively. However, all carbon supplements improved the production of xylanase and α-amylase, except of lactose decreased α-amylase production. While most of carbon supplements exhibited repressive effect on the production of total soluble carbohydrates except of sucrose, a little increase in the yield of total reducing sugars was detected (Figure 6b). The same carbon supplements except starch caused repressive effect on pectinase production by *B. licchenformis* [5]. Supplementation of carbon sources increased α-amylase production by *B. cereus* during SSF using wheat bran [22]. *B. thermooleovorans* is reported to prefer starch, glucose, lactose, maltose and maltodextrins as carbon sources for α-amylase secretion [42,43]. In contrast, carbon sources such as glucose, maltose and starch did not enhance α-amylase production by thermophilic *B. coagulans* in solid-state fermentation using wheat bran [29]. Xylanase production by *Bacillus* sp. AR-009 grown on wheat bran was repressed upon addition of lactose, glucose and sucrose [3].

Studies on supplementation of nitrogen sources such as ammonium sulphate, ammonium nitrate, ammonium chloride, yeast extract or urea at 1% concentration to the wheat bran showed various effects on pectinase, xylanase and amylase production by *B. megatherium* (Figure 7a). Among the nitrogen sources, yeast extract induced a significant repression to all enzyme productivity. While sodium nitrate inhibited α-amylase production from 123 to103 units/g, it enhanced xylanase production from 216 to 355 units/g. Urea on the other hand inhibited production of xylanase and enhanced α-amylase. Also, a significant enhancement of pectinase activity was obtained by ammonium chloride supplementation from 496 to 610 units/g. While all nitrogen sources inhibited the production of total soluble carbohydrates, yeast extract, urea, ammonium sulphate and ammonium chloride enhanced the productivity of the total reducing sugars (Figure 7b). Hashemi et al. [4] reported that ammonium nitrate was the best inducer for α-amylase secretion followed by yeast extract 'by *Bacillus* sp. On the other hand, the limitation or starvation of nitrogen during the fermentation resulted in the limited growth of *B. subtilis* and the enhancement of α-amylase production [44]. For *Bacillus* sp. AR-009, yeast extract increased xylanase production [3]. Maximum production of pectinase was achieved when yeast extract was used in fermentation medium [5].

Conclusions

In conclusion, the production of pectinase, xylanase and amylase and saccharification content (total soluble carbohydrates and reducing sugars) by a newly local isolat *B. megatherium* using wheat bran in SSF will have several advantages. The optimization of enzyme production and sccharification content required only adjustment of incubation time and temperature, moisture content and initial pH. Wheat bran supplied enough nutrients without any need for addition of supplements of carbon and nitrogen sources. All these combined together could greatly reduce the overall cost of production of enzymes and saccharification content by *B. megatherium*. In the future, the reducing sugars will be used for hydrogen production.

Competing interests

The authors declare that they have no competing interests.

Authors' contributions

El-R, MS, AA, GM, II and AL-H performed all experiments and read and approved the final manuscript.

Acknowledgement

This Project was funded by the King Abdulaziz City for Science and Technology (KACST) under grant number 11-ENE1527-03. The authors, therefore, acknowledge with thanks KACST for support for Scientific Research. Also, the authors are appreciating the kind cooperation provided by the Deanship of Scientific Research (DSR), King Abdulaziz University.

Author details

[1]Chemistry Department, Faculty of Science, King Abdulaziz University, P.O. Box 80203, Jeddah 21589, Saudi Arabia. [2]Biochemistry Department, Faculty of Science, King Abdulaziz University, Jeddah, Kingdom of Saudi Arabia. [3]The Center of Excellence for Advanced Materials Research, King Abdulaziz University, Jeddah 21589, Saudi Arabia. [4]Biology Department, Faculty of Science, King Abdulaziz University, Jeddah, Kingdom of Saudi Arabia.

References

1. Howard RL, Abotsi E, Jansen van REL, Howard S: **Lignocellulose biotechnology: issues of bioconversion and enzyme production.** *Afric J Biotechnol* 2003, **2:**602–619.
2. Mansfield SD, Mooney C, Saddler JN: **Substrate and enzyme characteristics that limit cellulose hydrolysis.** *Biotechnol Prog* 1999, **15:**804–816.
3. Gessesse A, Mamo G: **High level xylanase production by an alkalophilic *Bacillus sp.* by using solid state fermentation.** *Enzyme Microb Technol* 1999, **25:**68–72.
4. Hashemi M, Razavi SH, Shojaosadati SA, Mousavi SM, Khajeh K, Safari M: **Development of a solid-state fermentation process for production of an alpha amylase with potentially interesting properties.** *J Biosci Bioeng* 2010, **110:**333–337.
5. Ur Rehman H, Qader SAU, Aman A: **Polygalacturonase: production of pectin depolymerising enzyme from bacillus licheniformis KIBGE IB-21.** *Carbohydr Polym* 2012, **90:**387–391.
6. Anuradha R, Suresh AK, Venkatesh KV: **Simultaneous saccharification and fermentation of starch to lactic acid.** *Process Biochem* 1999, **35:**367–375.
7. Kapdan IK, Kargi F: **Review biohydrogen production from waste materials.** *Enzyme Microb Technol* 2006, **38:**569–582.
8. Sun Y, Cheng J: **Hydrolysis of lignocellulosic materials for ethanol production: A review.** *Bioresour Technol* 2002, **83:**1–11.
9. Bhat MK, Bhat S: **Cellulose degrading enzymes and their potential industrial applications.** *Biotechnol Adv* 1997, **15:**583–620.
10. Beg QK, Kapoor M, Mahajan L, Hoondal GS: **Microbial xylanases and their industrial applications: a review.** *Appl Microbiol Biotechnol* 2001, **56:**326–338.
11. Hoondal GS, Tiwari RP, Tiwari R, Dahiya N, Beg QK: **Microbial alkaline pectinases and their industrial application: A review.** *Appl Microbiol Biotechnol* 2002, **59:**409–418.

12. Sivaramakrishnan S, Gangadharan D, Nampoothiri KM, Soccol CR, Pandey A: **Alpha-amylases from microbial sources—an overview on recent development.** *Food Technol Biotechnol* 2006, **44:**173–184.
13. Lonsane BK, Ramesh MV: **Production of bacterial thermostable α-amylase by solid-state fermentation: a potential tool for achieving economy in enzyme production and starch hydrolysis.** *Adv Appl Microbiol* 1990, **35:**1–56.
14. Pandey A: **Recent process developments in solid-state fermentation.** *Process Biochem* 1992, **27:**109–117.
15. Couto SR, Sanroman MA: **Application of solid-state fermentation to food industry - a review.** *J Food Eng* 2005, **22:**211–219.
16. Pandey A: **Aspects of fermenter design for solid-state fermentations.** *Process Biochem* 1991, **26:**355–361.
17. Sukumaran RK, Singhania RR, Mathew GM, Pandey A: **Cellulase production using biomass feed stock and its application in lignocellulose saccharification for bio-ethanol production.** *Renew Energ* 2009, **34:**421–424.
18. Mohamed SA, Al-MalkiL AL, Khan JA, Kabli SA, Al-Garni SM: **Solid state production of polygalacturonase and xylanase by *Trichoderma* species using cantaloupe and watermelon rinds.** *J Microbiol* 2013, **51:**605–611.
19. Rahardjo YSP, Sie S, Weber FJ, Tramper J, Rinzema A: **Effect of low oxygen concentrations on growth and α-amyase production of *Aspergillus oryzae* in model solid-state fermentation systems.** *Biomol Eng* 2005, **21:**163–172.
20. Ustok FI, Canan Tari C, Gogus N: **Solid-state production of polygalacturonase by *Aspergillus sojae* ATCC 20235.** *J Biotechnol* 2007, **127:**322–334.
21. Senthilkumar SR, Ashokkumar B, Raj KC, Gunasekaran P: **Optimization of medium composition for alkali-stable xylanase production by Aspergillus fischeri Fxn 1 in solid-state fermentation using central composite rotary design.** *Bioresour Technol* 2005, **96:**1380–1386.
22. Singh RK, Mishra SK, Kumar N: **Optimization of α-amylase production on agriculture byproduct by *Bacillus cereus* MTCC 1305 using solid state fermentation.** *Res J Pharm Biol Chem Sci* 2010, **1:**867–876.
23. Panwar D, Srivastava PK, Kapoor M: **Production, extraction and characterization of alkaline xylanase from Bacillus sp. PKD-9 with potential for poultry feed.** *Biocatal Agric Biotechnol* 2014, **3:**118–125.
24. Teather RM, Wood PJ: **Use of congo red-polysaccharide interactions in enumeration and characterization of cellulytic bacteria from the bovine rumen.** *Appl Environ Microbiol* 1982, **43:**777–780.
25. Bergey JG, Holt NR, Krieg PHA: *Bergey's Manual of Determinative Bacteriology*, Lippincott Williams. 9th edition; 1994. ISBN ISBN 0-683-00603-7.
26. Miller GL: **Use of dinitrosalicylic acid reagent for determination of reducing sugar.** *Anal Chem* 1959, **31:**426–429.
27. Dubois M, Gilles KA, Hamitton JK, Rebers PA, Smith F: **Colorimeteric method for determination of sugars and related substances.** *Anal Chem* 1956, **28:**350–356.
28. Rajashri DK, Anandrao RJ: **Optimization and scale up of cellulase-free xylanase production in solid state fermentation on wheat bran by *Cellulosimicrobium* sp. MTCC 10645.** *Jordan J Biol Sci* 2012, **5:**289–294.
29. Babu KR, Satyanarayana T: **α-Amylase production by thermophilic *Bacillus coagulans* in solid-state fermentation.** *Process Biochem* 1995, **30:**305–309.
30. Thiago LR, Kellaway RC: **Botanical composition and extent of lignification affecting digestibility of wheat and oat straw and pastalum hay.** *Animal Feed Sci Technol* 1982, **7:**71–81.
31. Mrudula S, Reddy G, Seenayya G: **Effect of substrate and culture conditions on the production of amylase and pullulanase by thermophilic *Clostridium thermosulforegenes* SVM17 in solid state fermentation.** *Malays J Microbiol* 2011, **7:**19–25.
32. Malathi S, Chakraborti R: **Productions of alkaline protease by a new *Aspergillus flavus* isolate under solid sustrate fermentation conditions for use as a depilation agent.** *Appl Environ Microbiol* 1991, **57:**712–716.
33. Unakal C, Kallur RI, Kaliwal BB: **Production of α-amylase using banana waste by *Bacillus subtilis* under solid state fermentation.** *Eur J Exper Biol* 2012, **2:**1044–1052.
34. Cordeiro CAM, Martins MLL, Luciano AB, da Silva RF: **Production and properties of xylanase from *Thermophilic Bacillus* sp.** *Braz Arch Biol Technol* 2002, **45:**413–418.
35. Ramesh MV, Lonsane BK: **Critical importance of moisture content of the medium in α-amylase by *Bacillus licheniformis* M27 in a solid-state fermentation system.** *Appl Microbiol Biotechnol* 1990, **33:**501–505.
36. Kim JH, Hosobuchi M, Kishimoto M, Seki T, Ryu DDY: **Cellulase production by a solidstate culture system.** *Biotechnol Bioeng* 1985, **27:**1445–1450.
37. Nagendra PG, Chandrasekharan M: **L-glutaminase production by marine *Vibrio costicola* under solid-state fermentation using different substrates.** *J Marine Biotechnol* 1996, **4:**176–179.
38. Feniksova RV, Tikhomirova AS, Rakhleeva BE: **Conditions for forming amylase and proteinase in surface culture of *Bacillus subtilis*.** *Mikrobiologia* 1960, **29:**745–748.
39. Lonsane BK, Ghildyal NP, Budiatman S, Ramakrishna SV: **Engineering aspects of solid state fermentation.** *Enzyme Microb Technol* 1985, **7:**258–265.
40. Murad HA, Saleem MME: **Utilization of uf-permeate for producing exopolysaccharides from lactic acid bacteria.** *Mansoura Univ J Agric Sci* 2001, **26:**2167–2175.
41. Kobayashi T, Koike K, Yoshimatsu T, Higaki N, Suzumatsu A, Ozawa T, Hatada Y, Ito S: **Purification and properties of a low-molecular weight, high-alkalinepectate lyase from an alkaliphilic strain of Bacillus.** *Biosci Biotechnol Biochem* 1999, **63:**65–72.
42. Narang S, Satyanarayana T: **Thermostable α-amylase production by an extreme thermophilic *Bacillus thermooleovorans*.** *Lett Appl Microbiol* 2001, **32:**1–35.
43. Arunava B, Pal SC, Sen SK: **Alpha amylase production in lactose medium by *Bacillus circulanse*.** *J Microbiol* 1993, **9:**142–148.
44. Mirminachi F, Zhang A, Roehr M: **Citric acid fermentation and heavy metal ions: Effect of iron, manganese and copper.** *Acta Biotechnol* 2000, **22:**363–373.

Increased ethanol accumulation from glucose via reduction of ATP level in a recombinant strain of *Saccharomyces cerevisiae* overexpressing alkaline phosphatase

Marta V Semkiv[1], Kostyantyn V Dmytruk[1], Charles A Abbas[3] and Andriy A Sibirny[1,2*]

Abstract

Background: The production of ethyl alcohol by fermentation represents the largest scale application of *Saccharomyces cerevisiae* in industrial biotechnology. Increased worldwide demand for fuel bioethanol is anticipated over the next decade and will exceed 200 billion liters from further expansions. Our working hypothesis was that the drop in ATP level in *S. cerevisiae* cells during alcoholic fermentation should lead to an increase in ethanol production (yield and productivity) with a greater amount of the utilized glucose converted to ethanol. Our approach to achieve this goal is to decrease the intracellular ATP level via increasing the unspecific alkaline phosphatase activity.

Results: Intact and truncated versions of the *S. cerevisiae PHO8* gene coding for vacuolar or cytosolic forms of alkaline phosphatase were fused with the alcohol dehydrogenase gene (*ADH1*) promoter. The constructed expression cassettes used for transformation vectors also contained the dominant selective marker *kanMX4* and *S. cerevisiae* δ-sequence to facilitate multicopy integration to the genome. Laboratory and industrial ethanol producing strains BY4742 and AS400 overexpressing vacuolar form of alkaline phosphatase were characterized by a slightly lowered intracellular ATP level and biomass accumulation and by an increase in ethanol productivity (13% and 7%) when compared to the parental strains. The strains expressing truncated cytosolic form of alkaline phosphatase showed a prolonged lag-phase, reduced biomass accumulation and a strong defect in ethanol production.

Conclusion: Overexpression of vacuolar alkaline phosphatase leads to an increased ethanol yield in *S. cerevisiae*.

Keywords: Baker's yeasts, Ethanol production, *PHO8*, Alkaline phosphatase, ATP content, Biomass accumulation

Background

Alcoholic fermentation represents the largest application of the yeast *Saccharomyces cerevisiae* in the field of industrial biotechnology. In 2011, worldwide fuel ethanol production reached 84.6 billion liters [1]. With increased ethanol use as a biofuel and continued growth in its production for distilled beverages, the worldwide annual production of ethanol will exceed in 2014 100 billion liters. Due to economic and environmental reasons, exponential growth in the production of fuel ethanol was observed over the past decade [2]. Though lignocellulose is considered to be one of the most promising feedstocks for production of fuel ethanol, its current industrial production relies heavily on fermentation of traditional feedstocks such as sucrose (derived primarily from sugarcane or sugar beets) and glucose obtained from starchy materials (corn, wheat, barley, potatoes etc.). Therefore the construction of the strains of *S. cerevisiae* with elevated ethanol production (yield and productivity) from glucose is of great academic and industrial interest. We decided to reach this goal by manipulation of the ATP content in the cell.

The yeast *S. cerevisiae* catabolizes glucose via the Embden-Meyerhof-Parnas (EMP) pathway which yield anaerobically 2 moles ATP per mole of consumed glucose. The efficiency

* Correspondence: sibirny@cellbiol.lviv.ua
[1]Institute of Cell Biology, NAS of Ukraine, Drahomanov Street, 14/16, Lviv 79005, Ukraine
[2]University of Rzeszow, Zelwerowicza 4, Rzeszow 35-601, Poland
Full list of author information is available at the end of the article

of this pathway for anabolic processes is low with a maximal biomass yield of around 7% and an ethanol yield in the range of 90% - 93% from the glucose consumed [3]. Even a slight shift of this ratio in favour of greater ethanol yield from dextrose, can provide an additional several millions litres of ethanol to the worldwide production annually. In contrast to *S. cerevisiae*, the bacterium, *Zymomonas mobilis*, ferments glucose through Entner-Doudoroff (ED) pathway. This pathway provides only 1 mole of ATP per mole of glucose, and consequently directs only 3% of glucose to cell biomass achieving ethanol yield of up to 97% of the possible theoretical value [4]. This indicates that lowering the ATP yield during alcoholic fermentation increases ethanol yield with reduced substrate conversion to cell mass. Fast fermentation of glucose to ethanol is another important advantage of *Z. mobilis* over *S. cerevisiae* [4,5]. Attempts to substitute *S. cerevisiae* by *Z. mobilis* for the production of industrial ethanol were considered to be a promising approach to increase ethanol yield. However, *Z. mobilis* has several serious drawbacks which hamper its industrial use and these consist of: (i) a very narrow substrate range (only glucose is efficiently fermented whereas sucrose fermentation is hardly proceeds with low yield), (ii) natural auxotrophy for lysine, methionine and some vitamins, (iii) non-GRAS status, which prevents using cellular biomass as a feed additive, (iv) requirement for a higher pH for growth [6,7]. Furthermore, the technology of yeast cell utilization for alcoholic fermentation is well developed whereas the use of bacterial cells for ethanol production is far less common. Thus, it looks like construction of the yeast strains which produce less ATP during alcoholic fermentation would be a better approach to increase ethanol yield. These new yeast strains would combine all of the possible advantages of yeast with the high ethanol yield of *Z. mobilis*. There could be several approaches to achieve this goal, for example: (i) substitution of EMP pathway in yeast by ED pathway from *Z. mobilis* or other bacteria possessing genes of the pathway; (ii) increasing the activity of the enzymes involved in generation of futile cycles; (iii) the introduction of heterologous genes encoding for energy-consuming plasma membrane glucose symporters or (iv) construction of recombinant strains with elevated activity of ATPase of other ATP-degrading enzymes [8].

The first approach to express *Escherichia coli* ED dehydratase and ED aldolase genes *edd* and *eda* in a phosphofructokinase deficient mutant of *S. cerevisiae* has been published [9]. The engineered strains grew and fermented glucose to ethanol, though activities of ED dehydratase and ED aldolase were not reported. It is known that quite often prokaryotic enzymes display low or no activity in *S. cerevisiae* hosts [10], which is most likely caused by improper folding or instability of the expressed bacterial protein in yeast. In addition, there are difficulties in NADP regeneration in the yeast engineered to possess ED pathway since NADPH produced in the glucose-6-phosphate dehydrogenase reaction, cannot be re-oxidized via alcohol dehydrogenase reaction and yeast does not have NADH/NADPH transhydrogenase activity [11,12].

Alternative approaches to lower ATP level in yeast cell and increase ethanol yield could rely on the activation of some cytosolic ATPases, other ATP hydrolyzing enzymes or via the induction of the some kinds of futile cycles to dissipate cellular pool of ATP, e.g. bacterial or modified yeast fructose-1,6-bisphosphatase [13-15]. There are also published data on decrease of cellular pool of ATP and activation of alcoholic fermentation through overexpression of the soluble part (F_1) of H^+-ATPase or a portion of F_1 exhibiting ATPase from different origin in *S. cerevisiae* [16]. Similar results were obtained after overexpression of *PHO5* coding for acid phosphatase which is non-specific enzyme hydrolyzing also ATP [15].

In earlier work, we carried out a successful attempt to decrease intracellular ATP level by overexpression of 5' part of the *S. cerevisiae SSB1* gene encoding cytosolic ATPase domain and by the heterologous gene *apy* encoding apyrase from *E. coli* [17]. In this paper, we describe a new method to lower cellular ATP and increase ethanol yield and productivity during glucose fermentation by overexpression of *S. cerevisiae* intact *PHO8* gene which codes for alkaline phosphatase.

Results and discussion

Overexpression of the vacuolar forms of alkaline phosphatase in BY4742 laboratory strain

S. cerevisiae unspecific alkaline phosphatase (phosphomonoesterase), located in the vacuole, encoded by the gene *PHO8*, catalyses the dephosphorylation of many different compounds including ATP [18,19]. Therefore this enzyme may operate as non-specific ATPase. We hypothesized that an increase in alkaline phosphatase activity will reduce intracellular ATP level and thus enhance ethanol production under anaerobic conditions. Several factors should be taken into account to increase specific activity of the target enzyme in yeast. Among these are the copy numbers used for the gene of interest, choice of promoter which controls expression level, stability of the synthetized mRNA, and enzyme feedback inhibition. We have demonstrated that the *ADH1* promoter for the gene encoding alcohol dehydrogenase that we used in this work, is very strong and is two-fold activated under the conditions of alcoholic fermentation [O. Kurylenko, K. Dmytruk, A. Sibirny, unpublished observations]. Therefore the *ADH1* gene promoter was used for *PHO8* gene overexpression. Additional activation of *PHO8* overexpression can be achieved by multicopy genomic integration. In order to insert a high copy number of the *PHO8* gene into the yeast genome, an integrative plasmid containing δ sequences was constructed. The

S. cerevisiae δ sequences are the long terminal repeats of the retro-transposons Tyl and Ty2 with an estimated 425 δ sequences dispersed throughout the yeast genome. It was shown that the use of vectors containing δ sequences can provide tandem multicopy integration in one or several sites of yeast genomic DNA via homologous recombination [20]. The δ sequences-based plasmid pUC57-delta1_2-ADHpr-PHO8-CYCt-kanMX harboring *PHO8* gene under the control of *ADH1* promoter was constructed as described in Materials and Methods and subsequently transformed into *S. cerevisiae* laboratory strain BY4742 (Table 1).

The specific alkaline phosphatase activity of the selected recombinant strains, derivative of BY4742, was assayed. Among 8 tested transformants, six strains were characterized by 30–40 times higher specific phosphatase activity relative to the wild-type strain BY4742 (Figure 1A, strains 1–5 and 8). The data confirm that the constructed expression vector provided multi-copy integration into the genome of *S. cerevisiae.*

Genomic DNA isolated from the recombinant strains was subjected to dot-blot hybridization to estimate plasmid copy number harbouring target gene, integrated into the genome. Gene *PHO8* was used as a probe (Figure 1B). The recombinant strains contained 1–9 additional copies of *PHO8* gene with a good correlation between *PHO8* gene copy number and specific alkaline phosphatase activity was observed.

Southern-hybridization was performed to analyse the vector integration pattern of the tested strains (Figure 2A). Total genomic DNA from wild type and recombinant strains were HindIII digested and hybridized with a labelled *PHO8* gene. The constructed δ sequences-based vector had tandem multi-copy integration in up to three sites of yeast genomic DNA in the so called "head-to-tail" conformation (Figure 2B), which is in good agreement with previously published results [20].

A selected recombinant strain with substantially enhanced level of specific alkaline phosphatase activity showed small defects in biomass accumulation and intracellular ATP level relative to that of the wild-type strain (Table 2). Ethanol productivity of the wild-type strain BY4742 was 0.79 g/L/h or 0.30 g/g of biomass/h, whereas the recombinant strain BY4742/Pho8vac, had a much higher productivity: 0.92 g/L/h or 0.34 g/g of biomass/h during glucose alcoholic fermentation. The *PHO8*-overexpressing recombinant strain accumulated 13% more ethanol as compared to the parental strain BY4742 (Table 2). The data provide proof to our initial hypothesis that the overexpression of alkaline phosphatase can be used to increase ethanol synthesis during glucose alcoholic fermentation.

Table 1 Strains and plasmids used in the present investigation

Strains	Genotype	Reference
BY4742	*MATα, his3Δ1, leu2Δ0, lys2Δ0, ura3Δ0*	[21]
BY4742/Pho8vac	BY4742, P_{ADH}-PHO8-kan^r	This study
BY4742/Pho8cyt	BY4742, P_{ADH}-PHO8_trunc-kan^r	This study
AS400	Wild type strain commercially used in bioethanol production at Archer Daniels Midland Company (Decatur, IL, USA)	This study
AS400/Pho8vac	AS400, P_{ADH}-PHO8-kan^r	This study

Overexpression of the vacuolar form of alkaline phosphatase in AS400 industrial strain

The vacuolar form of alkaline phosphatase was also expressed in the industrial ethanol-producing strain AS400. The plasmid pUC57-delta1_2-ADHpr-PHO8-CYCt-kanMX was used for transformation of the strain AS400. Selected transformants AS400/Pho8vac had 2.8-fold increase in the specific activity of alkaline phosphatase as compared to the parental strain AS400 (Table 2). The recombinant strain AS400/Pho8vac with the highest specific activity of alkaline phosphatase and the parental AS400 strain were tested for the efficiency of alcoholic fermentation in a mineral YNB medium supplemented with 20% glucose. Biomass and ethanol accumulation was measured in transformant AS400/Pho8vac and the parental strain AS400 (Figure 3). Based on the results, it was determined that the transformant AS400/Pho8vac accumulated 6% more ethanol relative to parental strain during fermentation producing 5.37 g/L/h of ethanol versus 5.07 g/L/h for AS400 strain (Figure 3). Due to slight difference in the biomass accumulation and complete glucose consumption during fermentation, differences in ethanol production between parental and recombinant strains may be caused by the decrease of byproducts accumulation like glycerol, acetate or reduced amount of reserve carbohydrates. The recombinant strain AS400/Pho8vac was also tested for the efficiency of alcoholic fermentation in a CSL medium supplemented with hydrolysed corn meal. This strain was shown to have substantially increased productivity of ethanol synthesis (3.75 g/L/h) as compared to that of the parental strain (3.3 g/L/h) (Table 2). Ethanol yield per gram of consumed glucose was found to be increased 13.6% (Table 2). Measurement of the productivity of ethanol synthesis per biomass was not carried out in CSL medium as it contains many small insoluble particles which hamper cell biomass measurement using direct dry weight or by optical density analyses. The recombinant strains tested were all also characterized by substantially lower intracellular ATP level relative to the parental strain AS400 (Table 2). We do not exclude the possibility that the drop in intracellular ATP level may have a negative impact on biomass accumulation during aerobic large-scale cultivation, but we assume that this effect will be insignificant, since biomass accumulation of recombinant strain

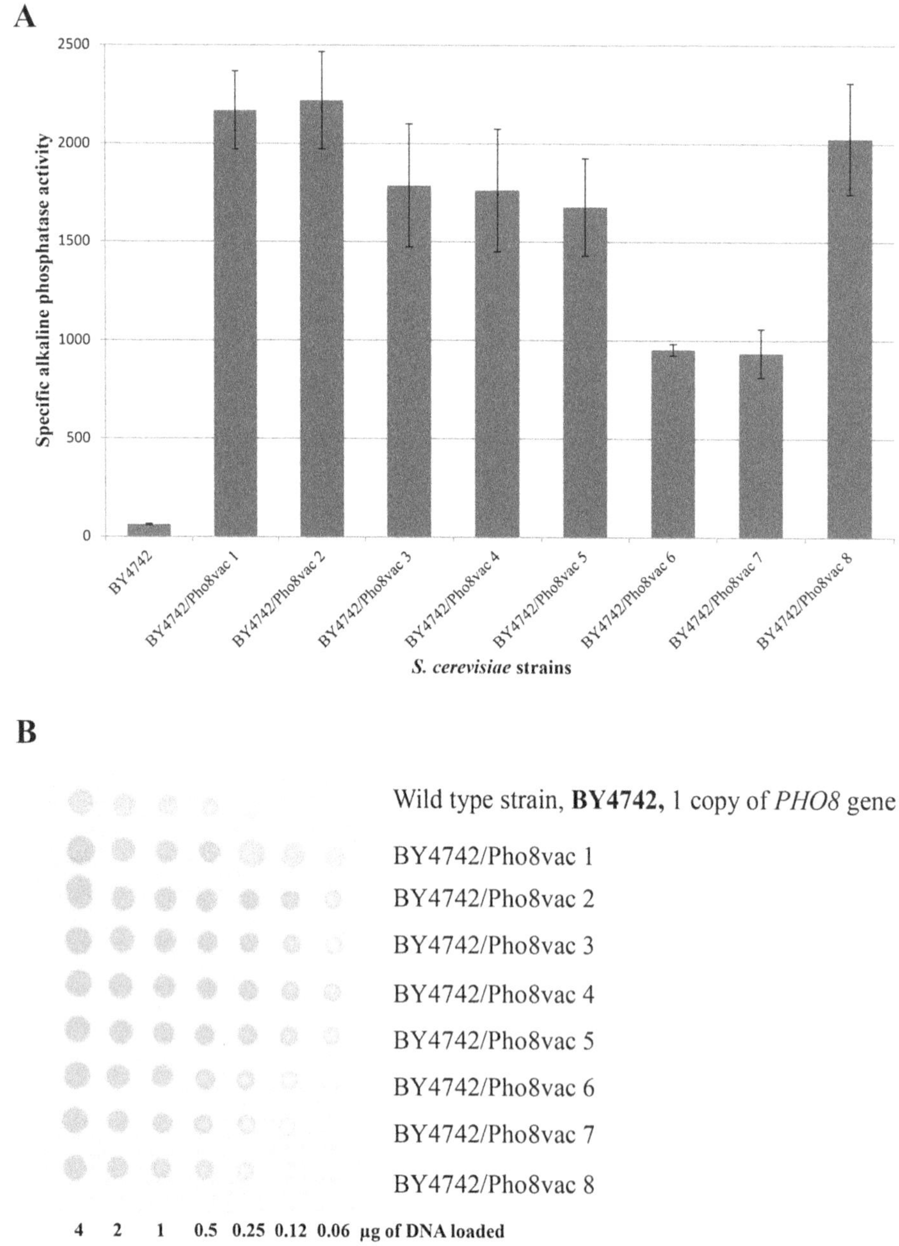

Figure 1 Specific alkaline phosphatase activity in nmoles of product/mg of prot.*min (A) and estimation of integrated expression cassette copy number by dot-blot hybridization (B) in the *S. cerevisiae* transformants and control strain. BY4742 – recipient strain; BY4742/Pho8vac – recombinant strains bearing plasmid pUC57-delta1_2-ADHpr-PHO8-CYCt-kanMX. Alkaline phosphatase activity is shown in nmoles of product/mg of prot.*min. For dot-blot hybridization genomic DNA was isolated from WT and recombinant strains, diluted and loaded onto nitrocellulose film. Amounts of loaded DNA is shown under the photo. *PHO8* gene was used as a probe.

AS400/Pho8vac didn't differ substantially from the parental strain AS400 in the experiments we carried out (Figure 3A).

Overexpression of the cytosolic form of alkaline phosphatase in BY4742

The results demonstrate that the overexpression of vacuolar alkaline phosphatase leads to an increase in ethanol production during alcoholic fermentation of glucose in laboratory and industrial strains of *S. cerevisiae*. This is due to the accelerated hydrolysis of substrate via increased activity from overexpression of vacuolar alkaline phosphatase. The target substrates are polyphosphates which are known to be the natural substrates of alkaline phosphatase [21]. As the synthesis of polyphosphates requires ATP, an increase in alkaline phosphatase activity may lead

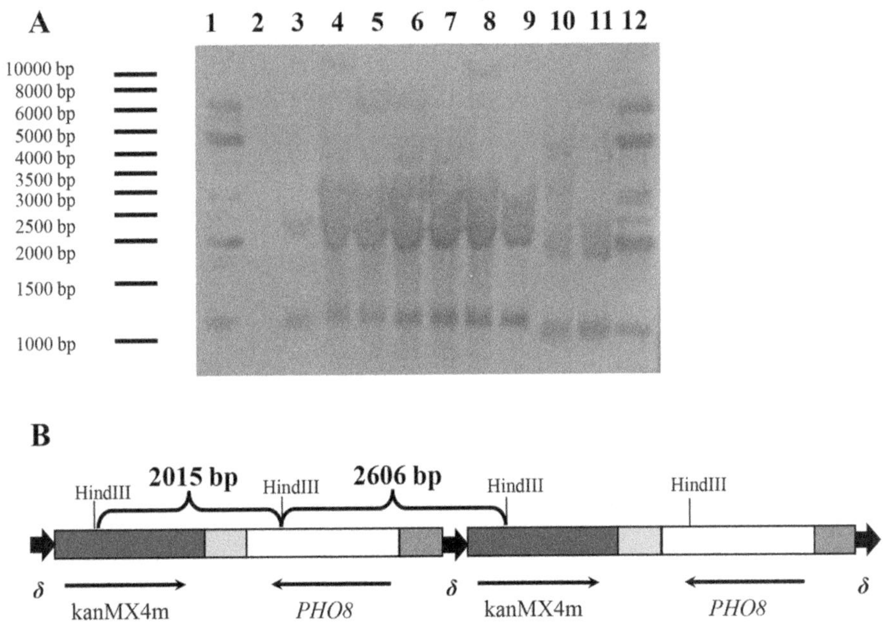

Figure 2 Analysis of expression cassette integration pattern. (A) Southern hybridization. *PHO8* gene was used as a probe. HindIII was used for genomic DNA restriction. 1, 12 – plasmid pUC57-delta1_2-ADHpr-PHO8-CYCt-kanMX; 2 – ladder; 3 - wild type strain, BY4742; 4–11 – BY4742/Pho8vac – recombinant strains containing vector pUC57-delta1_2-ADHpr-PHO8-CYCt-kanMX; **B**) "head-to-tail" conformation of vector integration.

to a futile cycle of ATP consumption which enhances need for ATP to replenish again the vacuolar polyphosphate pool [22]. We decided to check the effects of overexpression of the truncated forms of alkaline phosphatase which lack putative vacuolar targeting signal. A recombinant strain expressing truncated version of *PHO8* gene was constructed. Alkaline phosphatase is synthesized as an inactive precursor containing a C-terminal propeptide that is afterwards cleaved from the protein in yeast vacuoles in a Pep4-dependent manner [19]. During vacuolar delivery, which shares the early stages with the secretory pathway, the precursor form of this enzyme is glycosylated in the endoplasmic reticulum. It has also been shown that the active form of alkaline phosphatase binds with the metal co-factor zinc in the vacuole rather than in earlier compartments of the secretory pathway [23]. A truncated form of this enzyme was constructed which lacks 60 initial amino acids that are responsible for transmembrane protein delivery; and additionally 22 terminal amino acids, that represent the C-terminal propeptide that is normally cleaved from the protein in the vacuole [19,23]. The DNA fragment containing the expression cassette of the truncated *PHO8* gene in part with selective marker flanked by δ sequences were used for transformation of BY4742 strain of *S. cerevisiae*. The specific alkaline phosphatase activity of the selected recombinant strain BY4742/Pho8-cyt was assayed and compared with that of the strain BY4742/Pho8vac. The intracellular alkaline phosphatase

Table 2 Growth rate, ATP level, AP activity, ethanol productivity and yield of *S. cerevisiae* transformants and control strains

Strain	Specific growth rate, g/L/h*	ATP, μmoles of ATP/mg dry cell weight**	Alcaline phosphatase activity, nmoles of product/mg of prot.*min.*	Ethanol productivity*		Ethanol yield g/g of consumed glucose*
				g/L/h	g/L/g of biomass/h	
BY4742	0.031 ± 0.002	7.95 ± 0.10	85.2 ± 3.3	0.79 ± 0.014	0.30 ± 0.006	0.379 ± 0.007
BY4742/Pho8vac	0.032 ± 0.001	7.83 ± 0.06	1948.3 ± 175.3	0.92 ± 0.019	0.34 ± 0.007	0.442 ± 0.009
BY4742/Pho8cyt	0.025 ± 0.001	7.56 ± 0.08	1832.4 ± 137.4	0.42 ± 0.004	0.18 ± 0.002	0.202 ± 0.002
AS400		8.93 ± 0.15	45.3 ± 2.5	3.30 ± 0.083		0.396 ± 0.010
AS400/Pho8vac		8.11 ± 0.18	125.3 ± 4.4	3.75 ± 0.075		0.450 ± 0.010

*Results of three independent tests.
**Results of two independent tests.

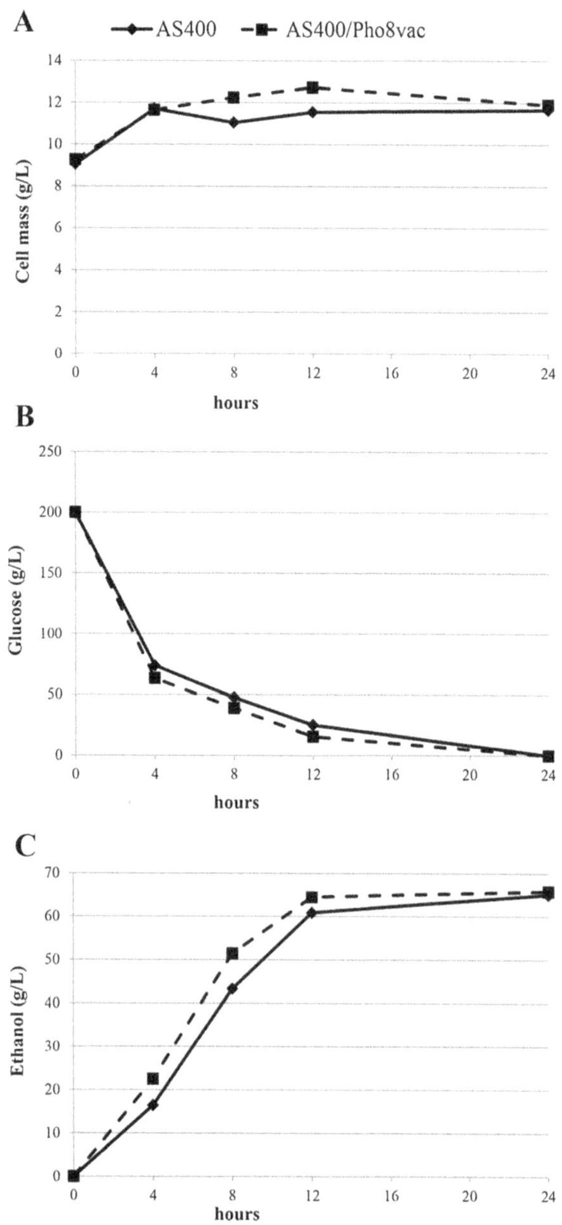

Figure 3 Growth, glucose consumption and ethanol production during alcoholic fermentation of recombinant strains overexpressing vacuolar form of alkaline phosphatase. Strains were cultured under semianaerobic conditions in YNB medium with 200 g/L D-glucose. Cultures were incubated at 30°C, shaking at 120 rpm using. AS400 – WT, recipient strain. AS400/Pho8vac – strain containing vector pUC57-delta1_2-ADHpr-PHO8-CYCt-kanMX. **A** – biomass accumulation (g of biomass/L of medium); **B** – glucose consumption (g of glucose/L of medium); **C** – ethanol production (g of ethanol/L of medium).

activity in the strain BY4742/Pho8cyt was at the same level as in the strain BY4742/Pho8vac overexpressing the intact form of *PHO8* (Table 2).

Growth kinetics of the constructed recombinant strains was measured. An equal amount of cellular biomass (0.015 g/L) of wild-type strain and recombinant strains were inoculated into the mineral media containing 10% glucose as a carbon source. The strain with truncated *PHO8* form was characterized by a longer lag-phase during the first day of cultivation (not shown), with lowered biomass accumulation and a reduced growth rate (Table 2). This recombinant strain (BY4742/Pho8cyt) had the highest specific activity of cytosolic form of alkaline phosphatase showed a decrease in the intracellular ATP level (Table 2).

Recombinant strain BY4742/Pho8cyt has produced significantly less ethanol during glucose alcoholic fermentation relative to recombinant strains with overexpressed native alkaline phosphatase and the wild-type strain BY4742, i.e. only about 0.42 g/L/h or 0.18 g/g of biomass/h (Table 2). We hypothesize that the cytosolic form of alkaline phosphatase decreases not only the level of ATP but also that of other sugar phosphate intermediates as well as regulators like fructose-2,6-bisphosphate, leading to detrimental effect on ethanol production. However, confirmation of this hypothesis requires additional experiments which are in progress.

Conclusion

This study established a new successful approach to increasing ethanol production from glucose in *S. cerevisiae* by homologous overexpression of *PHO8* gene due to multiple genomic insertion events. The overexpression of intact *PHO8* gene encoding vacuolar form of alkaline phosphatase led to decline in intracellular ATP level and biomass accumulation and up to 13% increase in ethanol productivity. By comparison, the overexpression of truncated form of *PHO8* gene encoding presumably cytosolic form of alkaline phosphatase caused a significant drop both in biomass accumulation and ethanol production. In conclusion, the overexpression of the intact vacuolar form of alkaline phosphatase results in overproduction of ethanol in laboratory and industrial strains of *S. cerevisiae*.

Methods

Strains, media, growth and fermentation conditions

The *S. cerevisiae* strain BY4742 (*MAT*α, *his3*Δ*1*, *leu2*Δ*0*, *lys2*Δ*0*, *ura3*Δ*0*; [24] and an industrial strain AS400 (obtained from Archer Daniels Midland Company, Decatur, IL, USA) (Table 1) were used for the expression of the intact or truncated version of *PHO8* ORF coding for alkaline phosphatase. For selection of yeast transformants on YPD, 200 mg/L of geneticin was added. When required, histidine (20 mg/L), leucine (60 mg/L), lysine (20 mg/L), or uracil (20 mg/L) were added.

S. cerevisiae strains were incubated at 30°C. Yeast strains were maintained in rich YPD (1% yeast extract, 1% peptone and 2% glucose) or mineral YNB (0.67%, yeast nitrogen base without amino acids, DIFCO, 0.5% ammonium sulfate, 2% glucose) media. For ethanol fermentation, all strains were tested with a YNB medium that was supplemented with 10% glucose or by using a medium containing corn steep liqour (CSL) supplemented with hydrolyzed corn meal as the primary carborn source.

The corn steep liquor (CSL) medium was prepared by combining two solutions that were prepared as follows. Solution 1 was prepared by adding 200 g of corn meal (Melvit, Poland) was mixed with 800 ml of water and pH adjusted to 6.0 using NaOH. The enzyme alpha-amylase (Liquizyme SC) was added at a rate of 0.1 units of per gram of meal and the slurry was liquefied by heating the mixture to 80°C for 30 minutes. To prepare Solution 2, 63 ml of CSL Concentrate (containing ~50% dry solids) was mixed with 137 ml of deionized water. The two solutions were autoclaved, cooled, combined and mixed. The enzyme glucoamylase (Liquizyme SC) was added at a rate of 0.2 units of glucoamylase per gram of meal to sterile flasks and the flasks were incubated at 28°C. Glucose concentration was measured in control flasks without yeast inoculation at 180–200 g/L in the CSL medium prepared.

Cells of BY4742 strain and its *PHO8*-expressing derivatives were grown in 100 ml of YPD medium in Erlenmeyer flasks (bottle size – 300 ml) overnight and then used to inoculate a 20 ml of YNB medium with 10% glucose in 50 ml Erlenmeyer flasks. An initial biomass concentration of 1.2 g (dry weight)/L was used for fermentation. Fermentation was carried out at a temperature of 30°C with limited aeration using a gyratory shaker at a setting of 120 revolutions/min.

For the *S. cerevisiae* AS400 strain and its *PHO8*-expressing derivatives, both YNB medium with 20% glucose and CSL medium supplemented with hydrolyzed meal were used in fermentation tests. An initial biomass concentration of 10 g (dry weight)/L was used for fermentation as this is the cell density normally used in industrial ethanol fermentation [25]. Strains were incubated under semi-anaerobic condition at 34°C for 2 days and samples were taken every 24 hours.

The *E. coli* DH5α strain (Φ80d*lacZ*ΔM15, *recA*1, *endA*1, *gyrA*96, *thi*-1, *hsdR*17(r_K^-, m_K^+), *supE*44, *relA*1, *deoR*, Δ(*lacZYA-argF*)U169) was used as a host for propagation of plasmids. Strain DH5α was grown at 37°C in LB medium as described previously [26]. Transformed *E. coli* cells were maintained on a medium containing 100 mg/L of ampicillin. The chromogenic substrates X-gal and IPTG (Fermentas, Vilnius, Lithuania) were used according to the manufacturer specifications.

DNA manipulations

Genomic DNA from *S. cerevisiae* strains was isolated using the Wizard® Genomic DNA Purification Kit (Promega, Madison, WI, USA). Plasmid DNA from *E. coli* was isolated using the Wizard® *Plus* SV Minipreps DNA Purification System (Promega). Taq and High Fidelity polymerase mix, T4 DNA ligase, T4 DNA polymerase and restriction enzymes were used according to recommendation of supplier (Fermentas). *S. cerevisiae* transformation was performed by standard protocol [26].

Plasmid construction

Expression cassette preparation. 154 bp 5'- and 180 bp 3'-parts of *S. cerevisiae* YJRWdelta12 sequence were amplified from genomic DNA of *S. cerevisiae* strain BY4742 using pairs of primers SM16/SM17 and SM18/SM19 (Table 3). Both delta sequences were fused via overlap PCR using primers SM16 and SM19, digested with EcoRI and HindIII and cloned into EcoRI/HindIII-linearized plasmid pUC57. The constructed plasmid was designated pUC57-delta1_2. Fragments of DNA having 807 bp and 269 bp bearing promoter of *ADH1* gene encoding alcohol dehydrogenase and terminator of *CYC1* gene encoding cytochrome C, were amplified from genomic DNA of BY4742 by pairs of primers Ko419/Ko420 and Ko453/Ko454, respectively. The *ADH1* promoter and *CYC1* terminator were fused via overlap PCR using primers Ko419 and Ko454 and SalI/XmaI cloned into the corresponding sites of the plasmid pUC57-delta1_2 resulting in the plasmid pUC57-delta1_2-ADHpr-CYCt. A 1470 bp DNA fragment corresponding to the selective marker *kanMX* providing resistance to geneticin was cut out from the plasmid pRS303K [27] with restriction endonucleases SacI and SmaI, blunt-ended and cloned into XbaI-digested and blunted plasmid pUC57-delta1_2-ADHpr-CYCt yielding the plasmid pUC57-delta1_2-ADHpr-CYCt-kanMX (Figure 4A).

Plasmid for expression of native PHO8. A 1701 bp DNA fragment bearing the ORF of *PHO8* gene coding for unspecific alkaline phosphatase was amplified from genomic DNA of the BY4742 strain using primers Ko508 and Ko509. This fragment was digested with BamHI and NotI and subcloned into BamHI/NotI digested plasmid pUC57-delta1_2-ADHpr-CYCt-kanMX. This plasmid was designated pUC57-delta1_2-ADHpr-PHO8-CYCt-kanMX (Figure 4B).

Plasmid for expression of truncated version of PHO8 gene. The cytosolic form of Pho8 lacking 60 initial amino acids which are responsible for transmembrane protein delivery; and 22 terminal amino acids composing the C-terminal propeptide that is normally cleaved from the protein in vacuole, was isolated. A 1452 bp DNA fragment corresponding to this truncated form, was amplified from genomic DNA of BY4742 strain using primers SM28 and SM29, digested with endonucleases BamHI and NotI and cloned into BamHI/NotI digested plasmid pUC57-delta1_2-ADHpr-PHO8-CYCt-kanMX.

Table 3 DNA oligonucleotides used in this study

Primer names	Sequence (5′-3′) restriction sites are italic	Restriction sites
SM16	CCG*GAATTC*[a]*GACGGGCAGTC*[b]TGTTGGAATAGAAATCAACTATC	EcoRI[a], AhdI[b]
SM17	CATCATTTTATATGTTTATATTCA*TCTAGA*[a]*CCCGGG*[b]*GTCGAC*[c] TTGATCCTATTACATTATCAATCC	XbaI[a], XmaI[b], SalI[c]
SM18	GGATTGATAATGTAATAGGATCAA*GTCGAC*[a]*CCCGGG*[b]*TCTAGA*[c]TGAATATAAACATATAAAATGATG	SalI[a], XmaI[b], XbaI[c]
SM19	CCC*AAGCTT*[a]*GACGGGCAGTC*[b]TGAGAAATATGTGAATGTTGAG	HindIII[a], AhdI[b]
SM28	CGC*GGATCC*[a]ATGTCTGCATCACACAAGAAGAAGAATGTC	BamHI[a]
SM29	TTT*GCGGCCGC*[a]TCAATCTGATGTGTGTTTGGTGTCCCTAATC	NotI[a]
Ko419	CGC*GTCGAC*[a]TTAATTAAAGTCCAATGCTAG	SalI[a]
Ko420	GATATCGACAAAGGAAAAGGG*GCGGCCGC*[a]*GGATCC*[b]*CTCGAG*[c]TGTATATGAGATAGTTGATTG	NotI[a], BamHI[b], XhoI[c]
Ko453	CAATCAACTATCTCATATACA*CTCGAG*[a]*GGATCC*[b]*GCGGCC GC*[c]CCCTTTTCCTTTGTCGATATC	XhoI[a], BamHI[b], NotI[c]
Ko454	CCCCC*CGGG*[a]GCAAATTAAAGCCTTCGAGC	XmaI[a]
Ko508	CGC*GGATCC*[a]ATGATGACTCACACATTACCAAGC	BamHI[a]
Ko509	TTT*GCGGCCGC*[a]TCAGTTGGTCAACTCATGGTAGTATTC	NotI[a]

[a,b,c]Restriction sites with corresponding restriction enzyme.

The constructed plasmid was designated as pUC57-delta1_2-ADHpr-PHO8_trunc-CYCt-kanMX (Figure 4C).

Selection of *S. cerevisiae* transformants

The vectors containing native or modified versions of the *PHO8* gene were digested with the AhdI. AhdI-fragments containing expression cassettes and selective marker flanked by δ-sequences, were eluted from an agarose gel and used for transformation of *S. cerevisiae* strains BY4742 and AS400. The transformants were selected on a solid YPD medium supplemented with 200 mg/L of geneticin. The selected transformants were stabilized by alternating cultivation in non-selective and selective media and examined by diagnostic PCR using a forward primer Ko419 specific to the *ADH1* promoter and a reverse one Ko509 specific to *PHO8* gene.

Biochemical methods and analyses

The specific activity of alkaline phosphatase was assayed in cell free extracts using the chromogenic substrate p-nitrophenyl phosphate as described elsewhere [18]. The assay was repeated three times and the measurement reported is an average of these determinations. The numbers reported use one unit of alkaline phosphatase activity

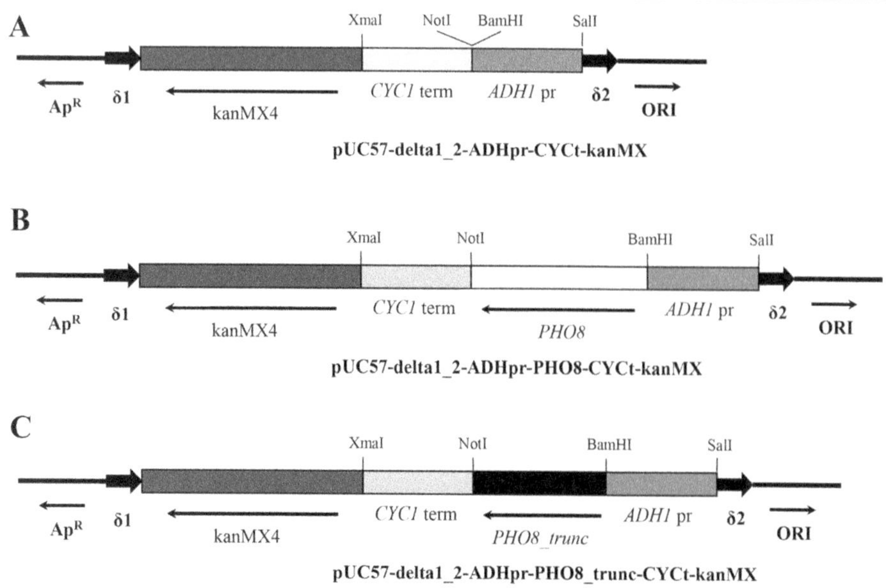

Figure 4 Linear schemes of plasmids used in this study: pUC57-delta1_2-ADHpr-CYCt-kanMX (A), pUC57-delta1_2-ADHpr-PHO8-CYCt-kanMX (B), pUC57-delta1_2-ADHpr-PHO8_trunc-CYCt-kanMX (C). δ elements are shown as thick black arrows. *ADH1* promoter, *CYC1* terminator and *kanMX4* gene are shown as chessboard-alike, doted and diagonal-hatched boxes, respectively. Unmodified *PHO8* gene ORF is shown as white box and truncated *PHO8* gene ORF is shown as black doted box. Origin of replication ORI and ampicillin resistance gene (*bla*) are shown as thin arrows.

as the amount of enzyme that liberates 1 μmol of p-nitrophenol per min under the assay conditions.

Measurements of ATP were carried out following extraction of this metabolite from the *S. cerevisiae* cells was using boiling ethanol and subsequent evaporation [28]. In order to optimize ATP assay condition for ethanol extraction, a wide range of yeast biomass from 2 to 15 mg was tested. Reproducible results were obtained during ATP extraction with 8–10 mg of *S. cerevisiae* cells using 1 ml of boiling ethanol. Extracted ATP was measured by coupled enzymatic reactions with hexokinase and glucose-6-phosphate dehydrogenase [29].

The concentration of ethanol in fermentation media was determined using alcohol oxidase/peroxidase-based enzymatic kit "Alcotest" [30,31]. The biomass was determined using turbidity with a Helios Gamma spectrophotometer (OD, 600 nm; cuvette, 10 mm) with gravimetric calibration. Glucose concentration was determined using the "Diagluc" assay kit (UBT, Lviv) [30]. All samples were assayed at least twice to ensure results are reproducible.

Competing interests

The authors declare that they have no competing interests.

Authors' contributions

MVS carried out gene engineering manipulations, enzyme activity assays and alcoholic fermentations. KVD participated in design of cloning and strain construction, analyzed the date and co-drafted the manuscript. CAA participated in plasmids sequencing, commented and approved the manuscript. AAS provided guidance and suggestions for experimental design and edited the manuscript. All authors have read and approved the manuscript.

Acknowledgement

This work was supported in part by Archer Daniels Midland Company (Decatur, IL); Polish grant of National Scientific Center (NCN) DEC-2012/05/B/NZ1/01657 awarded to AAS and FEMS Fellowship grant # UA-SMU2012-2 awarded to MVS.

Author details

[1]Institute of Cell Biology, NAS of Ukraine, Drahomanov Street, 14/16, Lviv 79005, Ukraine. [2]University of Rzeszow, Zelwerowicza 4, Rzeszow 35-601, Poland. [3]Archer Daniels Midland Co Research Center, 1001 N Brush College Rd, Decatur, IL 62521, USA.

References

1. Renewable Fuels Association: *Acelerating Industry Innovation – 2012 Ethanol Industry Outlook.* Washington, DC: Renewable Fuels Association; 2012:3, 8, 10, 22–23.
2. Schubert C: **Can biofuels finally take center stage?** *Nat Biotechnol* 2006, **24:**777–784.
3. Ingledew WM: *Alcohol production by Saccharomyces cerevisiae: a yeast primer, in the alcohol textbook.* 3rd edition. UK: Nottingham University Press; 1999.
4. Sprenger GA: **Carbohydrate metabolism in *Zymomonas mobilis*: a catabolic highway with some scenic routes.** *FEMS Microbiol Lett* 1996, **145:**301–307.
5. Panesar PS, Marwaha SS, Kennedy JF: ***Zymomonas mobilis*: an alternative ethanol producer.** *J Chem Technol Biotechnol* 2006, **81:**623–635.
6. Jeffries TW: **Ethanol fermentation on the move.** *Nature* 2005, **23:**40–41.
7. Bai FW, Anderson WA, Moo-Young M: **Ethanol fermentation technologies from sugar and starch feedstocks.** *Biotechnol Adv* 2008, **26:**89–105.
8. De Kok S, Kozak B, Pronk J, Van Maris A: **Energy coupling in *Saccharomyces cerevisiae*: selected opportunities for metabolic engineering.** *FEMS Yeast Res* 2012, **12:**387–97.
9. Lancashire WE, Dickinson JR, Malloch RA: **DNA encoding enzymes of the glycolytic pathway for use in alcohol producing yeast.** *US Patent* 1998, **5786186:**A1.
10. Hahn-Hagerdal B, Karhumaa K, Jeppsson M, Gorwa-Grauslund MF: **Metabolic Engineering for Pentose Utilization in *Saccharomyces cerevisiae*.** *Adv Biochem Engin Biotechnol* 2007, **108:**147–177.
11. Leskovac V, Trivic S, Peric D: **The three zinc-containing alcohol dehydrogenases from baker's yeast, *Saccharomyces cerevisiae*.** *FEMS Yeast Res* 2002, **2:**481–494.
12. Jeffries TW, Jin Y-S: **Metabolic engineering for improved fermentation of pentoses by yeasts.** *Appl Microbiol Biotechnol* 2004, **63:**495–509.
13. Navas MA, Cerdan S, Gancedo JM: **Futile cycles in *Saccharomyces cerevisiae* strains expressing the gluconeogenic enzymes during growth on glucose.** *Prot Natl Acad Sci USA* 1993, **90:**1290–1294.
14. Navas MA, Gancedo JM: **The regulatory characteristics of yeast fructose-1,6-bisphosphatase confer only a small selective advantage.** *J Bacteriol* 1996, **178:**1809–1812.
15. Rogers DT, Szostak JW: **Strains of yeast with increased rates of glycolysis.** *US Patent* 1993, **5268285:**A1.
16. Jensen P, Snoep J, Westerhoff H: **Method of improving the production of biomass or a desired product from a cell.** *US Patent* 2006, **20060094078:**A1.
17. Dmytruk KV, Semkiv MV, Sibirny AA: **Ethanol yield and reduction of biomass accumulation in the recombinant strain of *Saccharomyces cerevisiae* overexpressing ATPase.** *US Patent Application* 2012, **0088290:**A1.
18. Kaneko Y, Akio Toh E, Oshima Y: **Identification of the Genetic Locus for the Structural Gene and a New Regulatory Gene for the Synthesis of Repressible Alkaline Phosphatase in *Saccharomyces cerevisiae*.** *Mol Cell Biol* 1982, **2:**127–137.
19. Klionsky DJ, Emr SD: **Membrane protein sorting: biosynthesis, transport and processing of yeast vacuolar alkaline phosphatase.** *EMBO J* 1989, **8:**2241–2250.
20. Lee FW, Da Silva NA: **Improved efficiency and stability of multiple cloned gene insertions at the δ sequences of *Saccharomyces cerevisiae*.** *Appl Microbiol Biotechnol* 1997, **48:**339–345.
21. Fernandez M, Gascon S, Schwencke J: **Some enzymatic properties of vacuolar alkaline phosphatase from yeast.** *Curr Microbiol* 1981, **6:**121–126.
22. Klionsky DJ, Herman PK, Emr SD: **The Fungal Vacuole: Composition, Function, and Biogenesis.** *Microbiol Rev* 1990, **54:**266–292.
23. Qiao W, Ellis C, Steffen J, Wu J-Y, Eide D: **Zinc status and vacuolar zinc transporters control alkaline phosphatase accumulation and activity in *Saccharomyces cerevisiae*.** *Mol Cell Biol* 1985, **5:**248–252.
24. Giaever G, Chu AM, Ni L, Connelly C, Riles L, Veronneau S, Dow S, Lucau-Danila A, Anderson K, André B, Arkin AP, Astromoff A, El-Bakkoury M, Bangham R, Benito R, Brachat S, Campanaro S, Curtiss M, Davis K, Deutschbauer A, Entian KD, Flaherty P, Foury F, Garfinkel DJ, Gerstein M, Gotte D, Güldener U, Hegemann JH, Hempel S, Herman Z, *et al*: **Functional profiling of the *Saccharomyces cerevisiae* genome.** *Nature* 2002, **418:**387–391.
25. Lin Y, Tanaka S: **Ethanol fermentation from biomass resources: current state and prospects.** *Appl Microbiol Biotechnol* 2006, **69:**627–642.
26. Sambrook J, Russell DW: *Molecular cloning: a laboratory manual.* 3rd edition. New York: Cold Spring Harbor Laboratory Press; 2001.
27. Taxis C, Knop M: **System of centromeric, episomal, and integrative vectors based on drug resistance markers for *Saccharomyces cerevisiae*.** *Biotechniques* 2006, **40:**73–78.
28. Entian K-D, Droll L, Mecke D: **Studies on rapid reversible and non-reversible inactivation of fructose-1,6-bisphosphate and malate dehydrogenase in Wild-type and glycolytic block mutants of *Saccharomyces cerevisiae*.** *Arch Microbiol* 1983, **134:**187–192.
29. Ano Y, Hattori T, Kato N, Sakai Y: **Intracellular ATP correlates with mode of pexophagy in *Pichia pastoris*.** *Biosci Biotechnol Biochem* 2005, **69:**1527–1533.
30. Gonchar MV: **Sensitive method for quantitative determination of hydrogen peroxide and oxidase substrates in biological samples.** *Ukr Biokhim Zh* 1998, **70:**157–163.
31. Gonchar MV, Maidan MM, Pavlishko HM, Sibirny AA: **A new oxidase-peroxidase kit for ethanol assays in alcoholic beverages.** *Food Technol Biotechnol* 2001, **39:**37–42.

5

Analysis, pretreatment and enzymatic saccharification of different fractions of Scots pine

Monica Normark[1], Sandra Winestrand[1], Torbjörn A Lestander[2] and Leif J Jönsson[1*]

Abstract

Background: Forestry residues consisting of softwood are a major lignocellulosic resource for production of liquid biofuels. Scots pine, a commercially important forest tree, was fractionated into seven fractions of chips: juvenile heartwood, mature heartwood, juvenile sapwood, mature sapwood, bark, top parts, and knotwood. The different fractions were characterized analytically with regard to chemical composition and susceptibility to dilute-acid pretreatment and enzymatic saccharification.

Results: All fractions were characterized by a high glucan content (38-43%) and a high content of other carbohydrates (11-14% mannan, 2-4% galactan) that generate easily convertible hexose sugars, and by a low content of inorganic material (0.2-0.9% ash). The lignin content was relatively uniform (27-32%) and the syringyl-guaiacyl ratio of the different fractions were within the range 0.021-0.025. The knotwood had a high content of extractives (9%) compared to the other fractions. The effects of pretreatment and enzymatic saccharification were relatively similar, but without pretreatment the bark fraction was considerably more susceptible to enzymatic saccharification.

Conclusions: Since sawn timber is a main product from softwood species such as Scots pine, it is an important issue whether different parts of the tree are equally suitable for bioconversion processes. The investigation shows that bioconversion of Scots pine is facilitated by that most of the different fractions exhibit relatively similar properties with regard to chemical composition and susceptibility to techniques used for bioconversion of woody biomass.

Keywords: Scots pine, Chemical composition, Dilute-acid pretreatment, Enzymatic saccharification

Background

Lignocellulosic resources, such as wood, can be converted to fuels and chemicals that serve as renewable alternatives to commodities produced from fossil resources, such as petroleum [1,2]. Potential lignocellulosic feedstocks include residues from forestry and agriculture, as well as energy crops that are dedicated for production of energy carriers, and which include both woody and non-woody plants.

Since liquid fuels need to be produced in large quantities, there are good reasons to investigate utilization of a wide variety of lignocellulosic feedstocks and not only those that with relative ease can be converted by pretreatment, saccharification, and microbial fermentation. Boreal forests that cover large areas of the northern parts of the Palearctic and Nearctic regions mainly consist of coniferous trees. For instance, data for Sweden during the period 2006–2010 indicate that 82% of the forest resources (given as the standing volume) consisted of softwood [3]. In rough terms approx. one half of that was Norway spruce (*Picea abies*) and the other half was Scots pine (*Pinus sylvestris*). Softwood, including various species of pine, is also common in many other forests in temperate and subtropical areas of the world. Although softwood is typically recalcitrant with regard to biocatalytic conversion processes, there are also clear advantages with using softwood as feedstocks [4,5]. Softwood is typically rich in glucan, mannan, and galactan [6,7], which can be converted to hexose sugars that are generally preferred in comparison to pentose sugars by microbes producing cellulosic ethanol and other fermentation products. Furthermore, several countries have well established systems for sustainable management of coniferous forests, where the main product is high-quality sawn timber, but where a

* Correspondence: leif.jonsson@chem.umu.se
[1]Department of Chemistry, Umeå University, Umeå SE-901 87, Sweden
Full list of author information is available at the end of the article

considerable fraction is left for other purposes as an abundant lignocellulosic resource that is available throughout the year.

Considerable efforts are currently devoted to investigate the recalcitrance of natural and genetically-engineered varieties of different plants that are intended to be used for saccharification and further on to production of fuels and chemicals by fermenting microbes [2]. With regard to softwoods that are primarily used for production of sawn timber, the varieties that are planted in managed forests today will provide the supply of lignocellulosic residues in the decades to come. Because of that, there are good reasons to investigate the recalcitrance of the softwood that is used today, as we can expect the same type of feedstock in the near future.

We have divided the lignocellulose of Scots pine, one of the most important commercial species of softwood, into seven different fractions. The seven fractions were analyzed with regard to chemical composition, susceptibility to thermochemical pretreatment, and susceptibility to enzymatic hydrolysis. Investigations in this area shed light upon the fundamental properties of the feedstock, and are relevant for the potential utilization of softwood for bioconversion to fuels and chemicals. In case differences between different fractions are discovered, there are also emerging possibilities to fractionate forest feedstocks in an industrial scale using advanced techniques [8]. Some of the fractions included in this study, such as bark and top parts, are nevertheless available today using conventional fractionation technology. Other fractions that were included in the study, such as juvenile and mature sapwood and heartwood, were of interest to compare mainly for fundamental reasons. Therefore, studies in this area are relevant both with regard to fundamental issues concerning bioconversion of lignocellulose, and for applications related to a transition towards a bio-based economy.

Results and discussion

Chemical composition of pine fractions

The chemical composition of fractions of Scots pine wood and bark was determined and the results are summarized in Table 1. The glucan content varied between 38 and 43%. The lowest value was found in knotwood, while heartwood exhibited slightly higher values (42-43%). All fractions had small amounts of arabinan (2%) and galactan (2-4%). The xylan content varied between 5 and 7%, and the mannan content was 10-14%. The total lignin content varied between 27 and 32%. Most of the lignin was Klason lignin, while a minor part was acid-soluble lignin (Table 1). The total lignin content of the sapwood (27-28%) was slightly lower than that of the heartwood (29-30%). This agrees with previous studies of softwood, such as Norway spruce [9]. The highest lignin content was determined for knotwood (32%) (Table 1). The extraction procedure was based on a 9:1 mixture of petroleum ether and acetone rather than on ethanol, since inefficient extraction may lead to an overestimate of the lignin content due to precipitation of extractives with the lignin [10]. However, it is still possible that the lignin content of the extractive-rich knotwood (9% extractives, Table 1) is somewhat overestimated. This notion is supported by the fact that the total mass balance for knotwood ended up at 105%, while all other fractions had a mass balance between 92 and 100% (Table 1). The ash contents of the pine wood fractions were low (0.2-0.3%), while the bark fraction, which also contained phloem tissue and some wood, had a higher ash content (0.9%).

As expected, the xylan content of the Scots pine fractions was low in relation to literature data for the xylan content of agricultural residues, such as sweet sorghum bagasse (17.2%), corn stover (18.9%), and wheat straw (18.4%) [6]. Prominent softwood hemicelluloses include galactoglucomannan and arabinoglucuronoxylan [7]. The total carbohydrate content of the Scots pine fractions varied between 61 and 66%, which could be compared with literature data for *Populus tremuloides* (58.3%), birchwood (57.9%), sugarcane bagasse (64.3%), wheat straw (57.1%), and corn stover (57.7%) (adapted from [6]). Sjöström [7] reported a lignin content of 27.7% for *P. sylvestris*, which corresponds well with the values of the present study. As expected, the lignin content of the pine fractions was higher than the lignin-content values reported for hardwood species, such as *Populus tremuloides* (16.7%) and birchwood (22.8%), and agricultural residues, such as sugarcane bagasse (18.6%) and wheat straw (17.6%) [6]. Knotwood is known to have a high content of extractives and attempts have been made to collect knotwood fractions for preparation of bioactive extracts [11]. Lestander et al. [12] reported ash contents of 0.3-0.4% in various types of wood (Scots pine, Norway spruce and birch), while the average ash content in the bark of Scots pine was 2.0%. The lower value of the ash content of bark reported in this study compared to the value reported by Lestander et al. [12] can be attributed to the presence of phloem tissue and some wood in the bark fraction of our study. The ash content of the pine fractions is very low compared to that of many agricultural residues considered for utilization in lignocellulosic biorefineries, for example sweet sorghum bagasse (4.2%), corn stover (4.9%), and wheat straw (5.8%) [6]. In a literature review by Tao et al. [13] covering 742 data objects it was shown that straw of herbaceous grasses contains in average more than 3% ash and up to about 18% in rice. Kenney et al. [14] have reported a typical ash content of 6-8% in a dataset of 840 samples from corn stover, Miscanthus and wheat straw. A high content of carbohydrates that can be hydrolyzed to hexose sugars and a low ash content are advantageous properties from a biorefining perspective.

Table 1 Chemical composition of fractions of Scots pine[a]

	Contents in % (w/w)										
Fraction	*Arabinan ± SD*	*Galactan ± SD*	*Glucan ± SD*	*Mannan ± SD*	*Xylan ± SD*	*Klason lignin ± SD*	*Acid soluble lignin ± SD*	*Total lignin ± SD*	*Extractives ± SD*	*Ash ± SD*	*Total[b]*
Juvenile heartwood	2.0 ± 0.1	3.1 ± 0.1	42.7 ± 3.9	11.8 ± 0.2	6.6 ± 0.1	27.9 ± 0.3	1.6 ± 0.1	29.5 ± 0.4	4.5 ± 0.5	0.2 ± 0.1	100
Mature heartwood	1.8 ± 0.1	3.0 ± 0.1	42.2 ± 5.2	12.1 ± 0.2	5.3 ± 0.1	27.7 ± 0.1	1.4 ± 0.1	29.1 ± 0.2	4.4 ± 0.6	0.2 ± 0.1	98
Juvenile sapwood	1.9 ± 0.1	2.8 ± 0.2	39.7 ± 0.1	10.5 ± 0.3	6.2 ± 0.3	25.8 ± 0.4	1.7 ± 0.1	27.4 ± 0.5	3.0 ± 0.2	0.3 ± 0.1	92
Mature sapwood	1.7 ± 0.1	2.3 ± 0.1	41.8 ± 1.6	14.0 ± 0.2	5.5 ± 0.1	26.9 ± 0.2	1.5 ± 0.1	28.4 ± 0.2	2.9 ± 0.1	0.2 ± 0.1	97
Bark	2.4 ± 0.2	2.8 ± 0.2	41.8 ± 2.0	11.7 ± 0.3	5.4 ± 0.1	28.2 ± 0.3	1.9 ± 0.1	30.0 ± 0.3	3.5 ± 0.2	0.9 ± 0.3	99
Top parts	2.0 ± 0.1	3.4 ± 0.1	41.4 ± 1.4	11.4 ± 0.1	6.6 ± 0.1	28.1 ± 0.2	1.6 ± 0.1	29.7 ± 0.2	3.3 ± 0.6	0.3 ± 0.1	98
Knotwood	2.2 ± 0.1	4.1 ± 0.3	38.2 ± 1.4	11.9 ± 0.1	6.6 ± 0.1	30.0 ± 0.3	1.6 ± 0.1	31.5 ± 0.3	9.2 ± 0.3	0.3 ± 0.1	105

[a]Yield in g 100 g^{-1} wood/bark. Mean values and standard deviations based on three replicates.
[b]Includes arabinan, galactan, glucan, mannan, xylan, total lignin, extractives, and ash.

Figure 1 shows the syringyl/guaiacyl (S/G) ratio of the different fractions from Scots pine. The determined S/G ratios varied between 0.021 and 0.025 but the variation between different fractions was not statistically significant ($p > 0.05$, Student's t-test). The result indicates a very high proportion of guaiacyl residues in the lignin of all of the fractions. Lignin from conifers consists mainly of guaiacyl units derived from the monolignol coniferyl alcohol, whereas hardwood contain varying ratios of syringyl and guaiacyl units [15]. Glasser and Glasser [16] reported an S/G ratio for Loblolly pine of 0.023, which correlates well with the S/G ratios of the Scots pine fractions (Figure 1). The lignin content and the S/G ratio have been studied in relation to sugar release in enzymatic saccharification of different varieties of *Populus* [17,18]. A high lignin content has been associated with a low level of sugar release. Reports differ with respect to the potential influence of the S/G ratio on sugar release.

Pretreatment and pretreatment liquids

Pretreatment was performed using sulfuric acid as catalyst, which is a suitable method for recalcitrant feedstocks such as softwood [5], and which is relevant for industrial-scale biorefining. The pretreatment was performed with varying severity: mild, medium and severe (Table 2). The resulting pretreatment liquids were analyzed with regard to the monosaccharide content using high-performance anion-exchange chromatography (HPAEC) (Table 3). Juvenile sapwood was used as the reference fraction in comparisons between the different fractions, due to that it was quantitatively dominating in this material. After mild pretreatment, the monosaccharide yields in the pretreatment liquid of all the different fractions followed the order glucose > mannose > galactose > arabinose, xylose. The total amount of sugar was higher in pretreatment liquids obtained after pretreatment with medium severity (Table 3). Except for the xylose yield of mature heartwood, the yields of all separate monosaccharides from all fractions were higher after pretreatment with medium severity than after pretreatment with mild severity. The pretreatment liquids obtained after pretreatment using severe conditions generally contained less sugar than the pretreatment liquids obtained after pretreatment using medium conditions (Table 3). However, the glucose yields from knotwood, juvenile heartwood and mature sapwood were slightly higher after severe pretreatment than after medium pretreatment. No xylose and very small amounts of arabinose could be detected in pretreatment liquids after severe pretreatment (Table 3), which indicates that the pentose sugars were sensitive to the harsh conditions and had been largely degraded to furfural and possibly further on to carboxylic acids [19].

Some aliphatic acids, such as formic acid, are formed during pretreatment through thermochemical degradation of carbohydrates, while acetic acid is formed by hydrolysis of acetyl groups in xylan [19]. Low yields of both acetic acid and formic acid were obtained in the pretreatment liquid as an effect of the dilute-acid pretreatment (Figure 2). The differences between the yields of the acids from the different fractions were small (Figure 2). However, there was a very clear difference between the formation of acetic acid and formic acid in the sense that the concentrations of formic acid were rather similar regardless of whether the pretreatment conditions were mild, intermediate or severe, while the concentrations of acetic acid increased very much when the pretreatment changed from mild to intermediate conditions (Figure 2). When the pretreatment changed from intermediate to

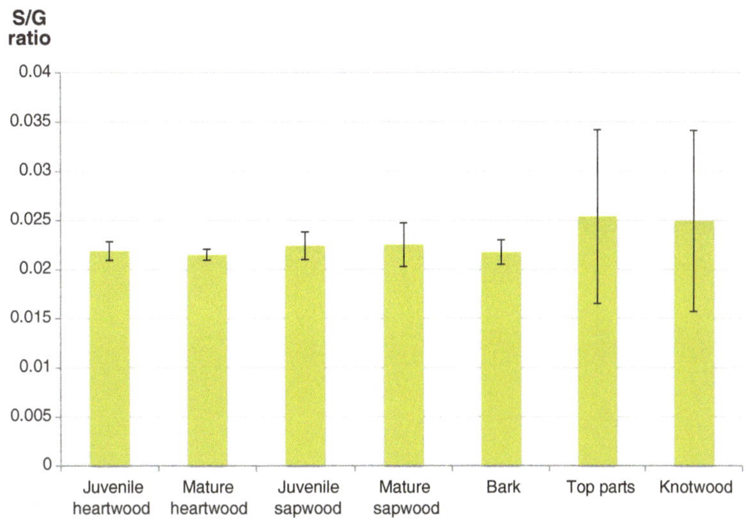

Figure 1 Estimation of the syringyl/guaiacyl (S/G) ratios of different fractions of Scots pine based on analysis using Py-GC/MS. The error bars show the standard deviations of six replicates.

Table 2 Different conditions used for pretreatment of fractions of Scots pine

	Combined severity (log Ro-pH)	Temperature (°C)	Time (min)	H_2SO_4 (% w/w)[a]
Mild	2.7	180	6.0	2.0
Medium	3.0	180	6.0	4.0
Severe	3.3	180	12.0	4.0

[a]The loading (in percent) refers to the weight of sulfuric acid in relation to the total weight of the liquid added to the milled wood.

severe, there was no corresponding increase in the concentrations of acetic acid (Figure 2). This suggests that a combined severity (CS) of 3.0 (Table 2) was sufficient for quantitative hydrolysis of the acetyl groups in the different pine fractions.

The effects of aliphatic acids (acetic, formic and levulinic acids) on the ethanol yield of *S. cerevisiae* were studied by Larsson et al. [20] who found that acid concentrations over 100 mM tended to result in decreased ethanol yields, while concentrations lower than 100 mM tended to increase the ethanol yield. The pretreatment liquids of the different pine fractions showed very low concentrations of both acetic acid (<1.3 g/L, which correspond to approx. 25 mM) and formic acid (<2.8 g/L, which correspond to approx. 50 mM) for all pretreatment conditions. These low concentrations would not be expected to reach an inhibitory level but rather stimulate the ethanol yield in an ethanolic fermentation with yeast.

Enzymatic saccharification

The susceptibility of untreated and pretreated pine fractions to enzymatic hydrolysis of cellulose was investigated using an analytical small-scale saccharification assay. Advantages with performing analytical saccharification in small scale include that large series of biomass samples can be processed in parallel, and that the number of replicates can be large enough to allow statistical analysis of the results.

Table 3 Sugar yields in pretreatment liquids[a]

	Pretreatment severity: mild				
Fraction	*Arabinose*	*Galactose*	*Glucose*	*Mannose*	*Xylose*
Juvenile heartwood	0.1 ± 0.0	0.5 ± 0.1	3.7 ± 0.4	1.3 ± 0.3	0.1 ± 0.0
Mature heartwood	0.1 ± 0.0	0.4 ± 0.1	3.3 ± 0.3	1.3 ± 0.4	0.1 ± 0.0
Juvenile sapwood	0.1 ± 0.0	0.5 ± 0.1	3.8 ± 0.2	1.4 ± 0.3	0.1 ± 0.1
Mature sapwood	0.2 ± 0.1	0.4 ± 0.1	4.0 ± 0.3	1.9 ± 0.2	0.1 ± 0.0
Bark	0.3 ± 0.1	0.6 ± 0.1	3.9 ± 0.5	1.7 ± 0.5	0.1 ± 0.0
Top parts	0.2 ± 0.1	0.8 ± 0.1	3.5 ± 0.3	1.7 ± 0.2	0.2 ± 0.3
Knotwood	0.2 ± 0.1	1.0 ± 0.2	3.4 ± 0.3	1.6 ± 0.4	0.1 ± 0.0
	Pretreatment severity: medium				
Fraction	*Arabinose*	*Galactose*	*Glucose*	*Mannose*	*Xylose*
Juvenile heartwood	0.5 ± 0.1	1.8 ± 0.3	13.0 ± 1.0	4.8 ± 0.6	0.8 ± 0.1
Mature heartwood	n.d.[b]	0.8 ± 0.5	9.7 ± 1.2	2.6 ± 1.7	n.d.
Juvenile sapwood	0.6 ± 0.1	1.5 ± 0.1	13.3 ± 1.9	5.2 ± 0.2	0.9 ± 0.1
Mature sapwood	0.4 ± 0.1	0.8 ± 0.4	8.8 ± 2.5	3.9 ± 1.5	0.5 ± 0.2
Bark	0.9 ± 0.2	1.3 ± 0.3	13.7 ± 1.7	4.1 ± 0.9	0.6 ± 0.3
Top parts	0.6 ± 0.2	2.1 ± 0.6	12.1 ± 3.4	4.8 ± 1.2	0.8 ± 0.3
Knotwood	0.7 ± 0.1	2.5 ± 0.1	11.2 ± 1.0	5.3 ± 0.2	0.9 ± 0.1
	Pretreatment severity: severe				
Fraction	*Arabinose*	*Galactose*	*Glucose*	*Mannose*	*Xylose*
Juvenile heartwood	0.3 ± 0.1	1.0 ± 0.1	15.9 ± 2.0	1.9 ± 0.4	n.d.
Mature heartwood	n.d.	0.5 ± 0.2	9.4 ± 1.9	0.9 ± 0.5	n.d.
Juvenile sapwood	0.2 ± 0.1	0.6 ± 0.1	10.9 ± 1.6	n.d.	n.d.
Mature sapwood	n.d.	0.5 ± 0.1	8.9 ± 2.4	1.3 ± 0.3	n.d.
Bark	0.3 ± 0.1	0.5 ± 0.1	9.8 ± 0.7	0.5 ± 0.1	n.d.
Top parts	n.d.	0.8 ± 0.3	10.2 ± 1.7	0.8 ± 0.4	n.d.
Knotwood	0.4 ± 0.1	1.3 ± 0.3	13.1 ± 1.7	1.9 ± 0.6	n.d.

[a]Yield in g sugar 100 g^{-1} wood (dry weight) in samples taken after 72 h and analyzed using HPAEC. Mean values and standard deviations based on three replicates.
[b]Not detected: yield <0.001 g sugar 100 g^{-1} wood (values under the lowest standard, 0.5 ppm, in the calibration curve).

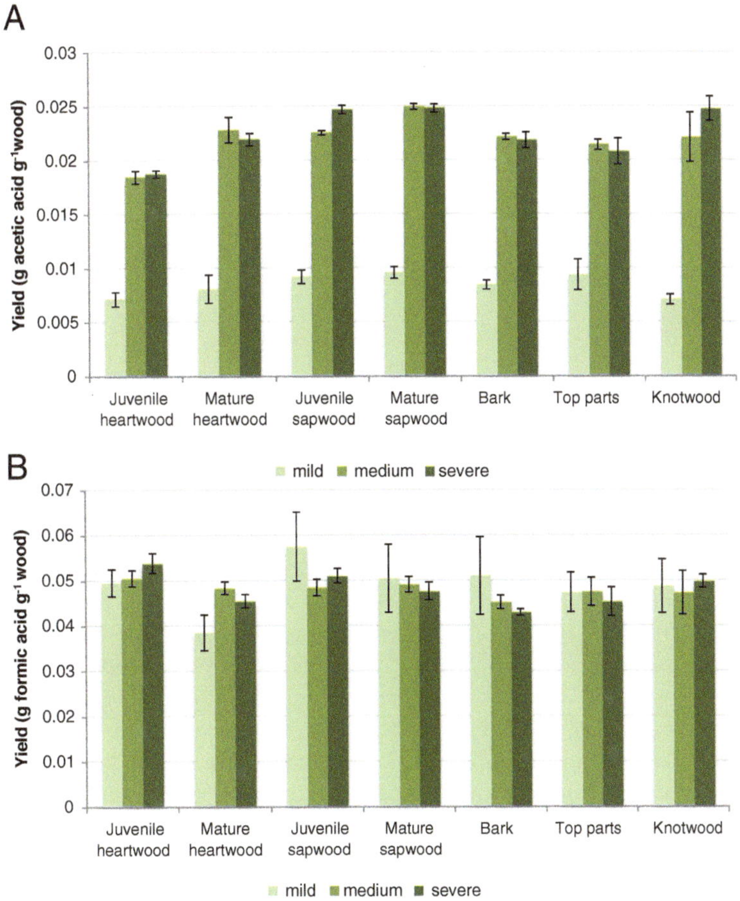

Figure 2 Mean values of the yields of (A) acetic acid and (B) formic acid expressed as g acid g^{-1} wood in the pretreatment liquid of fractions of Scots pine. The different pretreatment conditions are specified in Table 2. The error bars show the standard deviations of three replicates.

Figure 3 shows the glucose production rates (GPR) after 4 h of hydrolysis of the seven pine fractions. Pretreatment always gave higher GPR values than hydrolysis without pretreatment. However, the bark fraction gave relatively high GPR even without pretreatment and the GPR did not improve significantly ($p > 0.05$, Student's t-test) with mild pretreatment. It is noteworthy that the highest GPR values were always achieved using intermediate pretreatment conditions, and that increasing the CS to 3.3 did not give any further improvement. Compared to the reference fraction (juvenile sapwood), mild pretreatment resulted in statistically significantly ($p < 0.05$, Student's t-test) lower GPR values for the juvenile heartwood and the bark fraction. No significant differences compared to the reference fraction could be detected for the intermediate pretreatment conditions. At severe pretreatment, significantly ($p < 0.05$, Student's t-test) lower GPR compared to the reference could be detected for both the bark fraction and the mature heartwood.

Table 4 shows the sugar yields after enzymatic hydrolysis of the pine fractions, with or without pretreatment. The pretreated fractions gave none or very low yields of the pentose sugars arabinose and xylose. This is expected, as the pretreatment targets hemicelluloses. However, as the yields of arabinose and xylose in the pretreatment liquids (Table 3) were higher after medium pretreatment than after mild pretreatment, it would have been theoretically possible to obtain pentoses after enzymatic hydrolysis of fractions pretreated under mild conditions, especially since enzymatic hydrolysis of fractions that were not pretreated did generate some arabinose and xylose (Table 4). The fact that the enzymatic hydrolysates obtained from fractions that were not pretreated always contained all five monosaccharides analysed (Table 4) shows that the enzymes needed to generate these sugars were available in the enzyme preparations added. The enzymatic hydrolysates of fractions that had been pretreated using mild conditions contained galactose and mannose, while the fractions that were pretreated using medium or severe conditions did not (Table 4). As the highest yields of galactose and mannose in pretreatment liquids were achieved after pretreatment using medium conditions

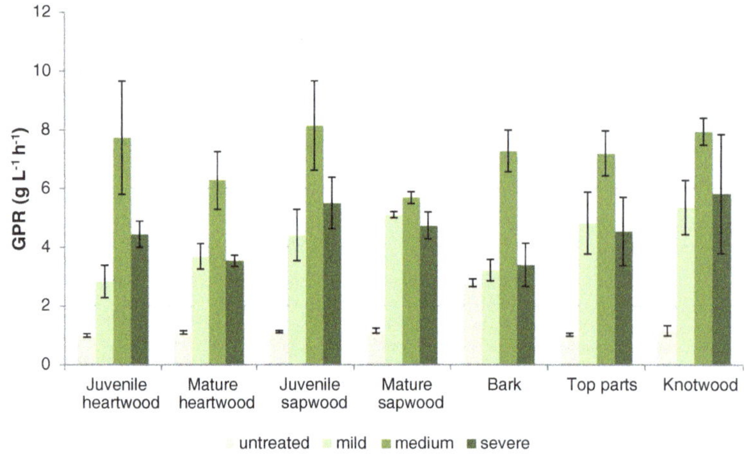

Figure 3 Mean values of glucose production rates (GPR) after 4 h of enzymatic hydrolysis of fractions of Scots pine, with or without pretreatment (pretreatment conditions as specified in Table 2). The error bars show the standard deviations of three replicates.

(Table 3), the fractions pretreated under medium and severe conditions were probably depleted of galactan and mannan, which explains that no or little galactose and mannose was generated in the enzymatic hydrolysis of the corresponding fractions. The glucose yield typically increased with increasing pretreatment severity (Table 4). For the bark fraction, the glucose yield was relatively high even without pretreatment and did not increase very much after pretreatment under mild or medium conditions (Table 4). Juvenile heartwood gave the lowest glucose yield without pretreatment or with mild pretreatment, but gave the highest glucose yield after severe pretreatment (Table 4). The medium pretreatment gave slightly higher glucose yields for sapwood fractions than for heartwood fractions, and the severe pretreatment gave slightly higher yields for juvenile fractions than for mature fractions, but these differences were not significant ($p > 0.05$, Student's t-test).

Figure 4 shows the combined glucose yield in pretreatment liquids and enzymatic hydrolysates after 72 h of hydrolysis. The yield without pretreatment was always much lower than the yield of pretreated pine wood and bark fractions. The glucose yield of all fractions increased with increasing severity of the pretreatment. There is a difference in that regard compared with the GPR values, which were based on samples taken after only 4 h of hydrolysis and for which the intermediate pretreatment gave higher values than those obtained for severe pretreatment. Without pretreatment all fractions showed significantly ($p < 0.05$, Student's t-test) lower glucose yield compared to the reference fraction, except the bark fraction, which gave 55% higher glucose yield. A possible explanation for this is that the morphology of the bark is different from that of the wood and provides a more porous structure that improves the enzymatic accessibility during the hydrolysis. With mild pretreatment conditions a significant ($p < 0.05$, Student's t-test) increase of glucose yield compared to the reference could be detected for the mature sapwood fraction. With medium pretreatment conditions, both mature sapwood and mature heartwood showed significantly lower glucose yield compared to the reference. With severe pretreatment conditions, the juvenile heartwood gave significantly higher glucose yield compared to the reference, whereas the bark fraction showed significantly lower yield.

Conversion of bark-rich softwood residues from hemlock using two-step acid hydrolysis was studied by Kim et al. [21], who found that using this approach it was possible to reach relatively high yields of hemicellulosic sugars, such as mannose, galactose, and xylose. Boussaid et al. [22] investigated steam pretreatment of wood chips of sapwood and heartwood of a 129-year-old Douglas fir. The sapwood fraction was found to be more susceptible to acid-catalyzed steam pretreatment and enzymatic hydrolysis at the medium severity level (195°C, 4.5% of SO_2, 4.5 min). The way the wood was fractionated was different in that study [22] compared to our study, in which the wood was divided to a larger number of fractions including fractions with juvenile and mature wood. It is nevertheless possible to distinguish a similar trend for the Scots pine fractions with regard to the glucose yields (g glucose per g wood) from the total (juvenile and mature fractions combined) sapwood and heartwood fractions after pretreatment at mild and medium severity. After pretreatment under severe conditions, the heartwood fractions of Scots pine instead gave higher glucose yield (0.59 g glucose per g wood) than the sapwood fractions (0.52 g glucose per g wood).

The fact that pretreatment under severe conditions resulted in the highest glucose yields for all fractions (Table 4, samples taken after 72 h) while the highest GPR values were achieved with material pretreated at medium

Table 4 Sugar yields after enzymatic hydrolysis[a]

	Untreated				
Fraction	*Arabinose*	*Galactose*	*Glucose*	*Mannose*	*Xylose*
Juvenile heartwood	0.2 ± 0.1	0.4 ± 0.1	3.0 ± 0.3	0.7 ± 0.1	0.3 ± 0.1
Mature heartwood	0.2 ± 0.1	0.4 ± 0.1	3.3 ± 0.3	0.8 ± 0.1	0.3 ± 0.1
Juvenile sapwood	0.5 ± 0.1	0.7 ± 0.1	4.1 ± 0.2	0.9 ± 0.2	0.4 ± 0.2
Mature sapwood	0.2 ± 0.2	0.4 ± 0.2	3.3 ± 0.1	0.9 ± 0.2	0.3 ± 0.2
Bark	0.2 ± 0.1	0.4 ± 0.1	6.4 ± 0.6	0.8 ± 0.1	0.3 ± 0.1
Top parts	0.2 ± 0.1	0.4 ± 0.1	3.0 ± 0.3	0.7 ± 0.1	0.3 ± 0.1
Knotwood	0.2 ± 0.1	0.4 ± 0.1	3.3 ± 0.3	0.8 ± 0.1	0.3 ± 0.1
	Pretreatment severity: mild				
Fraction	*Arabinose*	*Galactose*	*Glucose*	*Mannose*	*Xylose*
Juvenile heartwood	n.d.[b]	0.58 ± 0.1	6.9 ± 2.6	0.7 ± 0.1	n.d.
Mature heartwood	n.d.	0.62 ± 0.1	8.4 ± 1.3	0.8 ± 0.1	n.d.
Juvenile sapwood	n.d.	0.65 ± 0.1	8.5 ± 1.0	0.8 ± 0.1	n.d.
Mature sapwood	n.d.	0.64 ± 0.1	10.6 ± 0.7	0.9 ± 0.2	n.d.
Bark	n.d.	0.63 ± 0.1	8.1 ± 1.3	0.8 ± 0.1	n.d.
Top parts	n.d.	0.62 ± 0.1	10.5 ± 2.3	0.8 ± 0.1	n.d.
Knotwood	n.d.	0.61 ± 0.1	10.6 ± 2.4	0.8 ± 0.1	n.d.
	Pretreatment severity: medium				
Fraction	*Arabinose*	*Galactose*	*Glucose*	*Mannose*	*Xylose*
Juvenile heartwood	n.d.	n.d.	10.1 ± 1.8	n.d.	n.d.
Mature heartwood	n.d.	n.d.	8.3 ± 0.7	n.d.	n.d.
Juvenile sapwood	n.d.	n.d.	12.3 ± 1.0	n.d.	n.d.
Mature sapwood	n.d.	n.d.	12.1 ± 2.4	n.d.	n.d.
Bark	n.d.	n.d.	7.5 ± 0.9	n.d.	n.d.
Top parts	n.d.	n.d.	9.2 ± 2.2	n.d.	n.d.
Knotwood	n.d.	n.d.	10.5 ± 1.4	n.d.	n.d.
	Pretreatment severity: severe				
Fraction	*Arabinose*	*Galactose*	*Glucose*	*Mannose*	*Xylose*
Juvenile heartwood	n.d.	n.d.	18.0 ± 1.4	n.d.	n.d.
Mature heartwood	n.d.	n.d.	15.8 ± 2.3	n.d.	n.d.
Juvenile sapwood	n.d.	n.d.	17.9 ± 2.5	n.d.	n.d.
Mature sapwood	n.d.	n.d.	14.6 ± 3.0	n.d.	n.d.
Bark	n.d.	n.d.	14.7 ± 0.3	n.d.	n.d.
Top parts	n.d.	n.d.	15.9 ± 2.0	n.d.	n.d.
Knotwood	n.d.	n.d.	16.6 ± 0.5	n.d.	n.d.

[a]Yield in g sugar 100 g^{-1} wood (dry weight) in samples taken after 72 h and analyzed using HPAEC. Mean values and standard deviations based on three replicates.
[b]Not detected: yield <0.001 g sugar 100 g^{-1} wood (values under the lowest standard, 0.5 ppm, in the calibration curve).

severity (Figure 3, samples taken after 4 h) suggests that the highest severity caused formation of something that retarded the initial cellulose hydrolysis rate. This effect cannot have been caused by sugars or other soluble compounds that would inhibit cellulases, since the pretreatment liquid, which contained the soluble compounds, was removed from the solid residue before the enzymatic hydrolysis. Acid-catalyzed dehydration of carbohydrates is responsible for formation of pseudo-lignin, an aromatic material with positive Klason lignin value that is derived from carbohydrates rather than from native lignin [23-27]. Pseudo-lignin is suggested to cause non-productive binding of enzymes, which will affect enzymatic hydrolysis in a negative way. It is possible that the highest severity resulted in formation of pseudo-lignin to an extent that slowed down the initial phase of the hydrolysis of cellulose, resulting in a lower GPR, but that it was possible for the enzymes to overcome this obstacle in the long run.

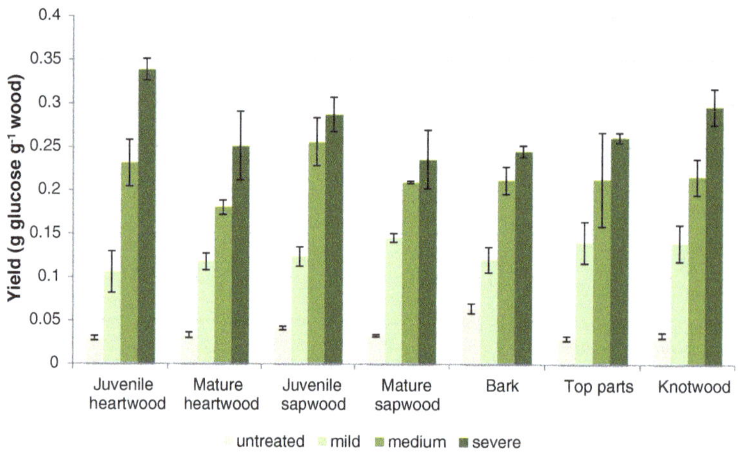

Figure 4 The combined glucose yield in pretreatment liquids and enzymatic hydrolysates. The figure shows mean values of glucose yields (expressed as g glucose per g wood after pretreatment and 72 h of enzymatic hydrolysis at 45°C in a orbital shaker set at 170 rpm). The error bars show the standard deviations of three replicate reactions for each pine fraction and pretreatment condition.

Evidently, there was also a positive effect of the severe pretreatment compared to the intermediate pretreatment, since the glucose yields after 72 h were higher. This could have been due to incomplete removal of hemicelluloses for the intermediate pretreatment, but the data do not support that hypothesis since the concentrations of hemicellulosic sugars were not higher in pretreatment liquid from the severe pretreatment than in pretreatment liquid from the intermediate pretreatment (Table 3), and since the enzymatic treatment did not release any hemicellulosic sugars from the material pretreated at medium severity (Table 4). Increased severity reduces the degree of polymerization (DP) of cellulose which could improve the digestibility [28], but differences in the DP may be difficult to relate to the susceptibility to enzymatic hydrolysis [29]. Assuming limited accessibility to cellulose to be the main obstacle for enzymatic hydrolysis [30], the effect could tentatively be attributed to improved enzyme accessibility caused by the more severe pretreatment producing a larger surface area by altering the physical shape of the pretreated material, i.e. by creating a more favorable topology.

Conclusions

Seven fractions of Scots pine were compared with respect to chemical composition and susceptibility to analytical-scale pretreatment and enzymatic saccharification. Chemical analysis showed that the fractions differed significantly as exemplified by the relatively high content of extractives of the knotwood fraction and the relatively high content of ash in the bark fraction. The investigation showed that the different fractions were relatively similar with regard to susceptibility to pretreatment and enzymatic saccharification, which is somewhat unexpected considering the variation between the fractions. The most obvious difference with regard to enzymatic saccharification was that without pretreatment the bark fraction was more susceptible than the other fractions. That different fractions exhibit similar susceptibility facilitates the utilization of forestry residues from Scots pine for bioconversion to fuels and chemicals.

Methods

Lignocellulosic material

Different fractions of Scots pine (*Pinus sylvestris*) were obtained from the lower parts (0–16 m height) of six trees with an age between 22 and 145 years, a height between 12.0 and 23.5 m, and a diameter at 1.3 m height between 12 and 35 cm. Each steam was separated into five fractions with a chain saw. The xylem wood (excluding bark and phloem wood) was divided into sapwood and heartwood. These fractions were further divided into juvenile and mature wood, which resulted in four fractions. The juvenile wood (inner part) was separated from the mature wood (outer part) at the 20th annual ring from the pitch. The heartwood and the sapwood were separated visually by their different coloration. The top parts of the pine stems (including bark) were collected as a separate fraction using the chain saw. The five resulting fractions were then chipped in a chipper (Edsbyhuggen 250H, Woxnadalens Energi AB, Edsbyn, Sweden). From the chips containing the top parts, two additional fractions, called knotwood (which had some juvenile sapwood attached) and bark (containing also phloem tissue and some attached wood) were obtained by visual sorting after drying. The juvenile sapwood fraction was used as reference in comparisons between the different fractions. The seven fractions and some of their properties have previously been described by Lestander et al. [8].

Milling and sieving

The pine fractions were first milled to a particle size of 1 mm [8], but were then milled once more using a Foss Tecator Cyclotech 1093 sample mill equipped with a 0.5 mm ring sieve. The milled material was then sieved in a Retsch Analytical sieve shaker AS 200 (Retsch, Haan, Germany) and the material obtained between 100 and 500 μm sieves was collected for each pine fraction.

Moisture content

The moisture content of each fraction obtained after milling and sieving was determined (four replicates) using a HG63 Halogen Moisture Analyzer (Mettler-Toledo, Greifensee, Switzerland).

Determination of the chemical composition of the pine fractions

The determination of extractives was performed as described by Lestander et al. [8] by extraction during 1 h using a Soxhlet system (Büchi Extraction System B-811, Büchi, Flawil, Switzerland) and 200 mL of a 9:1 mixture of petroleum ether (Petroleum Benzene, Merck, Darmstadt, Germany) and acetone for extraction of 3 g of wood. The extracted samples were then dried in air at room temperature for about 16 h until the weight of the samples was stable. The weight of the extracts was measured using a balance (Mettler Toledo XS204) and expressed as percentage of dry substance. For determination of carbohydrates and lignin, the pine fractions were first extracted, to achieve a more accurate estimation. The extraction procedure used was the same as described above, but with the exception that 15 cycles (1–1.5 h) were used in the extraction step instead of a specific extraction time (1 h). The contents of structural carbohydrates (arabinan, galactan, glucan, mannan, and xylan) and lignin (acid soluble and acid insoluble) of extracted wood and bark fractions were determined essentially using a procedure, NREL/TP-510-42618, described by Sluiter et al. [31]. Apart from using a different extraction method, the procedure used also differed with regard to the method used for determination of the concentration of monosaccharides generated by the hydrolysis. The concentrations of the monosaccharides were determined using High-Performance Anion-Exchange Chromatography (HPAEC) as described in the section "Analysis of hydrolysates" below. The determination of the ash content in the different pine fractions was performed using the Swedish Standard method SS-EN 14775.

Pyrolysis - gas chromatography/mass spectrometry (Py-GC/MS) analysis

The seven pine fractions were analyzed using Py-GC/MS at the Cell Wall and Carbohydrate Analytical Facility of the Umeå Plant Science Center (UPSC) (Umeå, Sweden). The instrument and the procedures used have been described previously [32]. The main objective of the Py-GC/MS analysis was to determine the syringyl-guaiacyl ratio (the S/G ratio) of the lignin of the different pine fractions.

Dilute-acid pretreatment

Enzymatic digestion of cellulose was performed with and without pretreatment of the pine fractions. The pretreatment was performed with dilute sulfuric acid using an advanced single-mode microwave instrument with stirring to achieve a more uniform temperature distribution in the reaction vessel.

The total mass of the reaction mixture was always 1000 mg. The amount of milled and sieved biomass in each reaction was 50 mg dry weight [5% (w/w)]. The pretreatment reactions were carried out using an Intiator 2.0 EXP instrument (Biotage, Uppsala, Sweden). The reaction vessels were glass vials (0.5-2 mL reaction size, Biotage) equipped with small (10 mm diameter) magnetic stirring bars (Biotage). Three different combinations of temperature, treatment time, and sulfuric acid concentration were used resulting in mild, medium and severe pretreatment conditions (Table 2). The pretreatment conditions can be described in terms of the resulting Combined Severity (CS) [33,34], which ranged between 2.7 and 3.3 (Table 2). For each pine fraction and pretreatment condition, reactions were performed in triplicates.

After pretreatment, the reaction mixture was transferred to pre-weighed 2-mL Sarstedt micro-centrifuge tubes using a 1-mL pipette tip with the end cut off to facilitate quantitative transfer of the content. The pretreatment reaction mixture was centrifuged for 15 min at 14,100 *g* (14,500 rpm) with a MiniSpin Plus centrifuge (Eppendorf, Hamburg, Germany). The solid phase and the liquid phase (referred to as the pretreatment liquid) were then separated. After separation, the pretreatment liquid was stored at −20°C awaiting analysis. The solid phase was washed two times with one mL Milli-Q water (Millipore, Billerica, MA) and one time with one mL sodium citrate buffer (50 mM citrate, pH 5.2). The centrifugation after each washing step was performed as described previously. The pH was checked with a pH indicator and, if necessary, a fourth washing step was performed (with the citrate buffer) so that the final pH was around 5.

Enzymatic hydrolysis

Enzymatic hydrolysis was carried out in parallell in analytical-scale 2-mL Sarstedt microcentrifuge tubes. The total mass of the reaction mixtures was 1000 mg. For each pine fraction and pretreatment condition, three replicates were processed. The lignocellulosic material in each reaction mixture consisted of 50 mg (dry weight) milled and sieved wood or bark, or, alternatively, the remaining solid fraction after pretreatment of 50 mg (dry weight) milled and sieved wood or bark. An enzyme cocktail consisting of 50 mg of a 1:1 mixture of Celluclast 1.5 L (a cellulase-rich liquid enzyme preparation from *Trichoderma reesei* ATCC 26921) and Novozyme 188 (a cellobiase-rich liquid enzyme preparation from *Aspergillus niger*) was added to each reaction mixture. The enzyme preparations were obtained from Sigma-Aldrich (St. Louis, MO, USA). Celluclast 1.5 L had a stated activity of 700 endoglucanase units (EGU)/g, while Novozyme 188 had a stated activity of 250 cellobiase units (CBU)/g. A sodium citrate buffer (50 mM citrate, pH 5.2) was added so that the final weight of the reaction mixture was 1000 mg. The tubes were incubated at 45°C for 72 h in an orbital shaker (Ecotron incubator shaker, Infors, Bottmingen, Swizerland) set at 170 rpm. The tubes were placed horizontally and were fastened with tape so that the screw caps were facing the same side of the incubator. Samples (10 μL) were withdrawn from the reaction mixtures after 0, 4, 48 and 72 h of incubation. At the end of the incubation, the tubes were centrifuged at 14,100 *g* (14,500 rpm) for 5 min and the supernatants were collected, stored at −20°C, and were then used for sugar analysis. The pellets, presumably consisting mostly of lignin and some remaining cellulose, were also collected and stored at −20°C.

Analysis of hydrolysates

The glucose concentrations in centrifuged samples taken after four h of incubation of enzymatic reaction mixtures were estimated using a glucometer (Accu-Check Aviva, Roche Diagnostics, Basel, Switzerland). Three replicates were analyzed for each pine fraction and pretreatment condition. The result was used to calculate the glucose production rate (GPR) expressed as g glucose L^{-1} h^{-1}.

The sugar content of pretreatment liquids and enzymatic hydrolysates (after 72 h reaction time) was analyzed by using HPAEC. Separation and analysis of monosaccharides was performed using an ICS-5000 system equipped with a CarboPac PA20 guard column (3 × 30 mm), a CarboPac PA20 analytical column (3 × 150 mm), and an electrochemical detector (all parts from Dionex, Sunnyvale, CA). The monosaccharides that were determined included arabinose, galactose, glucose, mannose, and xylose. Prior to analysis, all samples were diluted using ultra-pure water and were filtered through a 0.20 μm syringe-driven filter unit with a nylon membrane (Millipore). The results are reported as mean values of triplicate reactions.

The contents of acetic acid and formic acid were analyzed using HPAEC and the Dionex ICS-5000 system equipped with an IonPac AG15 guard column (4 × 50 mm), an IonPac AS15 analytical column (4 × 250 mm), and a conductivity detector (all parts from Dionex). Prior to analysis, all samples were diluted using ultra-pure water and filtered through a 0.20 μm syringe-driven filter unit with a nylon membrane (Millipore). The reported values are based on analyses of three replicate reactions.

Competing interests

The authors declare that they have no competing interests.

Authors' contributions

TAL and LJJ conceived and designed the study. The main part of the experimental work was carried out by MN under the supervision of SW. MN and LJJ wrote the manuscript, which was revised by SW and TAL. All authors read and approved the final manuscript.

Acknowledgements

We thank Carina Jonsson for help with ash analyses. This work was supported by the Bio4Energy research initiative (www.bio4energy.se) and by grants from the Kempe Foundations, the Swedish Energy Agency (35367-1), and the Swedish Research Council (621-2011-4388). We are grateful to the Plant Cell Wall and Carbohydrate Analytical Facility of UPSC/SLU, supported by Bio4Energy and TC4F, for analysis using Pyrolysis - Gas Chromatography/Mass Spectrometry (Py-GC/MS).

Author details

[1]Department of Chemistry, Umeå University, Umeå SE-901 87, Sweden.
[2]Department of Forest Biomaterials and Technology, Swedish University of Agricultural Sciences, Umeå SE-901 83, Sweden.

References

1. Sims REH, Mabee W, Saddler JN, Taylor M: **An overview of second generation biofuel technologies.** *Bioresour Technol* 2010, **101**:1570–1580.
2. Pu Y, Kosa M, Kalluri UC, Tuskan GA, Ragauskas AJ: **Challenges of the utilization of wood polymers: how can they be overcome?** *Appl Microbiol Biotechnol* 2011, **91**:1525–1536.
3. Skogsstyrelsen: *Skogsstatistisk Årsbok*. Jönköping; 2012.
4. Mabee WE, Gregg DJ, Arato C, Berlin A, Bura R, Gilkes N, Mirochnik O, Pan X, Pye EK, Saddler JN: **Updates on softwood-to-ethanol process development.** *Appl Biochem Biotechnol* 2006, **129–132**:55–70.
5. Galbe M, Zacchi G: **Pretreatment of lignocellulosic materials for efficient bioethanol production.** *Adv Biochem Eng Biotechnol* 2007, **108**:41–65.
6. Saddler JN: *Bioconverison of Forest and Agricultural Plant Residues*. Wallingford: CAB International; 1993.
7. Sjöström E: *Wood Chemistry: Fundamentals and Applications*. 2nd edition. San Diego: Academic Press; 1993.
8. Lestander TA, Geladi P, Larsson SH, Thyrel M: **Near infrared image analysis for online identification and separation of wood chips with elevated levels of extractives.** *J Near Infrared Spec* 2012, **20**:591–599.
9. Bertaud F, Holmbom B: **Chemical composition of earlywood and latewood in Norway spruce heartwood, sapwood and transition zone wood.** *Wood Sci Technol* 2004, **38**:245–256.
10. Burkhardt S, Kumar L, Chandra R, Saddler J: **How effective are traditional methods of compositional analysis in providing an accurate material balance for a range of softwood derived residues?** *Biotechnol Biofuels* 2013, **6**:90.
11. Willför SM, Ahotupa MO, Hemming JE, Reunanen MH, Eklund PC, Sjöholm RE, Eckerman CS, Pohjamo SP, Holmbom BR: **Antioxidant activity of knotwood extractives and phenolic compounds of selected tree species.** *Agric Food Chem* 2003, **51**:7600–7606.
12. Lestander TA, Lundström A, Finell M: **Assessment of biomass functions for calculating bark proportions and ash contents of refined biomass fuels derived from major boreal tree species.** *Can J For Res* 2012, **42**:59–66.
13. Tao G, Lestander TA, Geladi P, Xiong S: **Biomass properties in association with plant species and assortments I: A synthesis based on literature data of energy properties.** *Renew Sust Energ Rev* 2012, **16**:3481–3506.
14. Kenney KL, Smith WA, Gresham GL, Westover TL: **Understanding biomass feedstock variability.** *Biofuels* 2013, **4**:111–127.

15. Fengel D, Wegener G: *Wood: Chemistry, Ultrastucture, Reactions.* Berlin: Walter de Gruyter; 1983.
16. Glasser WG, Glasser HR: **The evaluation of lignin's chemical structure by experimental and computer simulation techniques.** *Pap Puu* 1981, **63:**71–74.
17. Davison BH, Sadie RD, Tuskan GA, Davis DF, Nghiem NP: **Variation of S/G ratio and lignin content in a *Populus* family influences the release of xylose by dilute acid hydrolysis.** *Appl Biochem Biotechnol* 2006, **129–132:**427–435.
18. Studer MH, DeMartini JD, Davis MF, Sykes RW, Davison B, Keller M, Tuskan GA, Wyman CE: **Lignin content in natural *Populus* variants affects sugar release.** *Proc Natl Acad Sci USA* 2011, **108:**6300–6305.
19. Jönsson LJ, Alriksson B, Nilvebrant N-O: **Bioconversion of lignocellulose: inhibitors and detoxification.** *Biotechnol Biofuels* 2013, **6:**16.
20. Larsson S, Palmqvist E, Hahn-Hägerdal B, Tengborg C, Stenberg K, Zacchi G, Nilvebrant N-O: **The generation of fermentation inhibitors during dilute acid hydrolysis of softwood.** *Enzyme Microb Tech* 1999, **24:**151–159.
21. Kim KH, Tucker M, Nguyen Q: **Conversion of bark-rich biomass mixture into fermentable sugar by two-stage dilute acid-catalyzed hydrolysis.** *Bioresour Technol* 2005, **96:**1249–1255.
22. Boussaid A-L, Esteghlalian AR, Gregg DJ, Lee KH, Saddler JN: **Steam pretreatment of douglas-fir wood chips: Can conditions for optimum hemicellulose recovery still provide adequate access for efficient enzymatic hydrolysis?** *Appl Biochem Biotechnol* 2000, **84–86:**693–705.
23. Brownell HH, Saddler JN: **Steam-explosion pretreatment for enzymatic hydrolysis.** *Biotechnol. Bioenerg. Symp.* 1984, **14:**55–68.
24. Schwald W, Brownell HH, Saddler JN: **Enzymatic hydrolysis of steam-treated aspen wood: influence of partial hemicellulosc and lignin removal prior to treatment.** *J Wood Chem Technol* 1988, **8:**543–560.
25. Jakobsons J, Hortung B, Erins P, Sundquist J: **Characterization of alkali soluble fraction of steam exploded birch wood.** *Holzforschung* 1995, **49:**51–59.
26. Sannigrahi P, Dong KH, Seokwon J, Ragauskas A: **Pseudo-lignin and pretreatment chemistry.** *Energy Environ Sci* 2011, **4:**1306–1310.
27. Kumar R, Hu F, Sannigrahi P, Jung S, Ragauskas AJ, Wyman CE: **Carbohydrate derived-pseudo-lignin can retard cellulose biological conversion.** *Biotechnol Bioeng* 2013, **110:**737–753.
28. Ucar G, Fengel D: **Characterization of the acid pretreatment for the enzymatic-hydrolysis of wood.** *Holzforschung* 1988, **42:**141–148.
29. Ramos LP, Nazhad MM, Saddler JN: **Effect of enzymatic hydrolysis on the morphology and fine structure of pretreated cellulosic residues.** *Enzyme Microb Technol* 1993, **15:**821–831.
30. Arantes V, Saddler JN: **Cellulose accessibility limits the effectiveness of minimum cellulase loading on the efficient hydrolysis of pretreated lignocellulosic substrates.** *Biotechnol Biofuels* 2011, **4:**3.
31. Sluiter A, Hames B, Ruiz R, Scarlata C, Sluiter J, Templeton D, Crocker D: *Determination of structural carbohydrates and lignin in biomass.* Golden, Colorado: National Renewable Energy Laboratory; 2008. NREL/TP-510-42-618.
32. Gerber L, Eliasson M, Trygg J, Moritz T, Sundberg B: **Multivariate curve resolution provides a high-throughput data processing pipeline for pyrolysis-gas chromatography/mass spectrometry.** *J Anal Appl Pyrol* 2012, **95:**95–100.
33. Overend RP, Chornet E: **Fractionation of lignocellulosics by steam-aqueous pretreatments.** *Phil Trans R Soc London, Series A* 1987, **321:**523–536.
34. Chum HL, Johnson DK, Black SK, Overend RP: **Pretreatment-catalyst effects and the combined severity parameter.** *Appl Biochem Biotechnol* 1990, **24/25:**1–4.

6

Production of recombinant human annexin V by fed-batch cultivation

Laura S Marder[1,2†], Juleane Lunardi[2,4†], Gaby Renard[4], Diana C Rostirolla[1,3], Guilherme O Petersen[1,2,4], José E S Nunes[4], Ana Paula D de Souza[5], Ana Christina de O Dias[4], Jocelei M Chies[4], Luiz A Basso[1,2,3], Diógenes S Santos[1,2,3] and Cristiano V Bizarro[1,2*]

Abstract

Background: Annexin V, a 35.8 kDa intracellular protein, is a Ca^{+2}- dependent phospholipid binding protein with high affinity to phosphatidylserine (PS), which is a well-known hallmark of apoptosis. Annexin V is a sensitive probe for PS exposure upon the cell membrane, and used for detection of apoptotic cells both *in vivo* and *in vitro*. Large-scale production of recombinant human annexin V is worth optimization, because of its wide use in nuclear medicine, radiolabeled with ^{99m}Tc, for the evaluation of cancer chemotherapy treatments, and its use in identification of apoptotic cells in histologic studies. Here we describe the high-yield production of a tag-free version of human annexin V recombinant protein by linear fed-batch cultivation in a bioreactor.

Results: We cloned the human *ANXA5* coding sequence into the pET-30a (+) expression vector and expressed rhANXA5 in batch and fed-batch cultures. Using *E. coli* BL21 (DE3) in a semi-defined medium at 37°C, pH 7 in fed-batch cultures, we obtained a 45-fold increase in biomass production, respective to shaker cultivations. We developed a single-step protocol for rhANXA5 purification using a strong anion-exchange column (MonoQ HR16/10). Using these procedures, we obtained 28.5 mg of homogeneous, nontagged and biologically functional human annexin V recombinant protein from 3 g wet weight of bacterial cells from bioreactor cultures. The identity and molecular mass of rhANXA5 was confirmed by mass spectrometry. Moreover, the purified rhANXA5 protein was functionally evaluated in a FITC-annexin V binding experiment and the results demonstrated that rhANXA5 detected apoptotic cells similarly to a commercial kit.

Conclusions: We describe a new fed-batch method to produce recombinant human annexin V in large scale, which may expand the commercial utilities for rhANXAV to applications such as *in vivo* imaging studies.

Keywords: Recombinant human annexin v, Fed-batch cultivation, Large-scale, Apoptosis detection

Background

Annexin V, formerly known as human placental anticoagulation protein, is a member of a family of calcium-dependent phospholipid binding proteins [1] that binds preferentially to phosphatidylserine (PS), a negatively charged phospholipid highly enriched in the inner leaflet of plasma membranes [2]. Annexin V shows minimal capacity to bind phospholipids that are constitutively present in the outer leaflet of plasma membranes, such as phosphatidylcholine and sphingomyelin [1]. During the early stages of apoptosis, cells expose PS on the surface, which is specifically recognized by phagocytic cells [3]. The exposed PS residues can be bound selectively by annexin V with nanomolar to picomolar affinity [4,5]. Because of the role of apoptosis in the pathophysiology of many diseases, there is a wide range of current and potentially new applications for annexin V as an apoptotic marker in clinical diagnosis [6]. Staining cells simultaneously with annexin V labeled with the fluorochrome fluorescein isothiocyanate (FITC) and the non-vital dye propidium iodide (PI) allows the discrimination among intact cells ($FITC^-/PI^-$) early apoptotic cells ($FITC^+/PI^-$) and late

* Correspondence: cristiano.bizarro@pucrs.br
†Equal contributors
[1]Centro de Pesquisas em Biologia Molecular e Funcional (CPBMF), Instituto Nacional de Ciência e Tuberculose (INCT-TB), Pontifícia Universidade Católica do Rio Grande do Sul (PUCRS), Av. Ipiranga 6681, 90619-900 Porto Alegre, Brazil
[2]Programa de Pós-Graduação em Biologia Celular e Molecular, PUCRS, Porto Alegre 90619-900, Brazil
Full list of author information is available at the end of the article

apoptotic or necrotic cells ($FITC^+/PI^+$) [7]. Moreover, recombinant human annexin V (rhANXA5) serves as an important *in vivo* diagnostic tool when labeled with different radionuclides, such as iodine-123 (^{123}I) and the metastable isotope technetium-99 (^{99m}Tc) providing a broad range of imaging applications in apoptosis research, as single-photon emission computed tomography and autoradiography to positron emission tomography [8]. In nuclear medicine, annexin V radiolabeled with ^{99m}Tc or ^{123}I is used to evaluate the efficacy of cancer therapy and disease progression or regression [9].

Because of its widespread use as a diagnostic tool, rhANXA5 is commercially produced in microorganisms such as *Escherichia coli* using recombinant DNA techniques [10-13]. To produce rhANXA5 in large scale in this study, we used *E. coli*, the most commonly used host for recombinant protein production [14]. In shake flask cultures, all components are added at the start of the cultivation, without the need to monitor or control any parameter such as pH or the level of dissolved oxygen, leading to slow growth and low recombinant protein production [15]. High cell-density culture techniques have been developed to improve productivity and to provide advantages such as reduced culture volume, enhanced downstream processing, reduced wastewater, lower production costs and reduced investment in equipment [14]. Fed-batch cultivation is an effective and simple method [16], allowing substantial concentrations of glucose, an inexpensive and readily usable carbon and energy source [17]. To our knowledge, there is currently no protocol available for a bioreactor-based, large-scale production of rhANXA5.

In this work, we describe a procedure for the production of 28.5 mg of homogeneous, nontagged rhANXA5 from 3 g wet weight of bacterial cells from bioreactor cultures. The identity and absence of host contaminants in purified rhANXA5 was confirmed by LC-MS/MS peptide mapping experiments. The molecular mass determination of intact rhANXA5 confirmed the integrity of the purified protein. Additionally, the produced rhANXA5 protein was shown to be functional in a bioassay for *in vitro* apoptosis/necrosis detection, in which it performed similarly to a commercially available kit.

Results and discussion

Cloning and expression of rhANXA5

The gene encoding human Annexin V, *ANXA5*, is located on human chromosome 4q27 locus and spans a region of DNA 29 kb in length containing 13 exons and 12 introns. *ANXA5* encodes a 35.8 kDa protein of 320 amino acid residues in length which is translated from a single mature transcript of approximately 1.6 kb [18]. The human *ANXA5* coding sequence was subcloned into the pET-30a (+) expression vector to generate the recombinant pET-30a (+):: ANXA5 plasmid. Both the sequence and the absence of PCR-introduced mutations in the *ANXA5* coding sequence were confirmed by automated sequencing.

The BL21 (DE3) and C41 (DE3) *E. coli* strains were transformed with the pET-30a (+)::ANXA5 construct by electroporation. The expression profiles were tested in lysogeny broth (LB) and in our semi-defined (SD) medium. The best results of soluble recombinant protein production from shaker cultivation were obtained using the BL21 (DE3) strain in SD medium induced with 1 mM IPTG at 37°C (Additional file 1: Figures S1 and Figure S2).

Bioreactor cultivation

The best growth conditions found in shaker cultivations were applied to the bioreactor batch and fed-batch cultures. We employed our SD medium and a temperature of 37°C in all experiments.

Transformed BL21 (DE3) *E. coli* was grown from a master cell bank (MCB) under batch cultivation in 1 L of SD medium at 37°C monitoring glucose consumption. Within 4 h, glucose was depleted from media, indicating this was the right time to start feeding. We compared DO-stat, pH-stat and linear ascending feeding strategies in uninduced cultures. A biomass concentration of 26.01 g (DCW) L^{-1} was attained in an uninduced culture using the linear ascending feeding profile [17]. Lower values for biomass concentration were obtained using DO-stat (20.15 g (DCW) L^{-1}) or pH-stat (11.59 g (DCW) L^{-1}) feeding strategies. Therefore, we selected the linear ascending feeding strategy for further fed-batch cultivations.

After 4 h of batch cultivation without feeding, the biomass concentration reached 4.71 g (DCW) L^{-1} (n = 4, SD = 0.6). We induced rhANXA5 expression in fed-batch cultivations with linear ascending feeding by adding IPTG to a final concentration of 1 mM. From three independent fermentations, we obtained a mean value of 27.48 g (DCW) L^{-1} (SD = 1.96) for the biomass concentration and a total protein content of 4.8 g L^{-1} (Figure 1). In contrast, we obtained a biomass concentration of 0.615 g (DCW) L^{-1} in shaker cultivation (OD_{600nm} = 1.84) after 6 h of culture. This represents a 45-fold increase in biomass concentration.

In an independent bioreactor cultivation, we performed densitometric analysis of culture samples (in triplicate) to calculate the product yield (g (rhANXA5) L^{-1}), the productivity (g (rhANXA5) L^{-1} h^{-1}) and the specific yield (g (rhANXA) g $(DCW)^{-1}$). rhANXA5 corresponded to 40.6% (mean value) of the total protein content (n = 3, SD = 0.03) (Additional file 1: Figure S4). Therefore, we obtained a product yield of 1.95 g (rhANXA5) L^{-1} of culture medium, a specific yield of 0.0715 g (rhANXA5) g $(DCW)^{-1}$ and a productivity of 0.065 g (rhANXA5) L^{-1} h^{-1}, considering the entire fermentation period (30 h).

Moreover, we can consider the entire procedure, from bioreactor cultivation to protein purification, and calculate

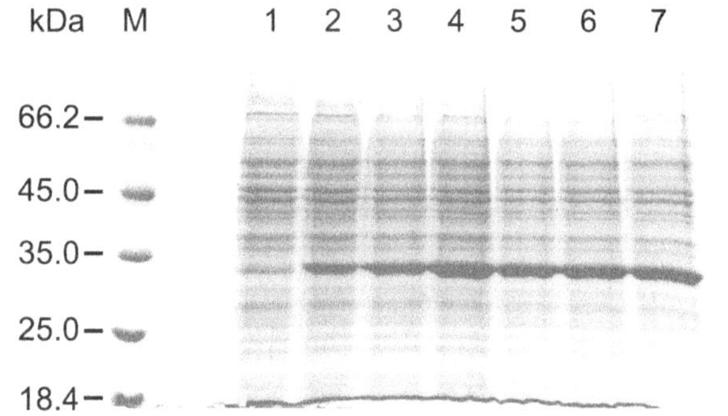

Figure 1 SDS-PAGE analysis of rhANXA5 expression in fed-batch cultivations. Recombinant *E. coli* BL21(DE3) cells were cultivated in semi-defined (SD) media for 30 h in fed-batch cultivations. Feeding started 4 h after the beginning of batch cultivation and rhANXA5 expression was induced after 18 h of cultivation by the addition of 1 mM IPTG to the cultures. M: Thermo Scientific™ Unstained Protein MW Marker; lane 1: sample collected immediately before IPTG induction (18 h culture); lanes 2–7: samples collected after IPTG induction from 20 h (lane 2), 22 h (lane 3), 24 h (lane 4), 26 h (lane 5), 28 h (lane 6), and 30 h (lane 7) cultures.

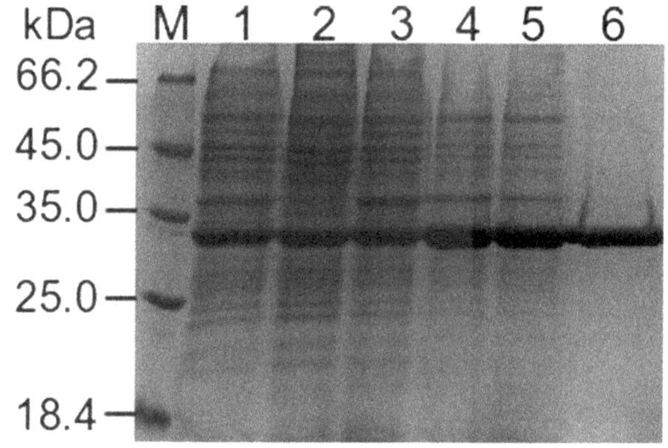

Figure 2 SDS-PAGE analysis of rhANXA5 purification steps. SDS-PAGE (12%) analysis of fractions from different steps of rhANXA5 purification. Each sample lane contains 8 μg of total protein. M corresponds to Thermo Scientific™ Unstained Protein MW Marker; lane 1: cells suspended in buffer A (50 mM Tris–HCl, pH 7.2, 10 mM $CaCl_2$) with no previous centrifugation step; lane 2: supernatant of cells in buffer A after sonication and centrifugation (for 20 min, 24,400 × *g*, 4°C); lane 3: cell pellet from the previous centrifugation suspended in buffer B (50 mM Tris–HCl, pH 7.2, 20 mM EDTA); lane 4: suspension from previous step dialyzed against 2 L of 20 mM Tris–HCl, pH 8.0 for 3 times, overnight in the first step, followed by two steps of dialysis 2 h each; lane 5: supernatant after centrifugation of dialyzed sample (for 30 min, 24,400 × *g*, 4°C); lane 6: homogeneous preparation of annexin V eluted from the Mono Q column.

the yield and productivity of homogeneous rhANXA5. From four independent purifications, we obtained a mean value of 28.5 mg of purified rhANXA5 from 3 g wet weight of cells (n = 4, SD = 6.7). In terms of volumetric yield, we obtained (mean value) 0.983 g (purified rhANXA5) L^{-1} (n = 4, SD = 0.23). The productivity over the entire fermentation (30 h) was 0.0361 g (purified rhANXA5) L^{-1} h^{-1} (n = 4, SD = 0.0084).

Purification

The overexpressed protein was purified by a single-step protocol consisting of a strong anion-exchange column (MonoQ HR16/10). Figure 2 shows the steps of rhANXA5 purification. The target protein eluted at approximately 190 mM of NaCl from a MonoQ HR 16/10 column (Additional file 1: Figure S3). The eluted protein was pooled and dialyzed against HEPES 100 mM NaCl pH 7.2, concentrated using an AMICON ultra-filtration membrane and stored at −80°C in 1 mL aliquots. This purification protocol yielded 28.5 mg of purified rhANXA5 from 3 g wet weight of cells (n = 4, SD = 6.7).

rhANXA5 identification by mass spectrometry

Homogeneous rhANXA5 samples were desalted and digested with trypsin, and the peptide mixtures were analyzed in LC-MS/MS peptide mapping experiments. A total of 320 spectra were identified with 29 different peptides derived from rhANXA5 protein. These peptides covered 80% of the rhANXA5 sequence.

Determination of rhANXA5 molecular mass

The spectra of intact rhANXA5 samples were recorded with a linear ion trap analyzer. Peaks spanning charge states 18+ to 39+ were detected (Figure 3a) and the spectra deconvoluted. We obtained a value of 35,804 Da for the average molecular mass of rhANXA5 (Figure 3b), consistent with the post-translational removal of the N-terminal methionine (theoretical average molecular mass of 35,937 Da with methionine and 35,805 Da without methionine).

FITC-annexin V binding test

To test the functional activity of purified rhANXA5, we assayed its ability to detect cells undergoing apoptosis. We treated B16F10 cells with cisplatin to induce apoptosis and after 6 or 12 h, we stained treated cells simultaneously with the non-vital dye PI and with rhANXA5 labeled with the fluorochrome FITC. Double staining allows the discrimination between intact cells ($FITC^-/PI^-$), early apoptotic cells ($FITC^+/PI^-$) and late apoptotic or necrotic cells ($FITC^+/PI^+$) [7]. The ability of FITC-labeled rhANXA5 to detect apoptotic cells was compared with a commercial kit from BD. We performed dose curve experiments with different concentrations of cisplatin and stained treated cells for 6 h (Figure 4a) or 12 h (Figure 4b) after treatment. We obtained similar results using our home-made kit (QuatroG) or the commercial kit at different times after treatment (6 h or 12 h) and concentrations of cisplatin (20, 40, 80 and 160 μg/mL), indicating that purified rhANXA5 is functionally active. Dot plot representations

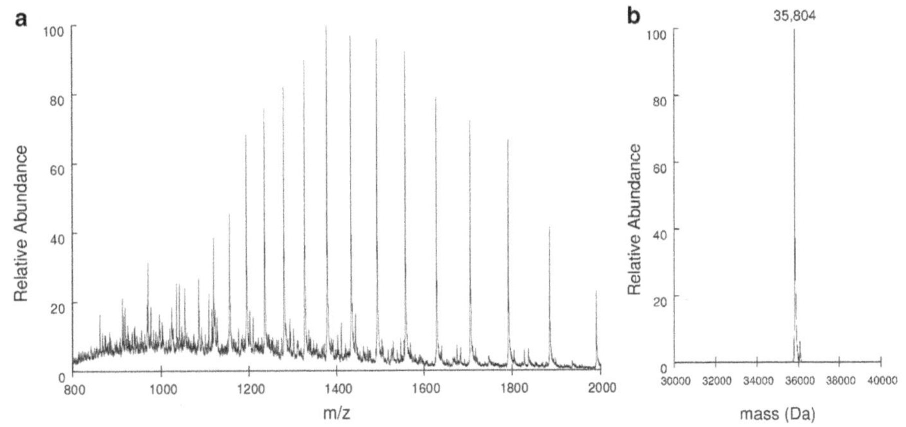

Figure 3 Determination of rhANXA5 molecular mass by mass spectrometry. a) ESI-FTMS spectra of rhANXA5 showing charge state distribution spanning from 18+ to 39+. **b)** Experimentally determined value of 35,804 Da for rhANXA5 average molecular mass after spectra deconvolution.

of flow cytometry experiments with cells stained 12 h after treatment clearly show differences in the distribution of FITC/PI cell populations between untreated cells (negative control), cells treated with 40 μg/mL and cells treated with 160 μg/mL of cisplatin (Figure 5).

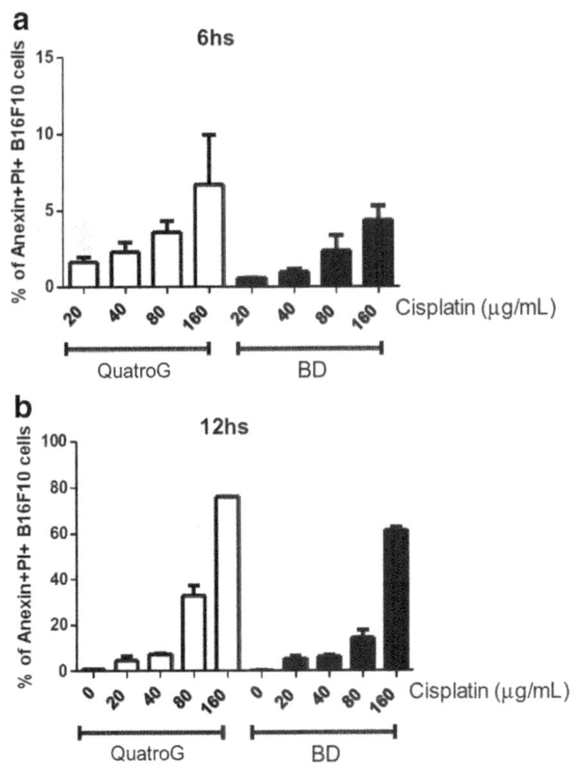

Figure 4 Analysis of apoptosis in B16F10 cells. Flow cytometric analysis of apoptotic cells stained with both FITC-labeled annexin V and PI using the BD Biosciences kit or a homemade kit (QuatroG) containing purified rhANXA5 conjugated with FITC. Apoptosis was induced by treating B16F10 cells with 20, 40, 80 or 160 μg/mL of cisplatin and double staining was performed after 6 h **(a)** or 12 h **(b)**.

Conclusions

Fed-batch cultivations represent an alternative to shaker cultivations, allowing control of process variables and improvement on biomass concentration and product yields [15]. A chimeric protein containing the C-terminus of hirudin fused to annexin V was previously expressed in large scale using fed-batch fermentation [19]. However, to the best of our knowledge, this is the first report of a scale up of rhANXA5 production.

Recombinant protein production methods must continuously be improved to meet commercial demands [14,20]. In this work, we produced 28.5 mg of purified recombinant human annexin V from 3 g wet weight of cells (n = 4, SD = 6.7) obtained in fed-batch cultures induced with IPTG. Moreover, this protocol generated a yield of homogeneous rhANXA5 (mean value) of 0.983 g (purified rhANXA5) L^{-1} (n = 4, SD = 0.23). The productivity over the entire fermentation (30 h) was 0.0361 g (purified rhANXA5) L^{-1} h^{-1} (n = 4, SD = 0.0084).

rhANXA5 could be commercially distributed to several research groups and to radiopharmaceutical companies. The use of fluorescence-labeled annexin V to identify apoptotic cells is currently limited to histologic and cell-sorting studies performed *in vitro*. Its use for *in vivo* imaging studies is hampered mainly by cost issues. Hence, production in large scale may expand the commercial utilities for rhANXA5, reducing costs and allowing a greater access of the scientific and physician community to this product.

Methods

Strains and plasmids

The *E. coli* strains BL21(DE3) and C41(DE3) were purchased from Novagen® (EMD Biosciences, Inc., Madison, WI, USA) and Lucigen Corporation (Middleton, WI, USA), respectively. The PCR-Blunt® cloning vector was

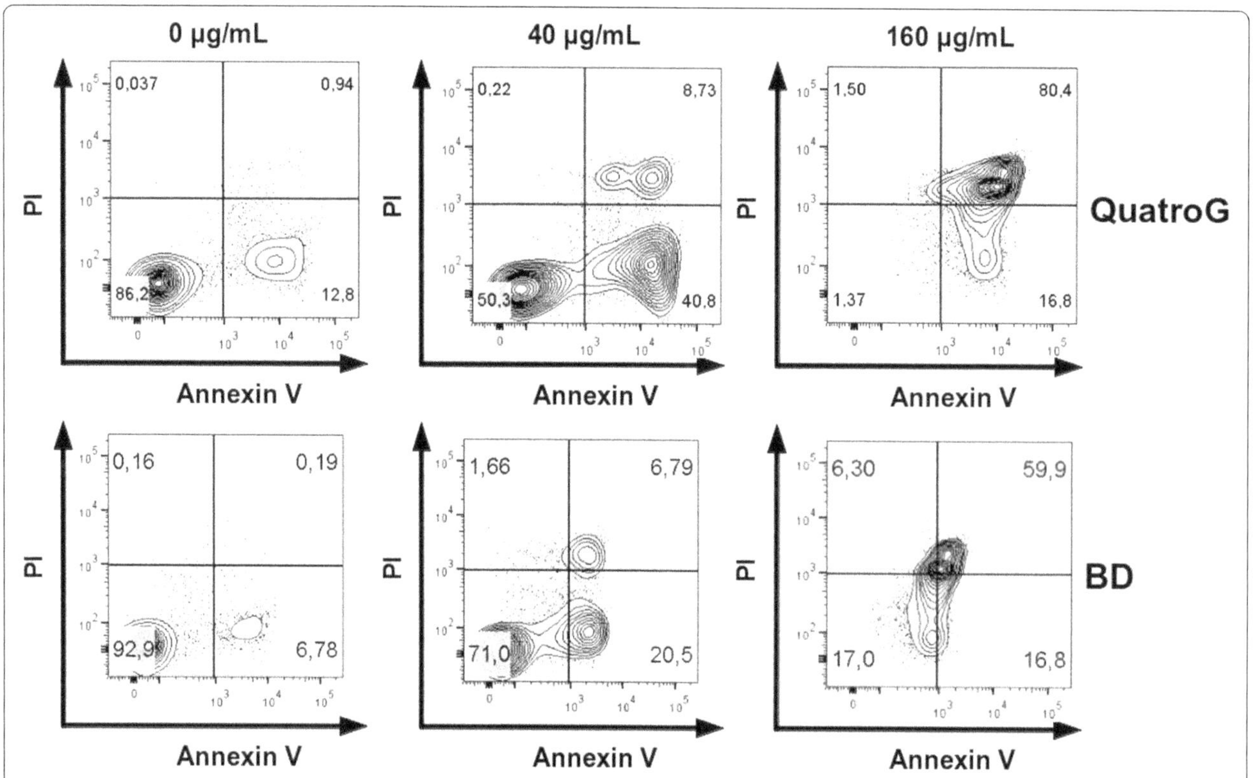

Figure 5 Representative dot plots of flow cytometric analysis. Dot plots of B16F10 cells stained with both FITC-labeled annexin V and PI using the BD Biosciences kit or a homemade kit (QuatroG) 12 h after inducing apoptosis with 0 µg/mL, 40 µg/mL or 160 µg/mL of cisplatin. The upper and lower right quadrants of each dot plot represents cells at late (PI and annexin V positive) and early (annexin V positive only) apoptosis.

purchased from Invitrogen® (Carlsbad, CA, USA) and the pET-30a (+) expression vector was from Novagen®.

Molecular cloning of rhANXA5

Oligonucleotides were designed based on the human *ANXA5* coding sequence in the GenBank (accession number: NM_001154.3) National Institute of Health (NIH) genetic sequence database [18]. Specific primers were designed to contain *Nde*I (forward primer: 5' GCG *CAT ATG* GCA CAG GTT CTC AGA GGC ACT 3') and *Hind*III (reverse primer: 5' GCG *AAG CTT* TTA GTC ATC TTC TCC ACA GAG C 3') restriction sites (italics). The human *ANXA5* gene sequence was PCR-amplified from a human blood cDNA sample.

The amplified *ANXA5* coding sequence was cloned into the PCR-Blunt® vector, cleaved with *Nde*I and *Hind*III restriction endonucleases (New England BioLabs®, Ipswich, MA, USA) and subcloned into the pET-30a (+) expression vector, previously digested with the same restriction enzymes, to generate the pET-30a (+)::ANXA5 plasmid. The cloned *ANXA5* sequence was confirmed by automated DNA sequencing.

Media preparation

Lysogeny broth (LB) (tryptone, 10 g L^{-1}; yeast extract, 5 g L^{-1}; NaCl, 10 g L^{-1}) was sterilized by autoclaving (30 min at 121°C). LB medium was used for shake-flask cultivation and for inoculum development of bioreactor cultivations.

Semi-defined medium (SD) [21] was used for shake-flask cultivations as well as fed-batch bioreactor cultivations, as both initial batch and also feeding media (with varied glucose and $MgSO_4$ concentrations). SD medium contains 0.5 g L^{-1} NaCl (a), 1 g L^{-1} NH_4Cl (b), 20 g L^{-1} yeast extract (c), 6 g L^{-1} Na_2HPO_4 (d), 3 g L^{-1} KH_2PO_4 (e), 1 µg L^{-1} thiamine (f), 1 mM $MgSO_4$ (g), 0.1% trace solution (h), and 0.1 mM $CaCl_2$ (i). For both shake-flask and bioreactor fed-batch cultivations (initial batch culture), SD medium was supplemented with glucose to a final concentration of 5 g L^{-1} (j). Components (a) to (e) were assembled and sterilized together by autoclaving (30 min at 121°C), while components (g) to (i) were sterilized separately by autoclaving and added under aseptic conditions. Thiamine solution (f) was filter-sterilized. Trace solution (h) contained 2.8 g L^{-1} $FeSO_4$, 2 g L^{-1} $MnCl_2$, 2 g L^{-1} $CaCl_2$, 0.26 g L^{-1} $CuCl_2$, and 0.3 g L^{-1} $ZnSO_4$. For feeding medium in bioreactor cultivations, SD medium was supplemented with glucose (final concentration of 300 g L^{-1}) and $MgSO_4$ (40 mM). To prepare this feeding medium, we mixed equal parts of a 2x concentrated SD medium and a 600 g/L glucose stock solution, both sterilized separately. SD medium used in bioreactor cultivations was also supplemented with 100 μL of Antifoam 204 (Sigma-Aldrich, São Paulo, SP, Brazil) per liter of culture. Both LB and SD media were supplemented aseptically with filter-sterilized kanamycin to a final concentration of 30 μg mL^{-1}.

Shake-flask cultivation

E. coli BL21(DE3) and C41(DE3) strains were transformed with the pET-30a (+) vector or the pET-30a (+)::ANXA5 recombinant plasmid by electroporation. Transformant colonies were selected on LB agar plates containing 30 μg mL^{-1} kanamycin [22]. Isolated colonies were selected and grown overnight (at 37°C and 180 rpm) in 5 mL of LB supplemented with 30 μg mL^{-1} kanamycin. To compare the expression profile with different strains (BL21(DE) or C41(DE3)) or media (LB or SD medium), saturated cultures were inoculated in 50 mL of LB or SD medium for shake-flask cultivations and grown at 37°C and 180 rpm to an optical density ($OD_{600\,nm}$) of 0.4–0.6. At this growth stage, we induced rhANXA5 expression by adding isopropyl-β-D-thiogalactopyranoside (IPTG) to a final concentration of 1 mM. At 3, 6, 9 or 24 h after induction, cells were harvested by centrifugation (11,800 *g*) for 30 min at 4°C, and the pellet was stored at −20°C. The expression of the recombinant soluble protein was confirmed by 12% sodium dodecyl sulfate-polyacrylamide gel electrophoresis (SDS-PAGE) visualized by Coomassie® Brilliant Blue R-250 staining. For biomass measurements from shake-flask cultures, isolated colonies were selected and grown overnight (at 37°C and 180 rpm) in 10 mL of LB supplemented with 30 μg mL^{-1} kanamycin and inoculated in 500 mL of SD medium. Inoculated cultures were grown (37°C and 180 rpm) to an optical density ($OD_{600\,nm}$) of 0.4–0.6, rhANXA5 expression was induced by the addition of IPTG to a final concentration of 1 mM, and cells were harvested after 6 h of IPTG induction.

Inoculum development for bioreactor cultivation

A master cell bank (MCB) of transformed *E. coli* BL21 (DE3) cells containing the pET-30a (+)::ANXA5 recombinant plasmid was prepared in 50% glycerol and stored at −80°C. For inoculum development, 150 μL of *E. coli* BL21(DE3) MCB cells (stored at −80°C) were grown overnight at 180 rpm and 37°C in 1 L flasks containing 250 mL of LB medium supplemented with 30 μg mL^{-1} kanamycin. The final optical density (OD_{600}) for each pre-inoculum culture was determined spectrophotometrically. For each experiment, we calculated and collected the initial volume of pre-inoculum culture needed to start bioreactor cultivation with an initial OD_{600} of 0.1. Collected aliquots were diluted in LB medium to a final volume of 100 mL before inoculation in 900 mL of SD medium for bioreactor cultivation.

Bioreactor cultivation

Batch and fed-batch culture experiments were conducted in a BIOSTAT® B Plus bioreactor (Sartorius Stedim, Goettingen, Germany) with two 2 L stirred tanks, filled with 1 L of SD medium each, at 37°C, pH 7.0 and supplemented with kanamycin. For pH control, 12% (v/v) ammonium hydroxide and 10% (v/v) phosphoric acid were employed. The bioreactor was equipped with two Rushton turbines and with agitation, aeration, temperature and pH controllers. A polarographic electrode was used to measure the dissolved oxygen concentration (DOC) in the culture. The pO_2, pH, stirrer speed, base and acid consumption and aeration rate were measured online and recorded by an external data acquisition and control system (Sartorius Stedim). Feeding was implemented using the bioreactor proprietary software micro-DCU system v. 0.63 (Sartorius Stedim) that allows controlling a peristaltic pump for feeding medium addition. The flow rate varied linearly from 0.066 mL min^{-1} (starting after 4 h of batch culture) to 0.594 mL min^{-1} after 26 h of feeding (30 h of bioreactor cultivation). In batch cultures, the DOC was maintained at 30% by cascading agitation (400–1000 rpm) with constant aeration rate (1vvm), and the process was finished when the biomass reached stationary phase.

Fed-batch cultivations were started as batch cultures with feeding starting at 4 h of cultivation (approximately OD_{600nm} 16.0) with SD as feeding medium (see subsection "Media preparation"). Different feeding strategies were tested. In DO-stat feeding fermentations, with DOC setpoint at 30%, the agitation rate was maintained at 800 rpm (after feeding initiation). In fed-batch fermentations with a linear ascending or pH-stat feeding, the DOC was maintained at 30% by cascading agitation (400–1000 rpm) with constant aeration rate (1 vvm).

For linear ascending feeding profile we used the following equation:

$$F = at + b$$

where F is the feeding rate (mL min^{-1}), t the cultivation time after initiation of the fed-batch culture (min) and, a and b are constants for the linear ascending feeding profile [23].

Analytical methods

Samples were withdrawn periodically for quantitative analysis along the cultivation. Cell growth was monitored by measuring the optical density at 600 nm ($OD_{600\,nm}$) in a spectrophotometer. One optical density unit was found to be equivalent to 0.3342 g L^{-1} of dry cell weight by gravimetric quantitation. Glucose concentration in the medium was measured with a glucose analyzer (model 2700 select, Yellow Springs Instruments, Yellow Springs, OH, USA). Acetate concentration was determined by high performance liquid chromatography (Äkta Purifier, GE HealthCare©, São Paulo, Brazil) equipped with an Aminex HPX-87H column (Bio-Rad Laboratories, Hercules, CA, USA), using 0.005 M H_2SO_4 as mobile phase and a UV-detector. The protein expression was analyzed by 12% SDS-PAGE stained with Coomassie® Brilliant Blue R-250 staining. The annexin V protein produced was quantified using the Qubit® Protein Assay Kit (Invitrogen™, Life Technologies, São Paulo, SP, Brazil) and a Qubit® 2.0 Fluorometer (Invitrogen™).

Purification

ANXA5 recombinant protein was purified using a Fast Performance Liquid Chromatography (FPLC) ÄKTA Purifier System (GE HealthCare©). All chromatographic steps were carried out at 4°C. Sample elution was monitored by UV detection at 215, 254 and 280 nm and fractions were analyzed by 12% SDS-PAGE. According to a previous report [24], frozen cells (3 g wet weight) were suspended in 30 mL of buffer A (50 mM Tris HCl, 10 mM $CaCl_2$ pH 7.2) and incubated with 1 mM of phenylmethanesulfonylfluoride for 30 min at 4°C. The cells were disrupted by sonication (eight pulses of 10") and centrifuged at 38,900 *g* for 30 min. The supernatant was discarded and the pellet was completely dissolved in 30 mL of buffer B (50 mM Tris HCl, 20 mM EDTA pH 7.2), stirred for 30 min at 4°C, and clarified by centrifugation at 38,900 *g* for 30 min at 4°C. The supernatant was dialyzed against 20 mM Tris HCl pH 8.0 (3 × 2 L, 3 h each). Residual precipitate was removed by centrifugation (38,900 *g* for 20 min) and the supernatant was loaded on a MonoQ HR 16/10 anion exchange column (GE Healthcare) previously equilibrated with 20 mM Tris HCl pH 8.0. Protein was eluted with 25% linear gradient of 20 mM Tris HCl, 1 M NaCl pH 8.0 at 1 mL min^{-1} flow rate. Homogeneous rhANXA5 was eluted at approximately 190 mM NaCl. Fractions containing homogeneous rhANXA5 were pooled, dialyzed against 20 mM N-2-hydroxyethylpiperazyne-N'-2-ethanesulfonic Acid (HEPES), 100 mM NaCl pH 7.2 and concentrated using an AMICON (Millipore Corporation, Bedford, MA, USA) ultra-filtration membrane (MWCO = 10 kDa), and stored at −80°C. Protein concentration was determined with Qubit® Protein Assay Kit using a Qubit® 2.0 Fluorometer.

rhANXA5 identification by mass spectrometry

rhANXA5 preparations (1 nmol) were desalted and subjected to proteolytic degradation using trypsin. The resulting peptides were separated by chromatography using 15 cm capillary columns (150 µm i.d., Kinetex C18 core-shell particles, Phenomenex, Inc., Torrance, CA, USA) and a nanoLC Ultra 1D plus equipment (Eksigent, Redwood City, CA, USA). Separated peptides were analyzed using an LTQ-Orbitrap hybrid mass spectrometer (Thermo Fisher Scientific Inc, Waltham, MA, USA). The chromatographic method used a step gradient from mobile phase A (0.1% formic acid in water) to mobile phase B (0.1% formic acid in acetonitrile): 0–2% B over 5 min; 2–10% B over 3 min; 10–60% B over 60 min; 60–80% B over 2 min; 80% B isocratic for 10 min; 80–2% B over 2 min; and 2% B isocratic for 8 min. We performed MS/MS fragmentation using collision-induced dissociation (CID) with an activation Q of 0.250, an activation time of 30.0 ms, and an isolation width of 1.0 Da. Using the Proteome Discoverer software (v. 1.3), we compared experimentally obtained MS and MS2 spectra with the *in silico* trypsin digestion of the human proteome. We allowed a precursor tolerance of 10 ppm, a fragment tolerance of 0.8 Da, static carbamidomethylation on cysteines, and oxidation on methionine residues. We restricted our analysis to matches with an Xcorr score > 2.0 for doubly charged ions and Xcorr score > 2.5 for triply charged ions.

Determination of rhANXA5 molecular mass

Purified rhANXA5 samples were desalted, reconstituted in acetonitrile 50%/MilliQ-water 49%/formic acid 1% and directly injected using a 500 µL syringe (Hamilton Company, Reno, NV, USA) in a static mode into an IonMax electrospray ion source. The electrospray source parameters were as follows: positive ion mode, 4.5 kV of applied voltage to the electrospray source, 5 arbitrary units (range 0–100) of sheath gas flow, 45.6 V of capillary voltage, 250°C of capillary temperature, and 238.8 V of tube lens voltage. Full spectra (600–2000 m/z range) were collected on a Thermo Orbitrap Discovery XL in profile mode using the linear ion trap analyzer (ITMS mode). The average spectrum was processed with the software MagTran [25] for charge state deconvolution.

FITC-annexin V binding test

To confirm the ability of rhANXA5 to detect cells undergoing apoptosis, we labeled rhANXA5 with the FluoroTag™ FITC Conjugation Kit (Sigma-Aldrich®). Our home-made kit also contained PI solution (Sigma-Aldrich®) and a 10 × binding buffer (0.1 M HEPES-NaOH pH 7.4, 1.4 M NaCl, 25 mM $CaCl_2$). B16F10 melanoma cells ($4X10^4$), a kind gift from Dr. Peter Henson (National Jewish Center for Immunology, Denver, CO, USA) were cultured in Dulbecco's Modified Eagle's Medium (DMEM) supplemented with 10% fetal bovine serum (FBS), gentamicin 80 mg L^{-1} (Novafarma, Anápolis, Brazil), and Fungisone 5 mg L^{-1} (Bristol Myers Squibb, New York, NY, USA). To induce apoptosis, the cells were treated with cisplatin (Libbs, Embu, SP, Brazil) at different concentrations (0, 40, 80 and 160 µg/mL) and after 6 or 12 h, the cells were stained with our home-made kit or with a commercial kit from BD Biosciences (FITC Annexin Apoptosis Detection Kit II, BD Biosciences, San Diego, CA, USA).

Cells were stained at a concentration of 10^5 cells mL^{-1} in 100 μL of 10X Binding Buffer using 5 μL of PI and 5 μL of rhANXA5 conjugated with FITC (home-made kit) or 5 μL of annexin-FITC from FITC Annexin Apoptosis Detection Kit II. All data were collected in a FACSCanto II flow cytometer (BD Bioscience) and analyzed using FlowJo software (Tree Stat, San Carlos, CA, USA).

Availability of supporting data

Supporting data are included in an additional file.

Additional file

Additional file 1: Figure S1. SDS-PAGE analysis (12%) of samples from shaker cultivations of *Escherichia coli* BL21(DE3) and C41(DE3) strains in lysogeny broth (LB) and in our semi-defined (SD) media. M - Thermo ScientificTM Unstained Protein MW Marker; Lanes 1 and 5 - pET-30a(+) (empty vector) without IPTG induction; lanes 2 and 6 - pET-30a (+)::ANXA5 without IPTG induction; lanes 3 and 7 - pET-30a (+) (empty vector) after 6h of IPTG induction; lane 4 - pET-30a(+)::ANXA5 after 3h of IPTG induction; lane 8 - pET-30a (+)::ANXA5 after 6 h of IPTG induction. **a)** C41(DE3) and LB. Overexpression of pET-30a (+)::ANXA5 in E. coli C41(DE3) strain using LB media. **b)** BL21(DE3) and LB. Overexpression of pET-30a (+)::ANXA5 in E. coli BL21(DE3) strain using LB media. **c)** C41(DE3) and SD. Overexpression of pET-30a (+)::ANXA5 in *E. coli* C41(DE3) strain using SD media. **d)** BL21(DE3) and SD. Overexpression of pET-30a (+)::ANXA5 in *E. coli* BL21(DE3) strain using SD media. **Figure S2.** Densitometric analysis of lanes 4 and 8 from Figure S1d. **Figure S3.** rhANXA5 purification in MonoQ HR16/10 column. **(a)** Eluted fractions from MonoQ HR16/10 were analysed by SDS-PAGE (12%). M corresponds to Unstained Protein MW Marker (Fermentas); Lane 1 corresponds to crude extract; Lanes 2-12 corresponds to MonoQ HR16/10 elution fractions. **(b)** Chromatogram of eluted fractions. rhANXA5 was eluted in fractions 118 to 125 (inlet). **Figure S4.** SDS-PAGE densitometric analysis of rhANXA5 expression in fed-batch cultivations. M: Thermo Scientific™ Unstained Protein MW Marker; Lanes 1–3: samples collected at 30 h of culture (triplicate). rhANXA5 corresponded to (1) 45.3%, (2) 38.6% and (3) 37.9% of total protein content.

Abbreviations

DOC: Dissolved oxygen concentration; FITC: Fluorescein isothiocyanate; HEPES: Hydroxyethylpiperazyne-N'-2-ethanesulfonic acid; LB: Lysogeny broth; MCB: Master cell bank; PI: Propidium iodide; PS: Phosphatidylserine; rhANXA5: Recombinant human annexin V; TB: Terrific broth.

Competing interests

The authors declare that they have no competing interests.

Authors' contributions

LSM carried out part of the experiments and contributed to the manuscript writing. JL carried out part of the experiments and contributed to the manuscript writing. GR helped in fermentation and protein purification. DCR helped in cloning, amplification and purification. GOP generated the FITC-rhANXA5. JESN assisted in the production of rhANXA5 in the bioreactor. APDS conducted the FITC-annexin V binding experiments. ACOD helped in cloning and expression. CVB performed the identification and molecular mass determination of rhANXA5 by mass spectrometry and contributed to the manuscript writing. JMC and LAB assisted in most of the experiments and DSS devised the experiments. All authors read and approved the final manuscript.

Acknowledgements

Financial support for this work was provided by the National Institute of Science and Technology on Tuberculosis (Decit/SCTIE/MS-MCT-CNPq-FNDCT-CAPES) and Millennium Initiative Program (CNPq), Brazil to DS and LB. DS (CNPq, 304051/1975-06) and LB (CNPq, 5201182/99-5) are research career awardees of the National Council for Scientific and Technological Development of Brazil.

Author details

[1]Centro de Pesquisas em Biologia Molecular e Funcional (CPBMF), Instituto Nacional de Ciência e Tuberculose (INCT-TB), Pontifícia Universidade Católica do Rio Grande do Sul (PUCRS), Av. Ipiranga 6681, 90619-900 Porto Alegre, Brazil. [2]Programa de Pós-Graduação em Biologia Celular e Molecular, PUCRS, Porto Alegre 90619-900, Brazil. [3]Programa de Pós-Graduação em Medicina e Ciências da Saúde, PUCRS, Porto Alegre 90619-900, Brazil. [4]Quatro G Pesquisa & Desenvolvimento, LTDA, Porto Alegre 90619-900, Brazil. [5]Instituto de Pesquisas Biomédicas (IPB), Laboratório de Imunologia Molecular, PUCRS, Porto Alegre 90619-900, Brazil.

References

1. Crompton MR, Moss SE, Crumpton MJ: **Diversity in the Lipocortin/Calpactin Family.** *Cell* 1988, **55**:1–3.
2. Schick PK, Kurica KB, Chacko GK: **Location of phosphatidylethanolamine and phosphatidylserine in the human platelet plasma membrane.** *J Clin Invest* 1976, **57**:1221–1226.
3. Fadok VA, Voelker DR, Campbell PA, Cohen JJ, Bratton DL, Henson PM: **Exposure of phosphatidylserine on the surface of apoptotic lymphocytes triggers specific recognition and removal by macrophages.** *J Immunol* 1992, **148**(7):2207–2216.
4. Tait JF, Gibson D, Fujikawa K: **Phospholipid binding properties of human placental anticoagulant protein-I, a member of the lipocortin family.** *J Biol Chem* 1989, **264**(14):7944–7949.
5. Andree HA, Reutelingsperger CP, Hauptmann R, Hemker HC, Hermens ET, Willems GM: **Binding of vascular anticoagulant alpha (VAC alpha) to planar phospholipid bilayers.** *J Biol Chem* 1990, **265**(9):4923–4928.
6. Blankenberg FG, Tait JF, Strauss HW: **Apoptotic cell death: its implications for imaging in the next millennium.** *Eur J Nucl Med* 2000, **27**(3):359–367.
7. Vermes I, Haanen C, Steffens-Nakken H, Reutelingsperger C: **A novel assay for apoptosis Flow cytometric detection of phosphatidylserine expression on early apoptotic cells using fluorescein labeled Annexin V.** *J Immunol Methods* 1995, **184**:39–51.
8. Lahorte CM, Vanderheyden JL, Steinmetz N, Van de Wiele C, Dierckx RA, Slegers G: **Apoptosis-detecting radioligands: current state of the art and future perspectives.** *Eur J Nucl Med Mol Imaging* 2004, **31**:887–919.
9. Yang DJ, Azhdarinia A, Wu P, Yu DF, Tansey W, Kalimi SK, Kim EE, Podoloff DA: **In Vivo and In Vitro Measurement of Apoptosis in Breast Cancer Cells Using ^{99m}Tc-EC-Annexin V.** *Cancer Biother Radiopharm* 2001, **16**:73–83.
10. Brumatti G, Sheridan C, Martin SJ: **Expression and purification of recombinant annexin V for the detection of membrane alterations on apoptotic cells.** *Methods* 2008, **44**:235–240.
11. Logue SE, Elgendy M, Martin SJ: **Expression, purification and use of recombinant annexin V for the detection of apoptotic cells.** *Nat Protoc* 2009, **4**:1383–1395.
12. Coxon KM, Duggan J, Cordeiro MF, Moss SE: **Purification of annexin V and its use in the detection of apoptotic cells.** *Methods Mol Biol* 2011, **731**:293–308.
13. Wang F, He XW, Yan HL, Huang JJ, Zhang Y, Jiang L, Gao YJ, Sun SH: **Non-fusion expression in Escherichia coli: Single-step purification of recombinant human annexin A5 for detection of apoptosis.** *Protein Expr Purif* 2006, **45**:80–87.
14. Lee SY: **High cell-density culture of *Escherichia coli*.** *Trends Biotechnol* 1996, **14**:98–105.
15. Krause M, Ukkonen K, Haataja T, Ruottinen M, Glumoff T, Neubauer A, Neubauer P, Vasala A: **A novel fed-batch based cultivation method provides high cell-density and improves yield of soluble recombinant protein in shaken cultures.** *Microb Cell Fact* 2010, **9**:1–11.
16. Kim BS, Lee SC, Lee SY, Chang YK, Chang HN: **High cell density fed-batch cultivation of Escherichia coli using exponential feeding combined with pH-stat.** *Bioprocess Biosyst Eng* 2004, **26**:147–150.
17. Luli GW, Strohl WR: **Comparison of growth, acetate production, and acetate inhibition of *Escherichia coli* strains in batch and fed-batch fermentations.** *Appl Environ Microbiol* 1990, **56**:1004–1011.

18. The National Center for Biotechnology Information (NCBI) - ANXA5 annexin A5 (*Homo sapiens* human). [http://www.ncbi.nlm.nih.gov/gene/308].
19. Yuan H, Yang X, Hua Z-C: Optimization of expression of an Annexin V-Hirudin chimeric protein in *Escherichia coli*. *Microbiol Res* 2004, **159**:147–156.
20. Choi JH, Keum KC, Lee SY: Production of recombinant proteins by high cell density culture of *Escherichia coli*. *Chemical Engineering Science* 2006, **61**:876–885.
21. Roth G, Nunes JES, Rosado LA, Bizarro CV, Volpato G, Nunes CP, Renard G, Basso LA, Santos DS, Chies JM: Recombinant *Erwinia carotovora* l-asparaginase II production in *Escherichia coli* fed-batch cultures. *Braz J Chem Eng* 2013, **30**(2):245–256.
22. Sambrook J, Russell DW: *Molecular Cloning: A Laboratory Manual*. 3rd edition. New York: CSHL Press; 2001:159–174.
23. Ayub MAZ: Fed-batch bioreactor process with recombinant Saccharomyces cerevisiae growing on cheese whey. *Braz J Chem Eng* 2006, **23**(4):435–442.
24. Wood BL, Gibson DF, Tait JF: Increased erythrocyte phosphatidylserine exposure in sickle cell disease: flow-cytometric measurement and clinical associations. *Blood* 1996, **88**:1873–1880.
25. Zhang Z, Marshall AG: A universal algorithm for fast and automated charge state deconvolution of electrospray mass-to-charge ratio spectra. *J Am Chem Soc Mass Spectrom* 1998, **9**:225–233.

Improving membrane based multiplex immunoassays for semi-quantitative detection of multiple cytokines in a single sample

Raffaele Altara[1*], Marco Manca[2], Marleen HM Hessel[1], Ben J Janssen[1], Harry H A Struijker-Boudier[1], Rob JJ Hermans[1] and W Matthijs Blankesteijn[1]

Abstract

Background: Inflammatory mediators can serve as biomarkers for the monitoring of the disease progression or prognosis in many conditions. In the present study we introduce an adaptation of a membrane-based technique in which the level of up to 40 cytokines and chemokines can be determined in both human and rodent blood in a semi-quantitative way. The planar assay was modified using the LI-COR (R) detection system (fluorescence based) rather than chemiluminescence and semi-quantitative outcomes were achieved by normalizing the outcomes using the automated exposure settings of the Odyssey readout device. The results were compared to the gold standard assay, namely ELISA.

Results: The improved planar assay allowed the detection of a considerably higher number of analytes (n = 30 and n = 5 for fluorescent and chemiluminescent detection, respectively). The improved planar method showed high sensitivity up to 17 pg/ml and a linear correlation of the normalized fluorescence intensity with the results from the ELISA (r = 0.91).

Conclusions: The results show that the membrane-based technique is a semi-quantitative assay that correlates satisfactorily to the gold standard when enhanced by the use of fluorescence and subsequent semi-quantitative analysis. This promising technique can be used to investigate inflammatory profiles in multiple conditions, particularly in studies with constraints in sample sizes and/or budget.

Keywords: Cytokines, Chemokines, Multiplex assay, Planar assay, Bead-based assay, Inflammatory mediators

Background

There is a large body of evidence supporting the use of inflammatory mediators as putative biomarkers for the early detection of a variety of diseases such as heart failure, cancer and infections [1-3]. Hence, determining changes in the levels of inflammatory markers in patients using multimarker-based strategies is expected to lead to better care [4].

In the last decade, multiplex cytokine arrays have become available in a bead-based (or suspension) format showing advantages over the more conventional immunoassays that rely on a solid support [5]. The principle of this assay is quite similar to that of the classic ELISA [6], but the primary antibodies are conjugated to color-coded micro beads rather than ELISA plates and the secondary antibody is attached to a fluorophore. The detection is based on dual-laser flow-based reading where a special instrument classifies the beads and subsequently determines the intensity of the fluorophore, the latter reflecting the concentration of the analytes bound to the beads. Notwithstanding the beauty of the bead-based technology, cases of a loss of sensitivity have been reported [7], resulting in certain proteins becoming undetectable in blood.

An alternative method for simultaneous detection and quantification is a planar assay known as fluorescence-linked immunosorbent assay (FLISA). This assay differs from the ELISA, in that the secondary antibodies are

* Correspondence: r.altara@maastrichtuniversity.nl
[1]Department of Pharmacology, Cardiovascular Research Institute Maastricht (CARIM), Maastricht University, Maastricht, The Netherlands
Full list of author information is available at the end of the article

biotin-labeled and the detection relies on the signal intensities emitted by the fluorophore labeled anti-biotin reporters. The sensitivity of this assay is higher than a traditional ELISA and it may be enhanced even further using tyramide signal amplification [8]. FLISA detection is usually suitable for concentrations that are normally present in the low pg/mL range, however it is recommended for the detection of limited number of targets as the optimization can be difficult using this format.

In this manuscript we report an improved planar multiplex assay (Proteome Profiler™). The assay is membrane-based and its protocol also resembles the ELISA principle. Primary antibodies are conjugated to a nitrocellulose membrane in confined spots. A sample pre-incubated with a secondary antibody is then added to the membranes. This assay has been marketed by the manufacturer as a qualitative assay; hence, the membranes should be processed according to a classic Western blot protocol and the results provide information of the presence of an analyte rather than about its amount.

In this manuscript, we present the fluorescent readout of the membranes which substantially increases the sensitivity of the procedure. Moreover, the introduction of an automated procedure to normalize the readout from the analytes allowed us to obtain semi-quantitative results. We evaluated the sensitivity and accuracy of the membrane-based immunoassay and compared the results with those obtained by ELISA. Moreover, we showed the applicability of this planar assay to mouse, rat and human samples.

Methods

Ethical disclosure

Animal research

This study was carried out in accordance with the recommendations in the Guidelines for the Care and Use of Laboratory Animals of the Dutch Institutes of Health. The protocol was approved by the Committee on the Ethics of Animal Experiments of the University of Maastricht. All surgery was performed under isoflurane anesthesia, and all efforts were made to minimize suffering.

Human research

The recruitment of healthy volunteers was performed according to the Dutch Medical Ethical Committee (protocol: METC 11-3-056) and in respect of the Declaration of Helsinki. The individuals in this manuscript have given written informed consent to publish these case details.

Transverse Aortic Constricted (TAC) Mouse

Pressure overload was induced in male C57BL/6 J mice (Harlan Laboratories, Boxmeer, The Netherlands; N = 5) as previously described [9]. Control male C57BL/6 J mice (Sham; N = 5) underwent the same surgical procedure without the actual tightening of the ligature. Animals were sacrificed four weeks after the Sham or TAC operation and blood samples were collected.

LPS treated rats

Male Wistar rats (N = 5) were treated, under isoflurane anesthesia, with LPS (Sigma-Aldrich) (0.5 mg/kg) intravenously to induce an acute systemic inflammatory response, and were sacrificed after 30 minutes. For the semi-quantitative test serum from the animals was pooled. A total amount of 1000 μL was used in the membrane-based assay, while a total amount of 50 μL of serum per well was analyzed as required by the ELISA manufacturer's instructions.

Sample preparation

Serum

Blood was sampled from the abdominal aorta during sacrifice. It was then allowed to clot for 4 hours at 4°C before centrifuging for 10 minutes at 2000 g. Serum was removed and immediately aliquoted and stored (for 6 months on average) at < -80°C.

Plasma

Blood was sampled from the abdominal aorta during sacrifice. Plasma was collected using EDTA as an anticoagulant. It was stored for 2 hours at 4°C and centrifuged for 10 minutes at approximately 2000 g. Plasma was removed and immediately aliquoted and stored (for 6 months on average) at < -80°C.

Membrane stripping

A pool of serum obtained from five TAC mice (200 μL for each) was tested according to the original Proteome Profiler™ Antibody Array based on chemiluminescence. Subsequently, stripping of the secondary antibody was performed by rinsing the membranes three times for 10 minutes with MilliQ water at 70°C. Then the new protocol for fluorescence detection, described in this manuscript (see below), was performed on the stripped membranes.

TNF-α detection via Enzyme Linked Immunosorbent Assay (ELISA)

Rat TNF-α Quantikine ELISA Kit was purchased from R& D Systems, Minneapolis MN, USA. The assay was performed following the manufacturer's instructions.

Proteome Profiler™ Antibody Array: membrane-based assay

Mouse, rat and human Cytokine Array Panel A (R& D Systems) were used according to the manufacturer's instructions (see the Extracellular Factors session on http://www.

rndsystems.com/product_detail_objectname_Proteome ProfilerArray.aspx) until step #9 (Incubate overnight at 2 - 8°C on a rocking platform).

The following modifications of the original protocol were made to adapt it for fluorescence read-out:

1. Carefully remove the membranes and place them into individual plastic containers. Gently wash the membranes with 20 mL of Wash Buffer 1× for 10 min on platform shaker. Repeat the procedure 3 times.
2. Discard the washing buffer and rinse the membranes with 20 mL of blocking solution [Odyssey Blocking Buffer (LI-COR Bioscience, Lincoln NE, USA) and PBS 1× - (1:1)] for 15 minutes on a rocking platform.
3. Every membrane is incubated with 5 mL blocking solution pre-mixed with 3.3 μL IRDye® 800CW Streptavidin (pre-diluted 1:100, LI-COR Bioscience) in 50 mL plastic tubes.
4. Tubes are then placed for 1 hr on a rolling bench protected from the light.
5. Wash the membranes 3 times with PBS 1× supplemented with 0.15%-Tween for 10 minutes.
6. Rinse 4 times with PBS 1×.
7. Leave the membranes to dry on paper in a dark place.
8. Scan with the LI-COR Odyssey infrared imaging scanner (Westburg BV, Leusden, The Netherlands).
   Resolution: 84 μm
   Quality: High
   Focus offset: 0.0 mm
   Intensity: 5, adjust as necessary. Avoid overexposure of the membrane. White dots (bleaching) should not appear on the final acquisition.
9. Background subtraction was performed by setting the negative controls of each independent membrane to 0.

NB. The Odyssey imaging scanner adjusts the exposure time automatically upon the signal intensity set. It has default scanning presets. The user can choose from 5 settings through the "quality" option: Lowest, Low, Medium, High and Highest. The settings indicated under point 8 are therefore necessary to have the correct setup, allowing normalization of the data later on.

Standard curve with human recombinant proteins

Recombinant Human IL-1 beta and Recombinant Human CXCL10/IP-10 from R& D Systems (R& D Systems™) were reconstituted according to the manufacturer's instructions. Following the reconstitution, the analytes were diluted to generate their respective stock solution.

The stock solutions were used to generate the following standard curves:

Concentrations (pg/mL): 1000, 500, 250,125, 62.5, 31.25.

Human IL-1 beta/IL-1 F2 Quantikine ELISA Kit and Human CXCL10/IP-10 Quantikine ELISA Kit (R& D Systems™) were used to quantify the analytes for further comparison with the Proteome Profiler™ outcome.

Statistical methods/data analysis

Mouse/rat data

All data included in this manuscript are expressed in units according to the international system. After background subtraction, polynomial curve fitting was used to infer the calibration curve for each membrane using Prism5 (GraphPad Software, Inc.; La Jolla CA, USA) and R [10]. Linear regression was used to test the colinearity between ELISA and the membrane-based assay read-outs. The results are illustrated by bar and scatter plots.

Human data

Raw data have been imported in R. Background was subtracted from raw data for each membrane. The readouts of the membranes have been tested for a linear relationship with the predicted concentrations by use of the MASS package ('Modern Applied Statistics with S' (4th edition, 2002)). Study of the quality of predictions was performed by gvlma (Global Validation of Linear Models Assumptions).

Results

Comparison of fluorescence and chemiluminescence detection

In Figure 1, two membranes that first were developed by chemiluminescence and then re-developed after stripping of the streptavidin-enzyme complex from the biotinylated secondary antibody are shown. The number of spots that could be detected, even visually, appeared to be 6-fold higher following the fluorescent (n = 30) compared to the chemiluminescent (n = 5) detection method. As a result, the membrane-based technique allowed the detection of the robust increase of at least 4 analytes (MIG, IP-10, MIP-2 and IL-17) out of the 30 detectable inflammatory mediators present in pooled serum samples of five TAC operated mice with similar hemodynamic characteristics.

Comparison of the membrane-based assay with fluorescence detection to ELISA

In order to prove the feasibility of the optimized membrane-based technology as a semi-quantitative assay, biological samples already containing high levels of cytokines were used for validation. For this purpose, a consecutive

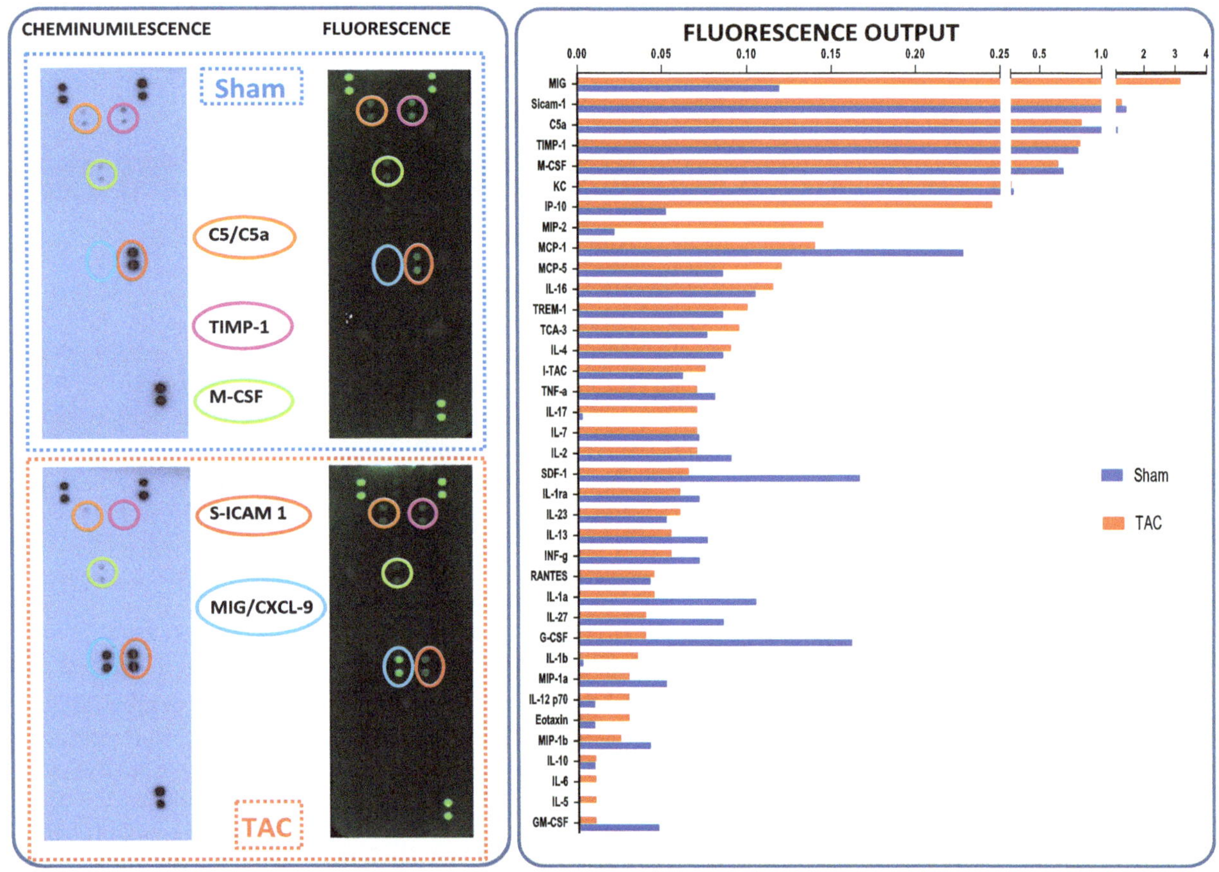

Figure 1 Comparison of chemiluminescence and fluorescence detection. In the left panel, chemiluminescence and fluorescence images, obtained from the same membrane after stripping, are represented. The latter method yielded a considerably higher number of detectable analytes that can be visualized (green spots). The fluorescence readout allowed to compare the circulating inflammatory profile of pooled samples (200 μL each) of 5 TAC operated mice to the relative shams (see graph in the right box).

dilution series of LPS-treated rat blood was created and TNF-α was measured using both the membrane-based assay and ELISA. As shown in Figure 2a, the fluorescence intensity (FI) decreased linearly with increased dilution. Moreover, the correlation of the fluorescence intensity with the serum concentration determined by ELISA was Pearson's $r = 0.91$ (Figure 2b).

Applicability to human cytokines

As proof of concept of the translatability of the fluorescent protocol to human research, a series of tests on consecutive dilutions of known cytokines in buffer were assayed. As shown in Figure 3, the Fluorescence Intensity (FI) readout of the membranes was almost co-linear with the predicted concentrations (Pearson's r value of 0.98 and 0.91 for IL-1β and IP-10, respectively). Testing the model (Predicted concentrations ~ Membrane readouts) fitting to detect biases and to check the overall quality of the distribution also gave promising results (Figure 4). The residuals vs. fitted values appeared to be skewed towards the low values. Overall, these results indicate that the residuals behave randomly, suggesting that the model fits the data well.

Discussion

In the present study, we show the superior sensitivity of fluorescence over chemiluminescence detection of a membrane-based assay for inflammatory mediators. Although this assay is intended for qualitative use by the manufacturer, we improved it using a fluorescence reporter system and a near-infrared imaging device to acquire the signal, and showed that the assay could be upgraded to yield semi-quantitative data (see Figure 5). We demonstrate that by normalizing the readouts to the background reference spots present on each membrane and using specific instrument settings the readout becomes semi-quantitative and shows a strong correlation with the outcomes of ELISA. Our adaptations make the membrane-based analysis suitable for use in situations where the amount of sample is limited, as information of ~30 analytes could be obtained from only 1 mL of serum.

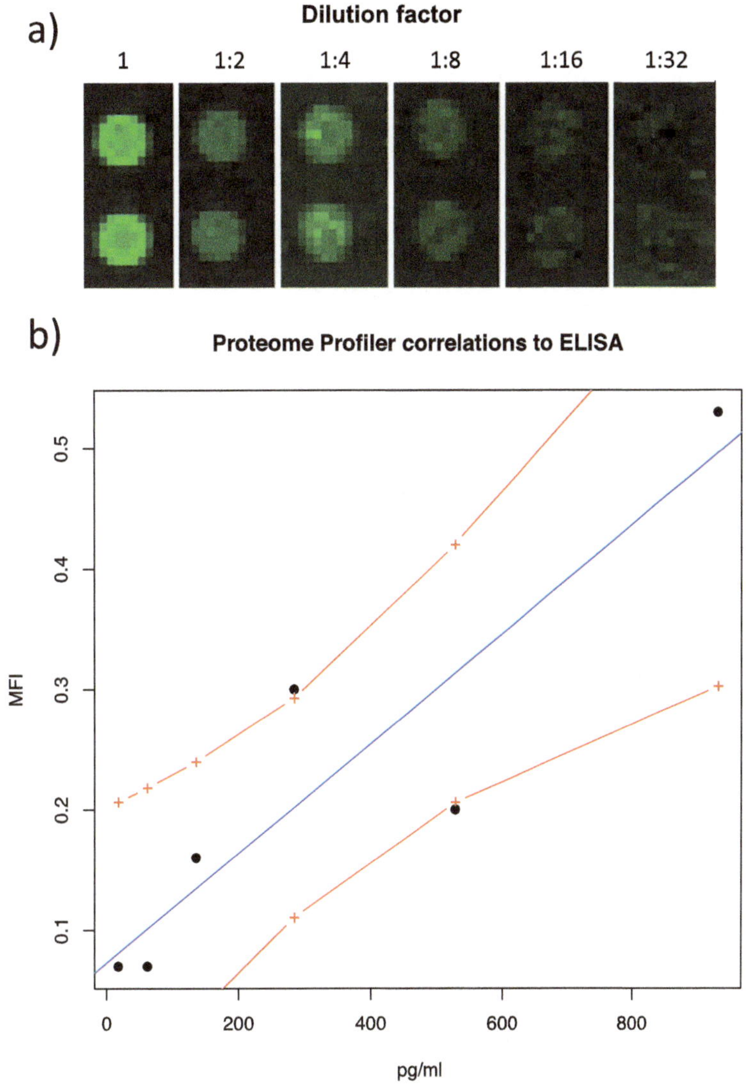

Figure 2 Membrane-based cytokine arrays as semi-quantitative assay. (a) TNF-α signal derived from the membrane-based assay for six progressive dilutions of a pool of plasma obtained from two LPS treated rats. The concentrations range from 930 to 18 pg/mL (from left to right). In **b**, the correlation between the membrane-based assay intensity and ELISA concentration detected for the same samples measured with the two different techniques is shown. The red dashed lines delimited the 95% confidence interval.

The improved membrane-based technique allowed the screening and detection of a large number of circulating inflammatory mediators in an in-vivo model for circulating cytokines/chemokines, which would be limited in conventional ELISAs due to sample availability and/or cost. Indeed, small animal models are known to yield low sample volumes, so to perform multiple ELISAs a large amount of animals would be necessary. Besides, to perform up to 40 classical ELISA investigations it would be rather laborious and costly.

In this study, a model of cardiovascular disease has been used to test the screening performance of the improved membrane-based assay on pooled samples obtained from mice. The TAC mouse model used is actually a known model for cardiac hypertrophy and it is conclusively proved that inflammation becomes activated [11,12] during cardiac remodeling. Moreover, the cardiac inflammatory mediators are released into the circulation [13], but the complete list of all the inflammatory mediators that are circulating in blood is still far from complete. In our investigation we obtained evidence that novel analytes (i.e. MIG and IP-10) can be found in the circulation of mice exposed to TAC for 4 weeks, giving to us a starting point for future investigations. However, our experimental approach (chemiluminescent detection- stripping –fluorescent detection) may have affected the actual inflammatory profile. Indeed, the efficiency of the stripping and the protein loss during that

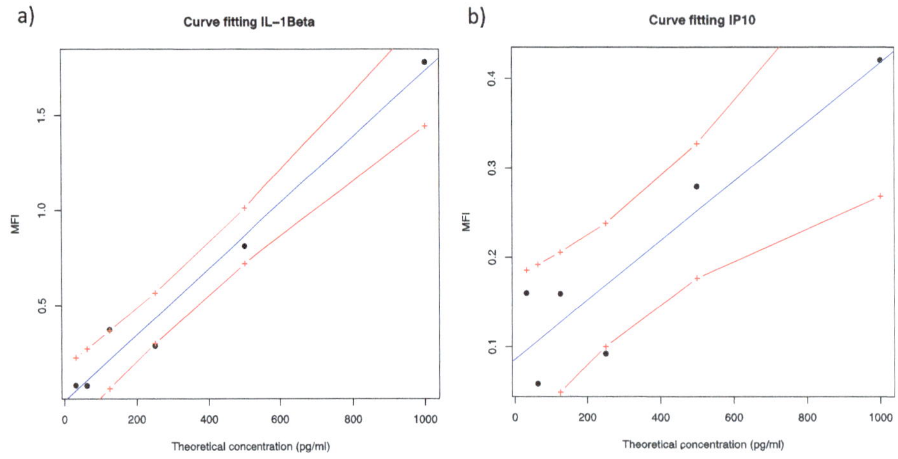

Figure 3 Curve fitting for IL-1β and IP-10. Visualization of the regression line (blue) fitted for the theoretical concentrations and membrane's fluorescence readouts of IL-1β **(a)** and IP-10 **(b)**. The dots are compactly distributed around the predicted line, and the unexplained error is small. The red lines delimited the 95% of CI. The equations of the curves are: **a)** $y = 561.73 \times + 8.97$ with Pearson's $r = 0.97$; **b)** $y = 1830.6 \times - 79.4$ with Pearson's $r = 0.91$. The red dashed lines delimited the 95% confidence interval.

passage could not be assessed. These results therefore need to be interpreted with caution, as the observed inflammatory profile may not reflect the actual one that is present in mice after 4 weeks of TAC.

In this study we are the first to introduce the versatility of the membrane-based assay, commercially known as Proteome Profiler, as a multiplex detecting assay for semi-quantitative analysis of inflammatory mediators. The use of the fluorescence detection protocol allowed us to accurately read the intensity of the spots at standardized exposure times (see materials and methods) which in turn allowed us to create a normalization protocol. The distinct advantages were that the measurements could be repeated without variation (data not shown) and the reading outcomes were comparable among the membranes, as the "development procedure"

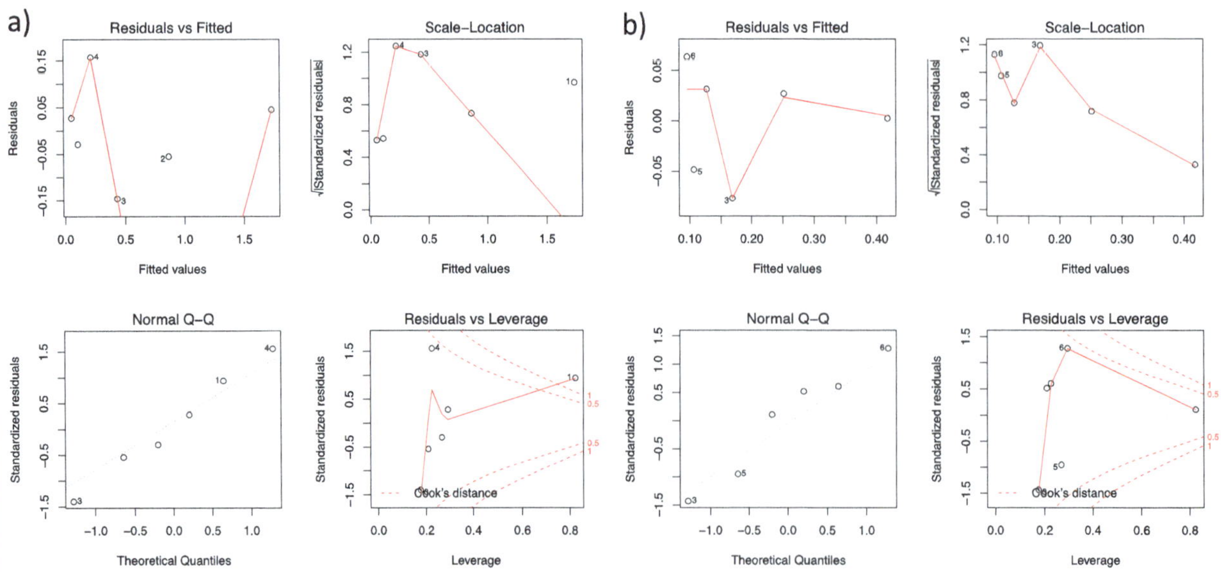

Figure 4 Model diagnostics of the curve fitting of IL-1β (a) and IP-10 (b)."Residuals vs. Fitted" and "Scale-Location" show that a trend appears in the residuals distribution for concentrations below 400 ng/mL. The "Q-Q plot" suggests that residuals are not normally distributed. Whilst this does not affect the goodness of the coefficient estimation, it impairs the t-distribution and thus renders the use of p-values meaningless. The Cook's distance suggests that 2 of the most diluted samples could be considered outliers in their influence on the model fitting. However, Cook's distance has no unambiguous interpretation and the best decision is to look at the original distribution, where the 2 incriminated points show no indication of anomalous position. The red dashed lines delimited the 95% of CI.

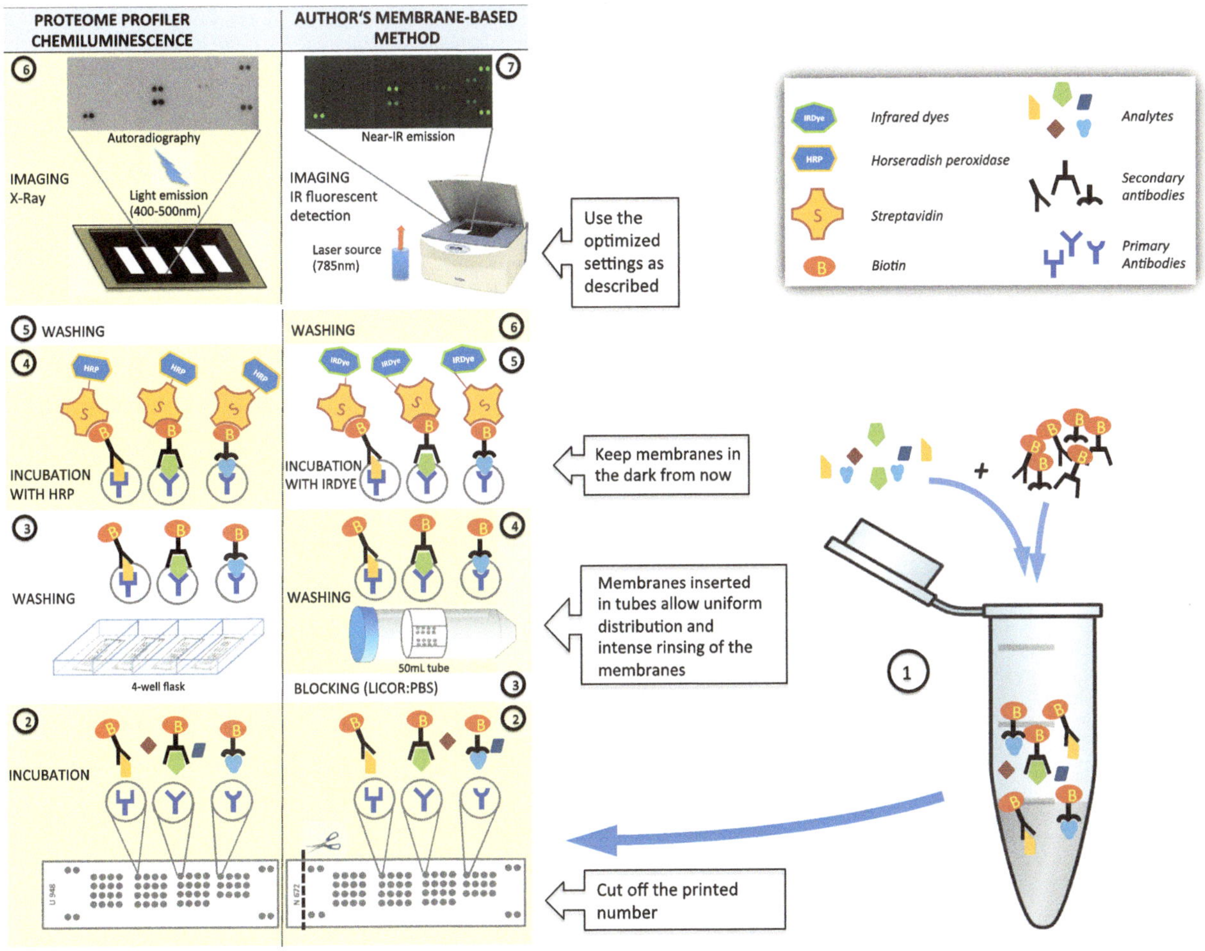

Figure 5 Schematic representation of the main differences between the Proteome Profiler intended for chemiluminescence and the fluorescent detection method and semi-quantitative analysis protocol proposed in this manuscript.

remained the same. This approach favored the construction of concentration curves with satisfactory results when it came to linearity (Figures 2 and 3).

One of the reasons that drove us to implement the membrane-based technique was to test whether it was possible to avoid the "hook effect". This effect is a phenomenon that can occur in multiplex immune-assays when the level of the analyte of interest is close to the lower detection limit; hence the signal to noise ratio is too low and the readouts corrected for the dilution factor result in an overestimation of the actual concentration. In previous experiments performed using the multiplex bead-based technology, we ran into such artifacts in several cases (data not shown). We observed that the detection of many inflammatory mediators in rodent blood was impaired and therefore this approach was only useful for the detection of analytes with high circulating levels.

To test the performance of our adapted membrane-based assay at low analyte concentrations, we tested our model (Predicted concentrations ~ Membrane readouts) fitting to detect biases and to check the overall quality of the distribution also at very low side of the curve (Figure 4). Below 200 pg/mL the residuals diverged, possibly showing a technical limitation of the membrane based assay. In fact, the residuals did not show any co-linearity with the fitted values, and the residuals versus leverage plots showed a well-behaved test. This behavior suggests that there is room for technical development to improve the reliability of the assay at low concentrations.

Conclusion

We present the potential use of a planar method as multiplex assay for the identification of a large panel of inflammatory mediators in blood. By switching from the chemiluminescent to the fluorescent readout and adapting our detection protocol, we showed that a membrane-based assay can yield semi-quantitative results for the detection of multiple cytokines and chemokines in mouse, rat and human blood samples. Applying this novel promising technique will allow scientists from several fields to extend their investigations in a prominent way, particularly when an inventory of multiple inflammatory markers is needed in a setting of low sample sizes and/or budget constraints. Considering the outcome of this study and what we known from the commercially available kits for cytokine/chemokine determination, we present a descriptive table that summarizes the major pros and cons of the assay with the respect to the two commonly alternative methods (see Additional file 1: Table S1).

Additional file

Additional file 1: Table S1. Advantages and disadvantages of the existing immunoassays kits for cytokines/chemokines determination.

Competing interests

The authors declare that they have no competing interests.

Authors' contributions

RA conceived the design of the study, carried out the immunoassays and part of the animal work and drafted the manuscript. MM participated in the design of the study and performed the statistical analysis. MHMH carried out the animal work and contributed with acquisition and interpretation of the data. BJJ participated in the design of the study and coordinated the animal work. HAJSB participated in the study design and coordination and critically revised the manuscript. RJJH conceived the design of the study, coordinated the project and critically revised the manuscript. WMB participated in the design of the study, coordinated the project and critically revised the manuscript. All authors read and approved the final manuscript.

Acknowledgements

The authors would like to thank Agnieszka E. Brouns-Strzelecka and Jacques J.M. Debets for their expert assistance with the animal work.
This research was performed within the framework of CTMM, the Center for Translational Molecular Medicine www.ctmm.nl, project TRIUMPH grant 01C-103, and supported by the Dutch Heart Foundation.

Author details

[1]Department of Pharmacology, Cardiovascular Research Institute Maastricht (CARIM), Maastricht University, Maastricht, The Netherlands. [2]Department of Pathology, Cardiovascular Research Institute Maastricht (CARIM), Maastricht University, Maastricht, The Netherlands.

References

1. Motilva V: **Inflammation and cancer: new targets and novel therapeutic approach.** *Curr Pharm Des* 2012, **18**(26):3829–3830.
2. Corsten MF, Schroen B, Heymans S: **Inflammation in viral myocarditis: friend or foe?** *Trends Mol Med* 2012, **18**(7):426–437.
3. Lam HS, Ng PC: **Biochemical markers of neonatal sepsis.** *Pathology* 2008, **40**(2):141–148.
4. Ahmad T, Fiuzat M, Felker GM, O'Connor C: **Novel biomarkers in chronic heart failure.** *Nat Rev Cardiol* 2012, **9**(6):347–359.
5. Leng SX, McElhaney JE, Walston JD, Xie D, Fedarko NS, Kuchel GA: **ELISA and multiplex technologies for cytokine measurement in inflammation and aging research.** *J Gerontol A Biol Sci Med Sci* 2008, **63**(8):879–884.
6. Lequin RM: **Enzyme immunoassay (EIA)/enzyme-linked immunosorbent assay (ELISA).** *Clin Chem* 2005, **51**(12):2415–2418.
7. Liu MY, Xydakis AM, Hoogeveen RC, Jones PH, Smith EO, Nelson KW, Ballantyne CM: **Multiplexed analysis of biomarkers related to obesity and the metabolic syndrome in human plasma, using the Luminex-100 system.** *Clin Chem* 2005, **51**(7):1102–1109.
8. Haab BB, Zhou H: **Multiplexed protein analysis using spotted antibody microarrays.** *Methods Mol Biol* 2004, **264:**33–45.
9. van Eickels M, Grohe C, Cleutjens JP, Janssen BJ, Wellens HJ, Doevendans PA: **17beta-estradiol attenuates the development of pressure-overload hypertrophy.** *Circulation* 2001, **104**(12):1419–1423.
10. Wass JA: **R: a language and environment for statistical computing.** *R Found Stat* 2012, **Version 2.11.1 (2010-05-31):**1.
11. Christia P, Bujak M, Gonzalez-Quesada C, Chen W, Dobaczewski M, Reddy A, Frangogiannis NG: **Systematic characterization of myocardial inflammation, repair, and remodeling in a mouse model of reperfused myocardial infarction.** *J Histochem Cytochem* 2013, **61**(8):555–570.
12. Coggins M, Rosenzweig A: **The fire within: cardiac inflammatory signaling in health and disease.** *Circ Res* 2012, **110**(1):116–125.
13. Braunwald E: **Biomarkers in heart failure.** *N Engl J Med* 2008, **358**(20):2148–2159.

8

Comparison of intracellular and secretion-based strategies for production of human α-galactosidase A in the filamentous fungus *Trichoderma reesei*

Wesley Smith, Jussi Jäntti, Merja Oja and Markku Saloheimo*

Abstract

Background: *Trichoderma reesei* is known as a good producer of industrial proteins but has hitherto been less successful in the production of therapeutic proteins. In order to elucidate the bottlenecks of heterologous protein production, human α-galactosidase A (GLA) was chosen as a model therapeutic protein. Fusion partners were designed to compare the effects of secretion using a cellobiohydrolase I (CBHI) carrier and intracellular production using a gamma zein peptide from maize (ZERA) which accumulates inside the endoplasmic reticulum (ER). The two strategies were compared on the basis of expression levels, purification performance, enzymatic activity, bioreactor cultivations, and transcriptional profiling.

Results: Constructs were cloned into the *cbh1* locus of the *T. reesei* strain Rut-C30. The secretion and intracellular strains produced 20 mg/l and 636 mg/l of GLA respectively. Purifications of secreted product were accomplished using Step-Tactin affinity columns and for intracellular product, a method was developed for gravity-based density separation and protein body solubilisation. The secreted protein had similar specific activity to that of the commercially available mammalian form. The intracellular version had 5-10-fold lower activity due to the enzymes incompatibility with alkaline pH. The secretion strain achieved 10% lower total biomass than either the parental or the intracellular strain. The patterns of gene induction for intracellular and parental strains were similar, whereas the secretion strain had a broader spectrum of gene expression level changes. Identification of the genes involved indicated strong secretion stress in the secretion strain and to a lesser extent also in intracellular production. Genes involved in the unfolded protein response (UPR) and ER-associated degradation were induced by GLA production, including; *hac1*, *pdi1*, *prp1*, *cnx1*, *der1*, and *bap31*.

Conclusions: Active human α-galactosidase could most effectively be produced intracellularly in *Trichoderma reesei* at >0.5 g/l by avoidance of the extracellular environment, although purification was challenging due to specific activity losses. Strain analysis revealed that in addition to the issues with secreted proteases, the processes of secretion stress including UPR and ER degradation remain as bottlenecks for heterologous protein production. Genetic engineering to eliminate these bottlenecks is the logical path towards establishing a strain capable of producing sensitive heterologous proteins.

Keywords: Therapeutic protein, Human α-galactosidase A, *Trichoderma reesei*, Protein body

* Correspondence: markku.saloheimo@vtt.fi
VTT Technical Research Centre of Finland, Espoo FI-02044 VTT, Finland

Background

Trichoderma reesei is a filamentous fungus known primarily for its use as a large scale producer of industrial proteins [1]. It is established within industry particularly for its ability to secrete high yields of active enzymes. This potential is an attractive benchmark to follow when considering the production of more sensitive heterologous proteins of significant value and therapeutic application, particularly in light of its compatible basis for humanised glycosylation engineering [2] and GRAS status of *T. reesei*. A limited number of mammalian proteins have been produced successfully in this system, including bovine chymosin [3] and antibody Fab fragments [4]. One salient feature of the host is its aggressive and opportunistic nature regarding nutrition uptake [5], with a broad use of complex carbon sources. Such opportunism often causes issues with the stability of heterologous proteins [6]; whereby the harsh extracellular proteolytic environment [7] directly degrades the sensitive recombinant products. Comparison with other potential systems for microbial production of high value proteins demonstrates not only the potential but also some clear limitations of *T. reesei* [8]. Optimisation of the production strategy is therefore one avenue that may be explored to bypass and also to define these limitations, leading to improved yields of high-value target proteins.

The most commonly used technique to enhance heterologous protein production in *T. reesei* is by fusion of the target protein with a secretion carrier, generally the catalytic domain and linker of the major cellulase enzyme cellobiohydrolase I. Carrier fusions of this type have been proposed to have a diversely positive effect on production levels by stabilising mRNA, improving intracellular trafficking in the secretory pathway, and proximity shielding against proteolysis [9]. An alternative strategy can be to direct intracellular accumulation of a target protein, previously highlighted by the hydrophobin fusion strategy leading to protein body (PB) formation [10,11]. Another self-assembling protein that can promote intracellular accumulation and PB formation is derived from the maize gamma zein storage protein [12], ZERA®. This fusion technique relies on ZERA's natural ability to accumulate in the ER and on ZERA-ZERA interactions driven by intermolecular disulphide bonds, recruiting the individual fusion proteins into PB structures. Protein body fusions of this nature have a tight conformation and are surrounded by the ER membrane, thereby insulating the foreign protein from other cell components and avoiding proteolysis to some extent during cell disruption and PB harvesting.

Human alpha galactosidase A (GLA) was chosen as a high value model protein for this study on the basis of its current use in the treatment of Fabry disease [13] by enzyme replacement therapy. GLA is currently produced commercially in Chinese Hamster Ovary (CHO) cells [14]. Fabry disease itself is an X-chromosome-linked lysosomal storage disease caused by loss of native GLA activity in affected individuals, allowing a toxic build-up of the substrate globotriaosylceramide. GLA in its mature form is a 46 kDa glycoprotein with four N-glycosylation sites and five disulphide bridges, and it is active as a 102 kDa homodimer. Enzymatic activity is thought to be independent of the glycosylation pattern, which is more likely to be important for structural integrity, solubility, targeting, and antigenic compatibility [15].

The aim of this study was primarily to investigate whether authentic and functional GLA can be produced in *T. reesei*, and to compare intra- and extracellular production strategies in an attempt to reach industrially significant production levels of active enzyme. A secondary aim following this comparison was to reveal the physiological consequences of the two production strategies. Transcriptional profiling and other techniques were used to elucidate the cellular mechanisms causing bottlenecks in production. Equivalence of a fungal version of human GLA in terms of yield and enzyme functionality would be a clear indication of the potential of *T. reesei* as an economically valid microbial system for the production of human therapeutic proteins.

Methods

Cloning and strain construction

Constructs for intracellular GLA production were made from a synthetic gene coding for ZERA-GLA [12], obtained from Geneart (Regensburg, Germany). GLA [NCBI Gene ID: 2717] was optimised to *the T. reesei* genomic codon bias and 40 bp homologous overlaps with the destination vector pHH01 [16] were included to excise a *Pac*I site upstream and downstream respectively; CATCTTTTGAGGCACAGAAACCCAATAGTCAACCGCGGAC, GACCTACCCAGTCTCACTACGGCCAGTGCGGCGGTATTGG. pHH01 is a *T. reesei* expression vector with a yeast backbone and sequences for amplification in *E. coli*, also containing an acetamidase selection marker, and a *cbh1* promoter and terminator with 3′ integration flanks to the *cbh1* locus in *T. reesei* (Figure 1). The secreted GLA construct was made by nested PCR cloning using Phusion HF polymerase (Thermo Fisher Scientific Inc. Waltham, MA, United States of America) according to the manufacturer's two step protocol conditions with a *cbh1* DNA template and the synthetic GLA plasmid. A forward oligo was used, overlapping pHH01 followed by a *cbh1* priming region (CATCTTTTGAGGCACAGAAACCCAATAGTCAACCGCGGACATCATGTATCGGAAGTTGGCCGTCATCTCGG) with a reverse oligo designed to introduce a KEX2 protease site between the native CBHI carrier and recombinant GLA (GCGCTGGCGGTGTAGGTGGTGCGCT

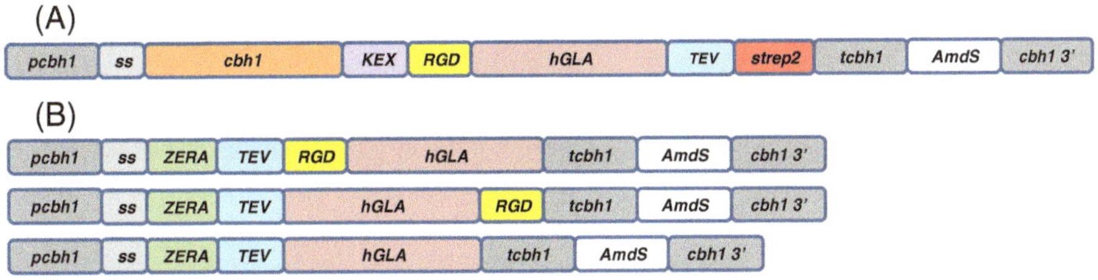

Figure 1 Map of DNA constructs. (A) The construct was designed to secrete the target protein human GLA (*hGLA*) under the *cbh1* promoter (*pcbh1*) with the aid of a CBHI secretion signal (*ss*) and CBHI carrier fusion (*cbh1*), which was to be proteolytically cleaved at a KEX2 site (*KEX*) on route to secretion. A TEV protease site (*TEV*) was also included for downstream polishing after purification by Strep Tactin affinity using a strep2 tag (*strep2*) **(B)** Three variations of GLA were designed with RGD integrin (*RGD*) targeting for intracellular accumulation by ER retention of the Zera® peptide system (*ZERA*). The *cbh1* locus was used as an integration site for all the constructs by homologous recombination using the *cbh1* promoter sequence and a section of DNA downstream of the *cbh1* region (*cbh1 3'*). The native terminator sequence (*tcbh1*) was followed by an *Aspergillus nidulans AmdS* (*AmdS*) marker gene for selective growth on acetamide. The size of the protein expressing region of the secretion construct was 2775 bp and 1632 bp for both RGD containing intracellular constructs whereas the control construct lacking RGD was 1563 bp.

TGTCCATGGGAGGTCCGGGAGAGCTTCCAGTG). After three polymerase chain reaction cycles an additional reverse oligo was added to the mix containing a GLA priming region with an extra terminal sequence to code for a strep2 tag with TEV protease site. (CCAATACCGCCGCACTGGCCTCACTTCTCGAACTGGGGGTGGCTCCAGCCCTGGAAGTAGAGGTTCTCGCCGCCACCGCCGAGGAGGTCCTTGAGGCTCA). A further 30 cycles were performed in order to anneal the entire nested product which contained 20 bp overlaps for each piece. Yeast homologous recombination cloning was used for all of the constructs. The expression vector pHH01 was linearised with a *Pac*I (New England Biolabs Inc. Ipswich, MA, USA) restriction enzyme between the *cbh1* promoter and terminator and transformed into *S. cervisiae* together with the ZERA-GLA and CBHI-GLA cloning fragments designed to replace the *Pac*I site and recircularise the plasmid. The transformation mixes were plated on SCD-Ura [17] for URA3 strain selection and the recombinant yeast plasmid DNA was then isolated and transformed to the *E. coli* DH5 alpha. Positive clones were identified by restriction enzyme digestions and DNA sequencing. The *T. reesei* expression cassettes shown in Figure 1 were excised from the cloning vectors using dual *PmeI* sites (Thermo) and transformed into *T. reesei.* Transformations were made using the Rut-C30 strain including a mus 53Δ [18] mutation to enhance genome-targeted integration according to the standard protoplast-based protocol [19]. Genomic Integration into the *cbh1* locus was confirmed by PCR screening from transformants successfully growing on Trichoderma Minimal Medium (TrMM) acetamide selection plates [19] after two successive streaks made on selective medium. Uninuclear clones were obtained by plating single colonies from spore suspensions onto selection plates and finally amplified for spore culturing stocks on potato dextrose DIFCO 213400 agar plates (Becton, Dickinson and Co. New Jersey, USA).

Culturing

Small scale 300 ml cultivations were performed in two litre shake flasks with TrMM buffered to pH 4.8 with 100 mM piperazine-n, n'-bis 3-propanesulfonic acid, and supplemented with 4% lactose Fagron 208090–0002 (Fagron Nordic A/S. Copenhagen, Denmark), 2% soluble spent grain (extracted from solid grain material (Harbro Ltd. Turriff, Aberdeenshire, United Kingdom) at pH 5.0 by adjustment with NaOH and using a 115°C heating step for 10 mins before centrifugation to obtain the soluble fraction), and 0.2% peptone. Five separate transformants were grown for each construct and compared for growth and expression levels in order to establish a representative clone for upscale culturing. Protease inhibition cultivations for the CBHI-GLA secretion strain were performed in Whatman™7701-5102 24 well culture plates (GE Healthcare UK Ltd. Little Chalfont, Buckinghamshire, United Kingdom) with the same medium in 4 ml volume including daily dosing of 10 nM pepstain A SIGMA P5318 (Sigma-Aldrich Co. LLC. St. Louis, MO, United States of America), 10 nM chymostatin SIGMA C7268, 5 mM EDTA, and 0.2 mg/ml trypsin inhibitor SIGMA T9128.

Larger scale bioreactor cultivations were performed in order to obtain sufficient material for purification of comparable test batches. 20 litre cultivations in a Braun Biostat C-DCU3 bioreactor (Sartorius AG. Göttingen, Germany) were started with a 10% inoculum, grown on 40 g/l lactose with 33% (v/v) of a soluble extract of spent grain. The bioreactor medium contained 60 g/l lactose with 40% (v/v) of the same spent grain extract. Cultivation conditions were controlled as follows; pH was maintained at 4.5 ± 0.5 using NH_4OH, foaming was controlled by automatic addition of Dow Corning 1500 antifoam agent when required (Dow Corning Corp. Midland, Michigan, United States of America), dissolved oxygen was maintained above 30% by agitation (400–550 rpm) and oxygen enrichment of the gas flow (0-66%) at a constant gas flow rate of 8 l/min, and temperature was controlled at 28°C. A fed-batch cultivation with lactose feeding (200 g/l) starting at the end of the batch phase was controlled by an algorithm based on the consumption of base for pH control [20,21].

Bioreactor cultivations for transcriptional profiling were performed in triplicate from a single inoculum per strain in one litre Sartorius Q DCU4 bioreactors using 4% lactose and 2% spent grain extract; all other parameters were controlled as in the previous batch method.

Protein gels and Western blot analysis

SDS PAGE samples were made with LI-COR 928–40004 protein sample loading buffer (LI-COR Biosciences Inc. Lincoln, NE, United States of America), a pre-cast Biorad Criterion gel system (Bio-Rad Laboratories Inc. Hercules, CA, United States of America) and blotted by a semi dry Biorad Transblot Turbo Western blotting system. Blots were then probed with primary antibody (anti ZERA rabbit αR8 1:7000 [12], Santacruz anti GLA H104 rabbit 25823 1:1000 (Santa Cruz Biotechnology Inc. Dallas, Texas, United States of America), iBA Strep-MAB classic 21507001 1:2000 (IBA GmbH. Göttingen, Germany), anti CBHI MAB261 1:1000 [22]) and quantified using a fluorescent secondary antibody: ODYSSEY IR Dye 680RD 92668071 1:10000/IRDye 800CW 92632710 1:10000 (LI-COR) and a LI-COR ODYSSEY CLX system for near infrared scanning and quantification). Quantifications were based on a moving average curve fit from five different dilutions of a protein standard. A representative batch of ZERA-GLA was chosen as a positive control and quantification standard by first quantifying against a commercial GLA from *E. coli* 12078-H08H (Sino Biological Inc. Beijing, China) using GelCode Blue G250 Coomassie stain (Thermo). Duplicate protein samples were used for each quantification. The densitometry calculations were processed with ImageQuant TL software (GE Healthcare) based on a best fit curve of five GLA standard dilutions. CBHI carrier quantities within the culture sample were determined against a pre-quantified batch of CBHI control protein (VTT Technical Research Centre of Finland, Espoo, Finland).

Purification of GLA

Mycelial biomass from ten litres of fermentation broth was harvested by vacuum filtration on double-layered 240 mm diameter Whatman GF/A filter paper (GE Healthcare) in a Büchner apparatus and washed with reverse osmosis water before being frozen at −20°C for storage. Many different buffer conditions and methods for cell breakage, PB extraction and solubilisation were tested both for protein and enzyme activity yield, as briefly outlined in Figure 2, before a standard protocol was determined. In this protocol 1 g of the cell material was resuspended in 10 ml of 10% sucrose, 10 mM NaH_2PO_4 (pH6.5), 1% Triton x100, 2 mM EDTA, 1 mM Phenylmethanesulfonyl fluoride Sigma P7626 and 10 μM Pepstatin A and homogenised by a 10 mm Polytron PT1200 disrupter probe (Kinematica AG. Luzern, Switzerland). The homogenous material was then subjected to three passes through an SLM Aminco FA030 French press (Horbia Instruments Inc. Irvine, CA, United States of America) at 20,000 psi. Initial separation of PBs from leftover biomass was performed in 50 ml tubes by centrifugation for 5 minutes at 200 × g. After separation from the mycelial pellet, the supernatant containing soluble cell extracts, small aggregates and the protein bodies was then used to enrich the protein bodies by centrifugation at 2000 × g for 10 minutes. Two washes were then performed on the pellet material using 10 mM NaH_2PO_4 (pH6.5) and 1% Triton-X100, to remove any endoplasmic reticulum membranes. PB-enriched aliquots were then made from one millilitre of original material and frozen for solubilisation method development.

The solubilisations were based on a controlled concentration ratio from disrupted biomass, e.g. 1 g of mycelial fresh weight was disrupted in 10 ml of breakage buffer, and PBs were then enriched into a small pellet and reconstituted in 1 ml of solubilisation buffer to a maximum concentration of approximately 2.5 mg/ml ZERA-GLA. A buffer containing 50 mM Bicine pH 8.2, 5% Glycerol, 150 mM NaCl, 16 mM Sarkosyl, 10 mM β-mercaptoethanol was used to resuspend the PB pellet before treatment in a Fritsch Laborette 17.202 sonic bath (Fritsch GmbH. Idar-Oberstein, Germany) for one minute. The suspended PB pellets were left to solubilise overnight at +8°C on a Brunswick TC7 roller drum (Eppendorff AG. Hamburg, Germany). The soluble fraction was recovered by centrifugation at 20,000 × g for 30 minutes. Proteolytic cleavage of the GLA enzyme from ZERA was achieved by use of a monomeric TEV protease Invitrogen AcTEV 12575 according to the manufacturer's instructions (Thermo).

Secreted GLA was purified using an IBA 2-1531-001 Strep-Tactin 1 ml flow column by adjustment of the sample pH to 7.5 using NaOH and elution by acid-driven dissociation in 50 mM sodium acetate activity buffer, pH 4.5 according to the manufacturer's instructions.

GLA activity assay

Activity measurements were made by mixing a 25 μl enzyme sample into 100 μl of the 2.46 mM fluorescing substrate 4-methyl-umbelliferyl-alpha-D-galactopyranoside Sigma M7633 in 10 mM acetic acid, pH 4.5. The enzyme reactions were continued for one hour on an orbital shaker at 37°C before being stopped by addition of 1.25 ml 200 mM glycine solution at pH 10.4. A standard curve was prepared of 0–500 ng/ml of 4-methylunbelliferone Sigma M1381. All samples were

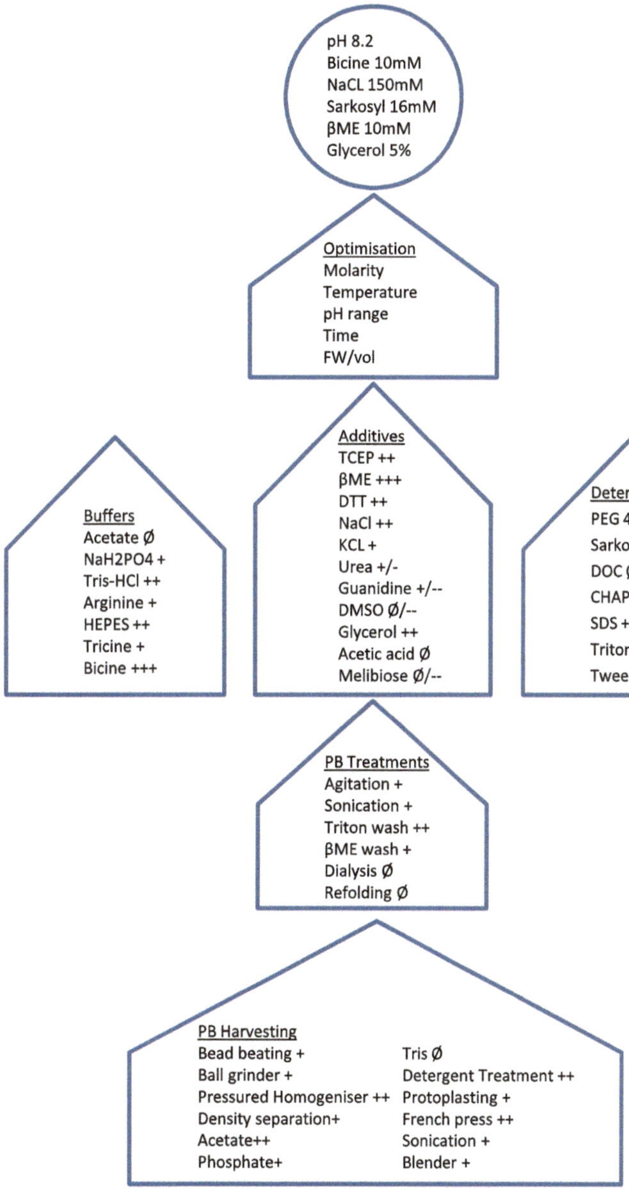

Figure 2 Schematic workflow of the protein purification protocol development. Biomass for the intracellular production strains was extracted by various PB harvesting techniques and eventually solubilised under many different conditions requiring stepwise optimisation. The effects at each step were measured by Western blotting for yield increase and by activity assay, to determine whether integrity of the GLA enzyme fusion was retained. The harvesting, PB treatment and solubilisation methods tested are shown in the arrows, and the final solubilisation conditions are shown in the circle on the top. A plus sign (+) represents an increase in active yield and two or three such signs show the approximate magnitude of this effect. Similarly one to three negative signs (–) denote the magnitude of active yield decrease. A null effect sign (Ø) was used where there was no apparent change in active yield. Some additives have both signs to show that the effect of active yield was dictated by a change in molarity of the additive.

analysed at an excitation wavelength of 365 nM and read at 450 nM on a Varioskan (Thermo). Activity units were expressed as μmol of 4-methylumbelliferone/h/mg protein.

RNA isolation and array hybridisation

Mycelial samples from the one litre bioreactors were collected by vacuum filtration on Whatman GF/A filter paper and then rinsed in 0.9% NaCl and frozen immediately in liquid nitrogen. RNA was isolated from 50 mg of the mycelia with RNeasy Plant RNA isolation kit (Qiagen GmbH. Hilden, Germany). RNA quality was assessed for purity and quantity using an Agilent RNA 6000 analyser (Agilent Technologies Inc. Santa Clara, CA, United States of America) The purified RNA was then converted into cDNA using an Invitrogen Superscript Double-Stranded cDNA Synthesis Kit (Thermo) prior to labelling with Cy3 One-Colour fluorescent dye (Roche

NimbleGen Inc. Basel, Switzerland) according to the manufacturer's instructions. The labelled cDNA was hybridised on an automated HS12 system (Roche) to custom-made four plex microarray slides (Roche) designed from the *T. reesei* genome version 2.0 [23,24]. Analysis was performed in an MS200 microarray scanner (Roche) to identify signal probe intensity of the six 50-74mer probe replicates designed from the known and predicted genes in *T. reesei.*

Microarray data analysis

R/Bioconductor (http://www.r-project.org) was used for all data analysis. The raw data was normalized with Robust Multichip Average (RMA) normalization [25]. The quality of the microarray data was assessed on the basis of a report from the arrayQualityMetrics–package [26] and by observation of the distribution of the $\log_2$ intensities on each array before and after normalization (Additional file 1A). Additionally, the similarity of replicate samples to each other was verified by plotting the arrays on a two dimensional display using principal component analysis (Additional file 1B).

Statistical differences in expression were analysed using linear modelling with tools of the limma package [27]. For each gene, a linear model was fitted by the least squares method and differential expression within pairs of experimental conditions was computed using an empirical Bayesian approach [28]. Genes with $\log_2$-scale fold change > +\-1 were considered to be up- or down-regulated, respectively. The significance test results were visualized as volcano plots showing the expression level difference ($\log_2$-transformed fold change) versus the significance of the change ($-\log_{10}$ of the p-value). The volcano plot shows, at a glance, the number of changes between the strains being compared. In a large and outspread volcano plot, genes in the top-left and top-right corners of the plot are most significantly down- or up-regulated. A small and shallow volcano indicates that there were no significant changes between the conditions, i.e. that the changes in the expression levels were small (below twofold) and insignificant (p-value >0.001).

Results

In order to compare the two different production strategies of human α-galactosidase in *T. reesei*, four expression constructs were made from a synthetic gene codon-optimised to *T. reesei* codon usage and transformed into the fungus (Figure 1). In one construct GLA was fused to the *T. reesei cbh1* catalytic domain and linker region in order to provide a secretion carrier. For the remaining constructs GLA was fused with ZERA peptide to promote protein body formation inside the cell. Three of the constructs had RGD peptides attached at either end of GLA to enhance its localisation in the lysosome for its use in enzyme replacement therapy for humans. The transformants obtained for each of these constructs were purified through single spore cultures, and integration of the expression construct to the *cbh1* locus was verified by PCR analysis.

Production of human α-galactosidase in *T. reesei*

Five individual transformants for each of the Intracellular ZERA-GLA variants and CBHI-GLA secretion strain were grown in replicate shake flask cultures. GLA production yields quantified by near infra-red Western blots were approximately 90 mg/l for all the intracellular ZERA-GLA variants, whereas the secreted CBHI-GLA product was barely above the limit of detection. No growth defects were observed and there was no significant variation in production levels between the replicates, and therefore one representative clone for each strain was chosen for up-scaled production. Secreted CBHI-GLA was deemed unsuitable for bioreactor studies due to its instability and low GLA yield. However, its production level was eventually improved to around 20 mg/l in 24-well plate cultures and to a total of 60 mg/l including CBHI carrier bound forms by addition of a cocktail of protease inhibitors (Figure 3A). The strain expressing ZERA-GLA with C-terminally attached RGD peptide was grown in a 20 litre batch cultivation and ZERA-GLA lacking RGD peptide was produced in a 20 litre fed batch cultivation based on the monitoring observations of the previous batch run.

Batch cultivation for ZERA-GLA produced 267 ± 18 mg/l and the subsequent fed-batch bioreactor run with extended lactose feeding increased the yield to 636 ± 41 mg/l. When including proteolytic cleavage products in the quantification, the total detected amount found in the cells was around 1200 mg/L (Figures 3B and C). In both the batch and fed-batch cultivations growth was similar to that of the parental strain and peak expression of ZERA-GLA was achieved at the end of the cultivations when CO_2 levels had decreased almost to the baseline after exhaustion of carbon source.

Purification and enzyme activity measurements of human GLA

In the ZERA peptide protein body purification concept the target protein is purified from lysed cell material based on the high density of the PBs. Developing a protocol for GLA isolation and solubilisation required extensive trials and optimisation (Figure 2), but ultimately a relatively robust protocol was established. In our initial purification trials we focused on optimising recovery of PB material from the mycelia. In the case of mild disruption methods such as the use of homogenisation blenders, PBs could not be extracted efficiently. In more harsh disruption methods such as grinding and bead beating, the PBs were also disrupted and consequently

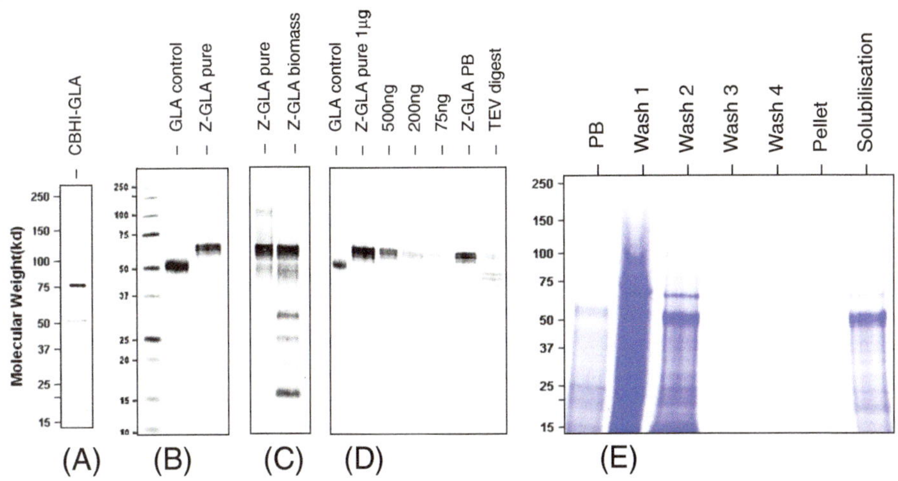

Figure 3 SDS-PAGE and Western blot analysis of the proteins produced. (A) Secreted CBHI-GLA purified via strep2 affinity, measured by strep-tactin-AP Western blot (100 ng loaded). The 50 kD band represents the GLA cleaved by KEX2 protease from the CBHI carrier. The larger band remains as a CBHI-GLA fusion. **(B)** Comparison of a solubilised batch of purified ZERA-GLA (Z-GLA pure) compared to commercial GLA (GLA control) (1 µg of each). The ZERA-GLA protein runs about 10 kD higher than commercial GLA as expected. **(C)** The same ZERA-GLA batch compared to ZERA-GLA contained in mycelium (Z-GLA Biomass) and extracted directly from filtered mycelia by boiling in protein sample loading buffer (1 µl of each). Detection by anti ZERA Western blot reveals the presence of several lower molecular weight degradation products contained within the cell, still retaining the ZERA peptide. **(D)** A TEV protease cleavage of GLA from the ZERA peptide (TEV digest) directly using a 10-fold digest buffer dilution of insoluble enriched protein body material (Z-GLA PB), compared to 0.5 µg of commercial GLA and 1 µg of the purified soluble form of ZERA-GLA. 5 µl were loaded for both the digest and starting PB material. The anti GLA Western blot shows that while cleavage is possible from insoluble PBs three lower molecular weight forms of GLA are present, perhaps due to endogenous protease. **(E)** A four step pellet wash purification of the ZERA-GLA PB material from a 5 µl sample of 1 g fresh weight diluted in 10 ml (PB) showing the soluble components removed from the PB material with a 5 µl sample of tenfold concentration increase (Wash). The PB pellet is then solubilised in the same volume and centrifuged leaving insoluble debris (Pellet) and the soluble ZERA-GLA (Solubilisation) (5 µl for both).

could not be recovered by density separation. Disruption in a French press with added Triton detergent allowed for sufficient disruption of the mycelial biomass without harming the PBs, which could subsequently be harvested efficiently by differential gravity separation. The remaining challenge of PB solubilisation was approached by testing a number of additives and detergents in a range of different buffers (Figure 2). These were then individually assessed for solubilisation yield and GLA activity and the best conditions were combined and optimised. During this method development phase for PB solubilisation it became apparent that loss of specific activity was a major challenge.

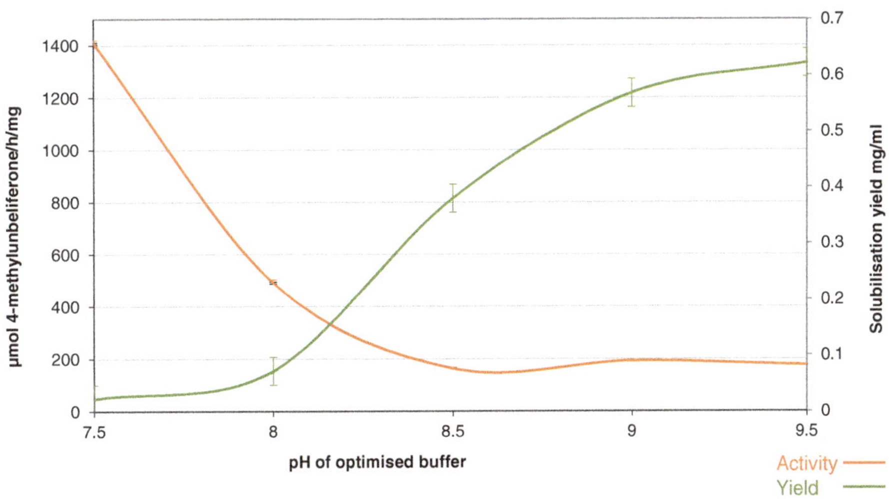

Figure 4 Enzyme activity in relation to recovered yield from insoluble material. A pre-pellet washed batch of PB material was solubilised over a neutral to alkaline pH range and tested for target protein concentration by quantitative Coomassie stain and GLA enzyme activity measurement. A specific activity loss trade-off vs. concentration yield, by solubilisation pH change was established to determine the optimum pH for the procedure.

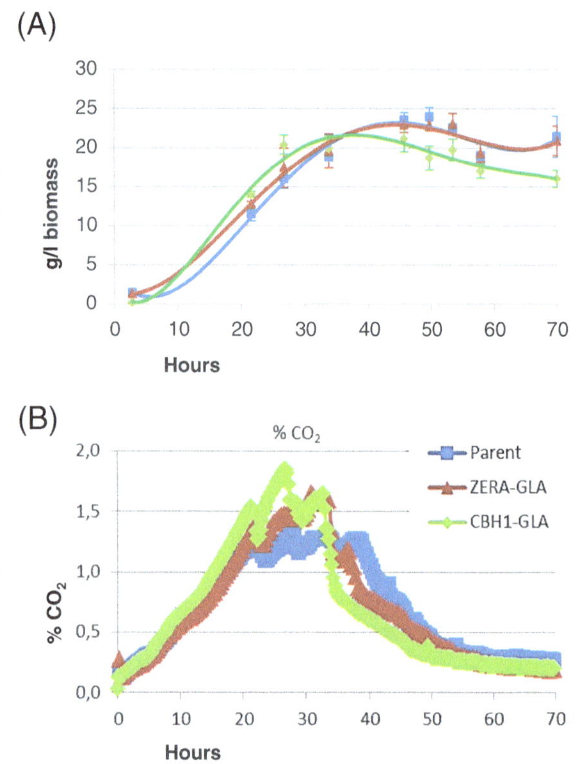

Figure 5 Bioreactor data. Batch bioreactor cultivations in triplicate. The three production strains were cultured in triplicate for the parental strain (Parental), intracellular production strain (ZERA-GLA) and secretion strain (CBHI-GLA) and then analysed by transcriptional profiling for physiological observations. **(A)** Cell dry weight determined over time in grams of dried biomass per litre of culture volume. **(B)** Carbon dioxide monitoring was determined over time and measured in percent dissolved CO_2.

Highly concentrated batches of >2 mg/ml of ZERA-GLA could be made in 50 mM Bicine buffer, pH 9.5, with the addition of 50 mM ionic surfactants, either SDS or Sarkosyl , 900 mM NaCl and 50 mM BME for reduction of the ZERA intramolecular disulphide bridges. This led to a clear protein enrichment of ZERA-GLA from cell lysate material (Figures 3D and E) with greater than 60% total recovery of material and >70% purity as determined by Coomassie staining. Unfortunately these conditions led to a major reduction of enzyme activity, with eventual total loss after one freeze thaw cycle. Optimisation of the pH and molarities of the solubilisation buffer components led to retention of a small proportion of the enzyme activity by trade-off with solubilisation yield (Figure 4). After optimisation, product concentrations of 0.1-0.2 mg/ml were observed with a 5-10-fold lower specific activity than the commercially available CHO-derived enzyme. Proteolytic release of GLA enzyme from the ZERA fusion could then be achieved at the TEV cleavage site using a monomeric TEV protease (Figure 3D); the specific activity after this polishing procedure remained unchanged.

The specific activity of CBHI carrier-secreted GLA collected by Strep tag purification was comparable to that of the commercial enzyme Replagal (results not shown). However, this method of production and purification was unreliable due to high levels of product degradation and loss. Secreted GLA became unstable during the Step-Tactin binding process when the acidic culture pH was increased to the neutral pH range of Strep-Tactin binding conditions, despite the addition of protease inhibitors.

Physiological consequences of the production strategies

In order to establish the specific physiological and transcriptional changes as a result of either secreted or intracellular production, bioreactor cultivations were performed in triplicate for each of the parental, GLA-secreting and intracellular production strains. Quantification of cell dry weights and CO_2 formation from the cultivations revealed that growth of the GLA-secreting strain peaked about 15 hours earlier than both the parental and intracellular ZERA-GLA strains (Figure 5). The secretion strain also produced about 10% lower peak biomass. However, according to both dry weight and CO_2 measurements the maximal growth rates of the three strains were rather similar.

The accumulation of intracellular ZERA-GLA was measured for all replicates (Figure 6). The best producer was the intracellular ZERA-GLA strain, at 150 mg/l. These production levels were compared over time, revealing an unexpected pattern of accumulation with two separate production peaks. ZERA-GLA came to an initial production peak at around 100 mg/L at 24 hours before decreasing to ca. 50 mg/L at 35–40 hours and finally recovering to 150 mg/L after 48 hours. Analysis of the secretion strain showed that despite the instability of GLA in culture medium, much higher quantities of the CBHI carrier fusion partner were accumulated. Levels of proteolytically cleaved CBHI carrier around 1.3 g/l were observed, precisely tenfold lower than the 13 g/l of endogenous CBHI that was produced in the parental strain (data not shown).

Transcriptional profiles for parental, secretion and intracellular production strains were determined by quantitative comparison of mRNA levels measured by microarray. The production strategies were compared at 22 hours during the mid-logarithmic growth phase and after 27 hours at the late-logarithmic growth phase of cultivation. To identify differentially expressed genes, we performed significance tests comparing two strains at a time. The significance test results were visualized as volcano plots showing the expression level difference between the two strains (log_2-transformed fold change)

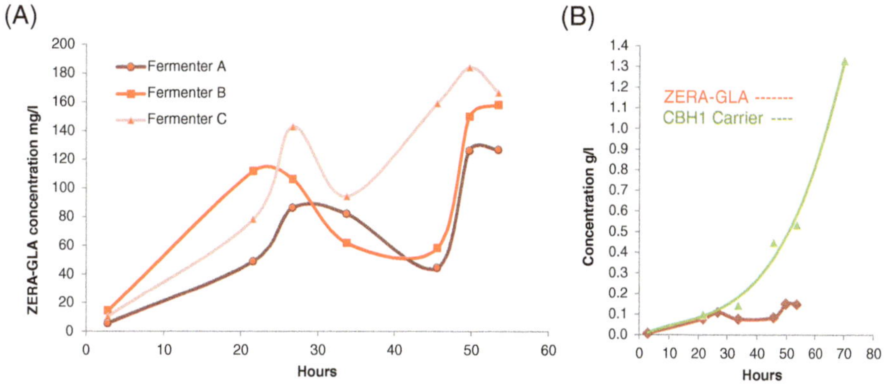

Figure 6 Total yield of target protein production during the bioreactor cultivations. (A) Triplicate cultivations of ZERA-GLA intracellular production showing total ZERA-GLA expressed over time by quantification from anti-ZERA Western blot analysis. Two peaks of ZERA-GLA production are apparent, separated by a decline in the ZERA-GLA level at approximately 30 to 45 hours of the culture. **(B)** A comparison of CBHI carrier produced as a result of CBHI-GLA production by secretion vs. intracellular ZERA-GLA production over time. By assessing the more stable CBHI carrier by quantitative anti CBHI Western instead of the unstable secreted GLA product it is clear to note that despite the eventual loss of secreted GLA through proteolysis more total GLA was actually produced than with the intracellular ZERA-GLA (quantified by anti ZERA Western).

versus the significance of the change ($-\log_{10}$ of p-value). These volcano plots (Figure 7A) showed that in the secretion strategy the mRNA transcript levels of many genes had changed compared to the parental strain, in most cases by an increase in gene transcription. The gene expression during the intracellular production strategy is relatively similar to that of the parental strain, the only significant shift in data spread being at 22 hours when a small number of up-regulated genes were over twofold induced. In general the transcriptional changes observed followed a similar pattern when progressing to the 27 hour time point, albeit at a reduced level of fold change. Further analysis of the secretion vs. intracellular strain at these relevant time points, and comparison of the genes with significant changes, demonstrated that similar changes had occurred for both forms of production, although the relative level of fold change in regulation was lower for the intracellular production and often did not exceed the twofold change threshold. The listing of genes up-regulated during GLA production by category shows that genes involved primarily in metabolism, transport, and translation were enriched (Figure 7B).

Genes related to the secretion pathway are obviously of special interest. A number of genes involved in UPR and protein folding or degradation in the ER (ERAD) are induced (Additional file 2). These induced genes include *hac1*, encoding a transcription factor of UPR, the protein disulphide isomerase *pdi1*, PDI-related *prp1*, the chaperone calnexin *cnx1*, and the ERAD pathway genes *der1* and *bap31*. This observation suggests that both the secretion strain and the intracellular strain suffer from secretion stress. As the extent of induction of these genes is higher in the secretion strain, it appears that this strain suffers more strongly from secretion stress than the intracellular production strain. This can be shown specifically by analysing the differences in gene induction; for example *hac1* and *der1* are above the twofold change threshold in the secretion strain at 2.23 and 2.02, but below the threshold in the intracellular strain at 1.45- and 1.73-fold.

Discussion

Human GLA was produced in *T.reesei* to >0.5 g/l levels with the aid of the ZERA fusion partner. However, several noteworthy limitations of this production strategy were revealed. Attempts to produce this enzyme by the classical CBHI carrier secretion method were initially unsuccessful. It was possible to obtain low mg/l yields by addition of a strong cocktail of protease inhibitors to the culture media and harvesting at an early time-point before significant proteolytic degradation occurred (data not shown). The fact that far higher quantities of CBHI carrier lacking the GLA fusion partner were observed from these cultures, coupled with the knowledge that CBHI fusion carriers have previously been successfully used for more proteolytically stable proteins such as Fab fragments [4], demonstrates the influence of the extracellular protease profile of *T.reesei* in the degradation of sensitive heterologous proteins. Traditionally it has been challenging to produce large quantities of mammalian proteins in filamentous fungi [8], and this was the driving force to investigate the possibilities of intracellular protein production using the ZERA PB storage system. It was therefore not surprising to note the relative success of intracellularly accumulated GLA, providing further evidence that it is indeed possible to express functional mammalian therapeutic proteins in *T. reesei* when degradation by secreted proteases can be avoided.

(A)

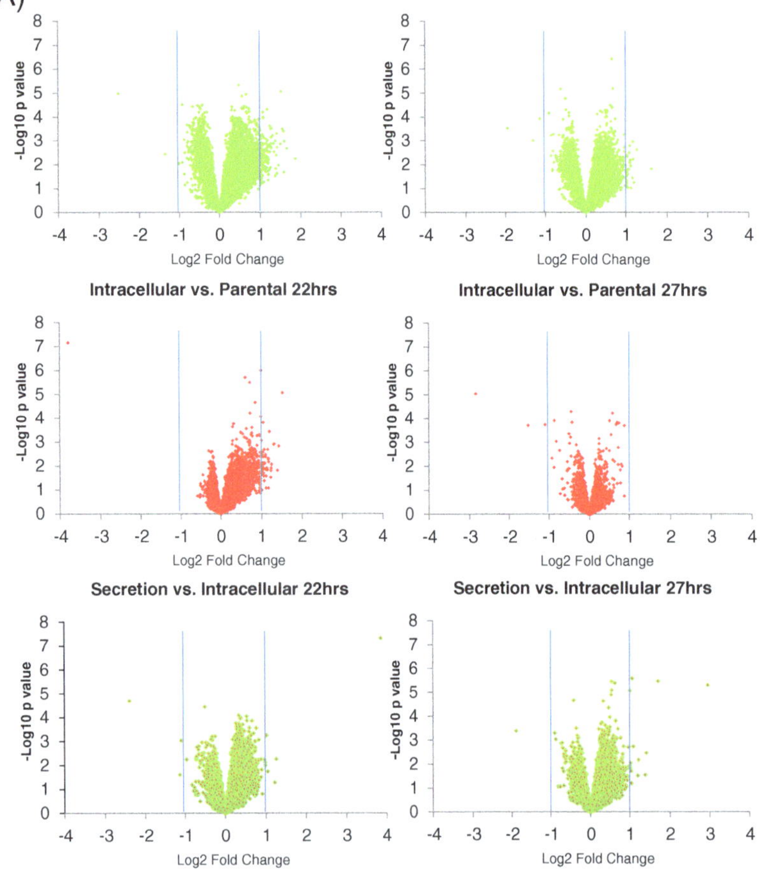

(B)

Gene Class	Total hits in all strains	Secretion vs. Parental	Intracellular vs. Parental	Mutual S vs. P & I vs. P
Metabolism	28	23	7	3
Mitochondrial	8	8	3	3
Glycoside Hydrolase	6	5	1	0
Carbohydrate Metabolism	2	2	1	1
Amino Acid Metabolism	3	2	1	0
Motility/Cell wall	5	4	2	1
Transport	10	8	1	0
Signalling	2	2	0	0
Regulatory	7	7	0	0
Nucleosome/DNA	4	4	2	2
Transcription/RNA	6	6	0	0
Translation	26	25	11	10
Secretory Pathway	7	6	3	2
Folding/UPR	6	5	3	2
Degradation/ERAD	4	3	3	2
Vacuolar	2	2	1	1
Secreted	6	4	4	3
Unknown	20	19	4	3
Total	152	135	47	33

Figure 7 Transcriptional profiling data comparing the production strategies. (A) Volcano plots of the combined triplicate RNA microarray data at the points of highest interest, 22 and 27 hours showing the spread of gene expression events in terms of up and down regulation between each production strategy comparison. **(B)** A table showing the transcriptional changes exceeding the 2-fold change threshold for each comparison and also listed by gene category throughout the cultivations. The total number of hits shows the combined number of transcriptional changes for all of the production strategies and the final column shows the mutual transcriptional changes observed between secretion and intracellular production. In the case of intracellular production most of these transcriptional changes observed are also found in the mutual category showing that many similar genes are highlighted between both types of production.

It is important to note at this point that despite the relative success of obtaining full length GLA from intracellular accumulation, there is still an intrinsic difference between secretion ability and cell storage capacity in *T. reesei*. Quantification of CBHI carrier levels observed from the bioreactor cultures revealed a near tenfold higher production level of CBHI carrier cleaved from secreted GLA than intracellular ZERA-GLA; 1300 mg/l and 150 mg/l, respectively. This shows that despite the low yield of actual GLA observed due to proteolysis, more GLA was in fact synthesized in the strain producing secreted GLA as compared with the strain producing intracellular ZERA-GLA. Furthermore, the levels of endogenous CBHI from the parental strain were a further tenfold higher at around 13 g/l (data not shown). Clearly there are not only issues with stability of GLA during secretion or after secretion but also common cellular mechanisms capable of limiting the basal expression or retention of this human enzyme.

When considering the quality of recombinant GLA enzyme, the primary concern is its ability to be purified while retaining relevant levels of specific activity. Despite the low yields of full size GLA obtained by carrier secretion a small quantity was purified, with specific activity close to that of the commercial form obtained from CHO. This was not an unexpected result considering that *T. reesei* is a known producer of high quality recombinant enzymes; however it was significantly more challenging to reproduce similar levels of specific activity from ZERA-GLA produced intracellularly. Establishing a purification method for intracellular ZERA-GLA required greater focus, having not been previously tested in any filamentous fungi. Extraction of intact PBs from the mycelial biomass required rather harsh yet well-defined disruptive forces so as not to also disrupt the PBs. Once the PBs were obtained, the methodology to enrich and wash to some extent by gravity-based separation was rather simple and robust. In order to test the enzyme in a more comparable way, solubilisation of the PBs would be required. As the results observed above demonstrate, it was possible to solubilise ZERA-GLA under several conditions but loss of specific activity occurred to some extent in every case. The most important criteria for solubilisation of the ZERA-GLA fusion were high pH with the addition of a reducing agent and ionic surfactants. Unfortunately, these factors resulted in significant loss of enzymatic activity. It was possible by adjustment of the molarity of the solubilisation buffer components, combined with optimisation of pH, to retain some of the activity, although at the cost of lower solubilisation yield. A specific activity to yield trade-off was established at around pH 8.2 although at this pH only 10-20% of the PB material was recovered, compared with almost 100% using highly basic pH. Cleavage of GLA from the ZERA fusion by TEV protease could be achieved, but enzymatic activity did not recover. These findings are probably due to the natural environment of GLA contained within the lysosome. With its low pH optimum of 4.5, the enzyme is likely to be innately less stable and less suitable for purification in basic conditions. In the comparison of these two production strategies it is relevant to note that both methods have certain advantages and also challenges. Moreover, selection of more compatible model proteins without such stringent requirements could lead to more encouraging results.

Despite the various challenges of both secreted and intracellular modes of production, it remains pertinent to investigate further the physiological and gene level regulation changes of both, in order to shed light on the cellular processes and production bottlenecks involved. As a general summary it can be concluded that production of GLA as a secreted product fused with the CBHI carrier caused a more pronounced stress effect with many more up-regulated genes, than the intracellular ZERA strategy. This correlated with the results based on cell growth, which not only peaked earlier during bioreactor cultivation but also reached a lower overall total biomass than the parental culture. By contrast, intracellular production strains grew in an almost identical manner to the parental control. Many of the genes up-regulated were identified in both strains, but the extent of the up-regulation was higher in the secretion strain. This could mean that in the secretion strain the foreign protein is detected more strongly as it travels through the whole secretion pathway. By contrast the ZERA-GLA protein is probably only present in high concentrations in the ER before being packed into protein bodies, and is therefore not detected with equal strength. It was interesting to observe that the production levels of ZERA-GLA during fermentation occurred in two distinct peaks (Figure 6B). An early peak occurred at the end of the logarithmic growth phase, dropping to around half during stationary growth, before recovering to a final and eventually higher peak at the end of the stationary phase. The second and highest peak in production appeared to occur when maximum biomass had been reached and presumably energy resources for growth were no longer in competition with target protein production. This phenomenon of a mid-point drop in total production yield suggests the existence of mechanisms capable of clearing the heterologous protein out from the ER. This conclusion is also supported by the fact that many of the significant transcriptional changes observed point towards secretion stress. It is also possible that during secretion stress further feedback controlling down-regulation of target protein biosynthesis is initiated, causing additional yield limitations [29].

Conclusions

Enzymatically active human GLA was successfully produced in *T. reesei* at detectable levels by CBHI carrier secretion and more significantly to >0.5 gram per litre levels by intracellular accumulation with the ZERA peptide fusion strategy. It was possible to partially purify GLA from both production modes, although loss of specific activity was a challenging aspect concerning purification of the ZERA fusion product. Transcriptional profiling showed that secretion stress is induced in *T. reesei* in both of the production modes, albeit more strongly by the secretion strategy.

Additional files

Additional file 1: Data quality of the microarrays. (A) Density histogram of rma normalized data. The variance of log_2 intensities of the normalised arrays intensity fall within acceptable limits for comparison. **(B)** Boxplot of rma normalized data. The appropriate similarity of replicate samples was verified using principal component analysis in this two dimensional display.

Additional file 2: List of significant genes in all of the comparisons. Listing of all the significantly changing genes found above the fold change threshold in all transcriptional profiling comparisons according to the *T. reesei* genome version 2.0 database.

Competing interests
The authors declare that they have no competing interests.

Authors' contributions
WS carried out the laboratory work, including method development and wrote the manuscript. JJ supervised the study, providing advice and criticism. MO performed statistical analysis of the transcriptional profiling data. MS designed the research plan, coordinated the study and supervised drafting of the manuscript. All authors read and approved the final manuscript.

Acknowledgements
Financial support for this work was provided by the Finnish Agency for Technology and Innovations (TEKES; decision number 40333/08) in the ERA-IB consortium project 'IMAPPROT' and by the Academy of Finland through its Centre of Excellence CoE Hyber 'Molecular Engineering of Biosynthetic Hybrid Materials research' (2014–2019). We thank Jussi Joensuu for fruitful discussions and proof-reading of the manuscript, and Michael Bailey for the language revision.

References

1. Cherry JR, Fidantsef AL: **Directed evolution of industrial enzymes: an update.** - *Curr Opin Biotechnol.2003 Aug;14(4)* (0958–1669 (Print); 0958–1669 (Linking)).
2. Stals I, Sandra K, Geysens S, Contreras R, Van Beeumen J, Claeyssens M: **Factors influencing glycosylation of *Trichoderma reesei* cellulases. I: Postsecretorial changes of the O- and N-glycosylation pattern of Cel7A.** *Glycobiology* 2004, **14**(8):713–724.
3. Harkki A, Uusitalo J, Bailey M, Penttila M, Knowles JKC: **A novel fungal expression system: secretion of active calf chymosin from the filamentous fungus *Trichoderma reesei*.** *Bio/Technol* 1989, **7**(6):596–603. and antibody Fab fragments.
4. Nyyssönen E, Penttilä M, Harkki A, Saloheimo A, Knowles JK, Keränen S: **Efficient production of antibody fragments by the filamentous fungus *Trichoderma reesei*.** *Nat Biotechnol* 1993, **11**(5):591–595.
5. Druzhinina IS, Seidl-Seiboth V, Herrera-Estrella A, Horwitz BA, Kenerley CM, Monte E, Mukherjee PK, Zeilinger S, Grigoriev IV, Kubicek CP: ***Trichoderma*: the genomics of opportunistic success.** *Nat Rev Micro* 2011, **9**(10):749–759.
6. Peterson R, Nevalainen H: **Trichoderma reesei RUT-C30–thirty years of strain improvement.** *Microbiology* 2012, **158**(Pt 1):58–68.
7. Kredics L, Antal Z, Szekeres A, Hatvani L, Manczinger L, Vágvölgyi C, Nagy E: **Extracellular proteases of *Trichoderma* species.** *Acta Microbiol Immunol Hung* 2005, **52**(2):169–184.
8. Corchero JL, Gasser B, Resina D, Smith W, Parrilli E, Vázquez F, Abasolo I, Giuliani M, Jäntti J, Ferrer P: **Unconventional microbial systems for the cost-efficient production of high-quality protein therapeutics.** *Biotechnol Adv* 2013, **31**(2):140–153.
9. Nyyssönen E, Keränen S: ***Trichoderma reesei*, a promising novel host for antibody production.** *Proc ACS Symp Ser ACS Publ* 1995, **604**(9):108–122.
10. Mustalahti E, Saloheimo M, Joensuu JJ: **Intracellular protein production in *Trichoderma reesei* (*Hypocrea jecorina*) with hydrophobin fusion technology.** *N Biotechnol* 2013, **30**(2):262–268.
11. Joensuu JJ, Conley AJ, Lienemann M, Brandle JE, Linder MB, Menassa R: **Hydrophobin fusions for high-level transient protein expression and purification in *Nicotiana benthamiana*.** *Plant Physiol* 2010, **152**(2):622–633.
12. Torrent M, Llompart B, Lasserre-Ramassamy S, Llop-Tous I, Bastida M, Marzabal P, Westerholm-Parvinen A, Saloheimo M, Heifetz PB, Ludevid MD: **Eukaryotic protein production in designed storage organelles.** *BMC Biol* 2009, **28**(7):5. *doi*:(1741–7007 (Electronic); 1741–7007 (Linking)).
13. Keating GM: **Agalsidase Alfa.** *BioDrugs* 2012, **26**(5):335–354.
14. Ioannou YA, Bishop DF, Desnick RJ: **Overexpression of human alpha-galactosidase A results in its intracellular aggregation, crystallization in lysosomes, and selective secretion.** *J Cell Biol* 1992, **119**(5):1137–1150.
15. Ioannou Y, Zeidner K, Grace M, Desnick R: **Human α-galactosidase A: glycosylation site 3 is essential for enzyme solubility.** *Biochem J* 1998, **332**:789–797.
16. Christianson TW, Sikorski RS, Dante M, Shero JH, Hieter P: **Multifunctional yeast high-copy-number shuttle vectors.** *Gene* 1992, **110**(1):119–122.
17. Sherman F: **Getting started with yeast.** *Meth Enzymol* 1991, **194**(3):21.
18. Steiger MG, Vitikainen M, Uskonen P, Brunner K, Adam G, Pakula T, Penttila M, Saloheimo M, Mach RL, Mach-Aigner AR: **Transformation system for *Hypocrea jecorina* (*Trichoderma reesei*) that favors homologous integration and employs reusable bidirectionally selectable markers.** *Appl Environ Microbiol* 2011, **77**(1):114–121. *doi*(1098–5336 (Electronic); 0099–2240 (Linking)).
19. Penttilä M, Nevalainen H, Rättö M, Salminen E, Knowles J: **A versatile transformation system for the cellulolytic filamentous fungus *Trichoderma reesei*.** *Gene* 1987, **61**(2):155–164.
20. Bailey M, Tähtiharju J: **Efficient cellulase production by *Trichoderma reesei* in continuous cultivation on lactose medium with a computer-controlled feeding strategy.** *Appl Microbiol Biotechnol* 2003, **62**(2–3):156–162.
21. Bailey MJ, Adamitsch B, Rautio J, Von Weymarn N, Saloheimo M: **Use of a growth-associated control algorithm for efficient production of a heterologous laccase in *Trichoderma reesei* in fed-batch and continuous cultivation.** *Enzyme Microb Technol* 2007, **41**(4):484–491.
22. Aho S, Olkkonen V, Jalava T, Paloheimo M, Bühler R, Niku-Paavola M, Bamford DH, Korhola M: **Monoclonal antibodies against core and cellulose-binding domains of *Trichoderma reesei* cellobiohydrolases I and II and endoglucanase I.** *Eur J Biochem* 1991, **200**(3):643–649.
23. Martinez D, Berka RM, Henrissat B, Saloheimo M, Arvas M, Baker SE, Chapman J, Chertkov O, Coutinho PM, Cullen D: **Genome sequencing and analysis of the biomass-degrading fungus *Trichoderma reesei* (syn. *Hypocrea jecorina*).** *Nat Biotechnol* 2008, **26**(5):553–560.
24. Grigoriev IV, Nordberg H, Shabalov I, Aerts A, Cantor M, Goodstein D, Kuo A, Minovitsky S, Nikitin R, Ohm RA, Otillar R, Poliakov A, Ratnere I, Riley R, Smirnova T, Rokhsar D, Dubchak I: **The genome portal of the Department of energy joint genome institute.** *Nucleic Acids Res* 2012, **40**(Database issue): D26–D32.
25. Irizarry RA, Bolstad BM, Collin F, Cope LM, Hobbs B, Speed TP: **Summaries of Affymetrix GeneChip probe level data.** *Nucleic Acids Res* 2003, **31**(4):e15.
26. Kauffmann A, Gentleman R, Huber W: **ArrayQualityMetrics–a bioconductor package for quality assessment of microarray data.** *Bioinformatics* 2009, **25**(3):415–416.
27. Smyth GK: **Limma: linear models for microarray data.** In *Bioinformatics and computational biology solutions using R and Bioconductor. Springer* 2005, 397–420
28. Smyth GK: **Linear models and empirical bayes methods for assessing differential expression in microarray experiments.** *Stat Appl Genet Mol Biol* 2004, **3**(1):3.
29. Pakula TM, Laxell M, Huuskonen A, Uusitalo J, Saloheimo M, Penttila M: **The effects of drugs inhibiting protein secretion in the filamentous fungus *Trichoderma reesei*. Evidence for down-regulation of genes that encode secreted proteins in the stressed cells.** *J Biol Chem* 2003, **278**(45):45011–45020.

Development of an enzyme linked immunosorbent assay for detection of cyathane diterpenoids

Tian Shen[1], Lena M Hof[1], Heike Hausmann[2], Marc Stadler[3] and Holger Zorn[1*]

Abstract

Background: So-called cyathane type diterpenoids are produced as secondary metabolites by basidiomycetes. Based on their antibacterial, fungicidal, and cytotoxic properties, cyathane type terpenoids represent interesting target compounds in fungal biotechnology.

Results: An indirect competitive enzyme linked immunosorbent assay has been developed for detection of cyathane type diterpenoids. Rabbit polyclonal antibodies were raised against a mixture of striatal A and B conjugated to bovine serum albumin. The conditions for direct attachment of the hapten striatal B to a solid phase by passive adsorption were optimized. The cross reactivities of the striatals A, C and D, of the striatins A and B, and of the erinacines C and P to striatal B were determined. The validation study showed that the ELISA was precise and sensitive. The average IC_{50} of striatal B was 36.0 ng mL^{-1} with an inter-assay coefficient of variation (CV) of 13.2% (n = 5). Recoveries from striatal B spiked samples in the assay were in the range of 97.3 – 125.9%. A good correlation between the striatal B concentration measured by the ELISA and by HPLC-DAD ($y = 1.1122x - 0.1585$, $R^2 = 0.9942$) was obtained from linear regression analysis. The suitability of the ELISA for detection of cyathane type diterpenoids in submerged cultures and fruiting bodies of *H. erinaceus* was studied. It showed cross reactivity with supernatants from submerged cultures and extracts thereof, but did not show cross reactivity with extracts from fruiting bodies.

Conclusions: The developed method is appropriate for qualitative and quantitative detection of cyathane diterpenoids in complex mixtures. Due to its high sensitivity and specificity, it represents an ideal screening method for discovering new cyathane diterpenoids and new potential producers of them.

Keywords: Cyathane diterpenoids, ELISA, *Hericium erinaceus*, Striatal, Erinacine

Background

Cyathine and allocyathine were the first reported cyathane diterpenoids discovered by Allbutt and Ayer from a static liquid culture of the fungus *Cyathus helenae* (H. J. Brodie) in the early 1970s [1,2]. They were proven to be active against actinomycetes, Gram-positive and Gram-negative bacteria, and some fungi, including dermatophytes. Afterwards, various structurally related compounds, so-called cyathane type diterpenoids, were isolated from different basidiomycetous cultures, e.g. striatals and striatins from *Cyathus spp.*, sarcodonins from *Sarcodon spp.*, and erinacines from *Hericium spp.* (Figure 1) [3-7]. All of these compounds share a cyathane skeleton consisting of five-, six-, and seven membered rings and possess antibacterial, fungicidal, and cytotoxic properties. Fungal extracts prepared from *C. striatus* showed significant inhibitory effects on the NF-κB activation pathway and might be applied for cancer therapeutics [8]. Erinacines promote nerve growth factor (NGF) synthesis, which suggests the application of *H. erinaceus* or its secondary metabolites for the treatment and prevention of dementia and further neurodegenerative diseases [7,9-12].

Because of their interesting biological activities and their high potential in medicinal and pharmaceutical applications, cyathane diterpenoids have attracted increasing interest in recent years. At present, the most common method for analysis of cyathane diterpenoids is high performance liquid chromatography (HPLC) coupled to diode array and mass spectrometric detection [13].

As a rapid, sensitive and cost effective method, an ELISA specific for cyathane type secondary fungal metabolites may serve as a complementary method, especially in

* Correspondence: holger.zorn@uni-giessen.de
[1]Justus Liebig University Giessen, Institute of Food Chemistry and Food Biotechnology, Heinrich-Buff-Ring 58, Giessen 35392, Germany
Full list of author information is available at the end of the article

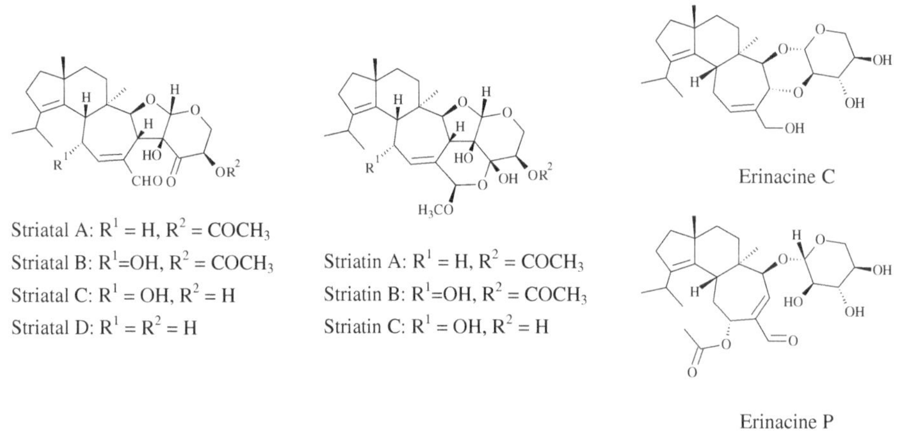

Figure 1 Structures of striatals, striatins and erinacines.

screenings for new producer strains. In the present study, an indirect competitive ELISA for the structure specific detection of cyathane type diterpenoids using polyclonal antibodies was developed, and its potential for the analysis of biological samples was proven.

Results

Optimum coating conditions

Striatal B was chosen as coating hapten in ELISA, because it is more polar than striatal A. In order to increase its solubility in aqueous solution, an equal volume of DMSO was added to the buffers. The optimum coating conditions (best signal-to-noise ratio and concentration dependence) were obtained by coating the plate in PBS/DMSO (1/1, v/v) buffer. More intense signals were obtained by coating at 24°C or 37°C, compared to coating at 4°C. Striatal B was adsorbed more efficiently by coating overnight than for 2 h. Based on these results, the optimum coating conditions for striatal B are summarized as follows: PBS/DMSO (1/1, v/v) buffer, and coating overnight at 24°C. The optimum concentration for coating of the hapten and the dilution of polyclonal antibodies (pAbs) were determined by checkerboard titration to be 5 μg mL^{-1} and 1:200, respectively, which were used in the following indirect competitive ELISA experiments.

Tolerance against organic solvents

Due to the poor solubility of cyathane diterpenoids in aqueous solutions, the tolerance against organic solvents used to dissolve these compounds was tested for assay optimization. An addition of 5% DMSO, acetonitrile or methanol, respectively, to the pre-incubation solution was found acceptable. Methanol could be used up to 25% without a notable negative effect. High percentages of DMSO up to 30% and of acetonitrile up to 20% resulted in considerable negative effects on the assay. An addition of DMSO >30% or acetonitrile >20% showed a sudden increase of signals. Since methanol reacts with striatals [4], an addition of 5% DMSO or acetonitrile was selected for pre-incubation in the following indirect competitive ELISA.

Cross reactivity

Figure 2 illustrates a typical standard calibration curve of striatal B. The specificity of the pAbs against striatal B was evaluated by determination of the cross reactivities of several structurally related compounds, such as striatals A, C, and D, striatins A and B, and erinacines C and P. The IC_{50} values are given in Table 1. All of the tested cyathane diterpenoids showed cross reactivities against striatal B. No color reaction was detected with non-coated plates as blanks.

Assay precision

Intra-assay and inter-assay precisions were determined for several concentrations of striatal B (Table 2). The intra-assay precision was calculated from the extinction difference at 450 nm and 630 nm within one microtiter plate. Less than 10% CV were obtained in every experiment, indicating an acceptable precision. The inter-assay precision was obtained from $\%B/B_0$-values of 8 plates on different days. Variations of less than 20% were observed. The average IC_{50} of striatal B standard was 36.0 ng mL^{-1} with an inter-assay CV of 13.2% (n = 5).

Recovery

To assess the assay accuracy, striatal B was added to the sample matrix in several concentrations, and the recoveries of the spiked samples were determined. In parallel, HPLC-DAD was used to analyze the spiked samples. The apparent recoveries of striatal B in the sample matrix as determined by indirect competitive ELISA ranged from 97.3% to 125.9%, while those obtained by HPLC-DAD varied from 95.8% - 110.2%. The limit of quantification was 0.02 μg mL^{-1} for ELISA, and 0.40 μg

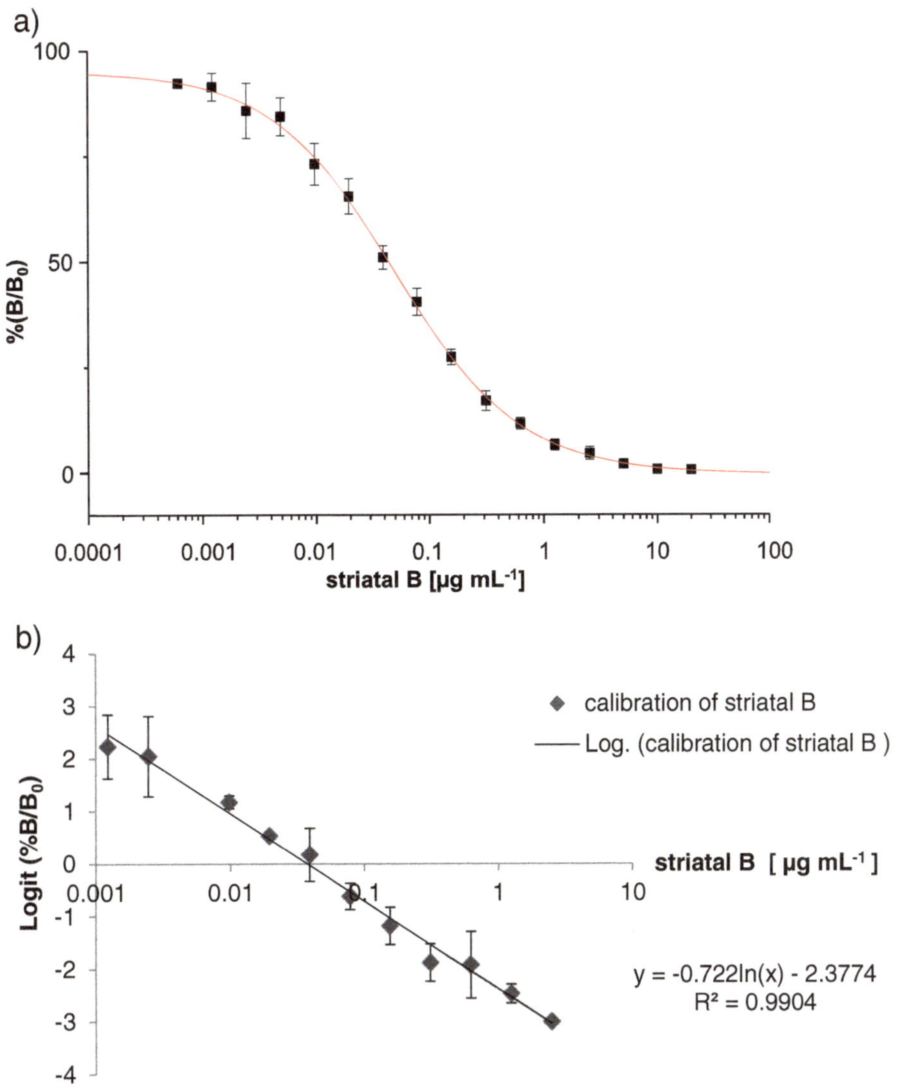

Figure 2 Calibration graphs for the quantification of striatal B. (a) Example of sigmoidal calibration curve for quantification of striatal B by indirect competitive ELISA. **(b)** Example of linearity of striatal B calibration curve. The linear working range was found to be 0.001 - 2.5 µg mL^{-1}, $y = -0.722 \ln(x) - 2.3774$, $R^2 = 0.9904$.

Table 1 Cross reactivities of structurally related compounds to striatal B

Compounds	Cross reactivity [%]*
Striatal B	100.0
Striatal A	39.8
Striatal C	77.8
Striatal D	74.4
Striatin A	24.1
Striatin B	109.3
Erinacine C	60.0
Erinacine P	21.1

*Related to molecular ratio.

Table 2 Intra- and inter-assay precision

Concentration [µg mL^{-1}]	Intra-assay (n = 3) CV [%]	Inter-assay (n = 8) CV [%]
0.63	2.7	16.3
0.32	3.9	14.8
0.16	1.7	16.3
0.08	2.4	8.3
0.04	4.5	10.4
0.0006	5.1	3.6

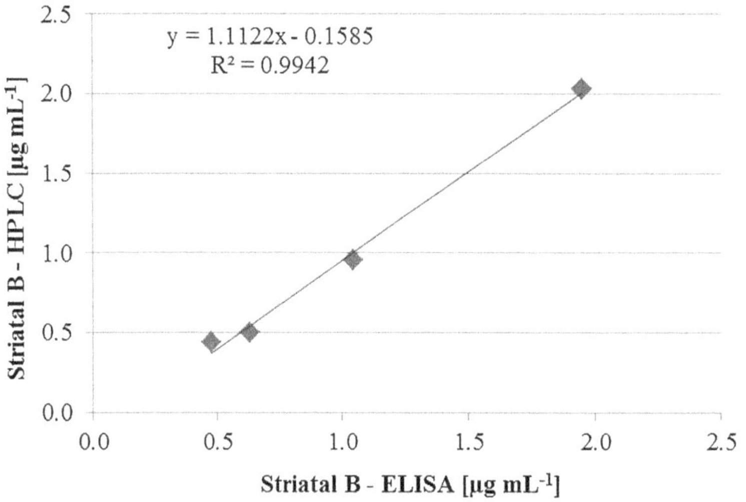

Figure 3 Correlation between striatal B concentrations in spiked samples measured by ELISA (x-axis) and by HPLC-DAD (y-axis).

mL^{-1} for HPLC-DAD. A good correlation between the striatal B concentrations measured by ELISA and by HPLC-DAD ($y = 1.1122x - 0.1585$, $R^2 = 0.9942$) was observed (Figure 3, Table 3).

Analysis of *H. erinaceus*

The indirect competitive ELISA was applied to samples prepared from submerged cultures as well as from fruiting bodies of *H. erinaceus*. The total concentrations of cyathane diterpenoids in supernatants and in extracts thereof are displayed in Figure 4. A similar change of % B/B_0-values was observed in both cases during the 8 culture days. %B/B_0-values in extracts of fruiting bodies from *H. erinaceus* were near to 100% and were independent from the sample concentration in the assay.

Discussion

In the present study, striatal B was directly attached as a low molecular weight hapten to microtiter plates. In many studies the use of protein conjugated haptens for immobilization was reported [14-16], as the ability of proteins to adsorb to the plastic surface of microtiter plates is much higher than that of low molecular weight haptens. However, the coupling ratio of haptens to carrier proteins may vary depending on the coupling reaction and a waste of valuable haptens during conjugate synthesis is often unavoidable. The involvement of carrier proteins results in several problems. Firstly, it may cause unspecific binding to pAbs and thus reduce specific signals. Secondly, due to the large molecular dimension of carrier proteins, a shielding or masking effect may occur that impedes the binding of pAbs to the conjugated haptens [17]. Some researchers reported about pre-treatments of the plate surface, e.g. by UV irradiation or reagents in order to realize a direct binding of small molecules [18-20]. In the present study, the target analyte (molecular weight <500 g mol^{-1}) was directly attached to microtiter plates without pre-

Table 3 Recovery of spiked striatal B in sample matrix

Spiked concentration [μg mL^{-1}]	ELISA (n = 3)		HPLC-DAD (n = 2)	
	Detected [μg mL^{-1}]	Recovery [%]	Detected [μg mL^{-1}]	Recovery [%]
10.010	out of calibration	-	10.50 ± 0.05	104.8%
5.005	out of calibration	-	5.26 ± 0.10	105.1%
2.002	1.949 ± 0.219	97.3%	2.03 ± 0.08	101.5%
1.001	1.048 ± 0.199	104.7%	0.96 ± 0.02	95.8%
0.501	0.630 ± 0.291	125.9%	0.50 ± 0.00	99.6%
0.400	0.478 ± 0.062	119.3%	0.44 ± 0.02	110.2%
0.200	0.252 ± 0.026	113.0%	n.d.	-
0.020	0.021 ± 0.001	102.8%	n.d.	-
0.001	n.d*	-	n.d.	-

*n.d: not detected.

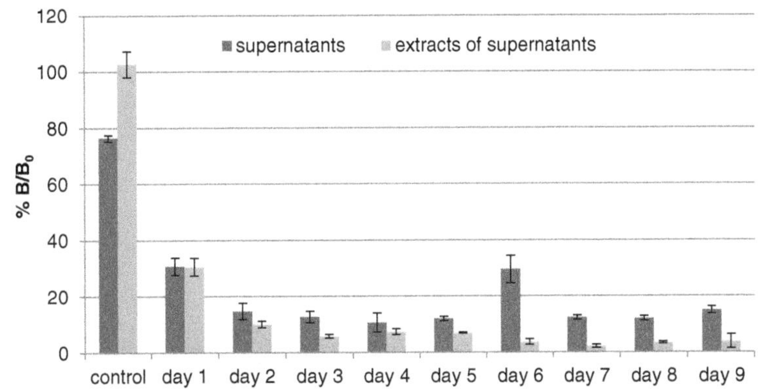

Figure 4 Analysis of supernatants of submerged cultures of *H. erinaceus* and extracts thereof during 8 culture days. Control: culture medium without *H. erinaceus*.

treatment of the plate surface. Significant differences between the various coating procedures were found during the study of coating conditions. The pH of the coating buffer significantly affected the coating efficiency. The final pH of the coating buffer PBS, potassium acetate and sodium carbonate after mixing with DMSO was 9.7, 5.8 and 11.1, respectively. An influence of the pH on the coating efficiency has been reported in several publications [21-23]. In addition, a decomposition of striatal B was detected in potassium acetate buffer in a time course study by HPLC analysis (data not shown). That might be a reason for the poor attachment of striatal B to the plate in this buffer system. For striatal B, coating at pH 9.7 was more efficient than at pH 11.1.

The presence of organic solvents may influence the performance of an ELISA. Methanol caused the least negative effect, and the use of low concentrations of acetonitrile or DMSO was also acceptable. Similar results have been reported e.g. by [17,23,24]. The tolerance against organic solvents allows for the analysis of samples from fungal cultures which require extraction with an organic solvent. The sudden increase of the signals in the presence of acetonitrile and DMSO in high concentrations indicated a false positive signal possibly caused by denaturation of the antibodies and thus an unspecific adsorption to the blocking proteins.

All of the examined cyathane diterpenoids showed cross reactivities against striatal B. As expected, striatin B with $IC_{50} = 109.3\%$ showed the highest cross reactivity. Striatin B and striatal B differ only in the aldehyde group at the C_7 ring, and this position of striatal B was used for BSA-conjugate synthesis for immunization. Therefore, striatin B and striatal B possess the same antigenic determinants. Erinacine P showed the lowest cross reactivity with $IC_{50} = 21.1\%$. Based on the cross reactivities and molecular structures it might be suspected that the resulting antiserum displays the highest affinity toward the cyathane ring moiety.

The recovery experiments suggested that the developed ELISA is accurate and significantly more sensitive than the HPLC-DAD method. It could be used for the estimation of the total cyathane diterpenoid concentrations of biological samples. The assay showed interferences with constituents of the culture media of *H. erinaceus*, as a matrix effect was observed with the control samples (cf. Figure 4, control). Therefore, the use of organic extracts is preferred for the analysis of cyathane type diterpenoids in fungal culture supernatants. The product concentrations could be estimated during the cultivation period. However, the total concentrations represent the sum of all compounds showing cross reactivity against striatal B. They may thus differ from the individual concentrations determined by HPLC.

No significant interferences ($\%B/B_0$-values close to 100%) were observed with crude extracts of fruiting bodies of *H. erinaceus*. Therefore, the secondary metabolites present in the fruiting bodies, e.g. hericenone and erinacerine, did not cross react with the polyclonal antibodies. These results are in good agreement with a review article about secondary metabolites from *H. erinaceus* [12]. Up to date, erinacines have been mostly reported from submerged cultures of *H. erinaceus*, and only traces of erinacines have been found in fruiting bodies [25]. In the present study, the trace concentrations of erinacines in the fruiting bodies were apparently below the detection limit of the ELISA.

The CV% values for the inter-assay precision might be caused by many factors, e.g. the position of the wells in the plate (edge effect), the day to day variation in reagent preparation, the temperature of buffers, the variability in washing procedures, pipetting errors and the loss of enzyme activity of the secondary antibodies after prolonged storage [16,26].

Conclusions

Polyclonal antibodies against striatal A/B were produced, and an indirect competitive ELISA was developed for determination of striatal B and structural related compounds. The method has shown satisfactory results concerning specificity, sensitivity and accuracy. The application to samples from submerged cultures and fruiting bodies of *H. erinaceus* indicated that the ELISA may be used for the estimation of total cyathane diterpenoid concentrations in complex mixtures. Due to its high sensitivity and specificity, it represents an ideal screening tool for discovering new potential producers of cyathane type diterpenoids. Further optimization will be needed to improve the inter-assay precision.

Methods

Chemicals and solutions

All reagents were of analytical grade unless specified otherwise. Ultra-pure water (produced by arium® 611 VF Water System, Sartorius, Göttingen, Germany) was used for all solutions, media, and HPLC eluents.

Acetonitrile (HPLC gradient grade), calcium carbonate, Edamin® K, 30% hydrogen peroxide (H_2O_2), and D-mannitol were purchased from Sigma Aldrich (Steinheim, Germany).

Salts, Agar-Agar, bovine serum albumin (BSA), EDTA, α-D(+)-glucose monohydrate, soy peptone, sucrose, 3,3',5,5'-tetramethyl benzidine (TMB), Tween® 20, yeast extract, and organic solvents were purchased from Applichem (Darmstadt, Germany) or from Roth (Karlsruhe, Germany).

Malt extract was obtained from Fluka, Neu-Ulm, Germany. Molasses was obtained from Südzucker, Offstein, Germany. Oatmeal was provided from Dr. Oetker, Düsseldorf, Germany.

Standards of erinacines and striatals, originally isolated from the two selected producer strains, were obtained from the library of pure natural products of InterMed Discovery, Dortmund, Germany (IMD, formerly Bayer Healthcare) and their purity and identity was confirmed by 2D-NMR and HR-MS, prior to their use in the experiments. Additionally, striatals A and B (for immunization) and striatals C and D, striatins A and B, and erinacines C and P (for determination of cross reactivities) were extracted and purified from submerged cultures of the basidiomycetes *Cyathus striatus* and *Hericium erinaceus*. The structures were confirmed by comparison to the pure reference compounds and by NMR analysis (cf. Additional file 1: Supplementary material).

Solutions for ELISA:

Coating buffers:

1. 10 mM phosphate buffered saline (PBS), pH 7.4 mixed with DMSO, 1/1, v/v.
2. 0.1 M potassium acetate buffer, pH 4.0 mixed with DMSO, 1/1, v/v.
3. 0.1 M sodium carbonate buffer, pH 9.6 mixed with DMSO, 1/1, v/v.

Washing buffer: 10 mM PBS, pH 7.4 with 0.05% Tween® 20 (PBST).

Blocking reagent (1% gelatin) was purchased from Roche Diagnostics (Mannheim, Germany). Secondary antibodies (goat anti-rabbit lgG peroxidase conjugate) were obtained from Merck (Darmstadt, Germany).

Color development: 0.12 mg mL^{-1} TMB and 0.05% H_2O_2 in 0.05 M sodium acetate solution, pH 4.5 (TMB solution). This solution was freshly prepared before use.

Stopping reagent: 0.5 M sulfuric acid.

Fungal strains

Cyathus striatus (STMA07048, isolated from basidiospores of a specimen from a trunk of Picea collected in July 1997 in Stelzenberg, Rheinland-Pfalz, Germany) and *Hericium erinaceus* (FU70034, isolated from basidiocarp tissue) for submerged culture were obtained from IMD. Prior to the experiments, the identity of both strains was confirmed by microscopic studies and by comparison of their ITS nrDNA sequences with reference data in Genbank. The fungi were maintained on a solid medium containing 20 g L^{-1} malt extract and 15 g L^{-1} Agar-Agar.They are deposited in the culture collection of IMD Natural Solutions GbR (formerly InterMed Discovery GmbH under liquid nitrogen. A duplicate strain and the corresponding specimen of the *C. striatus* material is also maintained at the personal herbarium and culture collection of Marc Stadler.

Fruiting bodies of *Hericium erinaceus* were obtained from a commercial provider www.pilzgarten.de, Helvesiek, Germany.

Culture media

Soy peptone, yeast malt, and sugar molasses media were prepared according to [25].

Instruments

HPLC-DAD

The HPLC system was from Merck Hitachi, Darmstadt, Germany, and comprised a pump (L-7100), an auto sampler (L-7200), an interface (D-7000), and a diode array detector (L-7455, 200 – 600 nm). Column: reversed phase, Nucleosil® 100–5 C_{18}, CC 125/3 mm with a guard column Nucleosil® 100–5 C_{18}, CC 8/3 mm (Macherey Nagel, Düren, Germany).

Flow rate: 0.4 mL min^{-1} for striatal and striatin analysis and 0.6 mL min^{-1} for erinacine analysis.

Eluent: Acetonitrile (A) and ultra-pure water (B).

Wavelength: 210, 233 and 254 nm.

Gradient for striatal and striatin: 50% A (0 min) – 94% A (14 min) – 94% A (17 min) – 100% A (19 min) – 100% A (25 min) – 50% A (30 min) – 50% A (40 min).

Gradient for erinacine: 30% A (0 min) – 50% A (15 min) – 50% A (16 min) – 100% A (23 min) – 100% A (38 min) – 30% A (43 min) – 30% A (47 min).

Preparative HPLC

A semi-preparative HPLC system (Young Lin, Hongye Anyang, Korea) equipped with quaternary pump (YL9110S) and dual wavelength UV/Vis detector (YL 9120S), combined with a fraction collector (CHF 122SC, Advantec, Osaka, Japan) was used. A column Kromasil 100 C_{18}, 7 μm, 250 × 20 mm (MZ Analysentechnik, Mainz) with a guard column Kromasil 100 C_{18}, 7 μm, 50 × 20 mm (MZ) was used to isolate the target compounds.

Flow rate: 15 mL min^{-1}.

Eluent: Acetonitrile (A) and ultra-pure water (B).

Gradient for striatals and striatins: 50% A (0 min) – 50% A (10 min) – 80% A (55 min) – 80% A (70 min) – 100% A (80 min) – 100% A (95 min). Wavelengths: 210 and 254 nm.

Gradient for erinacines: 25 A% (0 min) – 25% A (10 min) – 48% A (35 min) – 48% A (45 min) – 55% A (52 min) – 55% A (60 min) – 100% A (90 min) – 100% A (110 min). Wavelengths: 210 and 233 nm.

ELISA

The ELISA was performed in microtiter plates Immuno Plate Maxisorp F96 (Nunc, Denmark). Incubation steps and color development were performed in an incubator INE 500 (Memmert, Schwabach, Germany). Washing was done by an eight-channel pipette Research Pro 50–1200 μL (Eppendorf, Wesseling-Berzdorf, Germany). The extinction was measured by a microtiter plate reader Synergy 2 (BioTek, Bad Friedrichshall, Germany).

Biosynthesis and isolation of cyathane diterpenoids

Submerged cultivation

Pre-cultures of *C. striatus* were grown submerged in 250 mL Erlenmeyer flasks containing 100 mL soy peptone medium at 24°C and 150 rpm for 7 days. After that, the mycelium was homogenized with an Ultra-Turrax (IKA, Staufen, Germany) at 10,000 rpm for 10 sec. 40 mL homogenized mycelium were inoculated into a new 1,000 mL Erlenmeyer flask containing 400 mL soy peptone medium and incubated at 24°C and 150 rpm for further 12 days.

Pre-cultures of *H. erinaceus* were grown submerged in 250 mL Erlenmeyer flasks containing 100 mL yeast malt medium at 24°C and 150 rpm for 7 days. Afterwards, 40 mL homogenized mycelium were inoculated into 400 mL sugar molasses medium in 1,000 mL Erlenmeyer flasks and incubated at 24°C and 150 rpm for further 4 days to obtain erinacine P or 12 days to obtain erinacine C.

Extraction and isolation

After submerged cultivation, the mycelia from 2.2 L culture media were separated from the supernatants by centrifugation at 2,880 *g* and 4°C for 10 min (Allegra® X-15R, BECKMAN COULTER™, Krefeld, Germany), and extracted twice with 400 mL ethyl acetate. The combined organic extracts were dried over sodium sulfate, and afterwards evaporated to dryness. A brown residue was obtained after evaporation, which was dissolved in acetonitrile and subjected to preparative HPLC. The extracts of *Cyathus striatus* contained the striatals A, B, C and D. The striatins A and B were prepared by conversion of the striatals A and B by stirring in methanol at room temperature overnight, respectively. The erinacines P and C were isolated from the extracts of *Hericium erinaceus*.

Production of polyclonal antibodies

The aldehyde function of the striatals and NH_2 functions of BSA were used for synthesis of striatal A/B (1/1, w/w)-BSA conjugates. The ratio of striatals to BSA was determined by MALDI-TOF-MS. About 17 striatal molecules were coupled to one BSA molecule (Eurogentec S. A, Seraing, Belgien). The production of rabbit pAbs against striatal A/B was performed by Eurogentec S.A. Two rabbits with ID SA6928 and SA6929 were immunized four times with 200 μg of the striatal A/B-BSA conjugate on day 0, 14, 28 and 56. An additional immunization was done on day 91with 400 μg striatal A/B-BSA conjugate. The final antisera were collected on day 115 and stored at −20°C without further purification until use. The pre-bleed sera of these two rabbits on day 0 before immunization were used as negative control in ELISA. The final antiserum from rabbit SA6928 was used in the following assays because of its higher specificity against striatal B compared to SA6929 (data not shown).

Development of an indirect competitive ELISA

Optimization of coating conditions

Striatal B was directly attached to the plate in several different concentrations. The optimal conditions (buffer, temperature and time) for direct coating of striatal B were studied by an indirect non-competitive ELISA. The incubation temperatures were 4°C, 24°C and 37°C, and the coating time was 2 h or overnight.

Checkerboard titration

The optimal dilution of pAbs and the optimal coating amount of striatal B were determined simultaneously by checkerboard titration (a two-dimensional titration method) by indirect non-competitive ELISA. Striatal B with different concentrations (0.04 to 20 μg mL^{-1}, two fold dilution) was coated to the microtiter plate and then bound to pAbs with serial dilutions (1:50 to 1:6,400, stepwise in two fold dilution).

Tolerance against organic solvents

The tolerance against organic solvents (DMSO, acetonitrile, methanol) used to dissolve the cyathane diterpenoids was examined by indirect competitive ELISA. Different proportions of organic solvent (5 - 50%, v/v, 0% as blank) were added to PBST with a final dilution of pAbs 1:200 and pre-incubated without competing molecules at 24°C for 1 h. The effects of organic solvents on the ELISA system were evaluated by comparing the measured extinctions to those of the blanks.

ELISA

Coating

100 μL of 5 μg mL^{-1} striatal B in coating buffer 1 were pipetted in wells of a microtiter plate. 100 μL of 1 μg mL^{-1} BSA diluted in PBS were pipetted in two wells as positive control. The incubation was performed overnight at 24°C.

Blocking

200 μL of 1% gelatin were added to the wells and incubated at 24°C for 3 h.

Pre-incubation

Serial dilutions of analyte were prepared in a concentration range of 0.0003 - 20 μg mL^{-1} in PBST containing 5% acetonitrile or DMSO and a final pAbs dilution of 1:200. The incubation was performed for 1 h at 24°C in 2 mL Eppendorf® cups.

Incubation with pAbs

100 μL pre-incubated analyte were added to each well (in triplicate) and incubated for 1 h at 24°C. For the indirect non-competitive ELISA, 100 μL pAbs with a 1:100 dilution in PBST were used without competitors instead of the pre-incubated analyte.

Incubation with secondary antibodies

100 μL secondary antibodies with a 1:5,000 dilution in PBST were added to each well and incubated for 1 h at 24°C.

Color development

100 μL of the TMB solution were added to each well and incubated for 15 min at 24°C. Without washing step, 100 μL of 0.5 M sulfuric acid were added to each well to stop the color development.

Measurement

The extinctions were measured immediately by a microtiter plate reader at 450 nm and 630 nm.

Unless otherwise specified, the wells were washed with 300 μL washing buffer PBST three times after each incubation step. The indirect non-competitive ELISA involved all of the steps of the indirect competitive ELISA, except for the pre-incubation of pAbs with competitors.

Data analysis

Striatal B standards and samples were analyzed in triplicates. The differences between extinctions at 450 nm and 630 nm were used for calculation and plotting the sigmoidal curves. %B/B_0-values for calibration curves were calculated as follows:

$$\% \frac{B}{B_0} = \frac{\Delta E - \Delta E_{excess}}{\Delta E_0 - \Delta E_{excess}} \times 100$$

ΔE: the extinction differences of samples at 450 nm and 630 nm.

ΔE_0: the extinction differences of the upper asymptote of the sigmoidal curve between 450 nm and 630 nm.

ΔE_{excess}: the extinction differences of the lower asymptote of the sigmoidal curve between 450 nm and 630 nm.

Cross reactivity (IC_{50})

The assay specificity was evaluated by obtaining sigmoidal curves for several structurally related compounds (striatals A, C and D, striatins A and B, and erinacines C and P) as competitors against the striatal B standard. The calibration curve of striatal B standard was measured individually for each plate. The IC_{50} were estimated by using a linearization of the calibration curves with a logit-transformation of %B/B_0-values:

$$logit\ (\%\ B/B_0) = ln\left(\frac{\%\ B/B_0}{100 - \%\ B/B_0}\right)$$

Cross reactivity (IC_{50}) = A/B × 100%.

A: concentration of striatal B standard at logit (%B/B_0) =0.

B: concentration of competitors at logit (%B/B_0) =0.

The linear regression of logit (%B/B_0) was used for determination of concentrations in unknown samples.

Recovery

Acetonitrile was spiked with striatal B at several concentrations (0.001 - 10 μg mL^{-1}) and analyzed by indirect competitive ELISA as well as by HPLC-DAD. The concentrations of striatal B of spiked samples were calculated from linear regression of logit (%B/B_0) in ELISA and linear calibration for striatal B in HPLC-DAD. The recoveries of striatal B were obtained by:

$$recovery\ [\%] = \frac{\text{detected concentration}}{\text{spiked concentration}} \times 100\%$$

Analysis of *H. erinaceus* samples

The supernatants and extracts of supernatants from submerged cultures, as well as extracts of fruiting bodies of *H. erinaceus* were analyzed by indirect competitive ELISA:

Sample preparation for pre-incubation of supernatants in ELISA

50 μL of supernatants from submerged cultures (culture day 1 to day 8) were added to 950 μL PBST with a final DMSO proportion of 5% and a final pAbs dilution of 1:200. Culture medium without *H. erinaceus* was used as negative control.

Sample preparation for pre-incubation of extracts of supernatants

20 mL of supernatants from submerged cultures (culture day 1 to 8) were extracted once with 20 mL ethyl acetate. 15 mL of the ethyl acetate phase was evaporated to dryness, and the residues were redissolved in 1 mL acetonitrile for ELISA and HPLC-DAD analysis. 5 μL of the extracts were added to 995 μL PBST with a final acetonitrile proportion of 5% and a final pAbs dilution of 1:200. Extracts of culture media without *H. erinaceus* were used as negative controls.

Sample preparation for pre-incubation of extracts of fruiting bodies

32.6 g fruiting bodies of *H. erinaceus* were homogenized and then extracted with 200 mL ethyl acetate under stirring overnight. The ethyl acetate phase was dried over sodium sulfate and evaporated to dryness after filtration. 1 mg crude extract was dissolved in 1 mL acetonitrile. Several concentration levels (50.0, 10.0, 5.0, 2.5, 1.0 and 0.5 μg mL^{-1}) of the extracts were used in the pre-incubation solution. Acetonitrile was used as blank.

Availability of supporting data

The data sets supporting the results of this article are included within the article and its additional file.

Additional file

Additional file 1: Supplementary material.

Competing interests

The authors declare that they have no competing interests.

Authors' contributions

TS purified the cyathane type diterpenoids, carried out the ELISA and drafted the manuscript. LH was involved in the studies on the cross reactivities, and HH was responsible for NMR analyses and data interpretation. MS provided the fungal strains and contributed to the discussion. HZ participated in the design of the study and helped to draft the manuscript. All authors read and approved the final manuscript.

Acknowledgement

The authors thank the BMBF and CLIB 2021 for financial support of the project (FKZ 0315404 A). Part of the work was supported by the excellence initiative of the Hessian Ministry of Science and Art which encompasses a generous grant for the LOEWE research focus 'Integrative Fungal Research'.

Author details

[1]Justus Liebig University Giessen, Institute of Food Chemistry and Food Biotechnology, Heinrich-Buff-Ring 58, Giessen 35392, Germany. [2]Justus Liebig University Giessen, Institute of Organic Chemistry, Heinrich-Buff-Ring 58, Giessen 35392, Germany. [3]Department Microbial Drugs, Helmholtz Centre for Infection Research, Inhoffenstrasse 7, Braunschweig 38124, Germany.

References

1. Allbutt AD, Ayer WA, Brodie HJ, Johri BN, Taube H: **Cyathin, a new antibiotic complex produced by *Cyathus helenae*.** *Can J Microbiol* 1971, **17:**1401–1407.
2. Ayer WA, Taube H: **Metabolites of *Cyathus helenae*. cyathin A_3 and allocyathin B_3, members of a new group of diterpenoids.** *Tetrahedron Lett* 1972, **13:**1917–1920.
3. Anke T, Oberwinkler F: **The striatins–new antibiotics from the basidiomycete Cyathus striatus (Huds. ex Pers.) Willd.** *J Antibiot* 1977, **30:**221–225.
4. Anke T, Rabe U, Schu P, Eizenhöfer T, Schrage M, Steglich W: **Studies on the biosynthesis of striatal-type diterpenoids and the biological activity of herical.** *Z Naturforsch C* 2002, **57:**263–271.
5. Shibata H, Tokunaga T, Karasawa D, Hirota A, Nakayama M, Nozaki H, Tada T: **Isolation and characterization of new bitter diterpenoids from the fungus *Sarcodon scabrosus*.** *Agric Biol Chem* 1989, **53:**3373–3375.
6. Shibata H, Irie A, Morita Y: **New antibacterial diterpenoids from the *Sarcodon scabrosus* fungus.** *Biosci Biotechnol Biochem* 1998, **62:**2450–2452.
7. Kawagishi H, Zhuang C: **Compounds for dementia from *Hericium erinaceum*.** *Drugs Future* 2008, **33:**149–155.
8. Petrova RD, Mahajna J, Reznick AZ, Wasser SP, Denchev CM, Nevo E: **Fungal substances as modulators of NF-κB activation pathway.** *Mol Biol Rep* 2007, **34:**145–154.
9. Kawagishi H, Shimada A, Hosokawa S, Mori H, Sakamoto H, Ishiguro Y, Sakemi S, Bordner J, Kojima N, Furukawa S: **Erinacines E, F, and G, stimulators of nerve growth factor (NGF)-synthesis, from the mycelia of *Hericium erinaceum*.** *Tetrahedron Lett* 1996, **37:**7399–7402.
10. Kawagishi H, Shimada A, Shirai R, Okamoto K, Ojima F, Sakamoto H, Ishiguro Y, Furukawa S: **Erinacines A, B and C, strong stimulators of nerve growth factor (NGF)-synthesis, from the mycelia of *Hericium erinaceum*.** *Tetrahedron Lett* 1994, **35:**1569–1572.
11. Kawagishi H, Simada A, Shizuki K, Ojima F, Mori H, Okamoto K, Sakamoto H, Furukawa S: **Erinacine D, a stimulator of NGF-synthesis, from the mycelia of *Hericium erinaceum*.** *Heterocycllic Commun* 1996, **2:**51–54.
12. Ma B, Shen J, Yu H, Ruan Y, Wu T, Zhao X: **Hericenones and erinacines: stimulators of nerve growth factor (NGF) biosynthesis in *Hericium erinaceus*.** *Mycology* 2010, **1:**92–98.
13. Krzyczkowski W, Malinowska E, Herold F: **Erinacine A biosynthesis in submerged cultivation of *Hericium erinaceum*: quantification and improved cultivation.** *Eng Life Sci* 2010, **10:**446–457.
14. Jahn C, Zorn H, Petersen A, Schwack W: **Structure-specific detection of plant cuticle bound residues of chlorothalonil by ELISA.** *Pestic Sci* 1999, **55:**1167–1176.
15. Xu DM, Yu XY, Liu YQ, Feng JT, Pan LG, Liu XJ, He J, Zhang X: **Development of an enzyme-linked immunosorbent assay for podophyllotoxin.** *Int Immunopharmacol* 2005, **5:**1583–1592.
16. Tothiam C, Phrompittayarat W, Putalun W, Tanaka H, Sakamoto S, Khan IA, Ingkaninan K: **An enzyme-linked immunosorbant assay using monoclonal antibody against Bacoside A_3 for determination of jujubogenin glycosides in *Bacopa monnieri* (L.) Wettst.** *Phytochem Anal* 2011, **22:**385–391.
17. Sathe M, Merwyn S, Ghorpade R, Agarwal GS, Rao MK, Rai GP, Kaushik MP: **Design and synthesis of immunoconjugates and development of competition inhibition enzyme-linked immunosorbent assay (CIEIA) for the detection of O-isopropyl methylphosphonofluoridate (sarin): an organophosphorous toxicant.** *J Hazard Mater* 2011, **192:**1720–1728.
18. Boudet F, Theze J, Zouali M: **UV-treated polystyrene microtitre plates for use in an ELISA to measure antibodies against synthetic peptides.** *J Immunol Methods* 1991, **142:**73–82.
19. Zouali M, Stollar BD: **A rapid ELISA for measurement of antibodies to nucleic acid antigens using UV-treated polystyrene microplates.** *J Immunol Methods* 1986, **90:**105–110.

20. Lagace J, Arsenault S, Cohen EA: **Alcian blue-treated polystyrene microtitre plates for use in an ELISA to measure antibodies against synthetic peptides.** *J Immunol Methods* 1994, **175:**131–135.
21. Geerligs HJ, Weijer WJ, Bloemhoff W, Welling GW, Welling-Wester S: **The influence of pH and ionic strength on the coating of peptides of herpes simplex virus type 1 in an enzyme-linked immunosorbent assay.** *J Immunol Methods* 1988, **106:**239–244.
22. Güll I, Wirth M, Gabor F: **Development of a sensitive and reliable ELISA for quantification of wheat germ agglutinin.** *J Immunol Methods* 2007, **318:**20–29.
23. Kim YJ, Cho YA, Lee H-S, Lee YT: **Investigation of the effect of hapten heterology on immunoassay sensitivity and development of an enzyme-linked immunosorbent assay for the organophosphorus insecticide fenthion.** *Anal Chim Acta* 2003, **494:**29–40.
24. Brun EM, Garcés-García M, Puchades R, Maquieira A: **Enzyme-linked immunosorbent assay for the organophosphorus insecticide fenthion. Influence of hapten structure.** *J Immunol Methods* 2004, **295:**21–35.
25. Bhandari DR, Shen T, Römpp A, Zorn H, Spengler B: **Analysis of cyathane-type diterpenoids from *Cyathus striatus* and *Hericium erinaceus* by high-resolution MALDI MS imaging.** *Anal Bioanal Chem* 2014, **406:**695–704.
26. Whittington RJ: **Evaluation of a simple method for improving the precision of an ELISA detecting antibody in serum.** *J Immunol Methods* 1992, **148:**57–64.

10

Characterization of anode and anolyte community growth and the impact of impedance in a microbial fuel cell

Diana Sanchez-Herrera[1], Daniella Pacheco-Catalan[1], Ruby Valdez-Ojeda[1], Blondy Canto-Canche[2], Xochitl Dominguez-Benetton[3], Jorge Domínguez-Maldonado[1] and Liliana Alzate-Gaviria[1*]

Abstract

Background: A laboratory-scale two-chamber microbial fuel cell employing an aerated cathode with no catalyst was inoculated with mixed inoculum and acetate as the carbon source.

Electrochemical impedance spectroscopy (EIS) was used to study the behavior of the MFC during initial biofilm (week 1) and maximum power density (week 20). EIS were performed on the anode chamber, biofilm (without anolyte) and anolyte (without biofilm). Nyquist plots of the EIS data were fitted with two equivalent electrical circuits to estimate the contributions of intrinsic resistances to the overall internal MFC impedance at weeks 1 and 20, respectively.

Results: The results showed that the system tended to increase power density from 15 ± 3 (week 1) to 100 ± 15 mW/m^2 (week 20) and current density 211 ± 7 (week 1) to 347 ± 29 mA/m^2 (week 20). The Samples were identified by pyrosequencing of the 16S rRNA gene and showed that initial inoculum (week 1) was constituted by *Proteobacteria* (40%), *Bacteroidetes* (22%) and *Firmicutes* (18%). At week 20, Proteobacterial species were predominant (60%) for electricity generation in the anode biofilm, being 51% Rhodopseudomonas palustris. Meanwhile on anolyte, *Firmicutes* phylum was predominant with Bacillus sp.

This study proved that under the experimental conditions used there is an important contribution from the interaction of the biofilm and the anolyte on cell performance. Table 1 presents a summary of the specific influence of each element of the system under study.

Conclusions:

- The results showed certain members of the bacterial electrode community increased in relative abundance from the initial inoculum. For example, Proteobacterial species are important for electricity generation in the anode biofilms and *Firmicutes* phylum was predominant on anolyte to transfer electron.
- R1 is the same in the three systems and no variation is observed over time.
- The biofilm makes a significant contribution to the charge transfer processes at the electrode (R2 and Cdl) and, consequently, on the performance of the anode chamber.
- The biofilm can act as a barrier which reduces diffusion of the anolyte towards the electrode, all the while behaving like a porous material.
- The anolyte and its interaction with the biofilm exert a considerable influence on diffusion processes, given that it presents the highest values for Rd which increased at week 20.

Keywords: Microbial fuel cell, Community growth and electrochemical impedance spectroscopy

* Correspondence: lag@cicy.mx
[1]Renewable Energy Unit, Centro de Investigación Científica de Yucatán A.C (CICY), Calle 43 No. 130 Col. Chuburná de Hidalgo, C.P. 97200 Mérida, Yucatán, México
Full list of author information is available at the end of the article

Background

Wastewaters contain dissolved organics that require removal before discharge into the environment. However, wastewaters are being recognized as a renewable resource for the production of electricity, fuels and chemicals. Bioelectrochemical wastewater treatment has, therefore, emerged as a potentially interesting technology for the production of energy [1]. Reducing the cost of the materials used in MFCs is essential for practical applications. The cathode accounts for the greatest percentage of the total capital cost, and cathode surface area and materials generally limit higher power production in MFCs. Therefore, it is important to identify low-cost materials and efficient cathode architectures in order to improve MFC cost effectiveness and performance [2].

The bioanode, a crucial component in bioelectrochemical systems (BESs), is composed of an anode biofilm and a conductive electrode. The main catalytic components of interest in anode biofilms are exoelectrogens, microorganisms that are capable of exocellular electron transfer [3,4]. In mixed-culture systems, exoelectrogens compete for electron donors with other functional groups such as fermenters, acetogens and methanogens [5,6]. The complexity of anode biofilms makes it hard to elucidate electrochemical mechanisms at the bioanode, but a precise understanding of exoelectrogenesis and competition in anode biofilms will aid in improving the performance of BESs. Several reviews provide insightful summaries and perspectives regarding bioanodes [7,8].

Electrochemical impedance spectroscopy (EIS) was used as a non-intrusive tool to identify and elucidate the electrochemical properties of redox mediators produced by microbes. EIS enabled the study of the individual contributions from different resistances. These include ohmic resistance (representing the resistance from solution, electrode materials and membrane), charge transfer and concentration (diffusion) resistance on the anode and cathode behavior of mediators and their impact on MFC impedance without the need to interrupt MFC operation [9,10].

With the help of equivalent electrical circuit fitting analysis, EIS can also provide quantitative estimates of the kinetic rate constants for the anodic and cathodic reactions, double layer capacitance at the electrode surface, and the diffusion coefficients of electro-active species in the bulk electrolyte [10,11].

The aim of this study was to characterize community growth on the anode and in the anolyte and the impact of electrochemical impedance in a microbial fuel cell with an initial mixed inoculum and its selection process over time.

Results and discussion

Chemical oxygen demand

COD removed in the first week was 93%. Meanwhile, in week 20 it was 86.5% due to increased planktonic cell concentration in the anolyte increasing the organic load and decreasing COD over time. This is similar to what Biffinger *et al.* [12] observed. The COD removed is found in the medium-high range (63% to 98%) of results obtained by other authors in the literature [13].

VFA concentrations at week 20 were 0.01 mM, 0.003 mM and 0.0009 mM for butyric, propionic and acetic acid, respectively. These values were insignificant with respect to the initial concentration of the electron donor at 73.14 mM of sodium acetate. The ease with which acetate was metabolized by exoelectrogenic microorganisms was also observed by Velasquez-Orta *et al.* [14], who obtained carboxylic acid concentrations of less than 0.05 mM after 30 weeks. The percentages of H_2 and CH_4 in the samples analyzed at week 20 of this study were less than 0.01%. This rules out a methanogenic biochemistry with a Coulombic efficiency of 60.6%. These values were lower than those reported by Jung and Reagan [8] in a PEM-type MFC with Pt catalyst at the cathode and acetate as the carbon source. They obtained 6% H_2 and CH_4 with a Coulombic efficiency of 6.4%.

Composition by pyrosequencing

The Shannon index of diversity (H') was determined for all samples. The H' value was higher for "initial inoculum" (4.5) than "biofilm" (2.8) and "anolyte" (2.9). This indicates, that the initial inoculum was higher diverse than the other samples. The bacterial richness by Chao estimate indicated that initial inoculum showed a higher number of species (339) than the biofilm (142) and anolyte (150) samples, as was indicated by Shannon index.

The rarefaction curves (Figure 1) indicated higher OTU (Operational Taxonomic Unit) number on initial inoculum sample (330) than in biofilm (123) and anolyte (123). In comparison to initial inoculum curve, the curve of biofilm and anolyte samples is flatter with increasing sampling effort, and therefore possesses lower bacterial species.

Initial inoculum (1 week)

A total of 3,764 bacteria were detected in this sample. The phylogenetic spectrum (Figure 2) was dominated by *Proteobacteria* (40%), *Methylobacter* sp. (15%) and *Syntrophus* sp. (6%) as the predominant species. Within *Bacteroidetes* (22%), *Cytophagia order* (9%) was dominant with *Firmicutes* (18%) and *Clostridiales* family (5%). *Proteobacteria* phylum has been widely detected in microbial fuel cell studies. For example, in floating microbial fuel cell (FMFC), a modification of MFC the bacterial community was dominated by γ- and β-*Proteobacteria* (with identity to *Methylobacter luteus* species) [15]. Bacterioidetes [16] such as *Firmicutes* [17] have been frequently detected as dominant in MFC reactors. Primary clarifier effluents sampled from MFC showed phylotypes

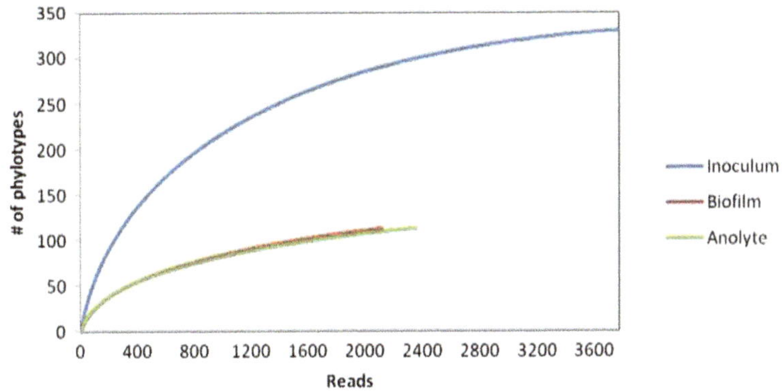

Figure 1 Rarefaction curves for each sample from MFC. Number of reads is shown on the x axis and number of OTUs at 95% sequence identity on the y axis.

relatively abundant in *Proteobacteria, Bacteroidetes* and *Firmicutes* phyla [16].

Biofilm (without anolyte - 20 weeks)

A total of 2,121 bacteria were detected in the sample collected from the electrode. The most abundant were *Proteobacteria* (60%), followed by *Bacteroidetes* (21%) and *Firmicutes* (17%). *Bacteroidete* identification in biofilm is consistent with many reports indicating its presence (Zhang *et al.* [18]; Zhang *et al.* [19]). This suggests it may play a critical role in electricity generation or efficient anode function (Yusoff *et al.* [20]). It is very interesting to note that of the 60% of *Proteobacteria* found 51% of these bacteria corresponded to *Rhodopseudomonas palustris* species. This bacterium is a member of the phototrophic purple non-sulfur bacteria which proliferates in different environments due to its versatile metabolism. It is photoautotrophic, photoheterotrophic, chemoheterotrophic and chemoautotrophic [21]. Its capability in electricity generation has been reported before in

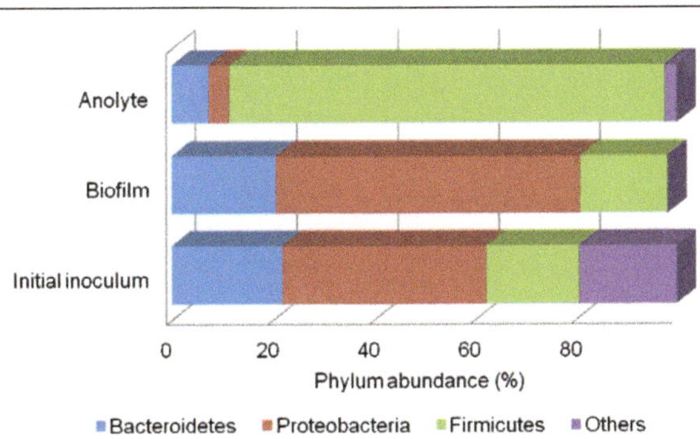

Figure 2 Taxonomic classification of pyrosequences from predominant bacterial communities of initial inoculum, biofilm, anolyte of MFC at the phylum level.

R. *palustris* DX-1 by Xing *et al.* [22]. This DX-1 strain can produce higher power densities when isolated than as mixed cultures in the same MFC. The genome completes of CGA009 strain of R. *palustris* have been sequenced [21]. U.S. Department of Energy has anticipated that genome sequence comparisons between DX-1 and strains of R. *palustris* will probably reveal key biochemical characteristics of strain DX-1 that are critical for its ability to generate power (Figure 2).

Anolyte (without biofilm - 20 weeks)

A total of 2,349 bacteria were detected in the sample taken from electrode. *Firmicutes* (86%) predominated in this sample, followed by *Bacteroidetes* (7%), *Proteobacteria* (4%) and *Lentispharaerae* (2%) (Figure 2). Predominance of *Firmicutes* at week 20 (140 days) is in accordance with studies by Ishii *et al.* [23]. The dominance of *Firmicutes* phylum was been reported in MFC at 79% in current production with acetate as the electron donor [17]. *Thermincola* sp. was the predominant species that generated current independent of an electron shuttle with acetate as an electron donor. In this study, *Bacillus* species for *Firmicutes* phylum were represented at 80%. In this respect, Nimjea *et al.* [24] demonstrated that the aerobic Gram-positive species *Bacillus subtilis* was able to grow anaerobically and produce a biofilm in a microbial fuel cell which generated a long-term power output. The electrochemical activity and the electron transfer mechanism were mainly due to excreted redox mediators in the broth solution and not to the membrane-bound proteins which were affected by physiological status.

Numerous reports indicate that *Firmicutes* are integral members of the MFC bacterial community, indicating their exocellular electron transfer (Choo *et al.* [25]; Rabaey *et al.* [26]). Thus, they dominated in the anolyte sample because of dependent electrode respiration (maybe by redox gradient-driven c-type cytochromes).

Power and current density

The maximum power and current densities were 100 ± 15 mW/m^2 and 347 ± 29 mA/m^2, respectively, at week 20. Values reported in the literature range from 80 to 1,330 mW/m^2 and from 0.55 mA/m^2 to 538 mA/m^2 in MFCs with no catalyst at the cathode. The open circuit potential stabilized at 704 mV (data not shown) [1,27,28].

Electrochemical impedance spectroscopy (EIS)

The EIS measurement of an individual electrode provides information that permits the analysis of electrochemical reactions on electrodes and bacterial metabolism, as well as surface and material properties of electrodes. These are critical to understanding the electricity-generating process and improving the power output of MFCs [29].

Anode chamber (Biofilm and Anolyte)

From Figure 3, it is clear that at both week 1 and week 20 an ohmic resistance (R1) is presented, followed by a semicircle at high frequencies, which corresponds to charge transfer processes on the surface of the electrode. The semicircle is characteristic of a single "time constant" [27]. Subsequently, a predominantly capacitive linear segment is observed (with medium to low frequencies) with a slight inclination. It is more evident in the diagram corresponding to week 20 due to the presence of other associated processes such as diffusion and/or charge transfer corresponding to the substrate. Generally, the bioelectrochemical substrate oxidation processes are slow, offer high impedance and are exhibited in the mid-to-low frequency domains [10]. For week 1, the charge transfer process of substrate oxidation is the rate limiting step since it is significantly slower than the mediator charge transfer process and oxygen reduction steps. The microbial growth on the anode has a beneficial effect on the kinetics of the bio-electrochemical reaction as it decreases the anode activation losses due to increased biocatalyst density [10]. Likewise, it is clear that a lower impedance response was present at week 20. If we consider impedance to be a vector with a specific magnitude [11], then from Figure 3 there is an improved performance in the anode chamber at week 20 (55.73 Ω) given that the impedance values are lower than week 1 (7,472 Ω) due to the presence of a biofilm. This development is similar to the report by Borole *et al.* [30] used a consortium enriched in a compact, flow-through porous electrode chamber coupled to an air-cathode. Anode impedance initially decreased from 296.1 Ω on day 24 (3.4 weeks) to 2.6 Ω after 6 months (24 weeks). Ramasamy *et al.* [10] reported that the anode impedance from two-chamber MFC on day 1 and week 3 were estimated to be 174 Ω and 32 Ω, respectively, indicating that the growth of the microbial biofilm was found to decrease the anode polarization resistance and facilitate the kinetics of the electrochemical reactions. For this study, the following elements were considered due to the similarity of

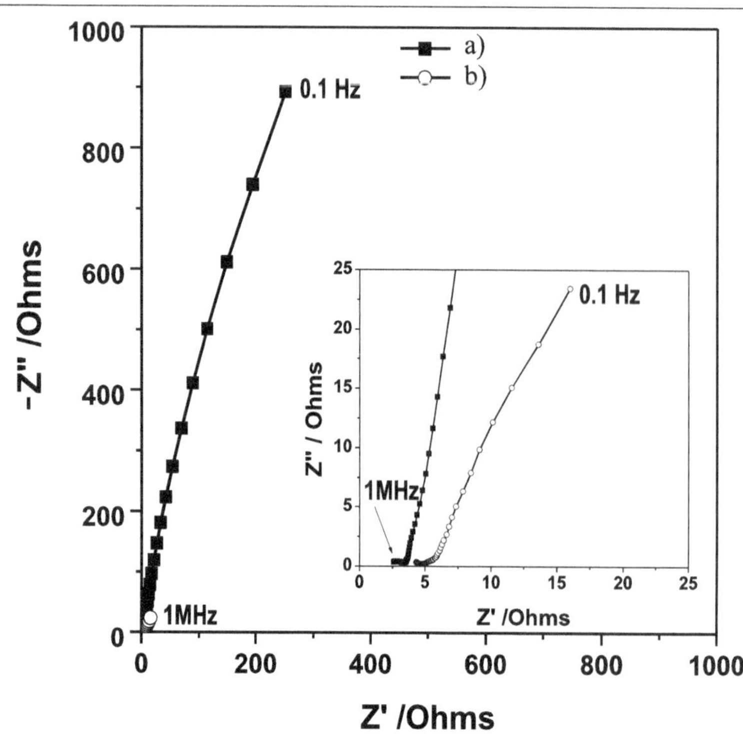

Figure 3 Nyquist diagrams corresponding to anode chamber at different times of growth: a) week 1 and b) week 20.

the behavior of the Nyquist diagrams in accordance with what Bisquert *et al.* [31] reported. Both a Warburg element (W1) and anomaly diffusion (M_a) were used to evaluate diffusion resistance (Rd) in terms of two parameters: Y_0 and B. Y_0 is the magnitude of the admittance at ω = 1 rad/s while B characterizes the time it takes for a reactant to diffuse through a thin fi w. The ratio B/Y_0 indicates the magnitude of diffusion resistance [31].

On analyzing the values obtained, the ohmic resistance (R1) was not observed to present significant changes between week 1 (R1 = 19.38 ± 5.48 Ωcm^2) and week 20 (R1 = 31.36 ± 9 Ωcm^2). This contrasted with the resistance values corresponding to the charge transfer phenomena in which a reduction was observed due to the presence of the biofilm (week 1, R2 = 66,860 ± 3.89 Ωcm^2; week 20, R2 = 434 ± 0.44 Ωcm^2). Given its high conductivity, it facilitated the mobility of electrons through the electrode/biofilm interface. This behavior is similar to what Manohar *et al.* [9] found, as well as other authors such as Malvankar *et al.* [32] and Borole *et al.* [30].

CPE values for week 1 and 20 were 1.58 E-03 ± 5.7 E-06 $\Omega^{-1}s^{\alpha}$ and 0.08 ± 1 $\Omega^{-1}s^{\alpha}$ with α = 0.9, respectively. In both systems, the incorporated CPE is considered to be a non-ideal capacitor. For determination, it was considered a faradaic system with time-constant interfacial capacitance RC [11].

$$C = \left[Q\left(\frac{1}{R1} + \frac{1}{R2}\right)^{(\alpha-1)}\right]^{1/\alpha} \qquad (1)$$

Q (S s^{α}) is the admittance when $\omega = 1,\ j = \sqrt{-1}$; α is an ideal constant and R1 and R2 are the ohmic and charge transfer resistances, respectively.

On comparing the capacitance values between week 1 Cdl = 8.70 E-04 ± 1.03 E-10 F and week 20 Cdl = 8.37 E-02 ± 4.68 E-02 F, an increase in capacitance was observed by two orders of magnitude due to the fact that the biofilm facilitates charge accumulation at the electrode interface. Although the values may vary based on the experimental conditions (electrode material, inoculum type, etc.) [33], the behavior is similar to that observed in previous works. For example, Borole *et al.* [30] reported their values of Cdl in the anode of 0.01 F during the first two months (8 weeks). Also, it increases by two orders of magnitude from 61 to 136 days (19.4 weeks), obtaining a value of 0.42 ± 0.04 F. Also, Ramasamy *et al.* [10] reported values for double layer capacitance from 0.5 mF (day 1) to 0.9 mF (day 12) for a surface of 15 cm^2 in anode, which is expected capacitance or carbon electrode. This confirmed that Cdl increased due to the stabilization of the system. As for the diffusion processes, these can occur when the solution species are diffused through the biofilm to the surface of the electrode [31]. At week 1, they are represented as a Warburg element (W1) (Figure 4) with linear, semi-infinite diffusion behavior (W1 = 6,485 ± 3,491.1Ω $s^{1/2}$) equivalent to Rd = 360.25 ± 19.95 Ωcm^2. At week 20, meanwhile, this element is replaced by a modified anomalous diffusion element (M_a) (Figure 4) equivalent to a linear transmission model reported by Bisquert *et al.* [31] with Rd = 36.18 ± 7.7 Ωcm^2. This decrease in resistance is caused by proton exchange due to the presence of the biofilm.

The model includes these elements, given that the impedance in the diffusion processes is defined as [34]:

$$Z(i\omega) \propto (i\omega)^{-\beta/2} (0 < \beta < 2) \qquad (2)$$

where, in the majority of cases, it is a Warburg-type impedance (β = 1). Anomalous diffusion (β ≠ 1) is characterized by a mean squared displacement of the diffusing particles that does not follow the ordinary linear law $r^2 \propto t$ but, more generally, has a power law dependence on time: $r^2 \propto t^{\beta}$ at low frequencies. The reason for this is the frequency transient time for a diffusing particle injected at x = 0 to cover a distance L. For $\omega >> \omega_d$ the particles will not sense the boundary at x = L so that the system will behave as semi-infinite.

The presence of the biofilm promotes anomalous diffusion processes, i.e. protonic diffusion which can be spatially restricted in either planar, cylindrical or spherical forms, producing variation in the limits of the diffusion region 0 < x < L present at low frequencies.

Many different mechanisms give rise to this anomalous behavior, including complex flows, structural complexity in the substrate of diffusion and in the diffusing tracers. Thus, no single theory of anomalous diffusion can account for all possible phenomenologies as in this case for the presence of biofilm.

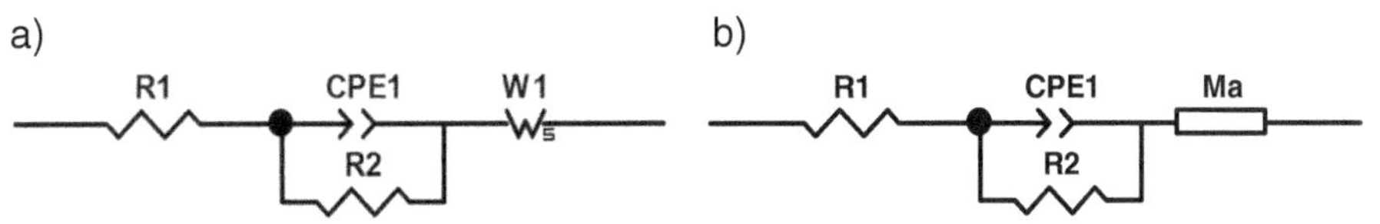

Figure 4 Equivalent circuits used for the analysis of impedance data for the three systems: anode chamber, biofilm and anolyte at a) week 1 and b) week 20.

Biofilm (without anolyte)

A more common "coating layer" on the electrodes of MFCs is a biofilm. Figure 5 shows the Nyquist diagram of the biofilm at weeks 1 and 20. First of all, there is a difference between the two systems at the same frequencies with week 20 presenting lower impedance, once again considering impedance to be a magnitude vector. This is due to the fact that at week 20 the biofilm had already achieved growth and stabilization, favoring processes which reduce the total resistance of the system [30].

Likewise, at both weeks 1 and 20, the final part of a semicircle is observed, corresponding to charge transfer processes on the surface of the electrode (high frequencies) followed by a linear segment with an incline angle that corresponds to diffusion processes, and finally a predominantly capacitive linear portion similar to the behavior presented by the anode chamber. Therefore, the equivalent circuits proposed for the anode chamber at weeks 1 and 20 were used (Figure 4).

On analyzing the data obtained, ohmic resistance was not found to present significant changes between week 1 (R1 = 36.00 ± 4.12 Ωcm^2) and week 20 (R1 = 35.99 ± 4.11 Ωcm^2). With respect to the CPE, as in the case of the anode chamber, the behavior is mainly non-ideal capacitance. The values obtained (Cdl) at weeks 1 and 20 were 10.71E-01 ± 1.37E-03 F and 8.14 E-04 ± 1.48 E-10 F, respectively. This allows for the inference that the capacitance contribution in the anode chamber can be attributed to the behavior of the biofilm as has been reported previously by other research groups [30].

The charge transfer resistance for week 1 presented higher values (R2 = 24,651 ± 799.38 Ωcm^2) compared to those at week 20 (R2 = 2,305.8 ± 12.4E-03 Ωcm^2) due to the interaction of the biofilm with the anolyte.

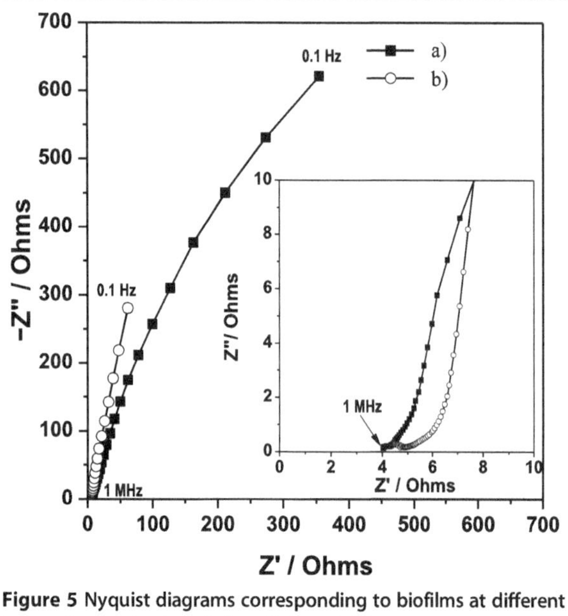

Figure 5 Nyquist diagrams corresponding to biofilms at different times of growth: a) week 1 and b) week 20.

Diffusion processes at weeks 1 and 20 were equivalent to Rd 32.51 ± 7.5 Ωcm^2 and 32.22 ± 8.93 Ωcm^2. These values were similar to those for the anode chamber at week 20, suggesting that the anolyte presents diffusion processes from the outset.

Anolyte (without biofilm)

Figure 6 shows the Nyquist diagram corresponding to the anolyte at weeks 1 and 20. In both cases, it was not possible to observe a variation in the impedance value, considering it to be a vector. The ohmic resistance is followed by an incomplete semicircle which indicates the presence of charge transfer processes. It is unlikely that any biochemically derived redox compounds, synthesized mediators, yield a complete faradaic response to an AC signal faster than 100 Hz. Hence, the reaction in the high frequency region depicts a fast electrochemical process such as oxidation of soluble metal ions in the growth medium.

The ohmic resistance values (R1) do not present significant changes due to the fact that they correspond to the resistance of the components of the system itself (weeks 1 and 20, 24.5 ± 15.55 Ωcm^2 and 42.9 ± 3.92 Ωcm^2, respectively).

The charge transfer resistance and diffusion resistance for week 1, R2 = 199 ± 3.89 Ωcm^2 and Rd = 389.3 ± 33.50 Ωcm^2, compared to week 20, R2 = 179 ± 47.8 Ωcm^2 and Rd = 3.75E + 03 ± 1.37E-03 Ωcm^2, show that the R2 values are similar, meaning that there is no contribution from the anolyte to the charge transfer processes. The anolyte without biofilm on the electrode is not redox-active toward acetate under these conditions. EIS of bare electrodes showed a very high charge transfer resistance [35].

When comparing the results of Rd obtained for the anolyte with those for the anode chamber and the biofilm, the anolyte at week 1 presents similar values to those for the anode chamber. However, at week 20 the value increases by an order of magnitude, whilst in the anode chamber it decreases by an order of magnitude because the diffusion resistance decreased considerably due to the presence of the biofilm.

CPE values for weeks 1 and 20 were 1.22E-03 ± 9.02E-05 (α = 0.6) and 1.40E-03 ± 1.18 E-03 (α = 0.6), corresponding to the capacitances (Cdl) 1.70E-05 ± 2.23E-08 F and 2.86E-05 ± 9.74E-06, respectively.

Although there is evidence in the literature for the presence of diffusion processes in this type of bioelectrochemical system indicating that the mass transfer limitations were insignificant and masked by the dominant kinetic limitations for the anode bio-electrochemical reaction, this study proved that under the experimental conditions used there is an important contribution from the

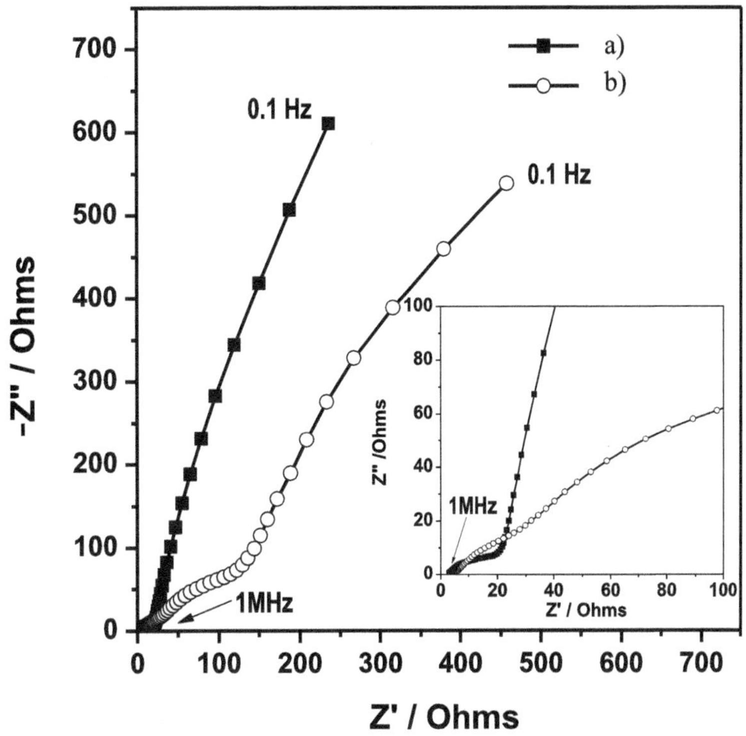

Figure 6 Nyquist diagrams corresponding to anolyte at different times of growth: a) week 1 and b) week 20).

interaction of the biofilm and the anolyte on cell performance. Table 1 presents a summary of the specific influence of each element of the system under study (A = highly influential; B = influential and C = uninfluential).

Conclusions

The Microbial Fuel Cell with no catalyst (granular carbon and stainless steel mesh collector) and mixed inoculum in this study showed certain members of the bacterial electrode community increased in relative abundance from the initial inoculum. For example, *Proteobacteria* species are important for electricity generation in the anode biofilms and *Firmicutes* phylum was predominant on anolyte to transfer electron. Likewise, the biofilm can act as a barrier which reduces diffusion of the anolyte towards the electrode, all the while behaving like a porous material. The anolyte and its interaction with the biofilm exert a considerable influence on diffusion processes, given that it presents the highest values for Rd. Rd increased at week 20.

Table 1 Influence of each of the elements corresponding to the phenomena in the MFC

Configuration	System	Charge transfer processes		Diffusion processes
	R1	R2	Cdl	Rd
Anode chamber	C	A	A	B
Anolyte	C	C	C	A
Biofilm	C	A	A	B

Methods

MFC preparation

Two MFCs were constructed from acrylic. The anode and cathode chambers were semicircular in shape with a capacity of 115 ml and a volume of 115 ml (Figure 7). Nafion® 117 supported between two acrylic sheets was used as a proton exchange membrane (Figure 7). The effective area of the previously activated membrane was 18.9 cm^2. The covers of each MFC featured ports for working, reference and counter electrodes for feeding and obtaining analysis samples, as well as for the oxygen diffuser at the cathode. The anode was carbon cloth (supplied by ElectroChem) with an area of 9 cm^2. The cathode used 30 g granular carbon with no catalyst, using a stainless steel mesh as the current collector (mesh size 400 × 400, alloy 316). Electrode connections to the exterior were made from stainless steel mesh (mesh size 400 x 400) with nylon thread covered with Termofit® (Figure 7).

Inoculum and carbon source

The inoculum was a mixed non-anaerobic consortium which consisted of 30 g/L deep soil, 300 g/L cattle manure, 150 g/L pig manure, 1.5 g/L sodium carbonate and 1 L water. The MFC was fed synthetic wastewater with acetate as a carbon source (6 g/L) with elements in g/L,

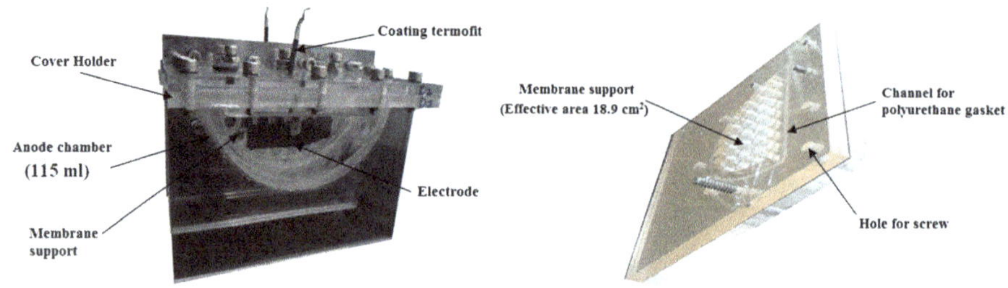

Figure 7 Assembled MFC and membrane support.

$NaHCO_3$ 1, Na_2CO_3 1, KH_2PO_4 0.2, NH_4Cl 0.1 and minerals in mg L-1, $ZnCl_2$ 10, $CaCl_2$ 10, $FeSO_4{\bullet}7H_2O$ 10, $CoCl_2{\bullet}6H_2O$ 5, $CuSO_4{\bullet}5H_2O$ 5, $NiCl_2{\bullet}6H_2O$ 20, $MnCl_2{\bullet}4H_2O$ 20 [36].

MFC configuration and operation

2 MFCs (semi-batch experiment with hydraulic retention time of 48 hours) were inoculated with 50 ml inoculum, 20 ml sodium phosphate buffer (Na_2HPO_4, 4.09 g/L and $NaH_2PO_4 \cdot H_2O$, 2.93 g/L) and 40 ml SW (Synthetic Wastewater) [37]. A control was evaluated under the same conditions.

The anode chamber was sparged with nitrogen gas to displace the oxygen present prior to closure and each time samples were taken. The pH was close to neutral. pH was adjusted with a $KHCO_3$ and K_2CO_3 (0.2 M) buffer in accordance with Puig *et al.* [38].

In the cathode chamber a potassium phosphate buffer (50 mM) as catholyte in accordance with Zhang *et al.* [39] was used, and the pH was maintained between 4 and 5 with the use of $KHCO_3$ and K_2CO_3 0.2 M and bubbled with filter-sterilized air. The working temperature was 27 ± 2°C.

Chemical analyses

The pH and temperature were measured with a Thermo Scientific Orion® multiparameter meter. COD was measured with the potassium dichromate in digestion solution technique (high range COD reagent from 0 to 15,000 ppm).

After a liquid sample was passed through a 0.22 μm pore membrane (type GV, Millipore), volatile fatty acids were analyzed using liquid chromatography. For the eluant, a mixture of an equal volume of 5 mM p-toluenesulfonic acid solution and 20 mM Bis-Tris solution containing 5 mM p-toluenesulfonic acid and 100 μM EDTA was used at 0.8 ml/min. The filtrate was acidified with concentrated HCl and short-chain alcohols [23] were analyzed using a gas chromatograph (GC - Perkin Elmer Clarus 500) with a flame ionization detector and an EC-1000 column (Altech).

Methane and hydrogen at the headspace of the anode chamber were measured using a gas chromatograph (GC - Perkin Elmer Clarus 500) equipped with a thermal conductivity detector and parallel packed columns (molecular sieve 5A) as described previously [40].

Bacterial composition identification

The DNA extraction protocol for identifying the bacterial composition of sludge used in microbial fuel cells was used as described previously by Canto-Canché *et al.* [41]. Metagenomic DNA was sent to the Research and Testing Laboratory facility in Lubbock, Texas (USA) for pyrosequencing of the 16S rRNA gene. The Research and testing analysis pipeline performed denoising and chimera checking. Sequences of ca. 398nt were submitted for analysis. Rarefaction curves were constructed by using the tool aligner, complete linkage clustering, and rarefaction of the RDP pyrosequencing pipeline. Shannon [42] and Chao indices [43] were calculated with the complete linkage clustering data.

EIS Experiments

Impedance measurements were taken on three different test cell configurations, as detailed in Table 2.

The EIS experiments were performed during biofilm growth at 1 and 20 weeks during closed circuit operation. Biologic potentiostat-AC frequency analyzer equipment was used for the EIS experiments, and the results

Table 2 Different test cell configurations used for EIS studies

Configuration	Working electrode	Reference electrode	Counter electrode	Resistive components in the configuration
Anode chamber	Anode	Ag/AgCl	Platinum mesh	Anode, membrane and anolyte
Biofilm (without anolyte)	Anode			Sterile buffer solution**
Anolyte (without biofilm)	Sterile Anode*			Sterile Anode*

*Carbon cloth.
**Per L of deionized water without acetate.

were analyzed using EC-Lab® V10.23 software by χ^2 minimization, obtaining values between 10^{-2}-10^{-3}. The resistance values were normalized based on the area of the electrode (9 cm^2). The frequency of the AC signal was varied from 0.1 Hz to 1 MHz with an amplitude of 10 mV. Impedance experiments were performed under galvanostatic closed circuit conditions at 0, 50, 100, 250, and 400 mA for the immature biofilm (week 1) and at 0 and 400 mA for the developed biofilm (week 20) for comparison purposes according to the methodology of Ramasamy *et al.* [10]. To ensure steady state during galvanostatic operation, the MFC was allowed to equilibrate for 10 min between each current setting before applying the AC signal. Experiments under potentiostatic control were performed utilizing a three-electrode arrangement consisting of the working electrode, a Ag/AgCl, sat. KCl, (197 mV vs. SHE) reference electrode and a counter electrode (platinum mesh).

The equivalent circuit of the anode chamber at weeks 1 and 20 contemplates the ohmic resistance, the charge transfer process and the double layer on the electrode, as well as diffusion and transfer processes from the electrolyte towards the surface of the electrode. These were, therefore, considered resistances. The constant phase element (CPE) is attributed to the heterogeneity of either the electrode or the reactions. The controlled diffusion (W) and anomaly (M_a) elements correspond to proton transfer processes (Rd) which, as will be explained later, are mainly attributed to the anolyte.

Abbreviations

MFC: Microbial fuel cell; COD: Chemical oxygen demand; EIS: Electrochemical Impedance Spectroscopy; VFAs: Volatile fatty acids; SW: Synthetic wastewater; SHE: Standard hydrogen electrode.

Competing interests

The authors declare that they have no competing interests.

Authors' contributions

DSH performed the experiments presented herein and helped draft the manuscript. DPC and XDB assisted in experimental design of the Electrochemical Impedance Spectroscopy. RVO and BCC supervised the bacterial composition identification analysis of microbial communities and assisted in drafting the manuscript. JDM developed the carboxylic acid analysis. LMAG conceived the study and financed, supervised the work, designed the microbial fuel cell and drafted the manuscript. All authors participated in the experimental design, evaluation of the data, reading and approval of final manuscript.

Acknowledgements

The authors thank the Mexican Council for Science and the financial support granted to carry out this study through the project SEP-CONACYT, number 106416 and Miguel Alonso Tzec Sima for his technical support.

Author details

[1]Renewable Energy Unit, Centro de Investigación Científica de Yucatán A.C (CICY), Calle 43 No. 130 Col. Chuburná de Hidalgo, C.P. 97200 Mérida, Yucatán, México. [2]Biotechnology Unit, CICY A.C, Mérida, Yucatán, México. [3]Separation and Conversion Technology, VITO-Flemish Institute for Technological Research, Boeretang 200, Mol 2400, Belgium.

References

1. Rozendal R, Hamelers H, Rabaey K, Keller J, Buisman C: **Towards practical implementation of bioelectrochemical wastewater treatment.** *Trends Biotech* 2008, **26**:450–459.
2. Zhang F, Cheng S, Pant D, Van Bogaert G, Logan B: **Power generation using an activated carbon and metal mesh cathode in a microbial fuel cell.** *Electrochem Commun* 2009, **11**:2177–2179.
3. Kim H, Park H, Hyun M, Chang I, Kim M, Kim B: **A mediator-less microbial fuel cell using a metal reducing bacterium, Shewanella putrefaciens.** *Enzyme Microb Technol* 2002, **30**:145–152.
4. Liu H, Cheng S, Logan B: **Power generation in fed-batch microbial fuel cells as a function of ionic strength, temperature, and reactor configuration.** *Environ Sci Technol* 2005, **39**:5488–5493.
5. Woodward L, Perrier M, Srinivasan B, Pinto R, Tartakovsky B: **Comparison of real-time methods for maximizing power output in microbial fuel cells.** *Industrial and Eng Chem Res* 2010, **49**:9222–9229.
6. Rismani-Yazdi H, Christy A, Dehority B, Morrison M, Yu Z, Tuovinen OH: **Electricity generation from cellulose by rumen microorganisms in microbial fuel cells.** *Biotechnol Bioeng* 2007, **97**:1398–1407.
7. Pham T, Aelterman P, Verstraete W: **Review Bioanode performance in bioelectrochemical systems: recent improvements and prospects.** *Trends in Biotechnol* 2009, **27**:168–178.
8. Jung S, Regan J: **Influence of External Resistance on Electrogenesis, Methanogenesis, and Anode Prokaryotic Communities in Microbial Fuel Cells.** *Appl Environ Microbiol* 2011, **77**:564–571.
9. Manohar A, Bretschger O, Nealson K, Mansfel F: **The use of electrochemical impedance spectroscopy (EIS) in the evaluation of the electrochemical properties of a microbial fuel cell.** *Bioelectrochemistry* 2008, **72**:149–154.
10. Ramasamy R, Ren Z, Mench M, Regan J: **Impact of initial biopelícula growth on the anode impedance of microbial fuel cells.** *Biotechnol Bioeng* 2008, **101**:101–108.
11. Orazem M, Tribollet B: *Electrochemical Impedance Spectroscopy.* USA: Wiley; 2008.
12. Biffinger J, Ray R, Little B, Fitzgerald L, Ribbens M, Finkel S, Ringeisen B: **Simultaneous analysis of physiological and electrical output changes in an operating microbial fuel cell with shewanella oneidensis.** *Biotech Bioeng* 2009, **103**:524–531.
13. Mohan S, Mohanakrishna G, Sharma P: **Composite vegetable waste as renewable resource for bioelectricity generation through non-catalyzed open-air cathode microbial fuel cell.** *Bioresource Technol* 2010, **101**:970–976.
14. Velasquez-Orta S, Curtis T, Logan B: **Energy from algae using microbial fuel cells.** *Biotech Bioeng* 2009, **103**:1068–1076.
15. Huang Y, He Z, Kan J, Manohar A, Nealson K, Mansfeld F: **Electricity generation from a floating microbial fuel cell.** *Biores Technol* 2012, **114**:308–313.
16. Ishii S, Suzuki S, Norden-Krichmar M, Nealson H, Gorby Y, Bretschger O, Sekiguchi Y: **Functionally Stable and Phylogenetically Diverse Microbial Enrichments from Microbial Fuel Cells during Wastewater Treatment.** *PLoS ONE* 2012, **7**(2):e30495.
17. Wrighton C, Agbo P, Warnecke F, Karrie W, Brodie L, DeSantis T, Hugenholtz P, Andersen G, Coates D: **A novel ecological role of the Firmicutes identified in thermophilic microbial fuel cells.** *Int Soc Microb Ecol* 2008, **2**:1146–1156.
18. Zhang Y, Min B, Huang L, Angelidaki I: **Electricity generation and microbial community response to substrate changes in microbial fuel cell.** *Biores Technol* 2011, **102**:1166–1173.
19. Zhang G, Zhao Q, Jiao Y, Wang K, Lee D, Ren N: **Efficient electricity generation from sewage sludge using bio-cathode microbial fuel cell.** *Water Res* 2012, **46**:43–52.
20. Yusoff M, Hu A, Feng C, Maeda T, Shirai Y, Hassan M, Yu C: **Influence of pretreated activated sludge for electricity generation in microbial fuel cell application.** *Biores Technol* 2013, **145**:90–96.
21. Larimer F, Chain P, Hauser L, Lamerdin J, Malfatti S, Do L, Land M, Pelletier D, Beatty J, Lang A, Tabita F, Gibson J, Hanson T, Bobst C, Torres J, Peres C, Harrison F, Gibson J, Harwood C: **Complete genome sequence of the metabolically versatile photosynthetic bacterium Rhodopseudomonas palustris.** *Nature Biotechnol* 2004, **22**:55–61.

22. Xing D, Zuo Y, Cheng S, Regan J, Logan B: **Electricity Generation by Rhodopseudomonas palustris DX-1.** *Environ Sci Technol* 2008, **42**:4146–4151.
23. Ishii S, Shimoyama T, Hotta Y, Watanabe K: **Characterization of a filamentous biofilm community established in a cellulose-fed microbial fuel cell.** *BMC Microbiol* 2008, **8**:6–15.
24. Nimjea V, Chen C, Chen C, Jeanc J, Reddya A, Fana C, Pand K, Liua H, Chena J: **Stable and high energy generation by a strain of *Bacillus subtilis* in a microbial fuel cell.** *J Power Sourc* 2009, **190**:258–263.
25. Choo Y, Lee J, Chang I, Kim B: **Bacterial communities in microbial fuel cells enriched with high concentrations of glucose and glutamate.** *J Microbial Biotechnol* 2006, **16**:1481–1484.
26. Rabaey K, Boon N, Siciliano S, Verhaege M, Verstraete W: **Biofuel cells select for microbial consortia that self-mediate electron transfer.** *Appl Environ Microbiol* 2004, **70**:5373–5382.
27. Rabaey K, Angenet L, Schröder U: *Bioelectrochemical Systems: From Extracellular Eclectron Transfer to Technological Application.* New York: IWA; 2010.
28. Yuan X, Song C, Wang H, Zhang J: *Electrochemical Impedance Spectroscopy in PEM Fuel Cells.* London: Springer; 2010.
29. Zhen Z, Mansfeld F: **Exploring the use of electrochemical impedance spectroscopy (EIS) in microbial fuel cell studies.** *Energy Environ Sci* 2009, **2**:215–219.
30. Borole A, Aaron D, Hamilton C, Tsouris C: **Understanding Long-Term Changes in Microbial Fuel Cell Performance Using Electrochemical Impedance Spectroscopy.** *Environ Sci Technol* 2010, **44**:2740–2745.
31. Bisquert J, Garcia-Belmonte G, Fabregat-Santiago F, Compte A: **Anomalous transport effects in the impedance of porous film electrodes.** *Electrochem Commun* 1999, **1**:429–435.
32. Malvankar N, Lau J, Nevin K, Franks A, Tuominen M, Lovley D: **Electrical Conductivity in a Mixed-Species Biofilm.** *Appl Environ Microbiol* 2012, **78**:5967–5971.
33. Ha P, Moon H, Kim B, Ng H, Chang I: **Determination of charge transfer resistance and capacitance of microbial fuel cell through a transient response analysis of cell voltage.** *Biosens Bioelectron* 2010, **25**:1629–1634.
34. Diard J, Le Gorrec B, Montella C: **Diffusion Impedances.** Hosted by Bio-Logic@ http://biologic.info.
35. Srikanth S, Marsili E, Flickinger M, Bond D: **Electrochemical characterization of geobacter sulfurreducens cells immobilized on graphite paper electrodes.** *Biotechnol Bioeng* 2009, **99**:1065–1073.
36. Alzate-Gaviria L, Sebastian P, Pérez-Hernández A: **Comparison of two anaerobic systems for hydrogen production from the organic fraction of municipal solid waste and synthetic wastewater.** *Int J Hydrogen Energy* 2008, **32**:3141–3146.
37. Phung N, Lee J, Kang K, Chang I, Gadd G, Kim B: **Analysis of microbial diversity in oligotrophic microbial fuel cells using 16S rDNA sequences.** *FEMS Microbiol Lett* 2004, **233**:77–82.
38. Puig S, Serra M, Coma M, Cabré M, Balaguer M, Colprim J: **Effect of pH on nutrient dynamics and electricity production using microbial fuel cells.** *Bioresource Technol* 2010, **101**:9594–9599.
39. Zhang L, Li C, Ding L, Xu K, Ren H: **Influences of initial pH on performance and anodicmicrobes of fed-batchmicrobial fuel cells.** *J Chem Technol Biotechnol* 2011, **86**:1226–1232.
40. American Public Health Association/American Water Works Association/ Water Environment Federation: *Standard Methods for the Examination of Water and Wastewater.* 21st edition. Washington, DC: American Public Health Association; 2005.
41. Canto-Canché B, Tzec-Simá M, Vázquez-Loría J, Espadas-Alvarez H, Chi-Manzanero B, Rojas-Herrera R, Valdez-Ojeda R, Alzate-Gaviria L: **Simple and inexpensive DNA extraction protocol for studying the bacterial composition of sludges used in microbial fuel cells.** *B Genetics and Molecular Res* 2013, **12**:282–292.
42. Shannon CE: **The mathematical theory of communication. 1963.** *MD Comput* 1997, **14**:306–317.
43. Chao A, Bunge J: **Estimating the number of species in a sochastic abundance model.** *Biometrics* 2002, **58**:531–539.

11

Characterization of the substitution pattern of cellulose derivatives using carbohydrate-binding modules

Laura von Schantz[1,4], Herje Schagerlöf[2,5], Eva Nordberg Karlsson[3] and Mats Ohlin[4*]

Abstract

Background: Derivatized celluloses, such as methylcellulose (MC) and hydroxypropyl methylcellulose (HPMC), are of pharmaceutical importance and extensively employed in tablet matrices. Each batch of derivatized cellulose is thoroughly characterized before utilized in tablet formulations as batch-to-batch differences can affect drug release. The substitution pattern of the derivatized cellulose polymers, i.e. the mode on which the substituent groups are dispersed along the cellulose backbone, can vary from batch-to-batch and is a factor that can influence drug release.

Results: In the present study an analytical approach for the characterization of the substitution pattern of derivatized celluloses is presented, which is based on the use of carbohydrate-binding modules (CBMs) and affinity electrophoresis. CBM4-2 from *Rhodothermus marinus* xylanase 10A is capable of distinguishing between batches of derivatized cellulose with different substitution patterns. This is demonstrated by a higher migration retardation of the CBM in acrylamide gels containing batches of MC and HPMC with a more heterogeneous distribution pattern.

Conclusions: We conclude that CBMs have the potential to characterize the substitution pattern of cellulose derivatives and anticipate that with use of CBMs with a very selective recognition capacity it will be possible to more extensively characterize and standardize important carbohydrates used for instance in tablet formulation.

Keywords: Substitution pattern, Methylcellulose, Hydroxypropyl methylcellulose, Application of carbohydrate-binding modules

Background

Cellulose is the most abundant natural biopolymer on earth [1]. It is primarily used to produce paper but as it is renewable and abundant, cellulose is also an attractive raw material for various industrial uses. However, native cellulose from plants and wood has a crystalline structure that complicates its use in bioprocesses. It consists of glucose saccharides bonded through glycosidic 1–4 β-bonds (Figure 1) to form linear polymer chains that have very strong inter-chain attractions making the complex insoluble [2]. Chemical modification of cellulose resulting in the incorporation of functional groups at the free hydroxyl groups along the cellulose chains improves the solubility of the carbohydrate [3].

* Correspondence: mats.ohlin@immun.lth.se
[4]Present Addresses: Alligator Bioscience AB, Lund, Sweden
Full list of author information is available at the end of the article

Derivatized cellulose is used in a wide range of industries and included in food, textile, healthcare and personal care products. In the pharmaceutical industry derivatized celluloses are often used as excipients [1-3]. A common formulation is mixing of the active substance with a polymer in a compressed tablet that upon hydration swells into a gel from which a controlled, prolonged delivery of the drug occurs [4,5]. Hydroxypropyl methylcellulose (HPMC) is the most frequently used polymer in such tablets [6], but other cellulose ethers with different patterns of substitutions are also commonly used (Figure 1), such as methylcellulose (MC), hydroxypropyl cellulose (HPC), hydroxyethyl cellulose (HEC) and carboxymethyl cellulose (CMC) [1,5,6].

The physicochemical properties of the derivatized polymer chains have been shown to affect their behavior and applicability as excipients in tablets [4]. An important factor affecting polymer behavior and drug release has been

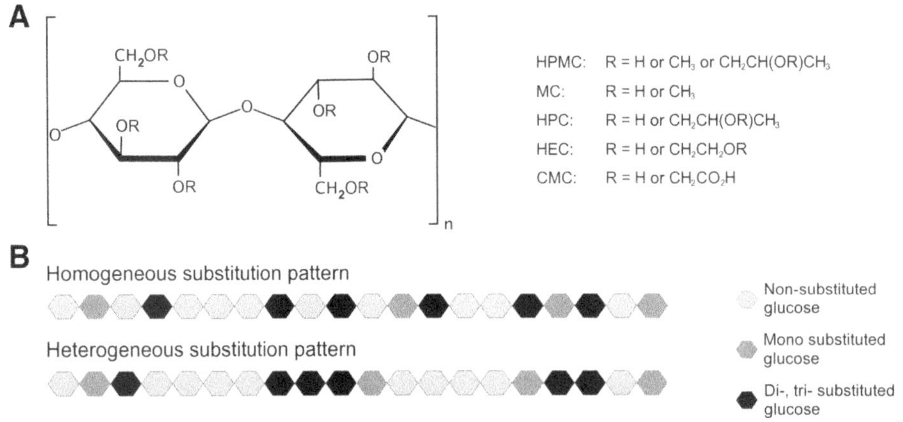

Figure 1 Chemical structure of different celluloses derivatives. A. Cellulose consists of glucose saccharides connected through glycosidic 1–4 bonds. In the generation of cellulose derivatives hydrogen atoms (annotated as R) on three hydroxyl groups on each glucose unit are susceptible to chemical derivatization. The possible substitutions for hydroxypropyl methylcellulose (HPMC), methylcellulose (MC), hydroxypropyl cellulose (HPC), hydroxyethyl cellulose (HEC) and carboxymethyl cellulose (CMC) are shown. **B**. Schematic illustration of substitution patterns in homogenously and heterogeneously substituted cellulose.

shown to be the average molecular weight that e.g. influences the viscosity of the dissolving tablets [7-10]. Other critical factors observed are the type and degree of substitution as well as the average number and position of the substituents in a single glucose unit [11-14]. In studies by Viridén et al. [15,16], tablets produced from seven different commercial HPMC batches were investigated. The batches had the same substitution grade and viscosity but behaved differently with respect to their capacity for gelation and polymer release and it was concluded that these differences were correlated to the substitution pattern of the HPMC batches. In a later study, Viridén et al. also showed that the drug release mechanism in HPMC tablets is controlled to a greater extent by erosion if the HPMC has a substitution pattern that is homogeneous, while it is largely controlled by diffusion if the HPMC has a more heterogeneous substitution pattern [17]. Thus it is important to be able to characterize the raw material used as excipient in tablets in order to minimize batch-to-batch variation in the tablet production process.

Though methods for determining the localization of substitutions within a glucose unit are well established [18-20], the characterization of the substitution pattern along the glucose backbone is more challenging and often involves a combination of several techniques. Many of the approaches require partial hydrolysis of the derivatized cellulose chains, which can be performed either randomly, by chemical strategies [21], or by using hydrolyzing enzymes that cleave the chains at specific regions [22]. After partial hydrolysis, the chemical structures of the resulting oligomers are often analyzed using methods such as mass spectrometry (MS) or nuclear magnetic resonance NMR [23]. MS analyses have been performed both with matrix-assisted laser desorption/ionization time of flight (MALDI-TOF) mass spectrometry [13] as well as with electrospray ionization (ESI), with or without separation with liquid chromatography [12,24,25]. Such analyses have in turn been performed either with or without prior isotopic labeling [26,27]. Other strategies differentiate the modified cellulose polymers by use of their cloud point, i. e. the gelation temperature, which partly depends on the substitution pattern [28,29].

Enzyme-aided characterization of cellulose derivatives is an attractive approach as it is based on specific protein-carbohydrate interactions [30]. Other proteins that can specifically recognize natural cellulose are carbohydrate-binding modules (CBMs). CBMs are defined as non-catalytic modules of carbohydrate active enzymes that have a continuous polypeptide chain and a separate fold and display carbohydrate-binding activity [31]. In the present study it was investigated if the specificity in natural and engineered CBMs can be used for characterization of the substitution pattern of cellulose derivatives by affinity electrophoresis (AE). Two natural CBMs with reported ability to bind to cellulose oligomers and regenerated cellulose were used, CBM4-2 [32] and CBM28 [33]. In addition, three mutants derived from CBM4-2 were used, of which A-6 was selected for its binding to Avicel [34], X-2 L110F for its binding to several polysaccharides including β-glucan and cellulose oligomers [35], and G-4 for its lack of binding to carbohydrates (i.e. it serves as a negative control) [34,36]. The mutants had been engineered for other studies and differ from CBM4-2 by a set of 7 to 9 mutations (Table 1) mostly found in the binding cleft. Also, because it is known that CBMs arranged in tandem often display a higher apparent affinity (avidity) for their target [37], a construct consisting of two connected CBM4-2 modules was included in the

Table 1 Amino acid differences existing between CBM4-2 and mutants thereof

	67	68	69	70	71	72	73	74	75	76	77	78	108	109	110	111	112	113	114
CBM4-2	N	P	W	D	I	E	A	T	A	F	P	V	Q	S	F	Q	E	Y	G
A-6	.	.	.	.	.	Q	.	.	.	H	.	.	.	.	Y	.	D	.	.
X-2 L110F	.	.	F	N	.	Q	.	.	.	L	.	.	.	.	.	D	.	.	.
G-4	.	.	F	.	.	D	.	.	.	L	.	.	.	.	L	E	H	.	.

	115	116	117	118	119	120	140	141	142	143	144	145	146	147	148	149	150	mutations outside the binding cleft
CBM4-2	R	L	H	E	Q	Q	V	I	R	A	P	I	H	F	G	Y	A	
A-6	.	.	L	Q	.	.	.	.	.	.	.	.	.	.	.	L	.	
X-2 L110F	.	.	.	E	.	.	.	.	.	.	.	.	.	.	.	.	.	**W91R**
G-4	.	.	F	.	.	.	.	.	.	.	.	.	Y	.	.	.	.	**E138A**

study. Altogether, these CBMs represent a collection of modules with different properties likely to represent different patterns of derivatized cellulose recognition.

Through these studies we are able to conclude that CBMs have the potential to discriminate between heterogeneously and homogeneously derivatized celluloses and envisage that such modules can be employed in the characterization of the substitution pattern of cellulose derivatives to be utilized for example in tablet formulations.

Results

Migration retardation of a set of CBMs was investigated by AE to evaluate if CBMs can distinguish between derivatized celluloses with a homogenous substitution pattern, i.e. randomly dispersed, from those with a heterogeneous substitution pattern, i.e. more clustered substitutions.

CBM4-2 and X-2 L110F distinguish between batches of MC with different substitution patterns

Non-denaturing electrophoresis demonstrated that CBM4-2 was retained more in the gels containing the heterogeneously substituted MC 1.78 or MC 1.80 than in the gel with homogeneously substituted MC 1.76 (Figure 2). CBM G-4, a carbohydrate non-binding negative control variant of CBM4-2, was as expected not retained in any of the gels. Of the other CBMs tested, A-6 and X-2 L110F were retained in all of the gels containing MC. X-2 L110F is retained more in the gels containing MC 1.80 and MC 1.78 than MC 1.76 although the difference in mobility is not as apparent as for CBM4-2. In contrast, A-6 is not able to make a distinction between the different MCs illustrating that by using different CBMs it is possible to get different views of the MCs.

CBM4-2 distinguishes between HPMC batches with different substitution pattern

Gels containing four types of HPMCs were also analyzed for their ability to interact with the CBMs. Of the four different batches included, two of the HPMCs are classified as homogeneously substituted and the other as heterogeneously substituted. In this experiment the CBMs investigated were a negative control (G-4), CBM4-2, X-2 L110F and CBM28 from *B. akibai* Cel5A. The retardations were modest yet a distinction was observed for batches with a heterogeneous versus a homogeneous substitution pattern (Figure 3). In gels containing heterogeneously substituted HPMC, CBM4-2 was retained and the protein band had a trailing shape. In contrast, in gels containing HPMC with a homogeneous substitution pattern, CBM4-2 was hardly retained at all.

Tandem CBM4-2 construct showed improved interaction with derivatized celluloses

In an effort to achieve better retardation a tandem CBM construct with possible higher apparent affinity to the cellulose derivatives due to avidity effects was generated. CBM4-2 was selected for this experiment as it appeared to be the CBM that is best capable of making a distinction between homogeneously and heterogeneously derivatized cellulose. The tandem construct consisted therefore of two CBM4-2 molecules linked together through a polypeptide linker. Indeed the difference in mobility for the tandem CBM in gels containing homogenously versus heterogeneously derivatized HPMC was more apparent than for the single CBM (Figure 3). It is thus possible to through engineering devise CBM constructs with enhanced ability to distinguish between batches of derivatized cellulose that

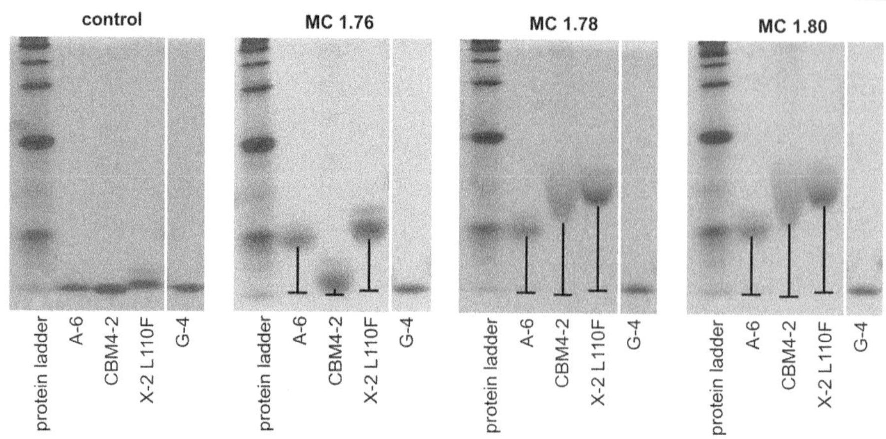

Figure 2 Affinity electrophoresis with three methylcellulose batches. The three batches have similar degree of substitution (1.76, 1.78 and 1.80 and denoted accordingly) however their substitution pattern differs. MC 1.76 is more homogeneously substituted while MC 1.78 and MC 1.80 have substitution patterns that are more heterogeneous. Three CBMs and a negative control (G-4) were run on native acrylamide gels containing one of each methylcellulose batch. CBM4-2 is more retained in the gels containing MC 1.78 or MC 1.80 compared to MC 1.76. A-6 and X-2 L110F are less discriminating and are retained in all gels containing methylcellulose. None of the CBMs were retained in the control gel containing no methylcellulose.

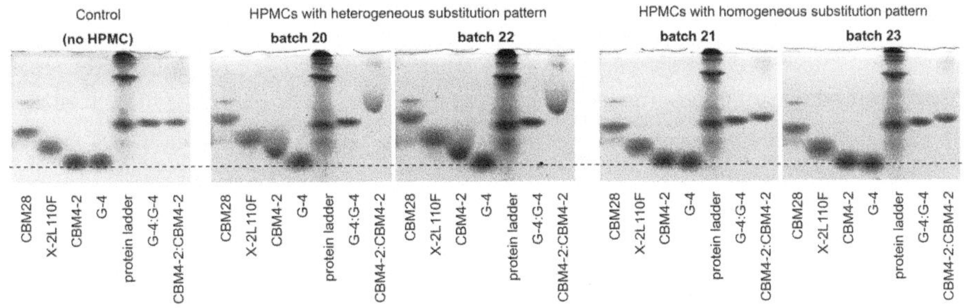

Figure 3 Affinity electrophoresis with four hydroxypropyl methylcellulose batches. Two of the batches (20 and 22) are known to have a heterogeneous substitution pattern and the other two (21 and 23) a homogeneous substitution pattern. Three CBMs (CBM28, X-2 L110F and CBM4-2) and a negative control (G-4) were run in 17% acrylamide gels containing no or 1 mg/ml HPMC. Also a tandem CBM construct of CBM4-2: CBM4-2 together with a tandem negative control (G-4:G-4) were included in the assay. The retardation of the CBMs is modest but differences in mobility can be seen for CBM4-2 and CBM28 in gels containing HPMC of heterogeneous substitution pattern as compared to in gels with HPMC of homogeneous substitution pattern. The difference in retardation is more distinct for the tandem CBM4-2 construct.

carry either evenly or clustered distribution of substituents along the glucosyl backbone.

Discussion

Proof-of-concept - CBM have the potential to characterize cellulose derivatives

Cellulases are commonly employed for the characterization of derivatized celluloses with respect to their substitution pattern. Many of the enzymes used, such as Cel5A, Cel7B and Cel45 from *Trichoderma reseei* [15,30], EGII from *Trichoderma longibrachiatum* [16], Cel5A from *Bacillus agaradhaerens* [24,30,38] or Cel5A from *Humicola insolens* [30] contain one or more carbohydrate-binding modules linked to either the C- or N-terminus of the catalytic module. The goal of this work was to investigate if CBMs on their own have the ability to make a distinction in the distribution pattern of cellulose derivatives based on their substitution pattern.

In this proof-of-concept study we show that CBMs indeed have the potential to be utilized in the characterization of chemically modified celluloses and can, due to the selective binding affinity displayed by the individual CBM, be used as tools to make a distinction between homogeneously and heterogeneously substituted MC and HPMC. The analysis was carried out using AE that provided a simple readout system. More extensive migration retardation of some CBMs, in particular CBM4-2, was observed in gels containing heterogeneously substituted cellulose derivatives. The molecular rationale for this behavior is not fully understood. The batches of cellulose derivatives with heterogeneous dispersed substituents might carry more binding epitopes and/or such of a nature that can create stronger binding affecting the overall CBM-cellulose polymer interaction, reflected in an increased retardation in the AE gels.

Visualization of CBM – carbohydrate interactions with affinity electrophoresis provides a rapid and easy read-out

Affinity gel electrophoresis is a technique that combines the visual simplicity of normal native gel electrophoresis with the specific separation of affinity interactions between various macromolecules. It has been widely used for the binding analysis of various interactions including those between antibody-hapten [39], lectin-glycoprotein [40] and between proteins and nucleic acids [41]. In 1977 AE was used to analyze the binding of a lectin against a soluble polysaccharide [42] and following the discovery of CBMs this protocol was adopted for studying the interaction of CBMs and polysaccharides [43]. Today AE is one of the most applied methods for rapid characterization of CBMs with unknown functions or for determining the role of a given amino acid after mutation [44]. In this study we show that this method not only provides information of the CBM but can also be used for the characterization of the polysaccharides. More specifically, the use of AE together with for example CBM4-2 has the potential to function as an initial screening method for the characterization of the substitution pattern of cellulose derivatives.

Future prospects

To establish a more detailed analysis of the substituent patterns of derivatized celluloses there are multiple possibilities to exploit in terms of a CBM/AE-based assay. An assay of this type may firstly rely on multiple readouts based not only on changes in CBMs' retardation at a fixed carbohydrate gel content but also on their migration behavior at different concentrations of the derivatized carbohydrate. The second important window of opportunity is to be found in our ability to expand the set of CBMs to include an even more selective binding capacity and to create a multiplexed assay employing multiple CBMs. Such CBM candidates may potentially be of natural origin and found linked to cellulases, but new CBMs can also be obtained through engineering, specifically designed to suit this application. Direct evolution and rational design are powerful approaches for the fine-tuning of protein properties, and a combination of the use of combinatorial libraries and phage-display technology has been successful for the generation of engineered CBMs [34] toward other carbohydrate polymers including xylans [45] and different molecular forms of xyloglucan [46,47]. In the present study CBMs engineered for such natural substrates were included but were unable or inefficient in their ability to discriminate between differentially substituted forms of cellulose. These results were not unexpected since the engineered CBM had not been specifically designed for this application but instead for recognition of other carbohydrate targets. In a future assay one could consider simultaneous use of CBMs that not only represent binders that discriminate substitutions, but also CBMs that specifically interact with these chemical groups, permitting a more complete mapping of the derivatized celluloses. We envisage that engineering of modules with predetermined tolerance for substitutions will be a valuable approach in the implementation of an assay for characterization of cellulose derivatives. It is expected, based on past experience on evolution of CBM binding properties [34,45,46,48], that an appropriately designed phage-display selection process will be able to identify variants from combinatorial libraries that are able to differentiate between differently substituted celluloses. Such engineering will improve assay functionality over the current proof of concept study that employed non-optimized CBMs only.

Conclusion

In conclusion, this study demonstrates that some CBMs have the ability to discriminate between batches of cellulose derivatives with differences in substitution homogeneity. We anticipate that CBMs with better capacity to characterize carbohydrate polymers can be identified from natural sources or created *in vitro* and that such modules can be implemented in a non-degrading screening assay that can complement today's standard analysis for use in quality control of batches of industrial cellulose derivatives.

Methods

Carbohydrate-binding modules

CBM4-2 from *Rhodothermus marinus* xylanase 10A and evolved CBM X-2 L110F [35] were produced in *E. coli* T7 Express (New England Biolabs, Ipswich, England) and purified using affinity chromatography columns on a Ni-NTA resin (Qiagen, Hilden, Germany) as previously described [35]. A codon-optimized gene (synthesized by GeneArt, Life Technologies, Carlsbad, CA, USA) coding for amino acids 571–762 of *Bacillus akibai* (also known as *Bacillus sp.* 1139) cellulase 5A, representing CBM28 [33], carried in the pET22b(+) vector (Novagen, Madison, WI, USA) was transformed into *E. coli* T7 Express. Protein production was made in 200 ml 2 x Yeast Tryptone broth supplemented with 100 μg/ml carbenicillin. The bacteria were cultivated at 37°C with shaking (210 rpm) until $OD_{600\ nm}$ reached 0.4. Protein production was induced by addition of 0.5 mM isopropyl-β-D-thiogalactopyranoside followed by a further cultivation at 30°C with shaking (210 rpm) for 4 hours. A-6, which was evolved from CBM4-2 and selected, using phage-display, for its ability to recognize Avicel [34], and G-4, which was also evolved from CBM4-2 but has no carbohydrate-binding activity and is used as negative control [34,36], were produced in *E. coli* BL21(λDE3) as previously described [34].

In an effort to create proteins with stronger binding to cellulose derivatives, we constructed and produced proteins consisting of two CBMs fused together via a peptide linker. Codon-optimized genes encoding such in tandem coupled CBMs carried in the pET22b(+) vector (Novagen) were purchased from GeneArt. The CBMs fused in tandem were produced in *E. coli* T7 Express as described above.

Cellulose derivatives

Three methylcelluloses were used in this study: MC 1.76 (viscosity type 4 cP), MC 1.78 (viscosity type 15 cP) and MC 1.80 (viscosity type 1500) from Shin-Etsu Chemical Co. (Tokyo, Japan) with similar degree of substitution (DS 1.76, 1.78 and 1.80, respectively) but with different molecular weights (20, 40 respectively 200 kDa) as analyzed by Melander et al. [49], Cohen et al. [24] and Fitzpatrick et al. [23]. The substitution patterns of the MCs have previously been characterized. MC 1.76 is more homogeneously substituted while MC 1.78 and MC 1.80 are more heterogeneously substituted [24]. Four hydroxypropyl methylcellulose (HPMC) batches were also analyzed. Of the HPMC used, batches 20 and 22 were from Dow Wolff Cellulosics (Bomlitz, Germany) and 21 and 23 from Shin-Etsu Chemical Co. All four batches had a similar degree of substitution and molecular weight but batches 21 and 23 were less susceptible to enzymatic degradation, performed with Celluclast (Novozymes A/S, Bagsvaerd, Denmark) and have therefore been classified as homogeneously substituted. Batches 20 and 22 on the other hand were more easily degraded thus classified as carrying a more heterogeneous substitution pattern.

Affinity electrophoresis

Affinity electrophoresis was used to determine the ability of soluble, charged, mobile CBMs to interact with largely immobile carbohydrates (and thus to become less mobile) under the influence of an electric field. Polyacrylamide gels for analysis of native proteins were cast according to the Ornstein-Davis method [50,51] with the important modification of the addition of polysaccharides. The retardation of the CBM (3 μg/lane) was investigated using gels containing MC or HPMC at a concentration of 3 and 1 mg/ml, respectively. The gels were polymerized in cassettes (Life Technologies) at an acrylamide/bisacrylamide concentration of 12.5/0.42% for gels containing MC and 17.5/0.58% for gels containing HPMC and run in a minigel system for 1 hour and 45 minutes. Staining of protein was done with SimplyBlue (Life Technologies) for 20 minutes, followed by destaining in water according to the manufacturer's instructions.

Abbreviations

CBM: Carbohydrate-binding module; AE: Affinity electrophoresis; HPMC: Hydroxypropyl methyl cellulose; MC: Methyl cellulose; HPC: Hydroxypropyl cellulose; HEC: Hydroxyethyl cellulose; CMC: Carboxymethyl cellulose; MS: Mass spectrometry; NMR: Nuclear magnetic resonance; MALDI-ToF: Matrix-assisted laser desorption/ionizaion time of flight); ESI: Electrospray ionization.

Competing interest

The authors declare that they have no competing interest.

Authors' contributions

LvS, HS and MO conceived the study; LvS, HS and MO contributed critical, unique research reagents; LvS performed the experiments; LvS, ENK, HS and MO analyzed the results; LvS wrote the manuscript; LvS, ENK, HS and MO approved the final version of the manuscript. All authors read and approved the final manuscript.

Acknowledgements

The study was supported by a grant from the Swedish Research Council (grant number: 621-2009-5152) and a grant from the European Commission (FP7, KBBE.2012.3.2-01, grant number 311932).

Author details

[1]Department of Immunotechnology, Lund University, Lund, Sweden. [2]Department of Biochemistry & Structural Biology, Lund University, Lund, Sweden. [3]Department of Biotechnology, Lund University, Lund, Sweden. [4]Present Addresses: Alligator Bioscience AB, Lund, Sweden. [5]Present Addresses: Camurus AB, Lund, Sweden.

References

1. Kamel S, Ali N, Jahangir K, Shah SM, El-Gendy AA: **Pharmaceutical significance of cellulose: a review.** *Express Polym Lett* 2008, **2**:758–778.
2. Medronho B, Romano A, Miguel MG, Stigsson L, Lindman B: **Rationalizing cellulose (in)solubility: reviewing basic physicochemical aspects and role of hydrophobic interactions.** *Cellulose* 2012, **19**:581–587.
3. Chang CY, Zhang LN: **Cellulose-based hydrogels: present status and application prospects.** *Carbohydr Polym* 2011, **84**:40–53.
4. Maderuelo C, Zarzuelo A, Lanao JM: **Critical factors in the release of drugs from sustained release hydrophilic matrices.** *J Control Release* 2011, **154**:2–19.
5. Tran PH, Tran TT, Park JB, Lee BJ: **Controlled release systems containing solid dispersions: strategies and mechanisms.** *Pharm Res* 2011, **28**:2353–2378.
6. Li CL, Martini LG, Ford JL, Roberts M: **The use of hypromellose in oral drug delivery.** *J Pharm Pharmacol* 2005, **57**:533–546.
7. Gao P, Skoug JW, Nixon PR, Ju TR, Stemm NL, Sung KC: **Swelling of hydroxypropyl methylcellulose matrix tablets. 2. Mechanistic study of the influence of formulation variables on matrix performance and drug release.** *J Pharm Sci* 1996, **85**:732–740.
8. Kavanagh N, Corrigan OI: **Swelling and erosion properties of hydroxypropylmethylcellulose (Hypromellos) matrices - influence of agitation rate and dissolution medium composition.** *Int J Pharm* 2004, **279**:141–152.
9. Reynolds TD, Gehrke SH, Hussain AS, Shenouda LS: **Polymer erosion and drug release characterization of hydroxypropyl methylcellulose matrices.** *J Pharm Sci* 1998, **87**:1115–1123.
10. Novak SD, Kuhelj V, Vrecer F, Baumgartner S: **The influence of HPMC viscosity as FRC parameter on the release of low soluble drug from hydrophylic matrix tablets.** *Pharm Dev Technol* 2013, **18**:343–347.
11. Adden R, Muller R, Mischnick P: **Analysis of the substituent distribution in the glucosyl units and along the polymer chain of hydroxypropylmethyl celluloses and statistical evaluation.** *Cellulose* 2006, **13**:459–476.
12. Enebro J, Momcilovic D, Siika-aho M, Karlsson S: **Liquid chromatography combined with mass spectrometry for the investigation of endoglucanase selectivity on carboxymethyl cellulose.** *Carbohyd Res* 2009, **344**:2173–2181.
13. Mischnick P, Momcilovic D: **Chemical structure analysis of starch and cellulose derivatives.** *Adv Carbohyd Chem Biochem* 2010, **64**:117–210.
14. Schagerlöf H, Richardson S, Momcilovic D, Brinkmalm G, Wittgren B, Tjerneld F: **Characterization of chemical substitution of hydroxypropyl cellulose using enzymatic degradation.** *Biomacromolecules* 2006, **7**:80–85.
15. Viridén A, Larsson A, Schagerlöf H, Wittgren B: **Model drug release from matrix tablets composed of HPMC with different substituent heterogeneity.** *Int J Pharm* 2010, **401**:60–67.
16. Viridén A, Wittgren B, Andersson T, Larsson A: **The effect of chemical heterogeneity of HPMC on polymer release from matrix tablets.** *Eur J Pharm Sci* 2009, **36**:392–400.
17. Viridén A, Larsson A, Wittgren B: **The effect of substitution pattern of HPMC on polymer release from matrix tablets.** *Int J Pharm* 2010, **389**:147–156.
18. Heinze T, Koschella A: **Carboxymethyl ethers of cellulose and starch - a review.** *Macromol Symp* 2005, **223**:13–39.
19. Mischnick P, Heinrich J, Gohdes M, Wilke O, Rogmann N: **Structure analysis of 1,4-glucan derivatives.** *Macromol Chem Physic* 2000, **201**:1985–1995.
20. Tezuka Y, Tsuchiya Y: **Determination of substituent distribution in cellulose acetate by means of a ^{13}C NMR study on its propanoated Derivative.** *Carbohyd Res* 1995, **273**:83–91.
21. Arisz PW, Kauw HJJ, Boon JJ: **Substituent distribution along the cellulose backbone in O-methylcelluloses using GC and FAB-MS for monomer and oligomer analysis.** *Carbohyd Res* 1995, **271**:1–14.
22. Saake B, Horner S, Kruse T, Puls J, Liebert T, Heinze T: **Detailed investigation on the molecular structure of carboxymethyl cellulose with unusual substitution pattern by means of an enzyme-supported analysis.** *Macromol Chem Physic* 2000, **201**:1996–2002.
23. Fitzpatrick F, Schagerlöf H, Andersson T, Richardson S, Tjerneld F, Wahlund KG, Wittgren B: **NMR, cloud-point measurements and enzymatic depolymerization: complementary tools to investigate substituent patterns in modified celluloses.** *Biomacromolecules* 2006, **7**:2909–2917.
24. Cohen AS, Nilsson C, Schagerlöf H, Tjerneld F, Gorton L: **Investigation of the enzyme Bacillus agaradhaerens Cel 5a as an analytical tool in mass spectral characterisation of methylcelluloses.** *Anal Chim Acta* 2006, **561**:16–24.
25. Cuers J, Unterieser I, Burchard W, Adden R, Rinken M, Mischnick P: **Simultaneous determination of substituent patterns in partially acid hydrolyzed O-Me/O-Me-d(3)-cellulose and quantification of the obtained oligomers by HPLC-ESI-MS.** *Carbohyd Res* 2012, **348**:55–63.
26. Adden R, Melander C, Brinkmalm G, Gorton L, Mischnick P: **New approaches to the analysis of enzymatically hydrolyzed methyl cellulose. Part 1. Investigation of the influence of structural parameters on the extent of degradation.** *Biomacromolecules* 2006, **7**:1399–1409.
27. Unterieser I, Mischnick P: **Labeling of oligosaccharides for quantitative mass spectrometry.** *Carbohyd Res* 2011, **346**:68–75.
28. Greiderer A, Steeneken L, Aalbers T, Vivo-Truyols G, Schoenmakers P: **Characterization of hydroxypropylmethylcellulose (HPMC) using comprehensive two-dimensional liquid chromatography.** *J Chromatogr A* 2011, **1218**:5787–5793.
29. Schagerlöf H, Johansson M, Richardson S, Brinkmalm G, Wittgren B, Tjerneld F: **Substituent distribution and clouding behavior of hydroxypropyl methyl cellulose analyzed using enzymatic degradation.** *Biomacromolecules* 2006, **7**:3474–3481.

30. Schagerlöf U, Schagerlöf H, Momcilovic D, Brinkmalm G, Tjerneld F: **Endoglucanase sensitivity for substituents in methyl cellulose hydrolysis studied using MALDI-TOFMS for oligosaccharide analysis and structural analysis of enzyme active sites.** *Biomacromolecules* 2007, **8**:2358–2365.
31. Boraston AB, Bolam DN, Gilbert HJ, Davies GJ: **Carbohydrate-binding modules: fine-tuning polysaccharide recognition.** *Biochem J* 2004, **382**:769–781.
32. Simpson PJ, Jamieson SJ, Abou-Hachem M, Karlsson EN, Gilbert HJ, Holst O, Williamson MP: **The solution structure of the CBM4-2 carbohydrate binding module from a thermostable Rhodothermus marinus xylanase.** *Biochemistry* 2002, **41**:5712–5719.
33. Boraston AB, Ghaffari M, Warren RAJ, Kilburn DG: **Identification and glucan-binding properties of a new carbohydrate-binding module family.** *Biochem J* 2002, **361**:35–40.
34. Cicortas-Gunnarsson L, Nordberg-Karlsson E, Albrekt AS, Andersson M, Holst O, Ohlin M: **A carbohydrate binding module as a diversity-carrying scaffold.** *Protein Eng Des Sel* 2004, **17**:213–221.
35. von Schantz L, Håkansson M, Logan DT, Walse B, Österlin J, Nordberg-Karlsson E, Ohlin M: **Structural basis for carbohydrate-binding specificity-A comparative assessment of two engineered carbohydrate-binding modules.** *Glycobiology* 2012, **22**:948–961.
36. Cicortas-Gunnarsson L, Dexlin L, Nordberg-Karlsson E, Holst O, Ohlin M: **Evolution of a carbohydrate binding module into a protein-specific binder.** *Biomol Eng* 2006, **23**:111–117.
37. Bolam DN, Xie H, White P, Simpson PJ, Hancock SM, Williamson MP, Gilbert HJ: **Evidence for synergy between family 2b carbohydrate binding modules in Cellulomonas fimi xylanase 11A.** *Biochemistry* 2001, **40**:2468–2477.
38. Momcilovic D, Schagerlöf H, Rome D, Jornten-Karlsson M, Karlsson KE, Wittgren B, Tjerneld F, Wahlund KG, Brinkmalm G: **Derivatization using dimethylamine for tandem mass spectrometric structure analysis of enzymatically and acidically depolymerized methyl cellulose.** *Anal Chem* 2005, **77**:2948–2959.
39. Tanaka T, Suzuno R, Nakamura K, Kuwahara A, Takeo K: **Thermodynamic analysis of the interactions of a mouse dinitrophenyl-specific myeloma protein, MOPC-315, with immobilized dinitrophenyl and trinitrophenyl ligands by affinity electrophoresis.** *Electrophoresis* 1986, **7**:204–209.
40. Nakamura S, Tanaka K, Murakawa S: **Specific protein of legumes which reacts with animal proteins.** *Nature* 1960, **188**:144–145.
41. Takeo K: **Affinity electrophoresis - principles and applications.** *Electrophoresis* 1984, **5**:187–195.
42. Horejsi V, Ticha M, Kocourek J: **Studies on lectins. XXXI. Determination of dissociation constants of lectin sugar complexes by means of affinity electrophoresis.** *Biochim Biophys Acta* 1977, **499**:290–300.
43. Tomme P, Boraston A, Kormos JM, Warren RAJ, Kilburn DG: **Affinity electrophoresis for the identification and characterization of soluble sugar binding by carbohydrate-binding modules.** *Enzyme Microb Tech* 2000, **27**:453–458.
44. Abbott DW, Boraston AB: **Quantitative approaches to the analysis of carbohydrate-binding module function.** *Methods Enzymol* 2012, **510**:211–231.
45. Cicortas-Gunnarsson L, Montanier C, Tunnicliffe RB, Williamson MP, Gilbert HJ, Nordberg-Karlsson E, Ohlin M: **Novel xylan-binding properties of an engineered family 4 carbohydrate-binding module.** *Biochem J* 2007, **406**:209–214.
46. Cicortas Gunnarsson L, Zhou Q, Montanier C, Nordberg Karlsson E, Brumer H III, Ohlin M: **Engineered xyloglucan specificity in a carbohydrate-binding module.** *Glycobiology* 2006, **16**:1171–1180.
47. von Schantz L, Håkansson M, Logan DT, Nordberg Karlsson E, Ohlin M: **Carbohydrate binding module recognition of xyloglucan defined by polar contacts with branching xyloses and CH-π interactions.** *Proteins* 2014, **82**:3466–3475
48. von Schantz L, Gullfot F, Scheer S, Filonova L, Cicortas Gunnarsson L, Flint JE, Daniel G, Nordberg-Karlsson E, Brumer H, Ohlin M: **Affinity maturation generates greatly improved xyloglucan-specific carbohydrate binding modules.** *BMC Biotechnol* 2009, **9**:92.
49. Melander C, Momcilovic D, Nilsson C, Bengtsson M, Schagerlöf H, Tjerneld F, Laurell T, Reimann CT, Gorton L: **Microchip immobilized enzyme reactors for hydrolysis of methyl cellulose.** *Anal Chem* 2005, **77**:3284–3291.
50. Ornstein L: **Disc electrophoresis - I Background and theory.** *Ann NY Acad Sci* 1964, **121**:321.
51. Davis BJ: **Disc electrophoresis - II Method and application to human serum proteins.** *Ann NY Acad Sci* 1964, **121**:404–427.

12

Improving the large scale purification of the HIV microbicide, griffithsin

Joshua L Fuqua[1,2*], Valentine Wanga[3] and Kenneth E Palmer[1,2*]

Abstract

Background: Griffithsin is a broad spectrum antiviral lectin that inhibits viral entry and maturation processes through binding clusters of oligomannose glycans on viral envelope glycoproteins. An efficient, scaleable manufacturing process for griffithsin active pharmaceutical ingredient (API) is essential for particularly cost-sensitive products such as griffithsin -based topical microbicides for HIV-1 prevention in resource poor settings. Our previously published purification method used ceramic filtration followed by two chromatography steps, resulting in a protein recovery of 30%. Our objective was to develop a scalable purification method for griffithsin expressed in *Nicotiana benthamiana* plants that would increase yield, reduce production costs, and simplify manufacturing techniques. Considering the future need to transfer griffithsin manufacturing technology to resource poor areas, we chose to focus modifying the purification process, paying particular attention to introducing simple, low-cost, and scalable procedures such as use of temperature, pH, ion concentration, and filtration to enhance product recovery.

Results: We achieved >99% pure griffithsin API by generating the initial green juice extract in pH 4 buffer, heating the extract to 55°C, incubating overnight with a bentonite $MgCl_2$ mixture, and final purification with Capto™ multimodal chromatography. Griffithsin extracted with this protocol maintains activity comparable to griffithsin purified by the previously published method and we are able to recover a substantially higher yield: 88 ± 5% of griffithsin from the initial extract. The method was scaled to produce gram quantities of griffithsin with high yields, low endotoxin levels, and low purification costs maintained.

Conclusions: The methodology developed to purify griffithsin introduces and develops multiple tools for purification of recombinant proteins from plants at an industrial scale. These tools allow for robust cost-effective production and purification of griffithsin. The methodology can be readily scaled to the bench top or industry and process components can be used for purification of additional proteins based on biophysical characteristics.

Keywords: Griffithsin, Tobacco mosaic virus, Protein purification, *N. benthamiana*, Bentonite, Microbicide, HIV

Background

Griffithsin (GRFT) is a high mannose targeting lectin that has shown promise as a microbicide candidate against viruses with oligomannose-rich envelope glycoproteins, such as HIV-1 or HSV-2 [1-4]. GRFT was originally isolated from a red alga (*Griffithsia* sp.) with native and recombinant forms showing similar antiviral activity against laboratory strains and primary isolates of HIV-1 [1,4]. Recombinant forms of GRFT have been successfully produced in both *E. coli* and *N. benthamiana*, with the latter used to produce multi-gram quantities of functional GRFT [4,5]. GRFT has shown to be a safe and effective microbicide against HIV, but needs to be evaluated in clinical trials [2,6]. The HIV epidemic is prevalent in many resource-limited or developing countries and the availability of indigenous manufacturing systems for GRFT in those areas could facilitate large-scale rollout of microbicide product. Cost and production complexity are factors that must be considered when manufacturing biologics for extremely cost-sensitive markets, such as for microbicides in developing economies.

This communication details our development of a process to reduce the production cost of GRFT from *N. benthamiana*. Currently published industrial scale GRFT manufacturing uses ceramic filtration followed by two-

* Correspondence: j.fuqua@louisville.edu; kenneth.palmer@louisville.edu
[1]Owensboro Cancer Research Program, 1020 Breckenridge St., Suite 201, Owensboro, KY 42303, USA
Full list of author information is available at the end of the article

stage chromatography [4]. These relatively complex methods require specialized equipment and increase production costs. The molecule's biophysical properties, and specifically the melting temperature, were taken into account when developing a new purification methodology. GRFT has a melting temperature of 78.8°C and remains functionally stable at temperatures not tolerated by many of the contaminating proteins [7,8]. There are three primary contaminants to contend with when trying to purify GRFT from *N. benthamiana*: two production host-associated, the small and large subunits of Ribulose-1,5-bisphosphate carboxylase/oxygenase (RuBisCO), and one contaminant from the TMV production system, TMV coat protein (CP). RuBisCO lacks the thermostability of GRFT at acidic pHs. Therefore, in these studies we tested heat treatment, under acidic conditions, of our crude plant extracts containing GRFT to determine if any of the major contaminants could be removed through precipitation/denaturation. The TMV CP has shown a propensity to self-aggregate under certain ionic, pH, and temperature conditions. Absorption by the clay material, bentonite, has been previously used to purify TMV particles and could potentially be a useful tool in removal of the coat protein contaminant [9]. Therefore purification methods utilizing altered heat, pH, ion concentrations, and bentonite filtration were all tested in an effort to develop a scalable purification method requiring less specialized equipment. We have developed a single-step-chromatography purification method that will decrease the cost and complexity of producing large quantities of GRFT, while increasing the protein recovery. The application of this knowledge on industrial size purification should further reduce GRFT cost of production and may be beneficial for production of other plant-made pharmaceuticals.

Results and discussion

Leaf punches were used to determine the pH and NaCl concentration parameters that were optimal for extraction of GRFT and minimal extraction of contaminating proteins. One dimensional SDS PAGE profiles showed the optimal extraction pH for GRFT to be pH 4.0. The addition of 100 mM NaCl showed minimal impact on the purity of GRFT in the small-scale samples, but from prior work it is understood that the addition of NaCl will improve purity as long as it is not deleterious to the protein of interest especially as scale is increased. Therefore all subsequent purifications used 100 mM sodium acetate and 300 mM NaCl pH4 as the extraction buffer. Analysis of the small scale protein extraction profiles by SDS-PAGE demonstrates little impact of temperature on the extract purity, but does show that GRFT maintains stability in temperatures ranging from 60 – 80°C. GRFT's thermostability in heated extracts was further assessed in larger scale extractions and Figure 1 shows clearly the impact that heat has on contaminating proteins. Nearly all contaminating proteins aggregated at temperatures above 50°C (Figure 1), increasing the extract purity 30 – 40% (Figure 1), while GRFT remains stable until the extract reaches 90°C.

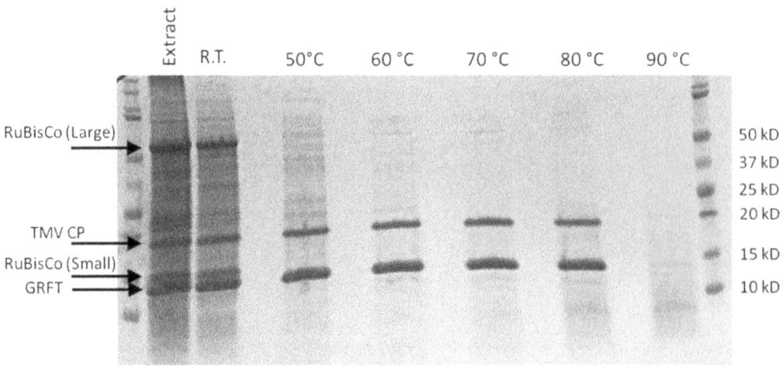

	Extract	R.T.	50°C	60°C	70°C	80°C	90°C
TMV CP Content (%)	100	81	56	62	71	54	6
GRFT Content (%)	100	82	81	87	93	88	14
GRFT Purity (%)	14	19	48	60	57	60	33

Figure 1 Removal of plant protein contaminants from GRFT Extract with Heat. Plants expressing GRFT through a TMV-based expression system were harvested at pH4 in sodium acetate buffer and the extract was divided into 7 aliquots. Each aliquot was incubated for 15 minutes at temperatures ranging from room temperature (24°C) to 90°C and centrifuged to remove any precipitant. Samples of the initial extract and all assessed temperatures were analyzed by SDS-PAGE with coomassie staining. The initial extract contains multiple contaminating proteins, which remain after incubation at room temperature. Once the extracts are heated to at least 50°C the majority of contaminating proteins precipitate except for the target protein GRFT and the TMV coat protein. The composition of the extract remains similar with temperature treatments ranging from 50 – 80°C, but at 90°C incubation nearly all proteins are precipitated. Table: Densitometry readings corresponding to SDS-PAGE shown in Figure 1.

The primary contaminating protein left in the GRFT extract after heating was the TMV CP. Therefore, methods for removing the TMV CP were examined. These included the addition of bentonite and/or $MgCl_2$, as well as pH and temperature adjustments. The more acidic extracts pH 4.0 and pH 6.5 showed increased activity of bentonite and $MgCl_2$ for removal of CP relative to the pH 7.0, 8.0, or 9.0 extracts, as long as extracts were incubated at 4°C (Figure 2). The pH 6.5 extract in the presence of bentonite and $MgCl_2$ reduced TMV CP levels by ~55% (Figure 2), but the pH 4.0 extract in the presence of bentonite and $MgCl_2$ reduced CP contamination by ~75%. Therefore, pH 4.0 and incubation at 4°C was used in subsequent purification with bentonite and $MgCl_2$, but the most effective concentrations of $MgCl_2$ and bentonite remained to be determined.

Multiple concentrations of bentonite and $MgCl_2$ were investigated in pH 4.0 buffer to determine the optimal concentration for removal of TMV CP. Bentonite was added to GRFT extracts at a range of 0 mg/mL to 10 mg/mL and $MgCl_2$ was added in a range of 0 – 0.1 M with each concentration tested in combination and separately to determine the most effective concentration to aggregate TMV. Figure 3 shows that 0.1 M $MgCl_2$ had a similar impact on TMV CP removal at all bentonite concentrations tested, with bentonite having a minimal impact on TMV levels in the absence of $MgCl_2$. Removal of 75 – 95% of TMV CP was observed with the addition of 0.1 M $MgCl_2$ (Figure 3), but bentonite in the absence of $MgCl_2$ only removed 10 – 30% of the CP contaminant. The addition of bentonite and $MgCl_2$ at their most effective concentrations,

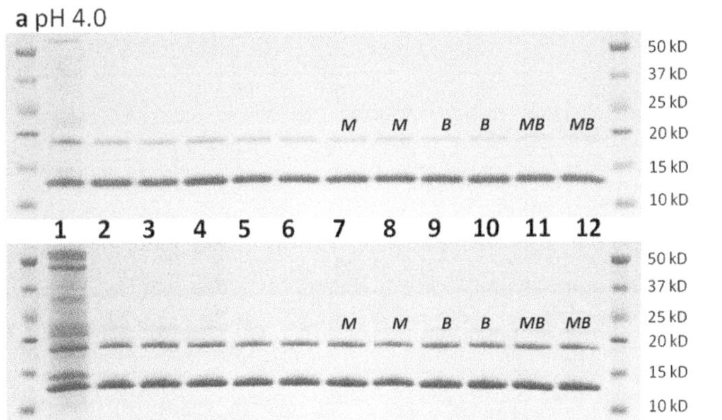

		1	2	3	4	5	6	7	8	9	10	11	12
		Extract	55°C	pH Adjust	Clarified	42°C incubation		42°C M(0.01M)		4°C B(1mg/mL)		4°C M(0.01M) B(1mg/mL)	
pH 4.0	TMV CP	123	70	67	100	52	54	38	44	32	34	27	19
	GRFT	70	77	74	100	82	80	83	81	81	85	83	80
pH 6.5	TMV CP	94	101	82	100	48	50	50	50	53	48	52	41
	GRFT	85	122	95	100	89	91	87	89	94	90	90	87
pH 7.0	TMV CP	241	98	103	100	102	81	108	110	114	107	118	104
	GRFT	125	98	106	100	107	90	106	109	108	109	92	102
pH 8.0	TMV CP	208	105	106	100	97	101	100	96	91	95	100	83
	GRFT	115	98	102	100	102	104	102	102	104	106	106	94
pH 9.0	TMV CP	162	96	103	100	90	87	97	81	93	93	90	91
	GRFT	104	92	101	100	93	96	102	92	99	99	97	99

Figure 2 Effects of pH on the activity of $MgCl_2$ and bentonite to remove the TMV coat protein. Extracts were harvested in sodium acetate (pH 4–5) or tris (pH7-9) buffer depending on the desired pH range and were visualized using coomassie stained SDS-PAGE. Sample 1 shows the proteins in the initial extract and sample 2 represents the proteins after the extract is heated to 55°C and centrifuged. Sample 3 represents the extract after it was adjusted to its corresponding pH **a)** pH 4.0 and **b)** pH 9.0. Sample 3 extracts were centrifuged to remove additional precipitate and represented in sample 4. Samples 5–12 represent various temperature, bentonite, or $MgCl_2$ treatments of sample 4 performed in duplicate. $MgCl_2$ (M), bentonite (B), and bentonite & $MgCl_2$ (MB) Samples 5 & 6 were incubated overnight at 42°C. Samples 7 & 8 were incubated overnight at 42°C after treatment with 0.01 M $MgCl_2$. Samples 9 & 10 were incubated overnight at 4°C with the addition of 1 mg/mL bentonite. Samples 11 & 12 were incubated overnight at 4°C after the addition of 1 mg/mL bentonite & 0.01 M $MgCl_2$. Table: Densitometry readings of SDS-PAGE represented in Figure 2 with values from three additional gels representing additional pH values that were explored (pH 6.5, 7.0, 8.0). The shaded area represents an extract condition achieving greater than 80% purity of GRFT.

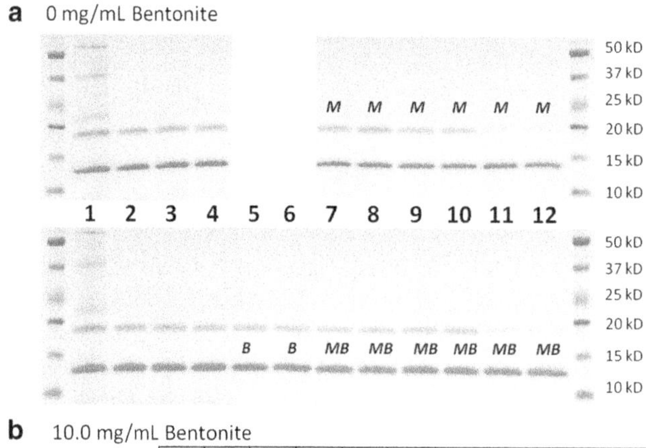

		1	2	3	4	5	6	7	8	9	10	11	12
		Extract	Clarified	0.0 M $MgCl_2$ 0 mg/mL Bentonite		0.0 M $MgCl_2$		0.001M $MgCl_2$		0.01M $MgCl_2$		0.1M $MgCl_2$	
0 mg/mL Bentonite	TMV CP	102	100	93	92	----	----	68	76	62	63	2	0
	GRFT	99	100	101	103	----	----	89	93	92	89	96	82
0.1 mg/mL Bentonite	TMV CP	103	100	95	87	92	81	93	75	70	83	14	7
	GRFT	88	100	91	99	91	85	94	86	99	94	94	86
1.0 mg/mL Bentonite	TMV CP	107	100	78	72	75	74	77	83	66	92	12	24
	GRFT	91	100	91	94	89	87	89	90	104	104	100	96
10.0 mg/mL Bentonite	TMV CP	122	100	103	89	89	70	77	91	97	97	28	20
	GRFT	100	100	101	97	99	90	96	99	99	105	101	99

Figure 3 Effects of $MgCl_2$ (0, 0.001 M, 0.01 M, 0.1 M) on the protein content of GRFT extracts treated with bentonite. Coomassie stained SDS-PAGE was used to see gross changes in purity of the GRFT extracts in both **a** and **b**. The initial GRFT extract is represented in sample 1 and subsequent heat & centrifugation step at sample 2. The supernatant was treated with multiple $MgCl_2$ concentrations (samples 5–12) in duplicate with the corresponding bentonite concentration 0 mg/mL bentonite **(a)** or 10 mg/mL bentonite **(b)**. The resulting supernatants of overnight treatment with $MgCl_2$ and/or bentonite are as follows; (3&4) 0 $MgCl_2$ & 0 bentonite, (5&6) 0 $MgCl_2$, (7&8) 0.001 M $MgCl_2$, (9&10) 0.01 M $MgCl_2$, (11&12) 0.1 M $MgCl_2$. Samples 11 and 12 both show a dramatic reduction in coat protein after treatment with 0.1 M $MgCl_2$ irrespective of the bentonite concentration. Table: Densitometry readings of SDS-PAGE represented in **a** and **b** including two additional bentonite concentrations not shown. The shaded area represents an extract condition achieving greater than 80% purity of GRFT.

10 mg/mL and 0.1 M, respectively, failed to be additive in their effects on TMV CP removal. The largest reduction in TMV CP levels was observed in the absence of bentonite with the addition of 0.1 M $MgCl_2$, reducing TMV CP contamination by greater than 95% (Figure 3). Bentonite has been proven to help in the clarification process of plant extracts and we continued to explore its impact on purification, but maintain that it has minimal impact on TMV CP removal from pH 4.0 extracts.

Since the goal was to provide an alternative, less expensive, and less complex purification method the steps outlined thus far were implemented on a larger scale with fewer variables explored. GRFT was extracted and aliquoted after temperature treatment. GRFT extract aliquots were untreated (control) or treated with 0.1 M $MgCl_2$ and/or 10 mg/mL bentonite and all were incubated overnight at 4°C. Extracts were compared via SDS-PAGE with densitometry and again showed that the greatest reduction of TMV occurred in the 0.1 M $MgCl_2$ treatment group (Figure 4A). The pellets correlating to the specific treatments contained large amounts of TMV CP likely through forced aggregation and removal during centrifugation. Minimal loss of functional GRFT is observed in any of the treatment groups when compared via HIV-1 gp120 binding ELISAs (Figure 4B). When western blots targeting TMV CP were quantified by densitometry a reduction of 65% was observed in extracts treated with $MgCl_2$, which is significantly decreased compared to the bentonite-only treated extract (Figure 4C). The table in Figure 4 shows GRFT purity in the four extracts ranges from 64 – 82% with the TMV content of extracts treated with $MgCl_2$ reduced by as much as 66%.

In an effort to achieve greater than 95% purity of GRFT a chromatography step was explored. Extracts were heated to 55°C and treated with $MgCl_2$ or $MgCl_2$ and bentonite overnight after which they were filtered through a 0.2 μm filter and loaded onto a Capto® MMC

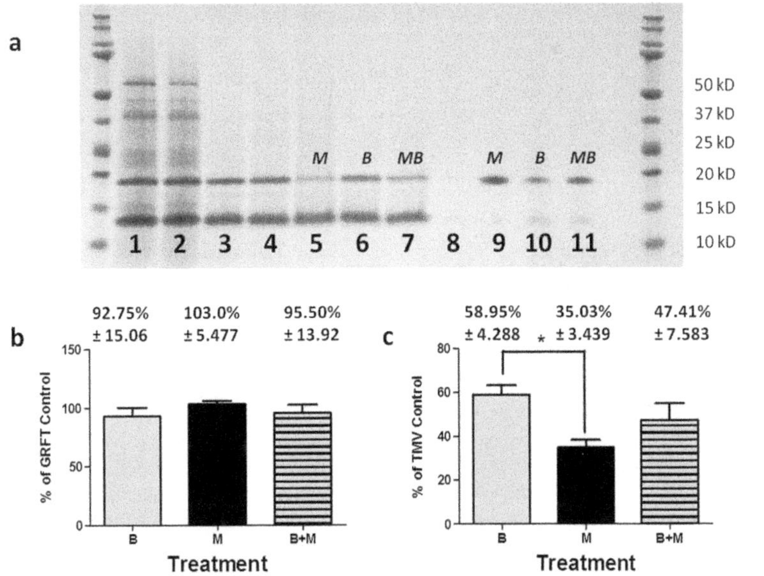

	1	2	3	4	5	6	7	8	9	10	11
	Extract	pH	Clarified	4°C	$MgCl_2$	Bentonite	$MgCl_2$ Bentonite	4°C	$MgCl_2$	Bentonite	$MgCl_2$ Bentonite
TMV CP	131	132	100	94	36	81	55	3	79	34	50
GRFT	119	111	100	96	95	97	82	3	2	6	4
Overall GRFT Purity	33	36	64	64	82	68	72	-	-	-	-

Figure 4 Summary of the effects of 0.1 M $MgCl_2$ and 10 mg/mL bentonite on GRFT extract purity. a) Coomassie stained SDS-PAGE showing the initial extraction of GRFT (1), subsequent pH adjustment to pH4 (2) and heating to 55°C followed by centrifugation (3). The resulting supernatant was untreated (control) or treated with $MgCl_2$ and/or bentonite and stirred overnight at 4°C. The following samples are the resulting supernatants; (4) Control –4°C, (5) 0.1 M $MgCl_2$, (6) 10 mg/mL Bentonite, (7) 0.1 M $MgCl_2$ & 10 mg/mL Bentonite . The following samples are the resulting pellets after treatment; (8) Pellet - 4°C, (9) Pellet - 0.1 M $MgCl_2$, (10) Pellet - 10 mg/mL Bentonite, (11) Pellet - 0.1 M $MgCl_2$ & 10 mg/mL Bentonite. **b)** Comparison of gp-120 binding GRFT concentrations relative to the control sample (4). Determined through a gp-120 binding ELISA and represented as a percentage of functional GRFT in the control extract. **c)** Comparison of the TMV coat protein concentration relative to the control extract. Determined through densitometry measures of TMV specific westerns and represented as a percentage of TMV in the control extract. b and c both analyzed by a one-way ANOVA with Bonferroni's multiple comparison test comparing all groups. * represents a post-hoc test with a p-value < 0.05. Table: Densitometry readings of SDS-PAGE represented in a. The shaded area represents an extract condition achieving greater than 80% purity of GRFT.

column. The Capto® MMC resin is a multimodal cation exchanger that does not require buffer exchange and was chosen to help minimize purification steps. After chromatography trace amounts of TMV can be seen in SDS gels and western blots in both the untreated-control and the $MgCl_2$ treated samples with no TMV contamination observed in the $MgCl_2$ and bentonite treated extract (Figure 5). Purity of all extracts after Capto MMC chromatography was greater than 97%, with the bentonite and $MgCl_2$ treatment group showing a purity of >99% by SDS-PAGE densitometry. The recovery levels and activity of the purified GRFT were assessed by gp120 binding ELISAs with very good recovery levels achieved in all treatments undergoing MMC chromatography. A minimum of 84.5% was recovered in the $MgCl_2$ treated extract after it was purified by chromatography and a maximum recovery of 92.1% was observed in the control extract that was purified by chromatography (Table 1). All extracts maintained an activity level at least 80% of GRFT purified by the previously published method [4], with GRFT purified by $MgCl_2$ & bentonite having 100.8 ± 4.33% of the activity (Table 1). In summary, purification of GRFT from *N.*

Table 1 Quantification of activity and recovery of GRFT from the plant extracts

Purification method	% gp-120 Binding	% GRFT Recovery
Heat and Chromatography	90.19 ± 1.53	92.13 ± 3.29
Heat, $MgCl_2$, Chromatography	80.25 ± 5.08	84.54 ± 6.84
Heat, $MgCl_2$ & Bentonite, Chromatography	100.8 ± 4.33	87.73 ± 4.82

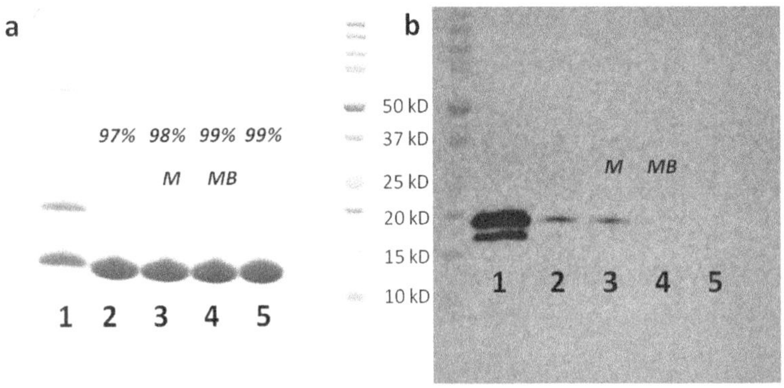

Figure 5 Purity and TMV CP content of GRFT extracts after Capto ™ MMC purification. SDS –PAGE **(a)** and TMV western blots **(b)** were used to visualize purity and TMV contamination. The initial extract (1) contains both GRFT and TMV CP as well as other plant protein contaminants. Lanes 2–4 were loaded with 15 μg of chromatography purified protein from the extracts that had previously been heated and treated with: control – no treatment (2), 0.1 M $MgCl_2$ (3), 0.1 M $MgCl_2$ & 10 mg/mL bentonite (4). Included for comparison was 15 μg of GRFT purified by the previously published methodology O'Keefe *et al.*, [4] (5). Above each lane is the densitometry determined GRFT purity.

benthamiana by heating to 55°C, overnight treatment with $MgCl_2$ & bentonite, and Capto® MMC chromatography recovered 88% of the GRFT with >99% purity and 100% activity of the original purification method.

Utilizing the information yielded from small scale experiments GRFT was purified at an industrial pilot scale (>5 kg of plant material) in a GMP ready facility. At this scale it was necessary to replace all centrifugation steps with a filter press because of the impracticality of centrifugation at scale. Ten liters of extraction buffer was added to 5 kg of *N. benthamiana* and the tissue was ground to yield 11.4 L of initial extract containing 2599.73 mg of GRFT for a total of 519.9 mg of GRFT per kg of plant material (Table 2). The extract was pH adjusted, heated, and filter pressed, which removed TMV CP and purified GRFT to >90% (Figure 6). The remaining small protein contaminants were removed by overnight incubation with 10 mg/mL of bentonite and 0.1 M $MgCl_2$ and subsequent filter pressing. The extract underwent filter press and sterile filtration simultaneously and at this stage was >99% pure, and sterile with only minimal color. It should be noted that the purification process could be halted at this point depending on the application of the process and necessity for >99% pure colorless product. Capto® MMC chromatography was used to remove the color and simultaneously affect a buffer exchange into phosphate buffered saline pH7.4. Purity was maintained with greater than 99% purity of the final product and recovery of 66–72.25%, determined by densitometry of SDS-PAGE (Figure 6) and gp120 ELISA respectively (Table 2). Because of the GMP-ready nature of the purification scheme, endotoxin levels were assessed and found to be 0.20 EU/mL of final product with a GRFT concentration of 1.20 mg/mL, which should be sufficiently low depending on the dosage and administration route (Table 3). In summary, the purification method developed for GRFT herein provided a >99% pure product with >66% recovery while maintaining low endotoxin levels at industrial pilot scale and could be further scaled without modification to the process.

Table 2 gp120 ELISA assessed GRFT recovery during industrial pilot purification

Sample	Green juice	Filter press #1	Filter press #2	Eluate
GRFT (mg)	2599.73	1816.34	1576.52	1543.57
% of GJ		79%	68%	66%

Conclusions

GRFT has been described as a promising topical microbicide component with potential to neutralize a broad range of viruses including HIV-1, Herpes simplex virus type 2, Japanese encephalitis virus, Hepatitis C virus, and SARS-CoV [1,2,10-12]. The safety and efficacy profile of GRFT as an HIV entry inhibitor was demonstrated in rabbits, on human cells, and in a sensitive murine HSV-2 susceptibilty assay [2,4,6]. A comprehensive set of highly promising data confirming the potential for use of GRFT in HIV prevention and even treatment, has led to considerable enthusiasm from policy makers and funders in bringing GRFT-based topical microbicides to clinical trial [2,4,6,13-16]. We have been motivated to improve and simplify the GRFT extraction protocol we initially developed [4] in anticipation of future need to manufacture extremely large amounts of this protein, preferably in geographic areas where the HIV-1 epidemic has the

Table 3 Endotoxin levels in industrial pilot purified GRFT

	GRFT Concentration (mg/mL)	Endotoxin (EU/mL)
GRFT Final	1.20	0.204

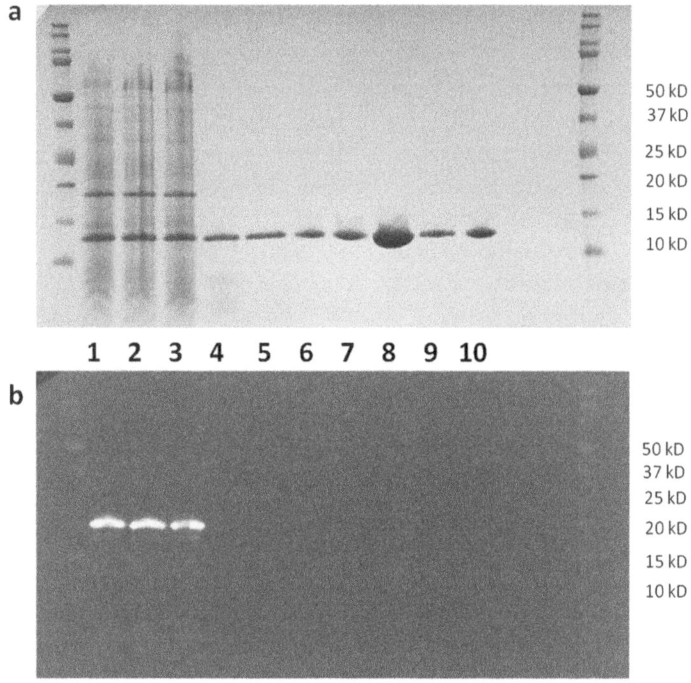

Lane #	1	2	3	4	5	6	7	8	9	10
GRFT Purity %	32	35	25	94	>99	>99	>99	>99	>99	>99
TMV CP %	17	19	13	ND	ND	ND	ND	ND	ND	ND

Figure 6 Analysis of industrial pilot scale purification of GRFT. Representation of the purity, by SDS -PAGE **a,** and TMV CP content, by western blot **b,** of GRFT during industrial pilot scale purification. Sample volumes loaded on the gel were normalized to the initial extraction volume. The initial extract (1) contains both GRFT and TMV CP as well as other plant protein contaminants. The extract was pH adjusted to pH4 (2)heated (3) and filter pressed(4). After which the extract was incubated overnight with $MgCl_2$ and bentonite and simultaneousely filter pressed and sterile filtered (5). Samples 6–7 are the GRFT final product At 1x volume, 2xvolume and 15 µg of protein. Included for comparison was 1.5 µg and 3.0 µg of GRFT purified by the previously published methodology O'Keefe *et al.,* [4] (5). The table represents the densitometric measures of GRFT and TMV protein content in the extracts.

most severe population-level impact. The infectious TMV vector system we use to produce GRFT requires only modest investment in plant controlled environment growth facilities, and does not require use of sophisticated *Agrobacterium* infiltration methods that are increasingly used in plant-made pharmaceuticals research. Beyond the additional regulatory concerns associated with *Agrobacterium*-launch vector systems, we have found that unacceptable levels of bacterial endotoxin contaminate GRFT extracts when these are used to express GRFT (unpublished data). Endotoxin removal after *Agroinfiltration* inserts yet another costly (and relatively inefficient) step in protein extraction, making the infectious TMV vectored expression approach more appealing for the extremely cost-sensitive manufacturing of topical microbicide components. Of course, the TMV vectored approach we use does present a downside in the need to efficiently remove the contaminating virus and CP from the extracts. The methodologies we describe in this communication show facile methods for reduction of TMV CP.

We previously produced GRFT in multi-gram quantities in *N. benthamiana*, but the purification methodology used required specialized ceramic filtration devices, two chromatography techniques, ultrafiltration-diafiltration and had relatively low recovery from infected extracts, 30% possibly increasing to 50% [4]. Of the purification techniques we describe here, using a combination of heat (55°C), $MgCl_2$ & bentonite, and a single chromatographic step were most effective. This three step approach produced GRFT with no loss in activity or purity, while recovering nearly 90% of the extracted GRFT. Each step targets various contaminating proteins within the plant extract. Heating the extract at 55°C removes the majority of the native plant proteins, which lack the thermostability of GRFT. Exposing the extract to a combination of $MgCl_2$ and bentonite removes the TMV CP contaminant, originating from the TMV expression vector. Bentonite and $MgCl_2$ work in tandem, with bentonite known to strip coat protein from the TMV RNA and potentially the $MgCl_2$ causing either structural shifts in the RNA causing aggregation or cleaving the RNA causing aggregation [9,17,18]. Either mechanism is plausible for the aggregation of TMV in the plant extract. Our data shows the necessity of both $MgCl_2$ and bentonite to be present for optimal TMV removal when coupling to a single chromatography step. Utilizing Capto® MMC resin allows for removal of trace protein impurities by binding GRFT to the resin and eluting based upon the isoelectric point of the protein with unobservable amounts of TMV left in extracts treated with bentonite and $MgCl_2$. Stripping of the CP from TMV likely reduces its ability to bind Capto® MMC resin and allows for less TMV contaminant binding and co-elution. All of the techniques, heating, cooling, pH adjustment, and chemical additives, used to purify GRFT are all facile, scaleable and amenable to technology transfer to facilities in resource poor countries. In an effort to demonstrate the scaleability and and amenability of technology transfer the process was performed at pilot scale in a GMP ready industrial facility with only slight reductions in the recovery, while providing pure active GRFT with low endotoxin levels. The methodology maintains increased recovery at scale, reducing per unit costs of GRFT, which should increase the availability of GRFT for HIV entry-inhibition in resource limited settings.

The techniques discussed here can be applied to other recombinant protein production methods to address similar protein contaminants such as native plant proteins, RuBisCO, and TMV CP. Recombinant proteins purified from transgenic or transiently expressing plants both can utilize portions of the methodology to remove contaminating proteins. It is possible that the manufacturing process could be further abbreviated with the use of transgenically expressed GRFT.

In conclusion, purification of GRFT at scale using this technique should provide multi-kilogram quantities of GRFT at a low cost with minimal product loss suitable for practical development of GRFT-based topical microbicides.

Methods

GRFT expression

GRFT was expressed in *N. benthamiana* using the same plasmid and methodology described in O'Keefe et al., [4]. Briefly, a TMV-based expression vector system containing the GRFT amino acid sequence was used to generate infectious RNA transcripts. Infectious transcripts were inoculated on *N. benthamiana* (~24 days post sowing) by gentle abrasion.

GRFT extraction – leaf punches

Ten days after inoculation, 8 mm leaf punches were randomly taken from eight *N. benthamiana* plants. Two leaf punches were homogenized in 200 μL of each extraction buffer. Extraction buffers ranged from pH 2–10 consisting of 100 mM buffering agent (glycine, sodium acetate, Tris, sodium phosphate) with and without 100 mM sodium chloride. Samples were centrifuged 15,000 × g for 15 minutes and the supernatant removed. The supernatants were then treated with the application of 60°C, 70°C, or 80°C for ten minutes and compared to an ambient temperature control sample for each buffer.

GRFT extraction – bench scale

Ten days after inoculation, leaf tissue was harvested and homogenized in the appropriate extraction buffer with a ratio of 1:2 (Plant tissue (g): Extraction buffer (mL)). Extraction buffer composition was either 100 mM sodium acetate, 300 mM sodium chloride, 20 mM ascorbic acid, 10 mM sodium metabisulfite for buffers with a pH range of 4.0 - 6.5 and 100 mM tris-hydrochloride, 300 mM sodium chloride, 20 mM ascorbic acid, 10 mM sodium metabisulfite for buffers in the pH range of 7.0 -9.0. Extract material was filtered through 4 layers of cheesecloth and 1 layer of miracloth, heated to the desired temperature for 15 minutes, and centrifuged at 15,000 × g for 15 minutes. In some experiments the purification process was halted at this point and the extract tested for purity by SDS PAGE, but in an effort to remove the TMV coat protein additional alterations were made to the extract. The semi-pure extract was treated with 0 – 0.1 M $MgCl_2$ and 0 – 10 mg/mL bentonite at varying pHs and then incubated at varying temperatures over night to determine if any combination of bentonite, $MgCl_2$, pH, and temperature could help remove the coat protein. After 16 hrs at their specified temperature all extracts were centrifuged at 15, 000 × g for 15 minutes and the supernatant removed. Pellets were resuspended in equivalent volumes of 2× SDS running buffer. Pellet and supernatant samples were analyzed by SDS-PAGE.

SDS-PAGE with densitometry

Sample buffer (4 X SDS) containing BME was added to all samples and samples were then boiled at 95°C for 10 minutes. Twenty microliters of sample was loaded on 12% Bis-Tris gels with the addition of 5 μL of protein ladder to the outer wells. Gels were ran at 200 V for 37 minutes then stained with coomassie stain for 1 hour and destained for 3 hours. Gels were imaged on a Kodak Image station 4000R Pro using Carestream SE M software. The software was then used to mark the lanes on the gell and find resulting bands. Bands were analyzed based on intensity and size. GRFT purity of each sample was determined based on the band sizes and strength of contaminating proteins in the same lane. The percentage of TMV CP or GRFT for each purification step was relative to the sample representing the step preceding the parameters that were investigated. Therefore all GRFT and TMV CP samples are represented as a percentage of the unmodified control.

GRFT gp120 ELISA

In a 96-well plate, wells were coated with 200 μL of 1 ng/mL gp120 (Protein Sciences, Meriden, CT) in PBS and incubated overnight at 4°C. Plates were washed with PBS-T and blocked with 5% PBS-T-M for 2 hours at room temperature. Wells were washed with PBS-T and incubated at room temperature for 2 hours with sample diluted 1:10 in PBS and further diluted in series 1:5 down the plate. Purified GRFT, uninfected plant extract, PBS sample and no gp120 coating controls were included in all ELISAs. Plates were washed and incubated for 1 hour at room temperature with RB anti-GRFT serum dilute 1:10,000 in 1% PBS-T-M. After washing the plate, the secondary antibody, GT anti-rabbit conjugated to HRP, was added to each well diluted 1:20,000 in 1% PBS-T-M. Plates were again washed and the ultra sensitive TMB substrate added to each well. After 3 minutes 0.2 M H2SO4 was added to stop the reaction and the colorimetric change was read at OD 405 nm on a Beckman- Coulter plate reader. The collected data was background subtracted and analyzed for the relative change in GRFT concentration between purification methods.

TMV western blot

Samples were ran on SDS-PAGE via methodology described earlier and then transferred to a PVDF membrane via BioRad transblot turbo system. Membranes were incubated for 1 hour with 3% PBS-T-M (phosphate bufferd saline with 0.5% Tween20 and 3% non-fat dry milk) at room temperature, washed with PBS-T three times and incubated overnight at4°C with rabbit antibody against TMV 1:5000 in 3% PBS-T-M. Blots were then washed with 3% PBS-T and incubated with goat anti-rabbit HRP conjugated antibody diluted 1:10,000 in 3% PBS-T-M for 1 hour at room temperature. Blots were washed and developed with ECL Prime development kit (GE Lifesciences). Blots were exposed on a Kodak Image Station 4000 R Pro using Carestream SE M software. Software was then used to identify and analyze visible bands based on intensity and size.

Chromatography and purity verification

GRFT extract was purified using an AKTA Purifier with a XK16 column packed with 10mLs of Capto™ MMC resin (GE Life Sciences). After overnight incubation the semi-pure extract was filtered through a 0.2 μm filter and loaded on to the column at a rate of 5 mL/minute. A two-step gradient of 90% and 100% phosphate buffered saline (137 mM NaCl, 2.7 mM KCl, 10 mM NaH_2PO_4, 2 mM KH_2PO_4) pH7.4 was used to elute GRFT from the column. Fractions (10 mL) were collected and pooled after being checked for the presence of GRFT through SDS-PAGE. The purity of the concentrated GRFT was verified by an overloaded SDS-PAGE and a western blot against the TMV CP. The glycan binding capabilities of the purified GRFT were determined through gp120 binding. The purified extract was compared to GRFT purified with the previously published methodology in over-loaded SDS-PAGE, western blots against TMV and gp120 ELISAs.

Industrial pilot scale purification

Five days after infectious transcript inoculation, plants were harvested and the rTMV was isolated. The isolated rTMV was used to inoculate approximately 600 plants through high velocity spraying with diatomaceous earth in a 100 mM phosphate buffer pH 7.4. Ten days after spray inoculation, five kg of leaf tissue was harvested and homogenized in extraction buffer (100 mM sodium acetate, 300 mM sodium chloride, 20 mM ascorbic acid, 10 mM sodium metabisulfite pH4.0) with a ratio of 1:2 (Plant tissue (g): Extraction buffer (mL)). The extract pH was adjusted back to pH 4.0 and heated to 55°C for 15 minutes. The heated extract was immediately filter pressed through a 0.3 μm filter (Ertel Alsop, M-853) with the addition of the filter aid Cell Pure® (30 g/L). The extract was incubated overnight in a holding tank with continuous stirring at 4°C with the addition of 10 mg/mL of bentonite and 0.1 M $MgCl_2$. After overnight incubation the extract was filter pressed again through a 0.3 μm filter and an inline sterilization filter (Sartorius, Sartopore 2). Capto® MMC chromatography was performed as described above, but the resin was packed in a sterilized AxiChrom 50/300 column. Purity was assessed via previously discussed methodology of gp120 ELISA, SDS-PAGE with densitometry, and TMV western blot with densitometry. Additionally, endotoxin levels were assessed with the methodology explained below.

Endotoxin analysis

Contaminating endotoxin levels in the final GRFT product purified at industrial pilot scale was analyzed with Charles River Laboratory Endosafe® PTS. Twenty-five μL of GRFT solution was added to each of the four wells on a Limulus Amebocyte Lysate Test cartridge and the endotoxin levels were analyzed accordingly.

Statistical analysis

All statistical analysis and summary graphical representations were performed on Graph Pad Prism® 5.0.

Competing interests

Kenneth Palmer is a member of Intrucept Biomedicine LLC that holds a license to make and sell Griffithsin from the United States Public Health Service.

Authors' contributions

JLF performed the purification studies, participated in the study design, and drafted the manuscript. VW performed preliminary purification studies and participated in study design. KEP conceived the study and participated in its design. All authors read and approved the final manuscript.

Acknowledgements

This research was made possible by support of NIH R-01 AI076169, NIH U19 grant AI113182, DoD grantW81XWH-10-2-0082 CLIN1, and the Helmsley Charitable Trust. Kenneth Palmer is the Helmsley Chair in Plant-based Pharmaceutical Research in the University of Louisville School of Medicine. The authors would like to acknowledge the contribution of Kentucky Bioprocessing Inc. in allowing use of their pilot scale facility.

Author details

[1]Owensboro Cancer Research Program, 1020 Breckenridge St., Suite 201, Owensboro, KY 42303, USA. [2]University of Louisville School of Medicine, James Graham Brown Cancer Center, 529 S Jackson Street, Louisville, KY 40202, USA. [3]Institute for Health Metrics and Evaluation, University of Washington, 2301 5th Ave, Suite 600, Seattle, WA, USA.

References

1. Mori T, O'Keefe BR, Sowder 2nd RC, Bringans S, Gardella R, Berg S, et al. Isolation and characterization of griffithsin, a novel HIV-inactivating protein, from the red alga Griffithsia sp. J Biol Chem. 2005;280(10):9345–53.
2. Nixon B, Stefanidou M, Mesquita PM, Fakioglu E, Segarra T, Rohan L, et al. Griffithsin Protects Mice from Genital Herpes by Preventing Cell-to-Cell Spread. J Virol. 2013;87(11):6257–69.
3. Emau P, Tian B, O'Keefe BR, Mori T, McMahon JB, Palmer KE, et al. Griffithsin, a potent HIV entry inhibitor, is an excellent candidate for anti-HIV microbicide. J Med Primatol. 2007;36(4–5):244–53.
4. O'Keefe BR, Vojdani F, Buffa V, Shattock RJ, Montefiori DC, Bakke J, et al. Scaleable manufacture of HIV-1 entry inhibitor griffithsin and validation of its safety and efficacy as a topical microbicide component. Proc Natl Acad Sci U S A. 2009;106(15):6099–104.
5. Giomarelli B, Schumacher KM, Taylor TE, Sowder 2nd RC, Hartley JL, McMahon JB, et al. Recombinant production of anti-HIV protein, griffithsin, by auto-induction in a fermentor culture. Protein Expr Purif. 2006;47(1):194–202.
6. Kouokam JC, Huskens D, Schols D, Johannemann A, Riedell SK, Walter W, et al. Investigation of griffithsin's interactions with human cells confirms its outstanding safety and efficacy profile as a microbicide candidate. PLoS One. 2011;6(8):e22635.
7. Moulaei T, Shenoy SR, Giomarelli B, Thomas C, McMahon JB, Dauter Z, et al. Monomerization of viral entry inhibitor griffithsin elucidates the relationship between multivalent binding to carbohydrates and anti-HIV activity. Structure. 2010;18(9):1104–15.
8. Ziolkowska NE, Shenoy SR, O'Keefe BR, McMahon JB, Palmer KE, Dwek RA, et al. Crystallographic, thermodynamic, and molecular modeling studies of the mode of binding of oligosaccharides to the potent antiviral protein griffithsin. Proteins. 2007;67(3):661–70.
9. Brakke MK, Van Pelt N. Influence of bentonite, magnesium, and polyamines on degradation and aggregation of tobacco mosaic virus. Virology. 1969;39(3):516–33.
10. O'Keefe BR, Giomarelli B, Barnard DL, Shenoy SR, Chan PK, McMahon JB, et al. Broad-spectrum in vitro activity and in vivo efficacy of the antiviral protein griffithsin against emerging viruses of the family Coronaviridae. J Virol. 2010;84(5):2511–21.
11. Takebe Y, Saucedo CJ, Lund G, Uenishi R, Hase S, Tsuchiura T, et al. Antiviral Lectins from Red and Blue-Green Algae Show Potent In Vitro and In Vivo Activity against Hepatitis C Virus. PLoS One. 2013;8(5):e64449.
12. Ishag HZ, Li C, Huang L, Sun MX, Wang F, Ni B, et al. Griffithsin inhibits Japanese encephalitis virus infection in vitro and in vivo. Arch Virol. 2013;158(2):349–58.
13. Hoorelbeke B, Xue J, Liwang PJ, Balzarini J. Role of the Carbohydrate-Binding Sites of Griffithsin in the Prevention of DC-SIGN-Mediated Capture and Transmission of HIV-1. PLoS One. 2013;8(5):e64132.
14. Meuleman P, Albecka A, Belouzard S, Vercauteren K, Verhoye L, Wychowski C, et al. Griffithsin has antiviral activity against hepatitis C virus. Antimicrob Agents Chemother. 2011;55(11):5159–67.
15. Xue J, Gao Y, Hoorelbeke B, Kagiampakis I, Zhao B, Demeler B, et al. The role of individual carbohydrate-binding sites in the function of the potent anti-HIV lectin griffithsin. Mol Pharm. 2012;9(9):2613–25.
16. Barton C, Kouokam JC, Lasnik AB, Foreman O, Cambon A, Brock G, et al. Activity of and effect of subcutaneous treatment with the broad-spectrum antiviral lectin griffithsin in two laboratory rodent models. Antimicrob Agents Chemother. 2014;58(1):120–7.
17. Sherwood JL, Fulton RW. The specific involvement of coat protein in tobacco mosaic virus cross protection. Virology. 1982;119(1):150–8.
18. Powell CA. The effect of cations on the alkaline dissociation of tobacco mosaic virus. Virology. 1975;64(1):75–85.

Selection of peptides binding to metallic borides by screening M13 phage display libraries

Martin Ploss[1], Sandra J Facey[1], Carina Bruhn[2], Limor Zemel[3], Kathrin Hofmann[2], Robert W Stark[3], Barbara Albert[2] and Bernhard Hauer[1*]

Abstract

Background: Metal borides are a class of inorganic solids that is much less known and investigated than for example metal oxides or intermetallics. At the same time it is a highly versatile and interesting class of compounds in terms of physical and chemical properties, like semiconductivity, ferromagnetism, or catalytic activity. This makes these substances attractive for the generation of new materials. Very little is known about the interaction between organic materials and borides. To generate nanostructured and composite materials which consist of metal borides and organic modifiers it is necessary to develop new synthetic strategies. Phage peptide display libraries are commonly used to select peptides that bind specifically to metals, metal oxides, and semiconductors. Further, these binding peptides can serve as templates to control the nucleation and growth of inorganic nanoparticles. Additionally, the combination of two different binding motifs into a single bifunctional phage could be useful for the generation of new composite materials.

Results: In this study, we have identified a unique set of sequences that bind to amorphous and crystalline nickel boride (Ni_3B) nanoparticles, from a random peptide library using the phage display technique. Using this technique, strong binders were identified that are selective for nickel boride. Sequence analysis of the peptides revealed that the sequences exhibit similar, yet subtle different patterns of amino acid usage. Although a predominant binding motif was not observed, certain charged amino acids emerged as essential in specific binding to both substrates. The 7-mer peptide sequence LGFREKE, isolated on amorphous Ni_3B emerged as the best binder for both substrates. Fluorescence microscopy and atomic force microscopy confirmed the specific binding affinity of LGFREKE expressing phage to amorphous and crystalline Ni_3B nanoparticles.

Conclusions: This study is, to our knowledge, the first to identify peptides that bind specifically to amorphous and to crystalline Ni_3B nanoparticles. We think that the identified strong binding sequences described here could potentially serve for the utilisation of M13 phage as a viable alternative to other methods to create tailor-made boride composite materials or new catalytic surfaces by a biologically driven nano-assembly synthesis and structuring.

Background

The diverse group of transition metal borides comprises compounds that exhibit or combine interesting physical properties such as semiconductivity, ferromagnetism, hardness, thermal and mechanical stability and oxidation resistance [1]. Some of the binary, ternary, and quaternary metal borides are even superconducting or represent some of the strongest permanent magnets [2,3]. Because of the high melting point of boron, the traditional synthesis of metal borides does not result in nanoscale or nanostructured products. As an alternative, a room temperature synthesis of nickel boride nanoparticles via hydrolysis of $NaBH_4$ has been reported [4-7]. Biological systems, on the other hand, can synthesise materials under mild and environmentally benign conditions [8]. Nickel borides like Ni_3B, Ni_2B or NiB are considered to exhibit catalytic activity in all kinds of reduction reactions, and hydrogenation reactions [9-15]. Like iron borides they are expected to exhibit paramagnetism, ferro- or superparamagnetism [16,17]. The electrical conductivities of different nickel borides vary with their boron content. Thus, the nickel-boron system represents an interesting class of

* Correspondence: bernhard.hauer@itb.uni-stuttgart.de
[1]Institute of Technical Biochemistry, University of Stuttgart, Allmandring 31, 70569 Stuttgart, Germany
Full list of author information is available at the end of the article

compounds for a model study on peptide-boride interactions and the potential of bio-modified inorganic solids. Until now nothing is known about the interaction between boron-containing materials and peptides, although peptides binding to solid surfaces have been extensively exploited for other chemical systems in nanoscale science because of their material selective properties [18-20]. To select for peptides that are capable of interacting with inorganic materials, phage peptide display is a powerful technique [21]. Phage display is a selection technique in which a combinatorial library of random peptides ($\sim10^9$) is expressed as a fusion with an M13 phage coat protein (usually p3), resulting in the display of the fused peptide on the surface of the phage particle. Selection of the desired peptides, which is usually a random 7- or 12-amino acid peptide, is achieved by multiple rounds of target binding, elution and amplification, a process known as biopanning (or panning). Because the DNA sequence for the displayed peptide is genetically fused to the p3 gene, the amino acid sequence of the phage-displayed peptide is readily obtained by sequencing the p3 encoding DNA.

Phage display has been used to identify peptides which specifically bind to metals like Ag, and Pd, and various inorganic materials like ZnO, SiO_2, TiO_2, ZnS, and CdS [22-29]. In the last few years, specific binding peptides to nickel have also been described [30-32]. Up to now, no specific binding peptides have been identified for borides in general, neither nanostructured nor bulk material, or more specifically nickel boride nanoparticles. The identification of such a selective binding peptide sequence would be a first, significant step towards peptide-modified boride materials and composites. Such linkers would help to make boride particles available for biological systems and allow for a tuning of the properties of borides using a biological peptide-enabled nanoassembly process.

Since to our knowledge no binding peptides to boride nanoparticles are described in the literature, our aim of this research was to identify peptides that bind specifically to metallic borides. This constitutes the first step in order to combine two different binding motifs e.g. a metallic and a ceramic boride into a single bifunctional phage or into a synthetic bifunctional peptide. This would be useful for the generation of new boride composite materials in which the physical properties of two different borides are unified which is until now not combinable. Peptides were selected after several rounds of screening with a commercial phage library. Several Ni_3B-binding peptides were selected using nanoparticles of amorphous and crystalline Ni_3B as targets. From the initial library, we identified ten phages that exhibit selective binding to either amorphous or crystalline or both using titer count analysis, fluorescence and atomic force microscopy.

Results and discussion

Isolation of Ni_3B-binding peptides

To identify Ni_3B-binding peptides, a commercial available M13 bacteriophage display library (Ph.D.-7; New England Biolabs) was screened against amorphous and crystalline Ni_3B nanoparticles. In the Ph.D.-7 library, random linear heptapeptides are displayed as N-terminal fusions to the minor coat protein p3 of M13. The M13 phage p3 constructs used in the screening selection have five copies of the peptide displayed on the one end of the phage particle. Five rounds of biopanning were conducted for each of the amorphous and crystalline Ni_3B substrates. Each biopanning round involved the steps of phage binding, removal of weak binding affinity phage by washing, and elution of bound phage by a rapid pH decrease. More stringent washing conditions were used in the additional rounds of panning. A 0.3% TBST wash solution was used to remove phage with low binding affinity to Ni_3B during the second round, whereas 0.5%, 0.6%, and 0.8% TBST wash solutions were used for the third, fourth, and fifth round, respectively. A total of 58 phage clones were selected and analysed by DNA sequencing after the fourth and fifth round of biopanning against amorphous and crystalline Ni_3B substrates. For amorphous Ni_3B substrates, 30 peptides were identified coding for 15 different peptide sequences (designated with an A for amorphous). A summary of the peptides identified from biopanning against amorphous Ni_3B is provided in Table 1. In the case of the biopanning against crystalline Ni_3B substrates, 28 peptides were isolated coding for 28 different peptide sequences (designated with a C for crystalline) shown in Table 2. The fact that all the peptides which were identified for crystalline Ni_3B are different, could be an indication of a more complex surface structure for crystalline Ni_3B nanoparticles than for amorphous Ni_3B nanoparticles. Interestingly, one peptide sequence, ANHQSAN, termed A6/C28, was isolated from both, amorphous and crystalline, Ni_3B substrates. While most of the peptides, especially identified for the crystalline Ni_3B substrate, were isolated only once, for the amorphous substrate the peptide A1 was isolated eight-times, A2 six-times, and A3 four-times, respectively.

Characteristics of the identified binding peptides

Due to the fact that cysteines interfere with the p3-mediated M13 infection process, no cysteines were present in the identified binding peptides [33]. The binding peptides showed approximately three-times more positively charged residues (K, R, and H) than negatively charged residues (D and E) (Table 3). Compared with the peptides binding to amorphous Ni_3B, the binding peptides to crystalline nickel boride show a two-times higher occurrence of the positively charged amino acids arginine and lysine, and a three-times higher occurrence

Table 1 Summary of the isolated binding peptides to amorphous Ni_3B

Peptide	Frequency	Sequence	Number of functional amino acid residues			p*I*
			Basic	Acidic	Hydrophobic	
A4	1/30	SEIVDNH	1	2	2	4.35
A1	8/30	TNLTLAS	0	0	3	5.19
A2	6/30	GALPNNL	0	0	3	5.52
A8	1/30	NVNSTSF	0	0	2	5.52
A9	1/30	SPDTVQK	1	1	1	5.55
A10	1/30	GNRLSAD	1	1	2	5.84
A7	1/30	LGFREKE	2	2	2	6.14
A11	1/30	TQVYHPM	1	0	3	6.40
A6	1/30	ANHQSAN	1	0	2	6.19
A5	1/30	TNSSFHK	2	0	1	8.44
A12	1/30	NTVIYQK	1	0	3	8.59
A13	1/30	HVQYWQF	2	0	3	8.75
A3	4/30	SLAVSRS	1	0	3	9.47
A14	1/30	VSVNSRT	1	0	2	9.72
A15	1/30	RLLNPWI	1	0	4	9.75

of histidine, respectively. Although the distribution of charged amino acids in the library is comparable in frequency, the increased abundance of these amino acids within the identified sequences could be possibly attributed to the surface composition of the nickel borides. The characterisation of the surface of different amorphous nickel borides by Okamoto *et al.* and Caputo *et al.* by X-ray photoelectron spectroscopy (XPS) and X-ray diffraction (XRD), respectively, revealed that the electron densities on nickel are increased by electron transfer from boron to the metal [6,34]. Based on the shift of the charge, specific binding peptides could therefore interact with the nickel boride substrates via electrostatic interactions. The majority of the peptides identified on Ni_3B prevalently possess alanine, leucine, proline, serine and threonine residues (Figure 1) which is a consequence of the amino acid distribution of the applied phage display library. In addition, the binding peptides identified on amorphous Ni_3B possess 2.5-times more asparagine residues as compared with the peptides identified on crystalline Ni_3B. This is vice versa with proline residues binding to crystalline Ni_3B. The strong presence of asparagine in the peptide sequences binding to amorphous Ni_3B, and asparagine and histidine in the peptide sequences binding to crystalline Ni_3B suggests that the binding between the Ni_3B and peptides could furthermore occur via hydrogen bonding interactions. The almost complete absence of tryptophan, phenylalanine, and methionine residues is also due to the low frequencies of these amino acids within the library. An amino acid position consensus was not evident from the 7-mer library screening of Ni_3B nanoparticles as substrates. However, besides the predominantly hydrophobic amino acids (A and L) and hydrophilic amino acids (S and T) the variety of amino acids with charged side groups implies that these groups could be essential in specific binding to both substrates via electrostatic interactions. The Ni_3B-binding peptides identified possess calculated theoretical isoelectric points (p*I*) ranging from strongly acidic (4.0) to highly basic (9.75). 60% of the peptides binding to amorphous Ni_3B show a p*I* value < 7.5 while 40% of the peptides have p*I* > 7.5. 57% of the peptides binding to crystalline Ni_3B show a p*I* value < 7.5 and 43% of the peptides have p*I* > 7.5. Therefore the identified peptides can be grouped into the categories acidic (p*I* 3.0 - 5.9; 35.7% of the peptides), neutral (p*I* 6.0 – 8.9; 52.4% of the peptides), and basic (p*I* 9.0 – 12; 11.9% of the peptides), respectively.

Relative binding affinity experiments

Because of the heterogeneity of all of the identified binding peptides and the absence of a distinctive binding motif, the binding strength of each of the 42 identified peptides to amorphous (Figure 2A) and crystalline (Figure 2B) Ni_3B was determined by titer assays. Since previously binding affinity experiments at pH 5 or pH 9 with a subset of the identified binding peptides showed no increase in binding affinity (data not shown) the relative binding affinity experiments were carried out at pH 7. As a control to verify that the peptide was the interactor of interest, M13KE wild-type (M13wt) phage without a random peptide insert were compared in the same manner as the peptide clones. As presented in Figure 2A, most of the phage clones, except C2 and C17, bind to amorphous Ni_3B with greater affinity than

Table 2 Summary of the isolated binding peptides to crystalline Ni_3B

Peptide	Frequency	Sequence	Number of functionnal amino acid residues			p*I*
			Basic	Acidic	Hydrophobic	
C15	1/28	LEQTPMF	0	1	3	4.00
C16	1/28	ELTQISS	0	1	2	4.00
C2	1/28	SDPQTHT	1	1	0	5.06
C3	1/28	TPPLLSP	0	0	2	5.19
C17	1/28	MNHAESY	1	1	2	5.22
C18	1/28	VPSLTPT	0	0	2	5.49
C4	1/28	VPIPYLP	0	0	4	5.49
C19	1/28	DPYNRIN	1	1	2	5.84
C20	1/28	RTFDAIS	1	1	3	5.84
C21	1/28	YELVLPK	1	1	4	6.00
C5	1/28	ETFPARG	1	1	2	6.10
C13	1/28	GPVNHQL	1	0	2	6.74
C22	1/28	LNHVLPA	1	0	4	6.74
C23	1/28	HAMRTEP	2	1	2	6.75
C6	1/28	ATSTAHA	1	0	3	6.79
C28	1/28	ANHQSAN	1	0	2	6.79
C10	1/28	SYTKLI-IL	2	0	3	8.33
C1	1/28	SPPKSNA	1	0	1	8.47
C9	1/28	SASKVHN	2	0	2	8.49
C24	1/28	SPSTHWK	2	0	1	8.49
C25	1/28	WNAKYTL	1	0	4	8.59
C12	1/28	YQWPAR	1	0	4	8.75
C26	1/28	GDPKAAR	2	1	2	8.75
C11	1/28	GDHSRHK	4	1	0	8.76
C14	1/28	AGLPKHQ	2	0	2	8.80
C27	1/28	STFNSRV	1	0	2	9.47
C8	1/28	VHTNPSR	2	0	1	9.73
C7	1/28	GASATRT	1	0	2	9.75

does the M13wt phage, which shows a binding affinity of 2 × 10^7 pfu/ml with amorphous Ni_3B. Compared with the M13wt phage, 9 out of 42 identified peptides (see arrows in Figure 2A) bound on average 100-times and 28 on average ten-times more efficiently to amorphous Ni_3B than the wild-type, verifying that binding to the substrate was a result of the peptide sequence and not due to non-specific phage coat protein interactions. The peptides A7, C12, and C15 (see arrows in Figure 2B) bound on average 1000-times more efficiently to crystalline Ni_3B as compared to the M13wt phage, which shows a binding affinity of 5 × 10^6 pfu/ml. Interestingly, the phage clone A7 identified for amorphous Ni_3B shows a higher binding affinity to crystalline Ni_3B than most of the Ni_3B-binding peptides (C1-C11, C13-C14, and C16-C28) identified for crystalline Ni_3B. In addition, the amorphous Ni_3B-binding peptides A8-A10 show a higher binding affinity to crystalline Ni_3B ($\geq 10^9$ pfu/ml) than several binding peptides found for crystalline Ni_3B specifically (i.e. C1-C11, C13-C14, C16-C24, and C27-28). The relative binding affinity experiments revealed a set of several strong binders ($>10^9$ pfu/ml) for amorphous (A1-A3, A7, C4, C9, C13, C15, and C24) and crystalline (A7, C12, and C15) substrates (Table 4). The phages displaying the peptides A7 and C15 emerged as the best binders for both substrates.

Competitive binding assays of the best binding peptides

The competitive binding assay represents a method to determine the most selective binding peptide sequence from a large subset of phage display results [35]. To identify a specifically recognised structural correlation between the surfaces of amorphous and crystalline Ni_3B nanoparticles, the individual phage clones of the nine

Table 3 Occurrences of amino acids of the isolated Ni_3B binding peptides

Amino acid	Averaged quantity of occurrences per heptapeptide	
	Amorphous	Crystalline
A	0.70	0.68
R	0.20	0.36
N	1.00	0.43
D	0.10	0.18
C	**0.00**	**0.00**
Q	0.20	0.25
E	0.10	0.21
G	0.27	0.21
H	0.17	0.50
I	0.10	0.14
L	1.20	0.50
K	0.13	0.29
M	0.03	0.11
F	0.10	0.14
P	0.30	0.82
S	1.00	0.75
T	0.73	0.68
W	0.07	0.07
Y	0.10	0.25
V	0.40	0.36

strongest binders (A1-A3, A7, C4, C9, C13, C15, and C24) to amorphous Ni_3B as determined by the relative binding affinity experiments were diluted into a single mini-phage library. The mini-phage library was then screened against amorphous or crystalline Ni_3B substrates. To receive better statistical results, the competitive binding assays were repeated three-times. After a single panning experiment, the strongly bound phages were isolated and a total of 58 phage clones were randomly selected and analysed by DNA sequencing to identify the strongest binding peptides. In Tables 5 and 6 the most abundant sequence indicates the strongest binding peptide to amorphous and crystalline Ni_3B, respectively. Among the 58 randomly chosen plaques, the phage displaying the A7 peptide LGFREKE has the strongest binding affinity to amorphous (seventeen-times, ~58%) and crystalline (thirteen-times, ~45%) Ni_3B. Interestingly, the amino acids lysine, arginine, and glutamic acid were shown to have a pronounced adhesion of their charged side groups to amorphous Si_3N_4 and SiO_2 under aqueous conditions [36]. It was also reported, that the surface of amorphous Si_3N_4 may present negative sites in solution [37]. Therefore substrates, possessing negatively surface charges in aqueous solution with pH values less than the intrinsic pK_a of the side chain of a basic amino acid, a peptide containing basic amino acids will be positively charged at these groups and will provide an attractive interaction to the given surface [36]. In aqueous solutions with higher pH values than the intrinsic pK_a of the side chain of an acidic amino acid, a peptide containing acidic amino acids will be negatively charged at these groups. Based on these facts, the best binder A7 (LGFREKE) possesses in solution at pH 7 an alternating series of positively and negatively charged side chains in its peptide sequence which could be responsible for an interaction between the charge depleted boron and the electrons accumulated on the nickel side of amorphous and crystalline Ni_3B nanoparticles. In contrast, the 7-mer peptide sequence TNLTLAS, termed A1, which emerged as the dominant amorphous Ni_3B binder in the initial screening, possesses no charged side chain. The lack of electrostatic interactions might contribute to the observed fact, that it does not show the highest frequency in the competitive assay.

Fluorescence microscopy of Ni_3B-M13 complex

The binding of phage displaying the A7 peptide LGFREKE to amorphous and crystalline Ni_3B nanoparticles was confirmed by fluorescence microscopy. Amplified phage displaying the LGFREKE peptide were incubated either with amorphous or crystalline Ni_3B nanoparticles and then washed to remove the nonspecifically bound phage particles. The samples were then dyed using a fluorescently tagged anti-M13 monoclonal antibody and washed again to eliminate nonspecifically bound anti-M13 antibodies. The fluorescence confocal images in Figure 3A and C show the LGFREKE displaying phage binding to the surface of amorphous and crystalline Ni_3B nanoparticles, respectively. Considerable fluorescence emits from the surfaces of the LGFREKE phage-Ni_3B nanoparticles. Although the relative binding affinity experiments showed an unspecific binding of the wild-type phage to the Ni_3B substrates, the control experiments (Figure 3E and G) showed no fluorescence. This could be due to the fact that the concentration of bound phage is below the detection limit.

AFM measurements

To verify the interaction between M13 phage displaying the A7 peptide on the p3 coat protein and Ni_3B particles, AFM measurements of phage bound to amorphous and crystalline Ni_3B particles were carried out (Figure 4). The modified M13 phage were incubated either with amorphous or crystalline nanoparticles as shown in Figure 4A and B, respectively. As shown in Figure 4A and B the height of the colour scale exceeds the height of the particles. This colour scale was chosen to show the 100-nm-thick particles together with the 6-nm-thick phage. Individual phage

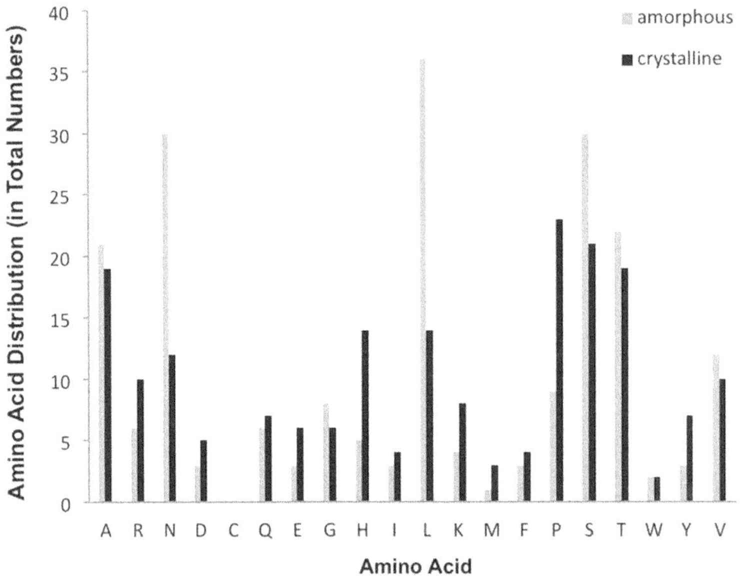

Figure 1 Amino acid distribution of the isolated peptide sequences. A total of 15 peptide sequences were analysed for amorphous Ni_3B (grey bars) and a total of 28 peptide sequences were analysed for crystalline Ni_3B (black bars).

are approximately 1 μm in length and 6.5 nm in height. This cross section and length of the modified phage correspond to that of M13 wt phage (data not shown). Some phage were oriented with their ends towards a nanoparticle (see arrows in Figure 4), which implies an affinity between the modified phage proteins and the particle. There are a few particles that seem to be attached to the unmodified side of the phage. Possible explanations for this observation include unspecific interaction or the presence of more than one phage, i.e. bundles of phage where the end of one phage coincides with the side of another phage. A small level of unspecific interaction between nanoparticles and phage cannot be avoided during preparation for AFM imaging. The phage oriented with their ends towards the nanoparticles, suggest that the peptide LGFREKE interacts with both amorphous and crystalline Ni_3B nanoparticles.

Conclusions

Metal borides are inorganic solids that are very interesting due to their high versatility in physical properties, as for example magnetism, electrical and thermal conductivity, hardness and catalytic activity, but at the same time they are difficult to prepare. Besides the application as catalysts in all kinds of reduction reactions, and hydrogenation reactions, metallic borides are employed for the generation of conductive ceramics, hard magnets, superconductors, and hard materials [9-15,38-40]. Typical synthesis routes employ high reaction temperatures and do not allow for a directed synthesis of particles that are well-defined in terms of size, shape, uniformity, phase purity, or crystallinity. Formation of metal boride nanoparticles in solvents has been described occasionally, but an interaction of borides with biological material or bio-molecules like peptides was unknown before [4,5]. To the best of our knowledge, the present study is the first to identify peptides that bind specifically to amorphous and to crystalline Ni_3B nanoparticles. Although all identified peptide sequences showed no significant binding motif, the analysis of the resulting Ni_3B-binding sequences found them to be enriched in charged amino acids. The identified binding sequences possessed approximately three times more positively charged than negatively charged amino acids which leads to the assumption of electrostatic interactions between the identified binding peptides and the Ni_3B substrates. Fluorescence microscopy and AFM studies directly confirmed the adhesion of the best binding phage displaying the peptide LGFREKE to amorphous and crystalline Ni_3B particles. The best binding peptide sequence could play an important role for the development of a new synthesis route of Ni_3B in the presence of phage displaying the peptide, or the synthetic peptide, respectively, which would represent a complete new approach. The synthesis of Ni_3B in the presence of LGFREKE could result in mono-disperse and peptide capped nanoparticles with well-defined and uniform sizes and shapes exhibiting new catalytic and magnetic properties. Furthermore, the utilisation of M13 phage as template for the Ni_3B mineralisation could potentially serve as a viable alternative to create tailor-made boride composite materials by a biologically driven nano-assembly synthesis and structuring for the generation of thermoelectric devices and field effect transistors.

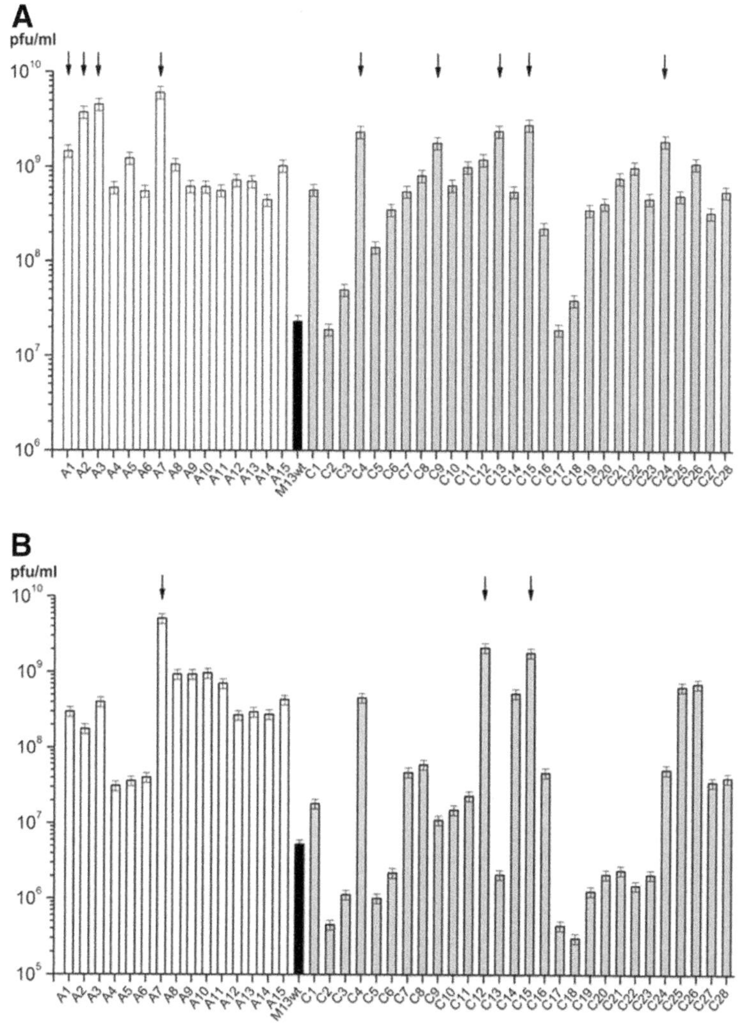

Figure 2 Evaluation of the binding strength of each of the 42 identified phage clones. The relative binding affinity of the phage clones to amorphous **(A)** and crystalline **(B)** Ni_3B nanoparticles was determined by titer assays at pH 7. The assay was repeated three-times for each clone and the elucidated phage amounts were arithmetically averaged. As a control, M13KE wild-type (M13wt) phage without a random peptide insert were compared in the same manner as the phage clones. Strong binding phage clones (> 10^9 pfu/ml) are indicated with an arrow. Clones identified on amorphous Ni_3B are designated with an A, clones identified on crystalline Ni_3B are designated with a C.

Methods

Phage and bacteria strains

Recombinant M13 phage from the Ph.D.-7 heptapeptide library (New England Biolabs, MA) were amplified and propagated in the *E. coli* host strain ER2738 (F', *lacI^q*, Δ(*lacZ*)*M15, proA$^+$B$^+$ zzf::Tn*10(Tet^R)/*fhuA2, supE, thi-1,* Δ(*lac-proAB*), Δ(*hsdS-mcrB*)5, [$r_k^-m_k^-$*McrBC*$^-$]), as described by the manufacturer. M13KE wild-type (M13wt) phage was used for the determination of unspecific binding to the Ni_3B targets. Bacteria cultures were maintained in lysogeny broth (LB) [41].

Peptide library screening

Ni_3B-binding peptides were identified using a commercially available M13 bacteriophage display library (Ph.D.-7; New England Biolabs) with random 7-mer peptides. The DNA sequences of the 7-mer peptides are N-terminal fusions to the minor coat protein p3 of the bacteriophage M13. A short linker sequence (GGGS) separates each random peptide from the minor coat protein p3. The phage display library consists of approximately 1.28 × 10^9 different peptides. The target binding, elution, and phage amplification steps were conducted according to the manufacturer's instructions. 20 mg of amorphous or crystalline Ni_3B nanoparticles were used as targets. The targets were washed three-times with Tris-buffered saline (TBS, 50 mM Tris–HCl, pH 7, and 150 mM NaCl) and blocked with TBS supplemented with 5 mg/ml of BSA for 1 hour. The blocked targets were then washed three-times with TBS. To each target, a total number of 2 × 10^{11} phage in

Table 4 Strong binders identified by relative binding affinity experiments

#	Peptide	Frequency	Phage titer (pfu/ml)	Sequence	pI
Ni_3B amorphous target					
1	A7	1/30	6.1×10^9	L G F R E K E	6.14
2	A3	4/30	4.6×10^9	S L A V S R S	9.47
3	A2	6/30	3.8×10^9	G A L P N N L	5.52
4	C15	1/28	2.8×10^9	L E Q T P M F	4.00
5	C13	1/28	2.4×10^9	G P V N H Q L	6.74
6	C4	1/28	2.4×10^9	V P I P Y L P	5.49
7	C24	1/28	1.9×10^9	S P S T H W K	8.49
8	C9	1/28	1.8×10^9	S A S K V H N	8.49
9	A1	8/30	1.5×10^9	T N L T L A S	5.19
Ni_3B crystalline target					
#	**Peptide**	**Frequency**	**Phage titer (pfu/ml)**	**Sequence**	**pI**
1	A7	1/30	5.1×10^9	L G F R E K E	6.14
2	C12	1/28	2.1×10^9	Y Q V V P A R	8.75
3	C15	1/28	1.8×10^9	L E Q T P M F	4.00

1 ml of TBS with 0.1% Tween-20 (0.1% TBST) were incubated for 1 hour. Non-specifically or weakly bound phage particles were removed by washing ten-times with 0.1% TBST. The remaining bound phage particles were eluted with a solution containing 0.2 M glycine-HCl, pH 2.2 and 1 mg/ml BSA. After centrifugation, the supernatant was removed from the target and neutralised with 1 M of Tris–HCl, pH 9. The eluted phage were amplified through infection into *E. coli* ER2738, followed by purification by precipitation with polyethylene glycol according to the manufacturer's instructions. This completed a round of panning. The entire biopanning process was repeated five rounds to enrich for tight binders. To inhibit the hydrophobic interactions between the phage library and the Ni_3B targets, the concentration of Tween-20 was successively increased in each panning step from 0.1%, 0.3%, 0.5%, 0.6%, and 0.8%, respectively. After the fourth and fifth round of screening, the eluted phage were diluted and titrated on LB agar plates containing 5-bromo-4-chloro-3-indolyl-β-D-galactopyranoside (Xgal) and isopropyl-β-D-thiogalactopyranosid (IPTG). A total of 58 randomly selected phage plaques, which appeared blue, were picked and analysed by DNA sequencing. The phage titer was determined by titration according to manufacturer's recommendations using appropriate dilutions of the precipitated phage.

Table 5 Competitive binding assay results for amorphous Ni_3B target

Peptide	Sequence	Frequency
A7	L G F R E K E	17/29
C24	S P S T H W K	3/29
C9	S A S K V H N	2/29
C15	L E Q T P M F	2/29
A1	T N L T L A S	1/29
A2	G A L P N N L	1/29
C4	V P I P Y L P	1/29
C13	G P V N H Q L	1/29
A3	S L A V S R S	0/29

Table 6 Competitive binding assay results for crystalline Ni_3B target

Peptide	Sequence	Frequency
A7	L G F R E K E	13/29
C9	S A S K V H N	5/29
C24	S P S T H W K	5/29
C13	G P V N H Q L	4/29
C4	V P I P Y L P	1/29
C15	L E Q T P M F	1/29
A1	T N L T L A S	0/29
A2	G A L P N N L	0/29
A3	S L A V S R S	0/29

Binding affinity assays

The binding affinity of each identified phage clone with the Ni_3B substrates was demonstrated by mixing equivalent phage amounts (1×10^{10} pfu/ml) of individual phage clones with either 20 mg of amorphous or crystalline Ni_3B nanoparticles in 1 ml of TBS containing 0.8% Tween-20. After 1 hour incubation at room temperature the unbound phage were removed by washing the substrates ten-times. Bound phage were then isolated using the standard acidic elution method as used in the screening procedure. For each clone, the bound phage amount was elucidated by titration according to manufacturer's recommendations using appropriate dilutions of the eluted phage. The binding strength of a single clone directly correlates with the phage titer of the eluted phage. The assay was repeated three-times for each clone and the elucidated phage titers were arithmetically averaged. M13KE wild-type phage, with no random peptides displayed on the p3 protein, was used as a control.

Competitive binding assays

The phages with the highest binding affinity to either amorphous or crystalline Ni_3B were evaluated by the method of a competitive binding assay. For this assay, the individual phage clones of the nine strongest binders on amorphous Ni_3B as determined by binding affinity assays, were separately amplified and diluted into a mini-phage library at a concentration of 1×10^7 pfu/μl per clone. Subsequently 1×10^{10} pfu/ml of this minimised phage library was applied to amorphous or crystalline Ni_3B nanoparticles

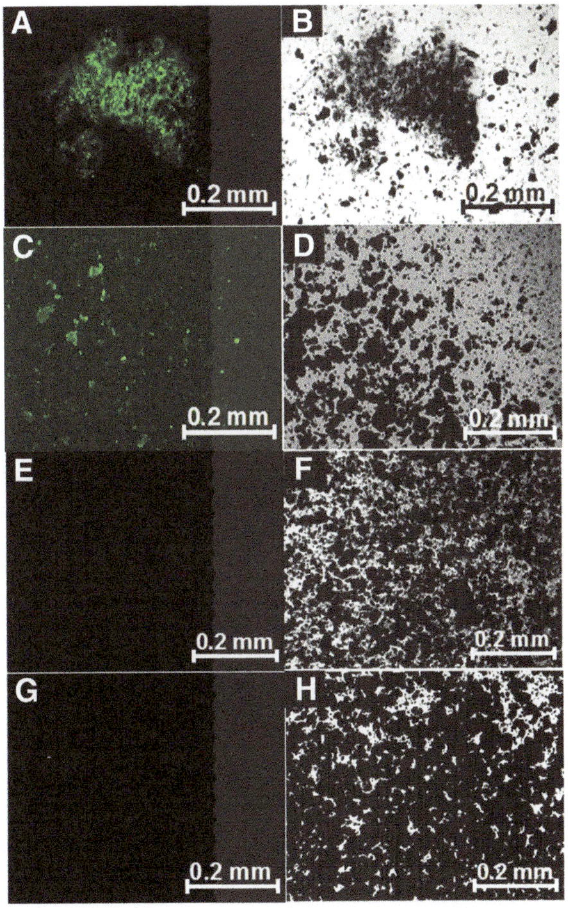

Figure 3 Analysis of the phage binding by fluorescence microscopy. Confocal fluorescence **(A, C)** and transmission optical microscopy **(B, D)** of LGFREKE phage bound to amorphous **(A, B)** and crystalline **(C, D)** Ni_3B nanoparticles. The samples were dyed using a fluorescently tagged anti-M13 monoclonal antibody. Confocal fluorescence **(E, G)** and transmission optical microscopy **(F, H)** of the control experiments of amorphous and crystalline Ni_3B nanoparticles, respectively, which were preincubated with M13KE wild-type (M13wt) phage before incubation with the fluorescently tagged antibody.

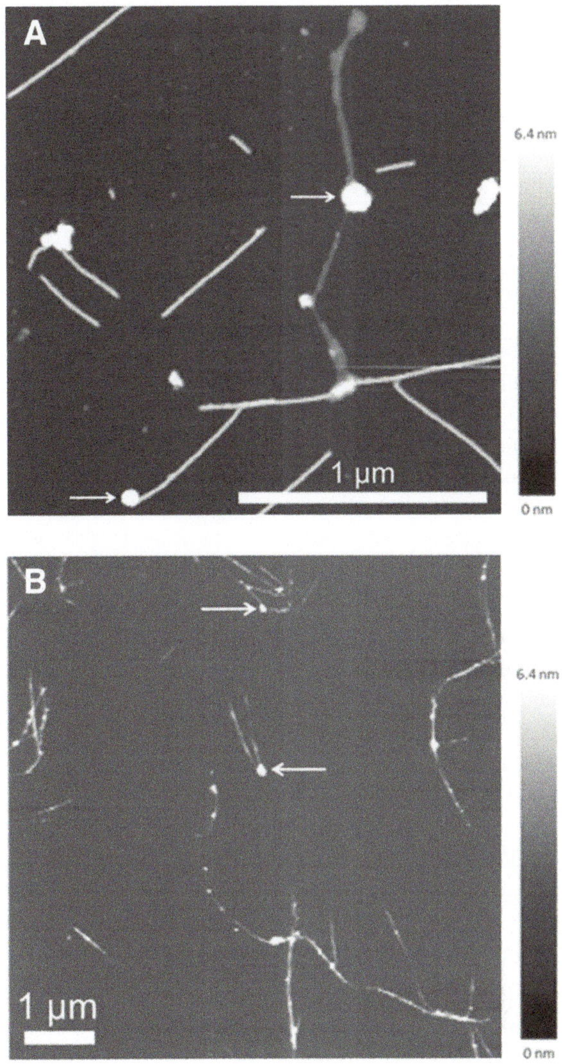

Figure 4 Visualisation of the phage binding by atomic force microscopy. Topographic AFM images of the binding of modified M13 phage (LGFREKE) to amorphous **(A)** and crystalline **(B)** Ni_3B nanoparticles. The AFM images of M13 phage displaying the A7 peptide on p3 showed the phage to bind to several amorphous and crystalline Ni_3B nanoparticles at the tip of p3, which are indicated by arrows. The colour scale exceeds the particle height to better display the particle cross section.

as described for the procedure of the binding affinity assays. After a single panning round, the weakly bound phages were removed by washing the substrates and the remaining strongly bound phages were isolated using the standard acidic elution method as used in the screening procedure. The eluted phages were quantified by titration according to the manufacturer's recommendations. The assay was repeated three-times for each substrate. A total number of 29 plaques were randomly picked for each substrate and sequenced to reveal the most strongly binding peptide sequences.

Synthesis and characterisation of amorphous and crystalline Ni_3B

The Ni_3B nanoparticles were synthesised from $NaBH_4$ (Alfa Aesar, 97%) and $NiCl_2 \times 6\ H_2O$ (Grüssing GmbH, 98%) in aqueous solution at about 0°C following a synthesis route described earlier [5]. According to this procedure amorphous Ni_3B is formed primarily, which is transformed into the crystallised Ni_3B sample at 490°C in vacuum (5×10^{-3} mbar, 18 h). The X-ray diffraction pattern of the annealed, crystalline sample allows for its identification as Ni_3B (see Additional file 1). In previous work, we have shown using methods like X-ray diffraction, energy-dispersive X-ray spectroscopy, X-ray absorption spectroscopy, scanning and transmission electron microscopy, and energy-loss electron spectroscopy that the corresponding amorphous phase actually consists of the same boride as the crystalline modification [5,7]. For the phage display the amorphous Ni_3B nanoparticles were used after extensive washing with deionized water and drying in a desiccator under vacuum. The crystalline nanoparticles were obtained from the amorphous product after the annealing process and used directly.

Fluorescence microscopy analysis

The binding of M13 phage to amorphous and crystalline nickel boride nanoparticles was characterised with a Zeiss Axiovert 200 inverted fluorescence microscope equipped with a Zeiss AxioCam HRm. The fluorescence images were taken using emission at 528 nm. M13 phage were incubated with 10 mg of amorphous and crystalline Ni_3B nanoparticles at a concentration of 3×10^{12} pfu/ml in 1 ml TBS supplemented with 0.8% Tween-20. After 2 h at room temperature, unbound phage were removed by washing the substrates three-times with TBS. Subsequently, the substrates were blocked with TBS supplemented with 2% BSA and 0.05% Tween-20 for 1 hour and washed three-times with TBS supplemented with 0.05% Tween-20. The samples were dyed using a FITC tagged anti-M13 monoclonal antibody against phage coat protein p8 (Acris Antibodies, Inc., Herford, Germany). Each target was incubated for 1 hour with 1 μg/ml of labeled monoclonal antibodies in TBS with 0.05% Tween-20. After removing unbound antibodies by washing the samples three-times with TBS containing 0.05% Tween-20, the samples were analysed by fluorescence microscopy.

Atomic force microscopy (AFM) analysis

Ni_3B nanoparticles, phages and compounds of phages with Ni_3B were characterised with an atomic force microscope (AFM). The measurements were carried out on a Dimension Icon AFM with a Nanoscope V controller (Bruker, Santa Barbara, California) in ambient conditions. The measurements were performed in tapping mode, to minimise the lateral interaction. Silicon cantilevers (PPP NCH, Nanosensors) with a nominal resonance frequency of 330 kHz, spring constant of 42 Nm^{-1}, and a nominal tip diameter of less than 10 nm were used. All images were taken at a resolution of 512×512 $pixel^2$. M13 phages were incubated with 10 mg of amorphous and crystalline Ni_3B nanoparticles at a concentration of 3×10^{12} pfu/ml in 1 ml TBS supplemented with 0.8% Tween-20. After 2 h, unbound phages were removed by washing the substrates three-times with TBS. Samples were prepared by dropping the sample suspension on a silicon wafer or freshly cleaved mica. The samples were dried in a stream of dry nitrogen. Image analysis and processing was performed with the Nanoscope Analysis software Version 1.40 (Bruker). A plane correction procedure and a line by line fit were used to compensate for the sample tilt.

Additional file

Additional file 1: X-ray powder diffraction profile. Calculated (solid blue line) and difference (solid red line) X-ray powder diffraction profiles for the Rietveld refinement of crystalline Ni_3B. Reflection positions are marked (blue bars).

Competing interests

The authors declare that they have no competing interests.

Authors' contributions

SJF and MP drafted the manuscript. CB and KH synthesised the Ni_3B substrates for the use in biopanning. MP identified and characterised the binding peptides, and visualised the adhesion of the phage by fluorescence microscopy. LZ did the AFM analysis of Ni_3B nanoparticles, phages and compounds of phages with Ni_3B. BA, SJF, RWS, and BH planned and designed this project. SJF, MP and BA read and approved the final manuscript. All authors read and approved the final manuscript.

Acknowledgements

This work was supported by the Deutsche Forschungsgemeinschaft (DFG) under grant HA 1251/2-1, STA 1026/5-1, and AL 536/11-1 as part of the Priority Programme SPP 1569 "Generation of multifunctional inorganic materials by molecular bionics".

Author details

[1]Institute of Technical Biochemistry, University of Stuttgart, Allmandring 31, 70569 Stuttgart, Germany. [2]Eduard-Zintl-Institute of Inorganic and Physical Chemistry, Technische Universität Darmstadt, Alarich-Weiss-Str. 12, 64287 Darmstadt, Germany. [3]Center of Smart Interfaces, Physics of Surfaces and Institute of Materials Sciences, Technische Universität Darmstadt, Alarich-Weiss-Str. 10, 64287 Darmstadt, Germany.

References

1. Schwarzkopf P, Kieffer R: *Refractory Hard Metals.* 1st edition. The Macmillan Company: New York; 1953:311–312.
2. Buzea C, Yamashita T: **Review of the superconducting properties of MgB_2.** *Supercond Sci Technol* 2001, **14:**R115–R146.
3. Herbst JF, Croat JJ: **Neodymium-iron-boron permanent magnets.** *J Magn Magn Mater* 1991, **100:**57–78.
4. Brown HC, Brown CA: **New, highly active metal catalysts for the hydrolysis of borohydride.** *J Am Chem Soc* 1962, **84:**1493–1494.
5. Kapfenberger C, Hofmann K, Albert A: **Room-temperature synthesis of metal borides.** *J Solid State Sci* 2003, **5:**925–930.
6. Caputo R, Guzzetta F, Angerhofer A: **Room-temperature synthesis of nickel borides via decomposition of $NaBH_4$ promoted by nickel bromide.** *Inorg Chem* 2010, **49:**8756–8762.
7. Rades S, Kornowski A, Weller H, Albert B: **Wet-chemical synthesis of nanoscale iron boride.** *Chem Phys Chem* 2011, **12:**1756–1760.
8. Chen CL, Rosi NL: **Peptide-based methods for the preparation of nanostructured inorganic materials.** *Angew Chem Int Ed* 2010, **49:**1924–1942.
9. Boar RB, Hawkins DW, McGhie JF, Barton DHR: **Some applications of nickel boride as a desulphurising agent. An improved route to triterpene 2-enes.** *J Chem Soc, Perkin Trans* 1973, **1:**654–657.
10. Sarma DN, Sharma RP: **Nickel boride reductive cleavage of allylic trimethylsilyl ethers.** *Tetrahedr Lett* 1985, **26:**371–372.
11. Seltzmann HH, Berrang BD: **Nickel boride reduction of aryl nitro compounds.** *Tetrahedr Lett* 1993, **34:**3083–3086.
12. Molvinger K, Lopez M, Court J: **Enantioselective borane reduction of ketones with oxazaborolidines boron-bound to nickel boride nanoparticles.** *Tetrahedr Lett* 1999, **40:**8375–8378.
13. Saxena I, Borah R, Sarma JC: **Reductive amination of aromatic aldehydes and ketones with nickel boride.** *J Chem Soc, Perkin Trans* 2000, **1:**503–504.
14. Brown CA: **Catalytic hydrogenation. V. The reaction of sodium borohydride with aqueous nickel salts. P-1 nickel boride, a convenient,**

highly active nickel hydrogenation catalyst. *J Org Chem* 1970, **35:**1900–1904.
15. Pavlik JW: **Catalytic hydrogenation using nickel boride.** *J Chem Educ* 1972, **49:**528.
16. Bratkovsky AM, Rashkeev SN, Wendin G: **Electronic structure, ferromagnetism, and EELS spectra of crystalline alloys FenB and NinB (n = 1, 2, 3): Aspects of universal behavior.** *Phys Rev B* 1993, **48:**6260–6270.
17. Schaefer ZL, Ke X, Schiffer P, Schaak RE: **Direct solution synthesis, reaction pathway studies, and structural characterization of crystalline Ni_3B nanoparticles.** *J Phys Chem C* 2008, **112:**19846–19851.
18. Sarikaya M, Tamerler C, Jen AKY, Schulten K, Baneyx F: **Molecular biomimetics: nanotechnology through biology.** *Nat Mater* 2003, **2:**577–585.
19. Baneyx F, Schwartz DT: **Selection and analysis of solid-binding peptides.** *Curr Opin Biotechnol* 2007, **18:**312–317.
20. Whyburn GP, Li Y, Huang Y: **Protein and protein assembly based material structures.** *J Mater Chem* 2008, **18:**3755–3762.
21. Smith GP, Petrenko VA: **Phage Display.** *Chem Rev* 1997, **97:**391–410.
22. Naik RR, Stringer SJ, Agarwal G, Jones SE, Stone MO: **Biomimetic synthesis and patterning of silver nanoparticles.** *Nat Mater* 2002, **1:**169–172.
23. Heinz H, Farmer BL, Pandey RB, Slocik JM, Patnaik SS, Pachter R, Naik RR: **Nature of molecular interactions of peptides with gold, palladium, and Pd-Au bimetal surfaces in aqueous solution.** *J Am Chem Soc* 2009, **131:**9704–9714.
24. Umetsu M, Mizuta M, Tsumoto K, Ohara S, Takami S, Watanabe H, Kumagai I, Adschiri T: **Bioassisted room-temperature immobilization and mineralization of zinc oxide - The structural ordering of ZnO nanoparticles into a flower-type morphology.** *Adv Mater* 2005, **17:**2571–2575.
25. Rothenstein D, Claasen B, Omiecienski B, Lammel P, Bill J: **Isolation of ZnO-binding 12-mer peptides and determination of their binding epitopes by NMR spectroscopy.** *J Am Chem Soc* 2012, **134:**12547–12556.
26. Naik RR, Brott LL, Clarson SJ, Stone MO: **Silica-precipitating peptides isolated from a combinatorial phage display peptide library.** *J Nanosci Nanotechnol* 2002, **2:**95–100.
27. Dickerson MB, Jones SE, Cai Y, Ahmad G, Naik RR, Kröger N, Sandhage KH: **Identification and design of peptides for the rapid, high-yield formation of nanoparticulate TiO_2 from aqueous solutions at room temperature.** *Chem Mater* 2008, **20:**1578–1584.
28. Lee SW, Mao C, Flynn CE, Belcher AM: **Ordering of quantum dots using genetically engineered viruses.** *Science* 2002, **296:**892–895.
29. Flynn CE, Mao CB, Hayhurst A, Williams JL, Georgiou G, Iverson B, Belcher AM: **Synthesis and organization of nanoscale II-VI semiconductor materials using evolved peptide specificity and viral capsid assembly.** *J Mater Chem* 2003, **13:**2414–2421.
30. Dong J, Chuan L, Zhang J, Xin ZT, Yang G, Gao B, Mao CQ, Liu NL, Shao NS, Fan M, Xue YN: **Selection of novel nickel-binding peptides from flagella displayed secondary peptide library.** *Chem Biol Drug Des* 2006, **68:**107–112.
31. Chung KCC, Cao L, Dias AV, Pickering IJ, George GN, Zamble DB: **A high-affinity metal-binding peptide from *Escherichia coli* HypB.** *J Am Chem Soc* 2008, **130:**14056–14057.
32. Zoroddu MA, Medici S, Peana M: **Copper and nickel binding in multi-histidinic peptide fragments.** *J Inorg Biochem* 2009, **103:**1214–1220.
33. Thomas CJ, Sharama S, Kumar G, Visweswariah SS, Surolia A: **Biopanning of endotoxin-specific phage displayed peptides.** *Biochem Biophys Res Commun* 2003, **307:**133–138.
34. Okamoto Y, Nitta Y, Imanaka T: **Surface characterisation of nickel boride and nickel phosphide catalysts by X-ray photoelectron spectroscopy.** *J Chem Soc, Faraday Trans* 1978, **1:**2027–2039.
35. Jaworski JW, Raorane D, Huh JH, Majumdar A, Lee SW: **Evolutionary screening of biomimetic coatings for selective detection of explosives.** *Langmuir* 2008, **24:**4938–4943.
36. Willett RL, Baldwin KW, West KW, Pfeiffer LN: **Differential adhesion of amino acids to inorganic surfaces.** *PNAS* 2005, **102:**7817–7822.
37. Amaral M, Lopes MA, Santos JD, Silva RF: **Wettability and surface charge of Si_3N_4-bioglass composites in contact with simulated physiological liquids.** *Biomaterials* 2002, **23:**4123–4129.
38. Fahrenholtz WG, Hilmas GE, Talmy IG, Zaykoski JA: **Refractory diborides of zirconium and hafnium.** *J Am Ceram Soc* 2007, **90:**1347–1364.
39. Brown D, Ma B-M, Chen Z: **Developments in the processing and properties of NdFeB-type permanent magnets.** *J Magn Magn Mater* 2002, **248:**432–440.
40. Nagamatsu J, Nakagawa N, Muranaka T, Zenitani Y, Akimitsu J: **Superconductivity at 39 K in magnesium diboride.** *Nature* 2001, **410:**63–64.
41. Bertani G: **Studies on lysogenesis.** *I. The mode of phage liberation by lysogenic Escherichia coli. J. Bacteriol* 1951, **62:**293–300.

Fermentation performance and physiology of two strains of *Saccharomyces cerevisiae* during growth in high gravity spruce hydrolysate and spent sulphite liquor

Emma Johansson[1,2*], Charilaos Xiros[1] and Christer Larsson[1*]

Abstract

Background: Lignocellulosic materials are a diverse group of substrates that are generally scarce in nutrients, which compromises the tolerance and fermentation performance of the fermenting organism. The problem is exacerbated by harsh pre-treatment, which introduces sugars and substances inhibitory to yeast metabolism. This study compares the fermentation behaviours of two yeast strains using different types of lignocellulosic substrates; high gravity dilute acid spruce hydrolysate (SH) and spent sulphite liquor (SSL), in the absence and presence of yeast extract. To this end, the fermentation performance, energy status and fermentation capacity of the strains were measured under different growth conditions.

Results: Nutrient supplementation with yeast extract increased sugar uptake, cell growth and ethanol production in all tested fermentation conditions, but had little or no effect on the energy status, irrespective of media. Nutrient-supplemented medium enhanced the fermentation capacity of harvested cells, indicating that cell viability and reusability was increased by nutrient addition.

Conclusions: Although both substrates belong to the lignocellulosic spruce hydrolysates, their differences offer specific challenges and the overall yields and productivities largely depend on choice of fermenting strain.

Keywords: Lignocellulosic material, Nutrients, Energy charge, Fermentation capacity, High gravity fermentation

Background

Lignocelluloses are a diverse group of substrates [1] including cellulose, hemicellulose, lignin and extractives [2], which produce inhibitors during hydrolysis. The quantity of inhibitor depends not only on the origin of the material but also on the pre-treatment and hydrolysis method. Fermentation inhibitors, which inhibit yeast metabolism, include 5-hydroxymethyl furfural (HMF), 2-furaldehyde (furfural), phenolic compounds and weak acids for review, see [3-6]. Reportedly, weak acids exert a growth inhibitory effect by inflow of non-dissociated acid into the cytoplasm of the microorganism [7]. HMF and furfural are consumed by *S. cerevisiae*, with consequent cost of ATP [8,5].

* Correspondence: emma.johansson@processum.se; christer.larsson@chalmers.se
[1]Department of Chemical and Biological Engineering, Chalmers University of Technology, 412 96 Göteborg, Sweden
[2]SP Processum AB, 891 22 Örnsköldsvik, Sweden

The inhibitory effect of phenolic compounds on yeast metabolism remains under investigation. A feasible suggestion is that phenolic compounds degrade the cell membrane integrity, reducing the membrane's efficacy as a selective barrier [9]. To release high amounts of monosaccharides for fermentation, lignocelluloses (on account of their recalcitrant nature) often require harsh pre-treatment conditions. In most cases, the quantity of inhibitors increases with the severity of pre-treatment. Therefore, scientists seek mild pre-treatments that maximize the saccharification yields.

Besides containing growth inhibitors, lignocelluloses are generally scarce in nutrients and nutrient supplementation is thought to increase fermentation performance [10-12]. Earlier studies on such raw materials have identified lack of nitrogen as a crucial limiting factor in fermentations. Nitrogen is especially important for fermentative performance because it promotes cell proliferation [13,14] and thereby ethanol yields, since the ethanol production rate is maximized in actively growing cells [15,16].

Efficient distillation energy requires both high productivity and high ethanol titres. A prerequisite of ethanol production is high-gravity substrate, which theoretically yields at least 40–50 g L^{-1} ethanol [17]. In practice, however, increasing the initial high dry matter content will also increase the concentrations of inhibitory compounds. The tolerance to inhibitors and the fermentation performance of cells depends on the nature of the substrate and the extent to which the fermenting organism is adapted to the specific challenges imposed by the substrate [18,19]. Substrate stress affects the energy metabolism of the yeast cells, since many stress responsive processes depend on ATP availability [20]. Some of the inhibitory compounds in lignocellulosic material appear to decrease the specific sugar uptake rate and the specific ethanol production rate [21,5], both of which are highly correlated with ATP production. Adenine nucleotides also participate in numerous intracellular reactions, and their intracellular concentrations may greatly affect metabolism and yeast cell performance [22]. The study of energy metabolism can therefore provide insights into the maintenance requirements of yeast cells in lignocellulosic fermentations.

To determine the re-usability of the yeast cells, we must know whether the cells can sustain high ethanol production rates under extended periods of stress and nutrient limitation. This is crucial for processes involving cell recycling or re-circulation. During cultivation, the ethanol production rate may not reflect the ethanol production capacity, because cell performance is reduced by the abovementioned substrate limitations. In this study, we measured the fermentation capacity of the cells after re-inoculation in a non-inhibitory nutritionally rich medium. Using this approach, we can detect whether the cells are irreversibly affected by the previous fermentation conditions.

To evaluate the different performance of yeast strains grown in different lignocellulosic substrates and the effect of nutrient addition, the fermentation performance, energy status and fermentative capacity of the strains were measured.

Methods

Microorganisms

Two yeast strains were used: *Saccharomyces cerevisiae* Thermosacc (Thermosacc) (Lallemand, USA) and *Saccharomyces cerevisiae* CCUG 53310 (CCUG) (Culture Collection University of Gothenburg, Sweden). Thermosacc is a commercial thermotolerant strain of *S. cerevisiae* developed to withstand the stress of industrial fermentation and higher concentrations of organic acids [23]. CCUG is an industrially harvested yeast strain, selected because it originates from spent sulphite liquor ethanol plant.

Media and chemicals

The fermentation media were a filtrated spruce dilute-acid hydrolysate (SH) with a water-insoluble solids (WIS) content of 20% (kindly provided by SEKAB E-Technology) and spent sulphite liquor (SSL) (kindly provided by Domsjö Fabriker, Aditya Birla). The concentrations of available hexoses, weak acids, HMF, furfural and phenolics in the spruce hydrolysate were 66.3 g L^{-1}, 8.0 g L^{-1}, 2.4 g L^{-1}, 1.9 g L^{-1} and 5.0 g L^{-1}respectively. In the spent sulphite liquor, the respective concentrations of the same compounds were 35.6 g L^{-1}, 6.4 g L^{-1}, 0.3 g L^{-1}, 0.2 g L^{-1} and 1.0 g L^{-1}. The spruce hydrolysate slurry was centrifuged and the supernatant was filtered through 0.2 μm pore size filters. The spent sulphite liquor was used without centrifugation and filtration. All chemicals were of analytical grade and were purchased from Sigma Aldrich (Sweden). Yeast extract was purchased from Becton Dickinson (Sweden).

Cell cultivations and fermentations

Cells were proliferated under aerobic conditions in 250 ml Erlenmeyer flasks using a defined media containing excess nutrients and vitamins [7]. The flasks were incubated at 30°C in an orbital shaker at 180 rpm. At the end of the aerobic growth phase (when the glucose was consumed) SH or SSL was added to the culture to a final concentration of 25% of the initial concentration (final volume ratio in the culture 1:4). The cultures were again aerobically incubated until all glucose was consumed. The fermentations were performed in batch mode in 300 ml Erlenmeyer flasks equipped with a glycerol loop to release CO_2 and exclude oxygen. The temperature was set to 30°C and agitation to 150 rpm. Prior to inoculation and filtration, the pH of all cultivations was adjusted to 5.5 with 5 M NaOH. As nutrient supplement, 1% (w/v) yeast extract was supplied according to the experimental design.

All experiments were performed in duplicate. CCUG is a flocculating strain, which affects the reproducibility of its results [24].

Fermentation capacity tests

In the fermentation capacity tests, a 10 ml sample was withdrawn from the lignocellulosic fermentation, and the cells were pelleted by centrifugation (SIGMA Laborzentrifugen GmbH, Osterode, Germany) at 4°C. The pelleted cells were inoculated in a nutritionally rich media containing 20 g L^{-1}glucose, 20 g L^{-1} peptone and 10 g L^{-1} yeast extract. Throughout the next 60 min, samples were regularly withdrawn (t = 10, 20, 40 and 60 min) and their ethanol content determined.

Cell viability test

Cell viability was evaluated by enumerating the colony forming units (CFU). Cells were grown on nutritionally rich non-inhibitory agar plates containing 20 g L^{-1} glucose, 20 g L^{-1} peptone and 10 g L^{-1} yeast extract.

Extraction of ATP, ADP, AMP and measurements of energy status

Samples (3 ml) were taken for ATP, ADP and AMP measurements and quenched as described in [25] in 17 ml pure methanol maintained at −40°C. The cells were pelleted in a centrifuge (SIGMA Laborzentrifugen GmbH, Osterode, Germany) at −20°C, 4000 *g* for 5 min, flash-frozen in liquid nitrogen and stored at −80°C until analysis. The ATP, ADP and AMP were then extracted according to [26]: 0.5 ml of 0.51 M trichloroacetic acid (TCA) containing 17 mM EDTA was added and the samples were incubated at 4°C for 15 minutes. The extracts were then centrifuged at 18078 *g* for 3 min and subsequently neutralized with 2 M Tris-base.

The energy charge was calculated as follows [27]:

$$EC = \left([ATP] + \frac{1}{2}[ADP]\right) / \left([ATP] + [ADP] + [AMP]\right).$$

Analytical procedures

The media were analysed by MoRe research using an Aminex HPX-87H column maintained at 45°C. Glycerol, ethanol and organic acids were detected by an RI detector, and a UV detector was used for HMF and furfural. The eluent was 10 mM H_2SO_4 and flow rate was 0.8 ml min^{-1}.

After the fermentation, ethanol and glycerol were analysed by HPLC using an RI detector and an Aminex HPX-87H column with a 30 mm × 4.6 mm Cation-H Bio-Rad micro-guard column maintained at 45°C. The eluent was 5 mM H_2SO_4 and flow rate was 0.6 ml min^{-1}. Hexoses were analysed by high performance anion exchange chromatography using an electrochemical detector and a 4 mm × 250 mm Dionex CarboPac PA1 column with a 4 mm × 50 mm guard column maintained at 30°C. Elution was performed at 1 ml min^{-1} using eluents A (300 mM NaOH) and B (100 mM NaOH + 85 mM sodium acetate). The adenine nucleotides were analysed by HPLC (Ultimate 3000, Dionex Corp., Sunnyvale, US) with an Luna® 5u C18(2) 100 Å LC column (150 mm × 4.6 mm) (Phenomenex Inc., Torrance, US) maintained at 20°C. The mobile phase was acetonitrile and tetrabutylammonium buffer (0.005 M tetrabutylammonium hydrogensulfate, 0.01 M Na_2HPO_4), pH 7.0. The acetonitrile gradient was as follows: t0 min 6%, t3 min 6%, t16 min 25%, t22 min 25%, t27 min 6%. The system was then equilibrated for 8 minutes to the initial conditions. The flow rate was 1 ml/min. All detections were performed with a photodiode array detector PDA-3000 (Dionex Corp., Sunnyvale, US) operated at 260 nm. Peak identities were confirmed by co-elution with standards and quantified by comparison with standard solutions of known concentrations.

Results and discussion

This study investigated the effects of fermenting strain, fermentation substrate and nutrient supplementation of the medium on the fermentative performance and physiological characteristics of *S. cerevisiae*. Effects were evaluated by ethanol production, cell growth, colony forming units, energy status, and fermentation capacity.

Importance of strain selection on the fermentation of lignocellulosic hydrolysates

Fermenting lignocellulosic hydrolysates and side streams into ethanol or other bulk chemicals is a demanding task for fermenting microorganisms. *S. cerevisiae* is considered the most robust microorganism for industrial scale fermentation in inhibitory media [28]. However, the fermentation performance of different *S. cerevisiae* strains critically depends on the culturing conditions [29-32]. In selecting an appropriate strain, we should consider the metabolic and physiological characteristics of the strain under the processing conditions (T and pH), and with regard to the ethanol, sugar and inhibitor concentrations in the fermentation medium. The two C6 fermenting strains of *S. cerevisiae* evaluated in the present study, namely, Thermosacc and CCUG, represent different strain selection strategies. Thermosacc is specifically designed to cope with certain challenges during fermentation, while CCUG has naturally evolved under the harsh industrial conditions prevailing in a sulphite mill.

As shown in Figure 1A, Thermosacc satisfactorily fermented SH and SSL under the prevailing conditions. After 22 h fermentation, the ethanol concentrations in SH- and SSL-containing media were 17 g L^{-1} and 9 g L^{-1}, respectively, corresponding to yields of 0.26 $g_{ethanol}$ and 0.25 $g_{ethanol}$ per g total hexoses. Although Thermosacc growth was limited in terms of cell dry weight (CDW) (data not shown), the viability of the cells (measured by the CFU, Figure 1) was high and remained more or less stable throughout the fermentation period. The strain harvested from the sulphite mill (CCUG) failed to ferment SH, but performed quite well in SSL, achieving an ethanol concentration of 11 g L^{-1} (a yield of 0.31 $g_{ethanol}$ g^{-1} total available hexoses) after 22 h fermentation. These results suggest that the CCUG strain has successfully adapted in the sulphite mill, but loses its adaptive benefits when exposed to a moderately different set of conditions. This hypothesis is consistent with earlier reports that stressed cells continuously and randomly produce genetic variants, which are immortalized as mutations only if they permit

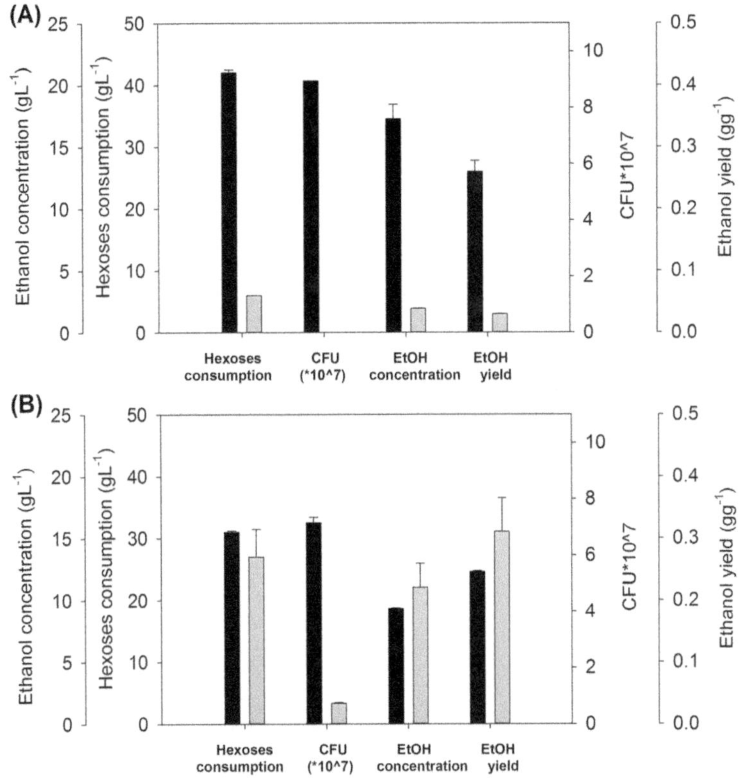

Figure 1 Hexose consumption, viability and ethanol production after 22 h fermentation. Fermentation substrate was **(A)** spruce hydrolysate **(B)** spent sulphite liquor. Black and grey bars denote the *S. cerevisiae* strains Thermosacc and CCUG, respectively. Error bars indicate the maxima and minima of two independent fermentations.

cell growth [33,34]. The low level of viable cells after 22 h fermentation in SH indicates that CCUG could not overcome the high toxicity of this foreign medium (Figure 1). However, it should be mentioned that viable cell counts of CCUG were comparably low in SSL, despite the strong fermentation performance in this medium.

Yeast performance in SH and SSL supplemented with yeast extract

The fermentation performance of yeast is varied by the complexity and variety of substances comprising lignocellulosic materials. The SH used in the present study is a high gravity substrate with 20% WIS content. The production of this hydrolysate also increases the concentration of inhibitory compounds and increases the stress factors affecting yeast performance, including osmolarity [35], sugar concentration [36] and ethanol concentration [37]. The concentrations of inhibitors such as acids, HMF, furfural and phenolics are elevated in SH while sulphite levels are higher in SSL (see Methods).

Supplementation with yeast extract (YE) significantly improved the fermentative performance of both yeast strains. Supplemented Thermosacc produced 25.4 g L^{-1} and 10.3 g L^{-1} ethanol from SH and SSL, respectively, corresponding to yields of 0.38 g g^{-1} hexoses and 0.27 g g^{-1} hexoses (based on total fermentable sugar contents in the media). The effect of yeast extract on SH fermentation by CCUG was even more impressive. The sugar uptake was doubled, while the ethanol production almost tripled (Figure 2A). As shown in Figure 2, YE addition affected all of the measured fermentation characteristics to similar extent. Increased ethanol production is partially attributable to the increased cell viability in the presence of yeast extract. However, in the CCUG strain, the higher conversion of hexoses into ethanol indicates that YE additive improved both cell viability and cell metabolism. These findings confirm earlier reports that yeast extract provides cellular building blocks such as amino acids, and also enhances the physiological status of the cells [38,39].

Yeast extract considerably enhanced the biomass formation and viability of Thermosacc in SH-containing medium. The biomass concentration of this organism reached 6 g L^{-1} CDW (data not shown) and the CFU count was 4.73×10^7 (Figure 2A). Nutrient addition also exerted a positive effect in SSL, although to lesser extent than in SH (Figure 2B). The different effects of nutrient addition to the two media probably reflect the higher maintenance requirement of cells fermenting SH. SH contains higher amounts of inhibitory compounds such as phenolics than SSL (5 g L^{-1} in SH versus 1 g L^{-1} in SSL). The

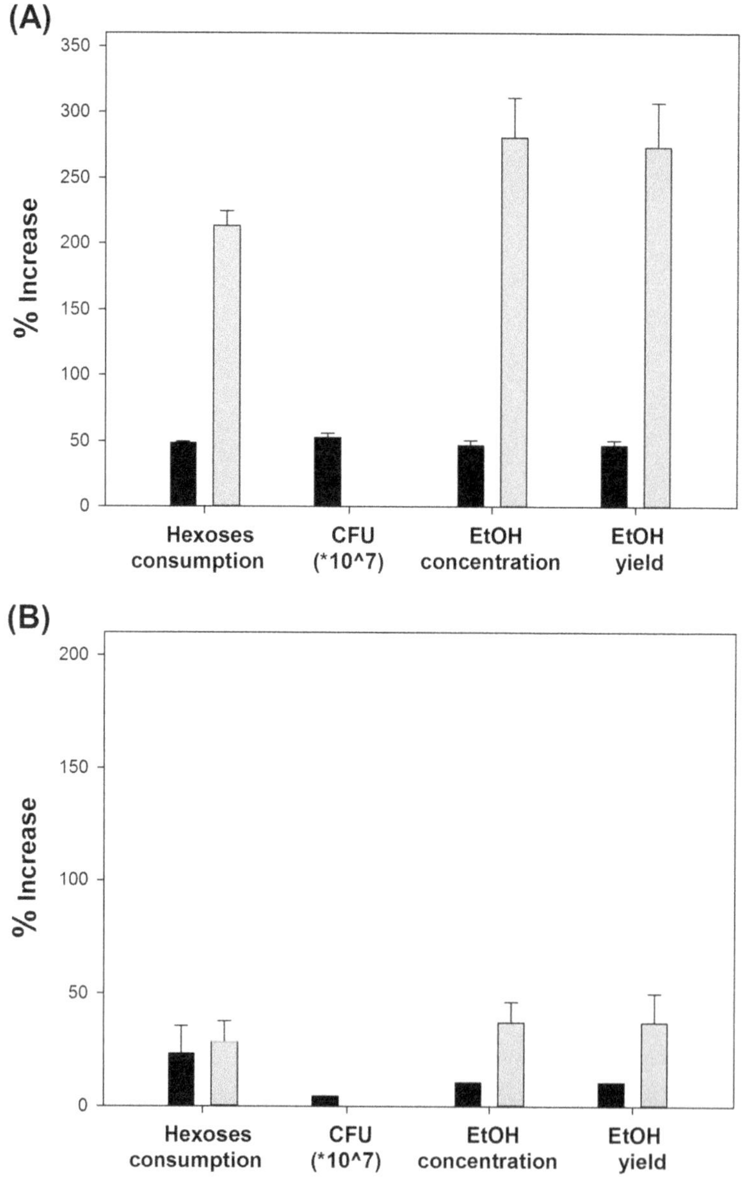

Figure 2 Increased fermentation performance in media supplemented with yeast extract. Black and grey bars indicate fermentations with *S. cerevisiae* strains Thermosacc and CCUG 53310, respectively. Fermentation substrate is **(A)** spruce hydrolysate **(B)** spent sulphite liquor. Error bars indicate the maxima and minima of two independent fermentations.

biomass production of CCUG was similar in both substrates. In this strain, yeast extract addition exerted a strong stimulatory effect despite the low cell viability. In all cases, the decreased concentrations of residual sugars and increased ethanol formation accompanied a higher biomass concentration, reflecting the improved physiological status of the cells.

The positive impact of supplementing lignocellulosic fermentations with yeast extract is most likely attributable to the complex mixture of vitamins, minerals and free amino nitrogens (FAN) present in yeast extract. All of these compounds are necessary for cell growth, which is correlated with ethanol production [15,16]. Ethanol production may be additionally enhanced by free amino nitrogens, whose potential role in the redox balance may lower the need for glycerol production [13].

Energy status of yeast cells cultivated in SH and SSL in the absence and presence of yeast extract

Living cells acquire energy through ATP. Besides satisfying a cell's maintenance energy requirements, ATP is required for growth and biomass formation. Under adverse conditions, maintenance energy requirements escalate to the extent that growth becomes ATP-limited [40].

Table 1 ATP, ADP and AMP concentrations throughout 24 h fermentation

Strain/condition	ATP			ADP			AMP			Energy charge
Fermentation time (h)	0	4	22	0	4	22	0	4	22	22
Thermosacc/SH	8.06 ± 0.08	8.02 ± 0.17	7.29 ± 0.16	0.18 ± 0.01	0.33 ± 0.06	0.15 ± 0.01	0.16 ± 0.03	0.15 ± 0.00	0.08 ± 0.01	0.98
Thermosacc/SH+	10.94 ± 0.43	11.47 ± 0.04	5.58 ± 0.45	0.55 ± 0.03	0.54 ± 0.01	0.19 ± 0.02	0.14 ± 0.03	0.07 ± 0.03	0.43 ± 0.09	0.92
Thermosacc/SSL	8.83 ± 0.25	5.39 ± 0.38	1.48 ± 0.17	4.76 ± 0.17	2.85 ± 0.55	1.10 ± 0.10	0.23 ± 0.05	0.21 ± 0.02	0.14 ± 0.00	0.75
Thermosacc/SSL+	12.86 ± 0.38	7.01 ± 0.92	1.75 ± 0.63	9.77 ± 0.08	6.40 ± 0.49	2.13 ± 0.57	0.14 ± 0.03	0.39 ± 0.07	0.38 ± 0.05	0.66
CCUG 53310/SH	7.28 ± 0.41	3.46 ± 0.63	2.11 ± 0.05	3.10 ± 0.67	2.4 ± 0.08	1.62 ± 0.19	3.21 ± 0.02	2.83 ± 0.06	2.17 ± 0.20	0.49
CCUG 53310/SH+	13.47 ± 0.24	5.64 ± 1.39	2.15 ± 0.03	6.73 ± 0.06	3.53 ± 0.54	1.67 ± 0.14	0.33 ± 0.01	4.99 ± 0.48	2.68 ± 0.29	0.46
CCUG 53310/SSL	4.95 ± 3.54	N.D.	7.42 ± 6.61	3.40 ± 1.79	N.D.	3.16 ± 2.13	0.05 ± 0.05	N.D.	0.08 ± 0.02	0.84
CCUG 53310/SSL+	5.77 ± 3.11	5.01 ± 0.42	6.26 ± 1.49	5.23 ± 1.65	3.46 ± 0.15	5.07 ± 1.81	0.08 ± 0.01	0.29 ± 0.10	1.59 ± 0.73	0.68

Concentrations are expressed in μmol (g DW)$^{-1}$ unless otherwise stated. SH = Spruce hydrolysate. SSL = Spent sulphite liquor. + indicates that the fermentation has been supplemented with yeast extract.
± indicates the maximum and minimum of two independent fermentations.

Concomitant with ATP reduction, AMP levels elevate, with potential decrease in energy charge. Nutrient-poor lignocellulosic substrates contain a cocktail of growth inhibiting substances (reviewed in [3]) that encourage such phenomena. From an ethanol production perspective, low ATP levels are not necessarily disadvantageous, since biomass formation can be restrained while glycolysis and ethanol flux is stimulated [41]. However, ATP levels must remain above a certain threshold to prevent glycolysis inhibition [42].

Cultivated in SH, the Thermosacc strain retained high ATP levels and low AMP levels (Table 1). In contrast, ATP levels temporally declined in the CCUG strain, while AMP accumulated. In SSL-cultivated Thermosacc, ATP levels dramatically decreased after 22 hours fermentation, while AMP levels did not appreciably increase. Again, CCUG was relatively resistant to SSL and retained its ATP levels after 22 hours fermentation (Table 1). These results are consistent with the fermentative performance of the two strains.

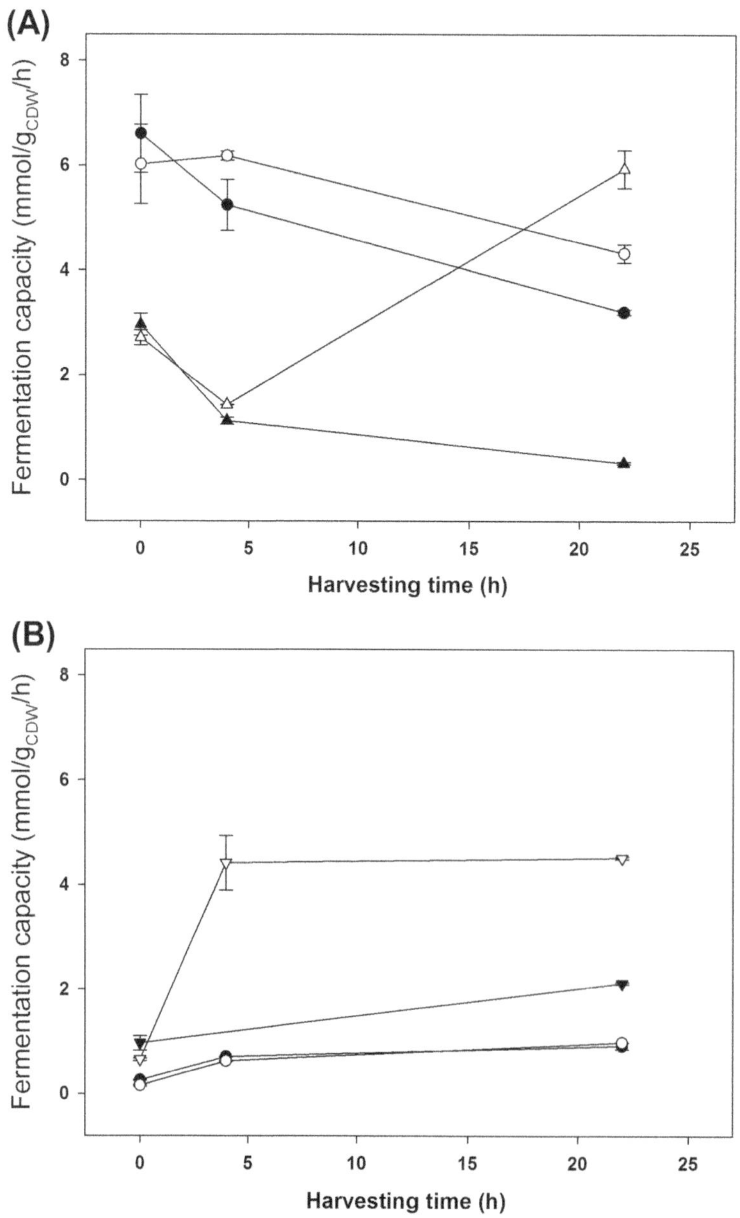

Figure 3 Fermentation capacity following re-inoculation in YPD media. Open and closed symbols represent cells harvested from nutrient-supplemented and non-supplemented fermentations, respectively. Circles and triangles denote the *S. cerevisiae* strains Thermosacc and CCUG 53310, respectively. **(A)** Fermentation capacity of cells harvested from spruce hydrolysate. **(B)** Fermentation capacity of cells harvested from spent sulphite liquor. Error bars indicate the maxima and minima of two independent fermentations.

Nutrient supplementation exerted little effect on energy charge in either medium. Throughout the SH fermentation, the energy charge in Thermosacc and CCUG was approximately 1 and below 0.5, respectively (Table 1). The low energy charge of CCUG reflects the low viability of the population, which may in turn reflect the high maintenance requirements of this strain. However, CCUG exhibited a higher energy charge in SSL medium, again suggesting that this yeast strain is better adapted to SSL and metabolically equipped for the challenges imposed by this substrate. This hypothesis is supported by the much higher titres and ethanol yields obtained by CCUG in SSL than in SH (Figure 1B). On the other hand, the energy charge of Thermosacc was slightly lower in SSL than in SH (Table 1). This reduction was accompanied by noticeable deterioration in fermentation performance. Hence, under these challenging conditions, maintaining high energetic status is crucial to sustaining a high fermentation rate. However, the absolute ATP levels were not obviously correlated with the rates of glycolysis and ethanol formation.

Changes in fermentation capacity during cultivation in SH and SSL in the absence and presence of yeast extract

The fermentation capacity of a cell quantifies the ability of the cell to produce ethanol in non-inhibitory nutrient rich media. This measure indicates whether the reduced ethanol production induced by cultivation in low-nutrient lignocellulosic media is restored when the cells are transferred to optimal, non-inhibitory conditions. By investigating how fermentation capacity relates to nutrient supplementation, we can assess whether yeast cells fermenting a lignocellulosic media can be revitalised and consequently reused.

Nutrient supplementation to the SH fermentation enhanced the fermentation capacity of Thermosacc, although this strain demonstrated reasonable fermentation capacity in the absence of additional nutrients (Figure 3A). The CCUG strain harvested from SH fermentations performed poorly when re-inoculated in non-inhibitory nutritional rich media, supporting the hypothesis that CCUG cannot ferment SH. Conversely, Thermosacc demonstrated very poor fermentation capacity in SSL, irrespective of nutrient addition, whereas CCUG responded very positively to nutrient-supplemented SSL. This suggests that CCUG is less affected by the conditions prevailing in SSL than Thermosacc (Figure 3B). The fermentation capacity of CCUG cells harvested from SSL fermentations also increased over time, although levels were initially low. In contrast, the SH fermentation capacity of CCUG generally deteriorated over time. Excepting Thermosacc fermenting SSL, the fermentation capacity was improved in cells harvested from nutrient-supplemented fermentations, indicating that cell viability and revitalizing ability is increased in nutrient-supplemented media.

Conclusion

This study highlights the importance of selecting a fermentation strain that is well adapted to the prevailing process conditions. The yeast strain CCUG, originally isolated from a sulphite mill, demonstrated stronger performance than Thermosacc in SSL media, while the opposite was observed in SH media. Nutrient supplementation did not moderate this difference, but markedly improved the performance of both strains in both substrates.

Competing interests

The authors declare that they have no competing interests.

Authors' contributions

EJ participated in the design of the study, carried out the fermentations, analysed the results and partially wrote the manuscript. CX participated in the experimental procedure, the HPLC analysis, result analysis and writing of the manuscript. CL conceived the study and participated in analysing the results and writing the manuscript. All authors read and approved the final manuscript.

Acknowledgments

Financial support by the Nordic Top-level Research Institute, the county administrative board of Västernorrland, SP Processum and the Kempe Foundation is greatly acknowledged. MoRe research is also acknowledged for performing analysis.

References

1. Jeon YJ, Xun Z, Rogers PL: **Comparative evaluations of cellulosic raw materials for second generation bioethanol production.** *Lett Appl Microbiol* 2010, **51**:518–524.
2. Sjöström E: *Wood chemistry fundamentals and applications.* Orlando: Academic Press, INC; 1981.
3. Sluiter JB, Ruiz RO, Scarlata CJ, Sluiter AD, Templeton DW: **Compositional analysis of lignocellulosic feedstocks. 1. Review and description of methods.** *J Agric Food Chem* 2010, **58**:9043–9053.
4. Palmqvist E, Hahn-Hägerdal B: **Fermentation of lignocellulosic hydrolysates II: inhibitors and mechanisms of inhibition.** *Bioresour Technol* 2000, **74**:25–33.
5. Almeida JRM, Bertilsson M, Gorwa-Grauslund MF, Gorsich S, Lidén G: **Metabolic effects of furaldehydes and impacts on biotechnological processes.** *Appl Microbiol Biotechnol* 2009, **82**:625–638.
6. Larsson S, Palmqvist E, Hahn-Hägerdal B, Tengborg C, Stenberg K, Zacchi G, Nilvebrant N-O: **The generation of fermentation inhibitors during dilute acid hydrolysis of softwood.** *Enzyme Microb Technol* 1999, **24**:151–159.
7. Verduyn C, Postma E, Scheffers WA, Dijken JP: **Effect of benzoic acid on metabolic fuxes in yeasts: a continous-culture study on the regulation of respiration and alcoholic fermentation.** *Yeast* 1992, **8**:501–517.
8. Jones RP: **Biological principles for the effects of ethanol.** *Enzyme Microb Technol* 1989, **11**:130–153.
9. Heipieper HJ, Weber FJ, Sikkema J, Kewelo H, de Bont JAM: **Mechanism of resistance of whole cells to toxic organic solvents.** *Trends Biotech* 1994, **12**:409–415.
10. Johansson E, Brandberg T, Larsson C: **Influence of cultivation procedure for Saccharomyces cerevisiae used as pitching agent in industrial spent sulphite liquid fermentations.** *J Ind Microbiol Biotechnol* 2011, **38**:1787–1792.
11. Jones AM, Ingledew WM: **Fuel alcohol production: appraisal of nitrogenous yeast foods for very high gravity wheat mash fermentation.** *Process Biochem* 1994, **29**:483–488.
12. Casey GP, Magnus CA, Ingledew WM: **High-gravity brewing: effects of nutrition on yeast composition, fermentative ability, and alcohol production.** *Appl Environ Microbiol* 1984, **48**:639–646.

13. Thomas KC, Ingledew WM: **Fuel alcohol production: effects of free amino nitrogen on fermentation of very-high-gravity wheat mashes.** *Appl Environ Microbiol* 1990, **56:**2046–2050.
14. Albers E, Larsson C, Lidén G, Niklasson C, Gustafsson L: **Influence of the nitrogen source on Saccharomyces cerevisiae anaerobic growth and product formation.** *Appl Environ Microbiol* 1996, **62:**3187–3195.
15. Devantier R, Scheithauer B, Granato Villas-Boâs S, Pedersen S, Olsson L: **Metabolite profiling for analysis of yeast stress response during very high gravity ethanol fermentations.** *Biotechnol Bioeng* 2005, **90:**703–714.
16. Bayrock DP, Ingledew WM: **Application of multistage continuous fermentation for production of fuel alcohol by very-high-gravity fermentation technology.** *J Ind Microbiol Biotechnol* 2001, **27:**87–93.
17. Galbe M, Sassner P, Wingren A, Zacchi G: **Process engineering economics of bioethanol production.** In *Adv Biochem Eng Biotechnol. Volume 108.* Edited by Olsson L. Heidelberg: Springer-Verlag; 2007:303–327.
18. Lau MW, Dale BE, Balan V: **Ethanolic fermentation of hydrolysates from ammonia fiber expansion (AFEX) treated corn stover and distillers grain without detoxification and external nutrient supplementation.** *Biotechnol Bioeng* 2008, **99:**529–539.
19. Alkasrawi M, Rudolf A, Liden G, Zacchi G: **Influence of strain and cultivation procedure on the performance of simultaneous saccharification and fermentation of steam pretreated spruce.** *Enzyme Microb Technol* 2006, **38:**279–286.
20. Gasch AP: **The Environmental Stress Response: A Common Yeast Response to Environmental Stresses.** In *Yeast Stress Responses. Volume 1.* Edited by Hohmann S, Mager P. Heidelberg: Springer-Verlag; 2002:11–70.
21. Sárvári Horváth I, Franzén CJ, Taherzadeh MJ, Niklasson C, Lidén G: **Effects of furfural on the respiratory metabolism of Saccharomyces cerevisiae in glucose-limited chemostats.** *Appl Environ Microbiol* 2003, **69:**4076–4086.
22. Nielsen J: **It is all about metabolic fluxes.** *J Bacteriol* 2003, **185:**7031–7035.
23. **Lallemand product sheet Thermosacc Dry.** [http://www.lallemandbds.com/wp-content/uploads/2012/12/2013_LBDS_Data-Sheet_Thermosacc-Dry_10080_Data_Rev.00.01.01.2013.pdf]
24. Purwadi R, Brandberg T, Taherzadeh: **A possible industrial solution to ferment lignocellulosic hydrolysates to ethanol: Continuous cultivation with flocculating yeast.** *Int J Mol Sci* 2007, **8:**920–932.
25. Canelas AB, Ras C, Pierick A, Dam JC, Heijnen JJ, Gulik WM: **Leakage-free rapid quenching technique for yeast metabolomics.** *Metabolomics* 2008, **4:**226–239.
26. Lundin A, Thore A: **Comparison of methods for extraction of bacterial adenine nucleotides determined by firefly assay.** *Appl Microbiol* 1975, **30:**713–721.
27. Atkinson DE: **The energy charge of the adenylate pool as a regulatory parameter. Interaction with feedback modifiers.** *Biochem* 1968, **7:**4030–4034.
28. Olofsson K, Bertilsson M, Liden G: **A short review on SSF - an interesting process option for ethanol production from lignocellulosic feedstocks.** *Biotechnol Biofuels* 2008, **1:**7.
29. Meneses FJ, Henschke PA, Jiranek V: **A survey of industrial strains of Saccharomyces cerevisiae reveals numerous altered patterns of maltose and sucrose utilisation.** *J Inst Brew* 2002, **108:**310–321.
30. Martin C, Jönsson LJ: **Comparison of the resistance of industrial and laboratory strains of Saccharomyces and Zygosaccharomyces to lignocellulose-derived fermentation inhibitors.** *Enzyme Microb Technol* 2003, **32:**386–395.
31. Modig T, Almeida JRM, Gorwa-Grauslund MF, Lidén G: **Variability of the response of Saccharomyces cerevisiae strains to lignocellulose hydrolysate.** *Biotechnol Bioeng* 2008, **100:**423–429.
32. Stanley D, Bandara A, Fraser S, Chambers PJ, Stanley GA: **The ethanol stress response and ethanol tolerance of Saccharomyces cerevisiae.** *J Appl Microbiol* 2010, **109:**13–24.
33. Foster PL: **Adaptive mutation - the uses of adversity.** *Annu Rev Microbiol* 1993, **47:**467–504.
34. Rosenberg SM: **Evolving responsively: adaptive mutation.** *Nat Rev Genet* 2001, **2:**645–645.
35. Blomberg A, Adler L: **Physiology of osmotolerance in fungi.** *Adv Microb Physiol* 1992, **33:**145–212.
36. Erasmus DJ, van der Merwe GK, van Vuuren HJJ: **Genome-wide expression analyses: metabolic adaptation of Saccharomyces cerevisiae to high sugar stress.** *FEMS Yeast Res* 2003, **3:**375–399.
37. Gray WD: **Studies on the alcohol tolerance of yeasts.** *J Bacteriol* 1941, **42:**561–574.
38. Alfenore S, Molina-Jouve C, Guillouet SE, Uribelarrea J-L, Goma G, Benbadis L: **Improving ethanol production and viability of Saccharomyces cerevisiae by a vitamin feeding strategy during fed-batch process.** *Appl Microbiol Biotechnol* 2002, **60:**67–72.
39. Jorgensen H: **Effect of nutrients on fermentation of pretreated wheat straw at very high dry matter content.** *Appl Biochem Biotechnol* 2009, **153:**44–57.
40. Gustafsson L, Ölz R, Larsson K, Larsson C, Adler L: **Energy balancing calculations as a tool to determine maintenance energy requirements under stress conditions.** *J Pure Appl Chem* 1993, **65:**1893–1898.
41. Larsson C, Nilsson A, Blomberg A, Gustafsson L: **Glycolytic flux is conditionally correlated with ATP concentration in Saccharomyces cerevisiae: a chemostat study under carbon- or nitrogen-limiting conditions.** *J Bacteriol* 1997, **179:**7243–7250.
42. Larsson C, Påhlman I-L, Gustafsson L: **The importance of ATP as a regulator of glycolytic flux in Saccharomyces cerevisiae.** *Yeast* 2000, **16:**797–809.

A novel serine hydroxymethyltransferase from *Arthrobacter nicotianae*: characterization and improving catalytic efficiency by rational design

Wei Jiang[1], Lin Chen[1], Nan Hu[2], Shaohui Yuan[1], Bin Li[1] and Ziduo Liu[1*]

Abstract

Background: Serine hydroxymethyltransferase (SHMT) is the key enzyme in L-serine enzymatic production, suggesting the importance of obtaining a SHMT with high activity.

Results: Here, a novel SHMT gene, *glyA*, was obtained through degenerate oligonucleotide-primed PCR and encoded a novel SHMT with 54.3% similarity to the known SHMT from *Escherichia coli*. The obtained protein *An*SHMT showed the optimal activity at 40°C and pH 7.5, and was more stable in weakly alkali conditions (pH 6.5-8.5) than *Hyphomicrobium methylovorum*'s SHMT (pH 6.0-7.5), In order to improve the catalytic efficiency of the wild type, the site-directed mutagenesis based on sequences alignment and bioinformatics prediction, was used and the catalytic efficiency of the mutant I249L was found to be 2.78-fold higher than that of the wild-type, with the replacement of isoleucine by leucine at the 249 position.

Conclusions: This research provides useful information about the interesting site, and the application of DOP-PCR in cloning a novel *glyA* gene.

Keywords: *Arthrobacter nicotianae*, SHMT, Characterization, Site-directed mutagenesis, Catalytic efficiency

Background

There have been only a few reports about the enzymatic characterization and the change of the enzymatic properties of SHMT by protein engineering [1,2]. Currently, L-serine production mainly relies on enzymatic conversion from glycine precursor plus a C_1 compound [3], and the key enzyme in L-serine enzymatic conversion is SHMT. The SHMT is coded by the *glyA* genes and act as the first enzyme in the assimilation of C_1 compounds through the addition of formaldehyde to glycine, producing the principal intermediate in the pathway, serine [4]. Therefore, it is necessary to obtain a SHMT with high activity of some new microorganisms and improve the catalytic efficiency through in vitro directed evolution.

SHMT (EC 2.1.2.1), a member of the α-class of pyridoxal phosphate enzymes, catalyzes the reversible interconversion of serine and glycine, changes the chemical bonding at the C^{α}-C^{β} bond of the serine side-chain mediated by the pyridoxal phosphate cofactor [5]. The crystal structures of SHMT from the human and rabbit confirm that SHMT belongs to the α-family of enzymes and shares a similar tertiary fold and mechanism [6,7]. SHMT is ubiquitous, highly conserved PLP-dependent enzyme with tetrahydrofolate (THFA) as the C_1 acceptor [8], purified from the animals [9,10], plants [11] and bacteria [12]. SHMT is widely used in the synthesis of the serine using glycine and formaldehyde, while the activity of the wild-type is not sufficient for industrial production [2].

The degenerate oligonucleotide-primed PCR (DOP-PCR) has been applied to the characterization of abnormal chromosomes and also in the cloning of new markers for specific chromosome regions [13]. Based on unknown genome information circumstances, using DOP-PCR to obtain *glyA* (encoding SHMT) can overcome the shortcomings of existing methods, such as cDNA library [14,15], southern-blot hybridization [16], direct PCR amplification according to the known genome information [17,18], and shotgun technology [19], for their waste time and material under the same conditions. Protein engineers redesign proteins in

* Correspondence: lzd@mail.hzau.edu.cn
[1]State Key Laboratory of Agricultural Microbiology, College of Life Science and Technology, Huazhong Agricultural University, Wuhan 430070, P. R. China
Full list of author information is available at the end of the article

order to improve their biomedical or industrial utility [20]. Although the crystal structures of SHMT from the human, rabbit and *Escherichia coli* (*E. coli*) are reported, little is known at present about the structure-function relationship among these enzymes [5-7]. And the exact active-site residues have never been confirmed using kinetic analyses of the wild-type and variant enzymes. Structure-based site-directed mutagenesis is usually applied to produce variants with dramatically improved specificities and the residues in or near the active site tend to be chosen as their special roles in the activity of the enzymes [21-23].

The objectives of the current study were to (i) clone the glyA gene from *Arthrobacter nicotianae* (*A. nicotianae*) using degenerate primers, (ii) study characteristics of recombinant SHMT by isolating the enzyme with high activity and (iii) improve its catalytic efficiency by site-directed mutagenesis.

Results

Cloning of *glyA* gene

Using degenerate primers, a fragment was obtained by DOP-PCR with the genomic DNA of *A. nicotianae* as the template. The open reading frame (ORF) of the *glyA* (1323 bp, GenBank: KF359496) fragment encoded a polypeptide of 440 amino acids, with a deduced molecular mass of 47.3 kDa. The phylogenetic analyses of the SHMT sequences produced a tree to further verify the evolutionary relationship among *An*SHMT and other known SHMTs, indicating that the *An*SHMT shared 54.3% amino acid identity with the known SHMT from *E. coli* (Figure 1).

Bioinformatic analysis of the amino acid sequence

Multiple sequence comparisons were performed by Clustal W Method, indicating that the conserved amino acid sequences and motifs are critical for the active site of SHMT (Figure 2). The highly conserved active site in all known SHMT enzymes T/ST/STTHKT/SL was also found in *An*SHMT (235–242) in the form of TSTTHKTL (Figure 2c), the putative active site [5,14,24]. The site of the site-directed mutagenesis underlined and marked by a five-pointed star was the close proximity to the highly conserved active site and the putative active site (TSTTHKTL) (Figure 2c). The deduced active site structure (TSTTHKTL) and the position of the mutant (249 isoleucine) were briefly presented the three-dimensional structure of *An*SHMT (Figure 3).

Based on multiple sequence comparisons, two well-known conserved sequences LTNKYAEGYPGRRYYGG (61–77) and GGHLTHG (134–140) (Figure 2a and 2b) were also detected in *An*SHMT [25]. The glycine rich region GQQGGP (268–273) (Figure 2c), was highly conserved sequence, significant homologous sequence, and proposed to be essential for PLP binding [24], the region of which was indicated on the three-dimensional structure of *An*SHMT (Figure 3).

Site-directed mutagenesis

A leucine residue was introduced into the 249 position using site-directed mutagenesis and conformed, followed by the test of its effects on enzymatic activity using purification I249L-SHMT.

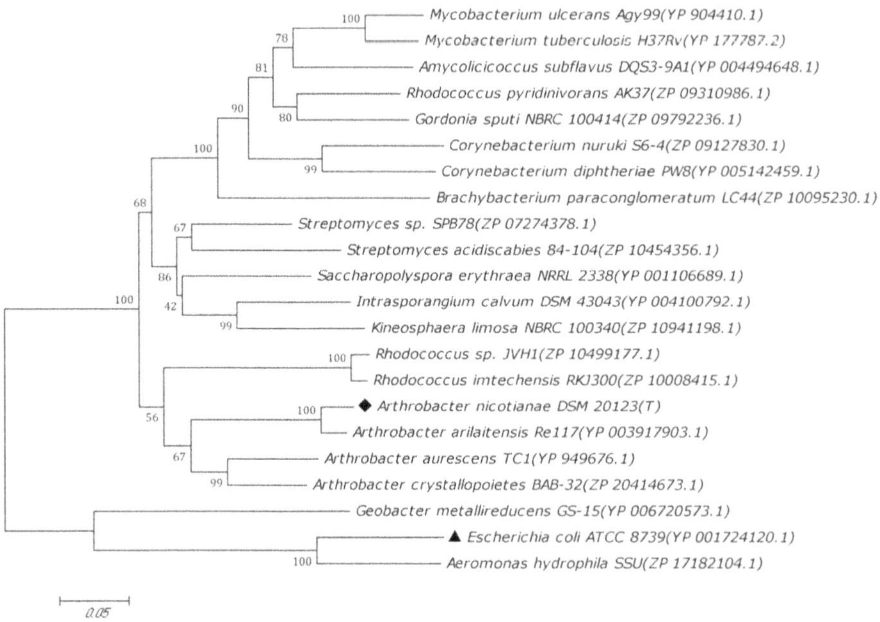

Figure 1 Phylogenetic tree of the AnSHMT. The phylogenetic tree was established using the program MEGA 5.05. The SHMT protein sequences were obtained from GenBank and PDB (http://www.rcsb.org/pdb/home/home.do), except for AnSHMT.

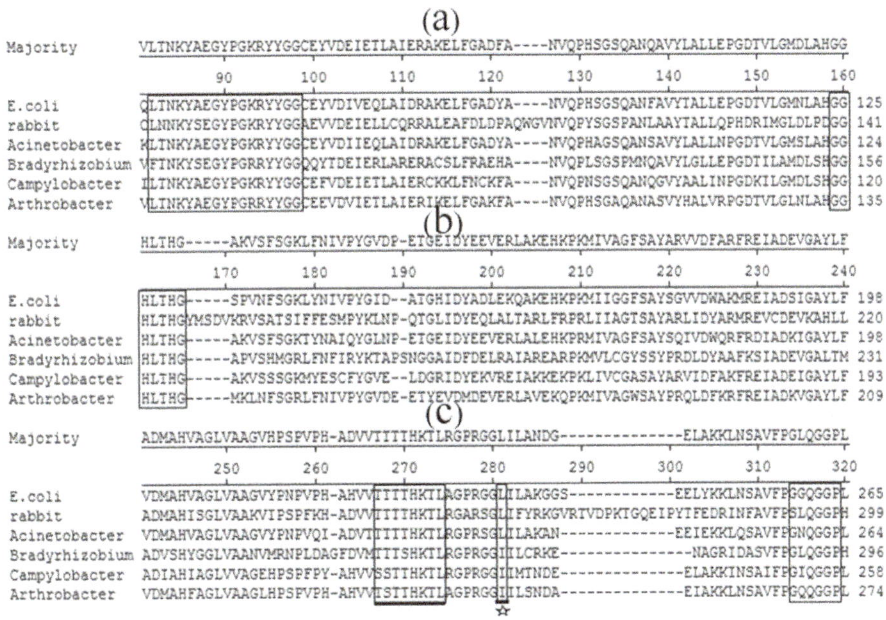

Figure 2 Multiple sequence comparison by ClustalW. (a) a part of the sequence alignment result (81-160 sites); **(b)** a part of the sequence alignment result (161-240 sites); **(c)** a part of the sequence alignment result (241-320 sites). Open boxes indicate the four highly conserved regions and the conserved active site is underlined. The site-directed mutagenesis site is underlined and is marked by five-pointed star. The amino acid sequences were obtained from NCBI database (http://www.ncbi.nlm.nih.gov/).

Expression, purification and activity of the *An*SHMT and the mutant

The *glyA* was cloned into the vector pGEX-6p-1 and expressed in DE3. The induced and non-induced recombinant bacteria (harboring pGEX-6p-*glyA*) and the induced control bacterium (harboring empty pGEX-6p-1 vector) were examined with SDS-PAGE (Figure 4c). After purification with Glutathione Sepharose 4B and digestion with 3C protease, the recombinant *An*SHMT and I249L-SHMT were harvested and resolved to a single band, with the purified *An*SHMT exhibiting the expected molecular mass (47.3 kDa) (Figure 4c).

The steady-state kinetic parameters of the wild-type enzyme and the I249L-SHMT were measured at 40°C at a DL-3-phenylserine concentration from 0.1–16.3 mg/ml (Table 1). Mutant I249L showed a 7% increase in Km and 2.97-fold increase in k_{cat}, resulting in approximately 2.78-fold increase in (k_{cat}/Km). The results indicated that the

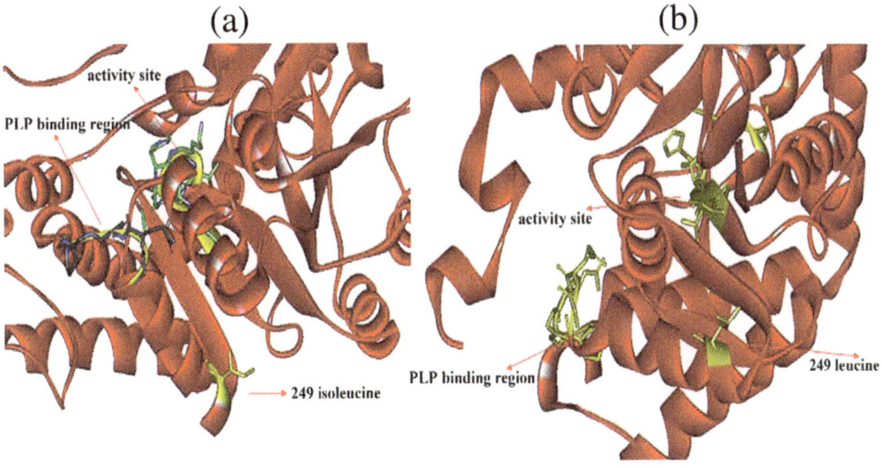

Figure 3 Modeled three-dimensional structure of the *An*SHMT. (a) the three-dimensional structure of the wild type enzyme; **(b)** the three-dimensional structure of the mutational enzyme. The catalytic site, PLP binding region and the site of site-directed mutagenesis were indicated on the three-dimensional structure, respectively. The three-dimensional structure was generated using the Swiss-Pdb viewer.

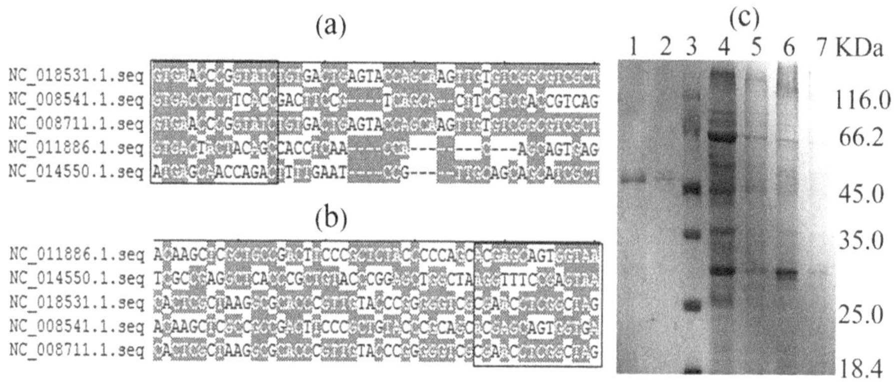

Figure 4 Design of both end degenerate primers and 12% SDS-PAGE analysis of the purified proteins. (a and b) The *glyA* gene sequences were obtained from NCBI database and belonged to the genus of *Arthrobacter* (GI: NC_008541; NC_008711.1; NC_011886.1; NC_014550.1; NC_018531.1). Degenerate primers were designed by the regions in the boxes. **(c)** 12% SDS-PAGE analysis of the purified proteins. The bands in the ellipses show GST (glutathione-S-transferase) and fusion protein (*An*SHMT and GST) respectively. Lane 1: purified I249L without GST. Lane 2: purified *An*SHMT (the wild-type enzyme) without GST. Lane 3: protein marker. Lane 4: recombinant bacterium (harboring pGEX-6P-*glyA*) induced by IPTG. Lane 5: recombinant bacterium (harboring pGEX-6P-*glyA*) non-induced by 0.1 mM IPTG. Lane 6: bacterium (harboring pGEX-6p-1) induced by IPTG. Lane 7: bacterium (harboring pGEX-6p-1) non-induced by 0.1 mM IPTG. The protein molecular weight ladder is Unstained Protein Molecular Weight Marker (Fermentas, Canada).

catalytic efficiency was considerably improved by replacing the isoleucine with leucine at the 249 position.

Effects of temperature and pH on enzyme activity

The maximal activity of the *An*SHMT was observed at 40°C, retaining over 50% of the maximal activity at temperatures from 30°C to 65°C (Figure 5a). The I249L-SHMT diaplayed the same optimum temperature (40°C) with the *An*SHMT (Figure 5a). After 1 h incubation under pH 7.5 (Figure 5c), the *An*SHMT retained over 35% of its maximal activity from 35°C to 45°C, but less than 15% at 55°C.

The *An*SHMT exhibited the optimal activity at pH 7.5, and was sensitive in low pH buffers, displaying less than 20% of its maximal activity at pH 5.5 and nearly no activity was detected below pH 2.5 (Figure 5b). Similarly, the I249L-SHMT also showed the maximal activity at pH 7.5, and retained over 85% of its maximal activity at pH 8.0 (Figure 5b). Without any stabilizer, the purified *An*SHMT showed significant stability under weakly acidic and alkaline environment (pH 6.5-8.5), retaining over 80% of the maximum activity over a pH range from 7.0 to 8.5 for 24 h at 4°C, and more than 70% of the maximal activity at pH 6.5. However, the *An*SHMT exhibited a rapid decrease in activity at pH 5.5 (Figure 5d).

Table 1 Steady-state kinetic parameters for the wild-type *An*SHMT and the mutant

Enzyme	*Km* (mM)	k_{cat} (min^{-1})	k_{cat}/Km ($min^{-1}\,M^{-1}$)
WT	57.86 ± 0.07	96.02 ± 0.12	1.66×10^3
I249L	61.95 ± 0.13	285.3 ± 0.15	4.61×10^3

Note. The data are the average of three replicates.

The effects of different metal ions or chemical reagents on *An*SHMT are presented in Table 2. The enzyme activated by Mg^{2+}, Ba^{2+} and EDTA, but could be strongly inhibited by Hg^{2+}, Cu^{2+}, Zn^{2+} and Fe^{2+}, and slightly inhibited by Co^{2+}, Mn^{2+}, DTT, and $CO(NH_2)_2$. Besides, the enzyme was not significantly influenced in activity by NH_4^+, Sl^{2+}, Ca^{2+}, Pb^{2+}, SDS and CTAB.

Discussion

The taxonomy of the bacterium *A. nicotianae* (ATCC 15236) was studied in 1982 [26]. While no paper on the characterization of the SHMT from *Arthrobacter* was reported. The full-length of the *glyA* gene from *A. nicotianae* was obtained by DOP-PCR [13] with moderate modifications, which represents a rapid, efficient, and species-independent technique for general DNA amplification. To the best of our knowledge, this is the first report on the application of this technique to clone the *glyA* gene, suggesting that its potential application in the cloning of other *glyA* genes. The *glyA* encoding a novel SHMT shared 54.3%, 53.6%, 50.9%, 41.2%, and 40.3% similarities to the known enzymes from *E. coli*, *H. methylovorum* GM2, *C. jejuni*, *B. japonicum*, and rabbit liver mitochondria, respectively.

Currently, L-serine is mainly produced through cellular or enzymatic conversion from the precursor glycine plus a C_1 compound, in which SHMT is the key enzyme [3,4]. *E. coli* SHMT forms a tight homodimer similar to that of AAT and other α-PLP enzymes, with the active site at the interface of the two monomers [5]. Directed evolution is a technique that can overcome the limitations of natural enzymes used as biocatalysts, for this technique does not only rely on a detailed understanding

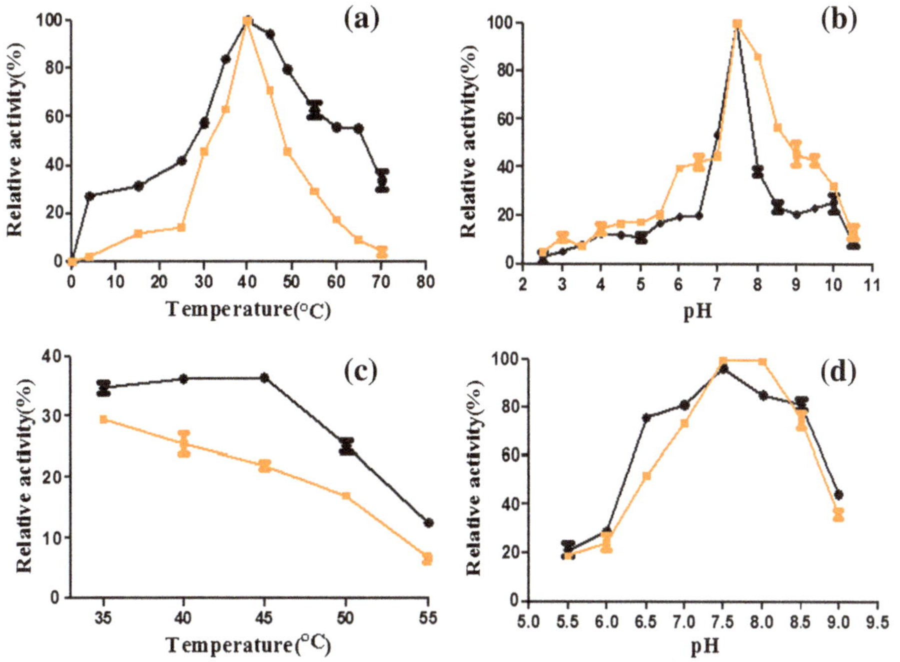

Figure 5 Effects of temperature and pH on activity. (a) Effect of temperature on the activity of the *An*SHMT and the I249L-SHMT. The optimal temperature was determined by measuring the activity at temperatures from 0 to 70°C. The maximal activity was taken as 100%. **(b)** Effect of pH on the activity of the enzymes. The activity was measured over a pH values ranging from 2.5 to 10.5, and the maximal activity was taken as 100%. **(c)** Effect of temperature on the stability of the *An*SHMT and the I249L-SHMT. At the optimal pH 7.5, the purified enzyme was pre-treated at a different temperature for 1 h. The activity of the enzyme without pre-incubation was defined as 100%. **(d)** The pH stability of the enzymes was determined by incubating the enzymes at a different pH at 4°C for 24 h. Then assays were conducted in the standard conditions and the enzyme activity without pre-treatment was taken as 100%. Error bars represent the standard deviation. Black circles represent the *An*SHMT (●), and black squares represent the I249L-SHMT (■).

of the relationship between enzyme structure and function, but on the simple powerful Darwinian principles of mutation and selection [27]. By targeting mutations to certain amino acids, we can map the enzyme active site, investigate mechanisms, and study structure-function relationships [28]. The site-directed mutagenesis methods are usually used to generate cloned DNAs with modified sequences to examine the importance of specific residues in the protein structure and function [29]. Therefore, studying the SHMT enzymatic properties and improving the catalytic efficiency can be available for insight into the industrial production of L-serine and further research.

To the best of our knowledge, this is the first report of the characterization of the SHMT which comes from *Arthrobacter*. Unlike *Hyphomicrobium methylovorum*'s SHMT (37°C) [4], the *An*SHMT exhibited the highest activity at 40°C in the assays of the enzymatic characteristics, and retained 84% and 94.2% of the maximal activity at 35°C and 45°C, respectively, and less than 40% of the activity after 1 h pre-treatment at 35°C and 45°C. Without any stabilizer, the *An*SHMT displayed higher pH stability profiles (pH 6.5-8.5) than that from *Hyphomicrobium methylovorum* (pH 6.0 and 7.5) under alkaline environment. The I249L-SHMT showed nearly no loss in activity after 24 h incubation under pH 8.0, while the *An*SHMT retained about 85% of the maximum activity. The *An*SHMT was not affected by EDTA, indicating that the metal ions were not appropriate to the enzyme. Additionally, the loss of activity of the enzyme in the presence of Hg^{2+} might be attributed to the interaction with thiol groups of Cys residue(s), dry residue(s), or carboxyl group (s) of amino acid(s) [30].

The replacement of isoleucine with leucine at the 249 position resulted in a 2.78-fold increase of the catalytic efficiency (k_{cat}/Km) in the mutant I249L over the wide type, namely a 7% increase in K_m and a 2.97-fold increase in k_{cat} indicating that the 249 site is vital to improve the enzyme activity. The increase of the 7% for the K_m may be due to the decreased of enzyme and substrate affinity with the replacement of isoleucine by leucine at the 249 position. The results of the site-directed mutagenesis performed by the sequences alignment and bioinformatics analysis indicate that the 249 site is vital to improve the enzyme activity. The improvement of the catalytic efficiency could be attributed to four factors: (i) based on the model structures, the mutant is too far away from the substrate to participate in direct

Table 2 Effects of metal ions and chemical reagents on the activity of purified *An*SHMT[a]

Reagents	Concentration (mM or %)	Relative activity (%)
control	0	100%
Hg^{2+}	1	-[b]
Co^{2+}	1	60.03 ± 0.27 c
K^{+}	1	100.44 ± 0.37
Na^{+}	1	100.62 ± 0.30
NH_4^{+}	1	95.12 ± 0.58
Sl^{2+}	1	94.44 ± 0.83
Mg^{2+}	1	108.38 ± 0.85
Ca^{2+}	1	91.38 ± 0.11
Ba^{2+}	1	102.58 ± 0.33
Cu^{2+}	1	8.95 ± 0.36
Zn^{2+}	1	2.91 ± 0.11
Mn^{2+}	1	60.81 ± 0.28
Fe^{2+}	1	-[b]
Pb^{2+}	1	98.90 ± 0.32
$CO(NH_2)_2$	1	60.59 ± 0.45
DTT	1	59.29 ± 0.08
CTAB	0.3%	97.05 ± 0.04
EDTA	1%	101.25 ± 0.65
SDS	5%	-[b]

[a]: all assays were performed in the standard conditions and the activity measured without additional reagents and ions was taken as 100%. The data are the average of three replicates; [b]: unmeasured data; [c]: Relative activity ± the standard deviation.

interaction, resulting in no mutation in the catalytic active site (Figure 3), which could be due to the possibility that the occurrence of most mutations in the active site would inactivate the enzyme, because SHMT is one of the most highly conserved proteins, and the mutations likely exert their effects by changing the conformation of the active site rather than by interacting directly with substrates [2]; (ii) the three-dimensional structure (Figure 3) of *An*SHMT shows that the isoleucine (249 site) is probably located at the end of an active channel and the PLP binding region at the other end of the active channel, so the change of the 249 site residue has no effect on the PLP binding progress; (iii) when PLP or other molecules bind the enzyme, the steric hindrance may be reduced by the residue and is altered by leucine; and (iv) as the relative abundance of isoleucine and leucine are 4.4 and 7.8, respectively, more steric hindrance is produced by the binding of the substrate to the enzymatic molecules through the cavity of the former. Further exploration concerning the structure-activity relationship of SHMT is necessary owing to the limited information about the relationship [5-7] and the interesting site would be useful for these work.

Conclusion

The *An*SHMT in the present study is a novel one from *A. nicotianae*, sharing 54.3% similarity to the known SHMT from *E. coli*. The enzyme showed better stability under weakly alkali conditions (pH 6.5-8.5) than *Hyphomicrobium methylovorum*'s SHMT (pH 6.0 and 7.5), and exhibited the optimal activity at pH 7.5 and 40°C. Through site-directed mutagenesis, the catalytic efficiency (k_{cat}/Km) of the mutant I249L was 2.78-fold higher than that of wide type. These characteristics provide useful information about the interesting site, and the application of DOP-PCR in cloning a novel *glyA* gene.

Methods

Reagents, bacterial strains, vectors and cultivation conditions

The restriction enzymes, T4 DNA ligase, *Taq* DNA polymerase, pMD18-T vectors, molecular weight marker, gel extraction kit and the DNA purification Kit used in the research were all from purchased from TaKaRa Co. (Dalian, China) and Qiagen Co. (Germany). The *Pfu* DNA polymerase, PreScission protease and the GST · Bind Resin were purchased from TransGen Co. (China), GE Healthcare Co. and Merck Co. (Germany), respectively. All oligonucleotide primers (Table 3) and fragments were synthesized and sequence by GenScript Co. (Nanjing, China). Unless otherwise stated, all the other chemicals used in this study arose out of analytical grade and purchased from Sinopharm Chemical Reagent (Wuhan, China).

The SHMT screening medium [31], consisting of 125 ml/L methanol, 5 g/L NaCl and 5 g/L glycine, was ready to detect the enzyme activity of the bacterium. *A. nicotianae*, with high SHMT activity used in this study, was isolated from the Nanhu-Lake (Wuhan, China) and preserved in our laboratory and was cultivated in Luria-Bertani (LB) medium at 37°C. *E. coli* DH5a and *E. coli* BL21 (DE3) strains were cultured in Luria–Bertani (LB)

Table 3 Primers used for plasmid construction and the site-directed mutagenesis

Primers	5′ to 3′	Purpose or references
DP-F	RTGAMCMCKNYATC	This study
DP-R	YTACCCSWGSTCSK	This study
An-F	CGGGATCC**ATG**AGCAACCAGACTTTTGAATC	This study
An-R	CGGAATTC**CTA**CTCGGAAACCTTTGGCAGGT	This study
I249L-F	CCGCGTGGCGGCCTGATTCTGTCGA	This study
I249L-R	CAGGCCGCCACGCGGACCAGCCAAC	This study

IUPAC ambiguity codes: M = A/C, R = A/G, W = A/T, S = G/C, Y = C/T, K = G/T and N = A/G/C/T. I indicates inosine. Restriction sites *Bam*H1 and *Eco*RI in primers *An*–F/R and the mutated site are underlined; Start and stop codons are marked in bold; site-directed mutagenesis site is underline in I249L-F and I249L-R.

medium containing ampicillin (100 μg/mL) and used as cloning and expression hosts, respectively. Plasmid pGEX-6P-1 was utilized for the preparation of the mutant and the purification of *An*SHMT and its mutant. The submitted work does not contain any experiments using animals or humans.

Gene cloning

The full ORF of *glyA* from *A. nicotianae* was obtained by DOP-PCR using both end degenerate primers with the following program: (i) 94°C for 5 min, (ii) 30 cycles of 94°C for 30 s, 44-48°C for 30 s, and 72°C for 1 min 30 s, and (iii) 72°C for 10 min, DP-F and DP-R (Table 3), which were designed according to the result of the multiple sequence alignment (Figure 4a and 4b) between *glyA* gene sequences and the genus of *Arthrobacter* obtained from the NCBI database. After amplification, the PCR product was purified by agarose gel electrophoresis, and was cloned into PMD18-T vector and sequenced. After that, the *glyA* was obtained from the chromosome DNA of *A. nicotianae* by PCR using the forward and reverse primers (An-F and An-R, Table 3) under the following conditions: 1 cycle at 94°C for 5 min; 30 cycles at 94°C for 30 s, 52°C for 30 s and 72°C for 1 min 30 s; and 1 final additional cycle at 72°C for 10 min. After double-digestion with *Bam*HI and *Eco*RI, the resulting PCR product was gel-purified and cloned into the *Bam*HI-*Eco*RI sites of pGEX-6p-1, generating pGEX-6p-*AnglyA* containing the glutathione-S-transferase (GST) tag. The construction accuracy was confirmed by sequencing at Nanjing GenScript Company.

Site-directed mutagenesis

The position of the site-directed mutagenesis was selected founded on the result of the multiple sequence alignment of SHMTs (Figure 2). The amino acid sequences which were reported to belong to *E. coli*, rabbit liver mitochondria, *A.radioresistens*, *B. japonicum*, *C. jejuni*, and *A. nicotianae*, respectively. The distinct amino acid residues close to activity site were observed and assumed to produce a negative effect on enzymatic activity [2]. As shown in Figure 2, an interesting phenomenon was detected in the site underlined and marked by a five-pointed star, indicating that it was isoleucine rather than leucine close to the sequence TSTTHKTL (Figure 2c), which was highly conserved in all known SHMT proteins [5,14,24].

The site-directed mutagenesis was done by PCR using the plasmid pGEX-6p-*AnglyA* as the template and primers designed from pairs of complementary oligonucleotides containing the desired mutant (Table 3). The PCR mixture (50 μl) was composed of a PCR buffer, 10 ng template pGEX-6p-*AnglyA*, 0.2 μM each oligonucleotide, 0.6 μM dNTP and 1 unit of FAST *Pfu* DNA polymerase (TransGen, China). The PCR was performed in the following three steps: (i) 97°C for 2 min; (ii) 20 cycles of 95°C for 20 s, 54°C for 20 s, and 72°C for 3 min 20 s; and (iii) 72°C for 7 min with the I249L-F and I249L-R (Table 3). The PCR product was gel-purified, treated with DpnI to eliminate methylated ancestral template through an overnight incubation at 37°C, and then transformed into *E. coli* DH5α competent cells. The recombinant plasmids were checked by DNA sequencing, and the successfully introduced desired mutation was designated as pGEX-6p-I249L.

Expression and purification of wild-type *An*SHMT and mutant enzyme

The recombinant plasmids pGEX-6p-*AnglyA* and pGEX-6p-I249L were transformed into DE3 for the SHMT expression. DE3 strain harboring the recombinant plasmid was inoculated into LB liquid medium with ampicillin (100 μg/mL) and incubated at 37°C overnight. Subsequently, the mixture was transferred into fresh LB liquid medium (1:100 dilution) containing ampicillin (100 μg/mL) and cultured at 37°C for 3–4 h. When the cells reached an optical density of 0.6-0.8 at 600 nm, protein expression was induced by adding IPTG to a final concentration of 0.1 mM and the mixture was incubated at 18°C for 12 h. After that, the cells were harvested and disrupted with High Pressure Homogenizer (NS100IL 2 K, Niro Soavi, Germany). Finally, the SHMTs were purified by glutathione-S-transferase (GST) Gene Fusion System (GE Healthcare, Sweden).

Measurement of SHMT activity

Standard enzyme activity of phenylserine degradation was measured with DL-3-phenylserine as substrate as previously described by Lee and Hsiao [32] with moderate modifications. Briefly, standard enzyme activity assay was performed by adding moderate purification enzyme into the 1 mL reaction system at pH 7.8 with sodium phosphate buffer containing 50 mM DL-3-phenylserine, 50 μM PLP, 1 mM Na_2EDTA (ethylene diamine tetraacetic acid) and 25 mM sodium sulfate. When cells were used in the reaction system, 0.03% (w/v) cetyltrimethyl ammonium bromide (CTAB) was added. The reaction was conducted for at 30°C, and 1 h later, and the production of benzaldehyde was measured by its strong absorbance at 279 nm [2]. One unit of SHMT activity was defined as the amount of enzyme that released 1 μmole benzaldehyde per hour (benzaldehyde as standard). Specific activity was reported in units/mg protein.

Biochemical characterization of the enzyme

The optimal pH of SHMT was determined at 30°C in different buffers at pH 2.5-10.5, namely 0.2 M Na_2HPO_4/ 0.1 M citric acid buffer (pH 2.5–8.0) and 50 mM glycine–

NaOH buffer (pH 8.0–10.5). The optimum temperature for the enzyme was measured by performing the SHMT activity assay for 1 h at temperatures from 0 to 70°C under pH 7.5. The thermal stability of the enzyme was determined under the optimal pH by pre-incubating the enzyme at temperatures from 35°C to 55°C for 60 min and then the residual activity was measured as described above. To determine the pH stability of the SHMT, the recombinant SHMT was incubated at 4°C for 24 h in different buffer systems (pH 5.5–9.0), and the residual activity was measured under standard assay conditions. The biochemical characterization of the I249L-SHMT was performed with the same methods.

Effects of metal ions and other chemical compounds on the enzyme activity were determined in the standard reaction system at 40°C for 1 h at pH 7.5.

Computer model generation

The position of site-directed mutagenesis was indicated on the three-dimensional structure of *An*SHMT, which was constructed from the known x-ray structure of *Burkholderia Pseudomallei Mycobacterium Tuberculosis T.Th.*Hb8 (PDB entry 3H7F, 2DKJ and 3ECD) using Swiss-Model, a knowledge-based protein modeling tool [33]. Furthermore, the PLP binding region and the active site were marked on the three-dimensional structure.

Competing interests

The authors declare that they have no competing interests.

Authors' contributions

WJ: designed and supervised the experiments, have performed gene cloning, expression in *E. coli*, rational design, enzyme characterization and drafted this manuscript. LC: have performed gene cloning, expression in *E. coli* and revised the manuscript. NH: have conceived the study, designed and supervised the experiments. SHY: have performed expression in *E. coli*, rational design and revised the manuscript. BL: have performed three-dimensional structure and revised the manuscript. ZDL: have conceived the study, designed and supervised the experiments, is a corresponding author. All authors have read and approved the manuscript.

Acknowledgement

Financial support for this work was provided by China National Natural Sciences Foundation (No. 51008012).

Author details

[1]State Key Laboratory of Agricultural Microbiology, College of Life Science and Technology, Huazhong Agricultural University, Wuhan 430070, P. R. China. [2]College of Biotechnology and Pharmaceutical Engineering, Nanjing Tech University, Nanjing 211800, P. R. China.

References

1. Izumi Y, Yoshida T, Miyazaki SS, Mitsunaga T, Ohshiro T, Shimao M, Miyata A, Tanabe T: **L-Serine production by a methylotroph and its related enzymes.** *Appl Microbiol Biotechnol* 1993, **39**(4):427–432.
2. Zuo ZY, Zheng ZL, Liu ZG, Yi QM, Zou GL: **Cloning, DNA shuffling and expression of serine hydroxymethyltransferase gene from < i > Escherichia coli</i > strain AB90054.** *Enzyme Microb Technol* 2007, **40**(4):569–577.
3. Peters-Wendisch P, Stolz M, Etterich H, Kennerknecht N, Sahm H, Eggeling L: **Metabolic engineering of Corynebacterium glutamicum for L-serine production.** *Appl Environ Microbiol* 2005, **71**(11):7139–7144.
4. Miyazaki SS, Toki S, Izumi Y, Yamada H: **Purification and characterization of a serine hydroxymethyltransferase from an obligate methylotroph, Hyphomicrobium methylovorum GM2.** *Eur J Biochem* 1987, **162**(3):533–540.
5. Scarsdale JN, Radaev S, Kazanina G, Schirch V, Wright H: **Crystal structure at 2.4 Å resolution of E. coli serine hydroxymethyltransferase in complex with glycine substrate and 5-formyl tetrahydrofolate1.** *J Mol Biol* 2000, **296**(1):155–168.
6. Renwick SB, Snell K, Baumann U: **The crystal structure of human cytosolic serine hydroxymethyltransferase: a target for cancer chemotherapy.** *Structure* 1998, **6**(9):1105–1116.
7. Scarsdale J, Kazanina G, Radaev S, Schirch V, Wright H: **Crystal structure of rabbit cytosolic serine hydroxymethyltransferase at 2.8 resolution: mechanistic implications.** *Biochemistry* 1999, **38**(26):8347–8358.
8. Blakley R: **The interconversion of serine and glycine: role of pteroylglutamic acid and other cofactors.** *Biochem J* 1954, **58**(3):448.
9. Jones-Mortimer M, Wheldrake J, Pasternak C: **The control of sulphate reduction in Escherichia coli by O-acetyl-L-serine.** *Biochem J* 1968, **107**(1):51.
10. Ulevitch RJ, Kallen RG: **Studies of the reactions of lamb liver serine hydroxymethylase with L-phenylalanine: kinetic isotope effects upon quinonoid intermediate formation.** *Biochemistry* 1977, **16**(24):5350–5354.
11. Rao DN, Rao NA: **Purification and regulatory properties of mung bean (Vigna radiata L.) serine hydroxymethyltransferase.** *Plant Physiol* 1982, **69**(1):11.
12. Barra D, Martini F, Angelaccio S, Bossa F, Gavilanes F, Peterson D, Bullis B, Schirch L: **Sequence homology between prokaryotic and eukaryotic forms of serine hydroxymethyltransferase.** *Biochem Biophys Res Commun* 1983, **116**(3):1007–1012.
13. Telenius H, Carter NP, Bebb CE, Ponder BA, Tunnacliffe A: **Degenerate oligonucleotide-primed PCR: general amplification of target DNA by a single degenerate primer.** *Genomics* 1992, **13**(3):718–725.
14. Garrow TA, Brenner A, Whitehead V, Chen X-N, Duncan R, Korenberg J, Shane B: **Cloning of human cDNAs encoding mitochondrial and cytosolic serine hydroxymethyltransferases and chromosomal localization.** *J Biol Chem* 1993, **268**(16):11910–11916.
15. Byrne PC, Sanders P, Snell K: **Nucleotide sequence and expression of a cDNA encoding rabbit liver cytosolic serine hydroxymethyltransferase.** *Biochem J* 1992, **286**(Pt 1):117–123.
16. Miyata A, Yoshida T, Yamaguchi K, Yokoyama C, Tanabe T, Toh H, Mitsunaga T, Izumi Y: **Molecular cloning and expression of the gene for serine hydroxymethyltransferase from an obligate methylotroph Hyphomicrobium methylovorum GM2.** *Eur J Biochem* 1993, **212**(3):745–750.
17. Schirch V, Hopkins S, Villar E, Angelaccio S: **Serine hydroxymethyltransferase from Escherichia coli: purification and properties.** *J Bacteriol* 1985, **163**(1):1–7.
18. Vidal L, Calveras J, Clapes P, Ferrer P, Caminal G: **Recombinant production of serine hydroxymethyl transferase from Streptococcus thermophilus and its preliminary evaluation as a biocatalyst.** *Appl Microbiol Biotechnol* 2005, **68**(4):489–497.
19. Hamilton BK, Hsiao H-Y, Swann WE, Anderson DM, Delente JJ: **Manufacture of L-amino acids with bioreactors.** *Trends Biotechnol* 1985, **3**(3):64–68.
20. Parikh MR, Matsumura I: **Site-saturation mutagenesis is more efficient than DNA shuffling for the directed evolution of β-fucosidase from β-galactosidase.** *J Mol Biol* 2005, **352**(3):621–628.
21. Wilks HM, Hart KW, Feeney R, Dunn CR, Muirhead H, Chia WN, Barstow DA, Atkinson T, Clarke AR, Holbrook JJ: **A specific, highly active malate dehydrogenase by redesign of a lactate dehydrogenase framework.** *Science* 1988, **242**(4885):1541–1544.
22. Chen R, Greer A, Dean AM: **Redesigning secondary structure to invert coenzyme specificity in isopropylmalate dehydrogenase.** *Proc Natl Acad Sci U S A* 1996, **93**(22):12171–12176.
23. Vick JE, Schmidt DM, Gerlt JA: **Evolutionary potential of (β/α) 8-barrels: in vitro enhancement of a "new" reaction in the enolase superfamily.** *Biochemistry* 2005, **44**(35):11722–11729.
24. Usha R, Savithri HS, Appaji Rao N: **The primary structure of sheep liver cytosolic serine hydroxymethyltransferase and an analysis of the evolutionary relationships among serine hydroxymethyltransferases.** *Biochim Biophys Acta* 1994, **1204**(1):75–83.

25. Hong MC, Wu ML, Chang MC: **Cloning and complete nucleotide sequence of Acinetobacter radioresistens CMC-1 AglyA gene encoding serine hydroxymethyltransferase.** *FEMS Microbiol Lett* 1999, **170**(2):413–418.
26. Stackebrandt E, Fowler V, Fiedler F, Seiler H: **Taxonomic Studies on Arthrobacter nicotianae and Related Taxa: Description of Arthrobacter uratoxydans sp. nov. and Arthrobacter sulfureus sp. nov. and Reclassification of Brevibacterium protophormiae as Arthrobacter protophormiae comb. nov.** *Syst Appl Microbiol* 1982, **4**(4):470–486.
27. Zhang Z-G, Yi Z-L, Pei X-Q, Wu Z-L: **Improving the thermostability of < i > Geobacillus stearothermophilus</i > xylanase XT6 by directed evolution and site-directed mutagenesis.** *Bioresour Technol* 2010, **101**(23):9272–9278.
28. Georgescu R, Bandara G, Sun L: **Saturation Mutagenesis.** In *Directed Evolution Library Creation.* Berlin: Springer; 2003:75–83.
29. Liu H, Naismith JH: **An efficient one-step site-directed deletion, insertion, single and multiple-site plasmid mutagenesis protocol.** *BMC Biotechnol* 2008, **8**(1):91.
30. McGavin M, Forsberg C, Crosby B, Bell A, Dignard D, Thomas D: **Structure of the cel-3 gene from Fibrobacter succinogenes S85 and characteristics of the encoded gene product, endoglucanase 3.** *J Bacteriol* 1989, **171**(10):5587–5595.
31. Patt T, Cole G, Hanson R: **Methylobacterium, a new genus of facultatively methylotrophic bacteria.** *Int J Syst Evol Microbiol* 1976, **26**(2):226.
32. Lee TK, Hsiao H-y: **Synthesis of l-tyrosine by a coupled reaction of serine hydroxymethyltransferase and β-tyrosinase.** *Enzyme Microb Technol* 1986, **8**(9):523–526.
33. Peitsch M: **Protein modeling by E-mail.** *Bio/Technol* 1995, **13**:658–660.

Improvement of an antibody-enzyme coupling yield by enzyme surface supercharging

Agneta A Prasse[1], Thomas Zauner[1], Karin Büttner[1], Ralf Hoffmann[1] and Thole Zuchner[1,2*]

Abstract

Background: Protein cross-coupling reactions demand high yields, especially if the products are intended for bioanalytics, like enzyme-linked immunosorbent assays. Amongst other factors, the coupling yield depends on the concentration of the proteins being used for coupling. Protein supercharging of enzymes can increase the solubility dramatically, which could promote enzyme-antibody coupling reactions. A highly soluble, supercharged variant of the enzyme human enteropeptidase light chain was created by a site-directed mutagenesis of surface amino acids, used for the production of an antibody-enzyme conjugate and compared to the wild type enzyme.

Results: Wild type and mutant enzyme could successfully be cross-coupled to an antibody to give antibody-enzyme conjugates suitable for ELISA. Their assay performances and the analysis of the enzyme activities in solution demonstrate that the supercharged version could be coupled to a higher extent, which resulted in better assay sensitivities. The generated conjugate, based on the supercharged enzyme, was feasible as a reporter molecule in a sandwich ELISA and allowed the detection of epidermal growth factor with a detection limit of 15.63 pg (25 pmol/L).

Conclusion: The highly soluble, surface supercharged, human enteropeptidase light chain mutant provided better yields in coupling the enzyme to an antibody than the wild type. This is most likely related to the higher protein concentration during the coupling. The data suggest that supercharging can be applied favourably to other proteins which have to be covalently linked to other polymers or surfaces with high yields without losses in enzyme activity or specificity.

Keywords: ELISA, Human enteropeptidase light chain, Protein-protein coupling, Surface supercharging

Background

The extraordinary feature of enzymes to catalyze reactions with a high selectivity is utilized in a broad field of applications, including the paper-, food- and pharmaceutical industry [1]. Besides this, enzymes are used as bioanalytical tools, such as in biosensors or as amplifying agents in immunoassays. One prominent application is the enzyme-linked immunosorbent assay (ELISA), where an antibody-enzyme conjugate is used for the detection and quantification of analytes. Due to its compatibility to complex samples, this approach became a method of high significance in routine analysis and research [2]. Most applications have in common, that the enzyme has to be immobilized. Whilst adsorption, entrapment and covalent-cross coupling procedures are feasible for continuous flow-through processes in industry [3], the latter is essential for the production of ELISA conjugates. The application in this type of immunoassay requires a synthesis protocol which ideally results in a conjugate with an enzyme:antibody ratio of at least 1:1, without causing side-products and, more importantly, remaining unconjugated antibody [4].

Different coupling strategies have been developed, especially for the popular reporter enzymes horseradish peroxidase, alkaline phosphatase and β-galactosidase. The coupling reaction for the first enzyme-antibody conjugate applied in ELISA was based on the homobifunctional linker glutaraldehyde [5,6]. Although the resulting enzyme-protein complex partially retains its enzymatic activity and immunological specificity, it suffers from the formation of

* Correspondence: Thole.Zuechner@octapharma.com
[1]Institute of Bioanalytical Chemistry, Faculty of Chemistry and Mineralogy and Centre of Biotechnology and Biomedicine, Universität Leipzig, Deutscher Platz 5, 04103 Leipzig, Germany
[2]Current address: Octapharma Biopharmaceuticals GmbH, Im Neuenheimer Feld 590, 69120 Heidelberg, Germany

inactive homo-conjugates and polymers [4]. Thus, heterobifunctional linker molecules have been developed, which can react to two diffeṫent functional groups, such as amino groups, *cis*-hydroxyl-groups in carbohydrates and thiol groups. As the coupling efficiency is also dependent on the amount of coupling sites on the proteins [7], the modification should occur to a large extend. Additionally, high concentrations of proteins will result in higher yields of coupling product, since diffusion- and orientation-controlled ligation processes become more likely. Although the commonly used enzymes in ELISA allow a satisfactory performance for most applications, the adoption of other enzymes might be interesting in terms of substrate diversity and sensitivity.

In 2007, Lawrence *et al.* presented a method to decrease protein aggregation tendencies by specific mutations of residues on protein surfaces. This specific replacement of surface amino acids with either acidic or basic amino acids was termed *supercharging*, as the net charge of the mutants were more positive or negative [8]. Supercharged proteins have been engineered to overcome limitations of macromolecule delivery into mammalian cells [9] and to increase thermal stability of antibody fragments [10]. This strategy might also be useful to improve the yield of protein-protein coupling reactions, especially the linking of enzymes to antibodies for immunological applications.

We adapted supercharging to the light chain of the human enteropeptidase in an earlier study and the resulting mutant, hEPl scC112S (N6D, G21D, G22D, C112S, N141D, K209E), showed a superior solubility and heat stability [11]. Here, we investigated the impact of enzyme supercharging on chemical crosslinking as exemplified for an antibody-enzyme conjugate. Therefore, hEPl scC112S and its wild type variant (hEPl wt) were conjugated to a secondary antibody and compared with respect to their assay performance.

Methods

Reagents

Enzymes (hEPl wt and hEPl scC112S) were expressed, refolded and purified in-house [11]. Polyclonal sheep-anti-mouse and donkey-anti-rabbit antibodies were from Dianova (Hamburg, Germany), monoclonal mouse-anti-EGF, EGF protein and polyclonal rabbit-anti-EGF were from Abcam (Cambridge, UK). Citric acid and Tween®20 were from Serva Electrophoresis (Heidelberg, Germany), Tris(hydroxymethyl)-aminomethan was from Applichem (Darmstadt, Germany). C_6-Succinimidyl-4-formylbenzoate and C_6-Succinimidyl-4-hydrazinonicotinate acetone hydrazone were from Acris Antibodies (Herford, Germany). Sodium chloride, hydrochlorid acid (conc.), $Na_2HPO_4x12H_2O$, KH_2PO_4, GD_4K-na and DMSO were purchased from Sigma-Aldrich (Taufkirchen, Germany) in the highest purities available. Roti®-Block was ordered from Roth (Karlsruhe, Germany). Bovine serum albumin standard was from Thermo Fisher Scientific (Rockford, USA). Deionized water was obtained from a Purelab Ultra water purification system (ELGA LabWater, Celle, Germany; 18.2 MΩ*cm).

Antibody-enzyme coupling

A 1.5-fold molar excess of enzyme (hEPl wt: 90 μg/mL or hEPl scC112S: 440 μg/mL) was coupled to a sheep-anti-mouse antibody (2.4 g/L) after dialysing the proteins against modification buffer (0.1 mol/L phosphate buffer, 0.15 mol/L NaCl, pH 8.0) separately for 16 hours at 4°C using a 12 kDa cut-off dialysis tube (Sigma-Aldrich). Afterwards, the antibody was mixed with 40 equivalents of C_6-succinimidyl-4-formylbenzoate (9 g/L in DMF), while enzymes have been modified with 40 equivalents of C_6-succinimidyl-4-hydrazinonicotinate acetone hydrazone (5.2 g/L in DMF). The reaction mixtures were incubated at room temperature for 3 hours and excess linker molecules were removed by dialysis against conjugation buffer (0.1 mol/L citric acid, 0.15 mol/L NaCl, pH 6.0) at 4°C. Antibody and enzyme solution were mixed and incubated at 4°C. After 20 hours, protein concentrations were determined using a Bradford assay according to standard protocols with bovine serum albumin as standard [12]. The solutions were used without further treatment.

The ELISA-conjugate was obtained by increasing the concentration of hEPl scC112S from 0.242 g/L to 3.6 g/L by lyophilisation from 5 mmol/L Tris-buffer (pH 8.0) and subsequent resolubilization of the enzyme in water. Coupling was performed as described above, using a donkey-anti-rabbit antibody (1.2 g/L) and a 10-fold molar excess of enzyme.

Enzyme assay

Kinetic characterization of the enzymes was performed in a 0.1 mol/L Tris–HCl buffer (pH 8.0, 10% DMSO) at room temperature using black 384-well plates (non-binding, Greiner Bio-one, Frickenhausen, Germany). GD_4K-na in concentrations varying from 3.7 to 947 μmol/L was cleaved by addition of hEPl wt and hEPl scC112S (final concentration 10 nmol/L). The linear increase of released 2-naphthylamine was recorded on a Paradigm fluorescence reader (Molecular Devices, Ismaning, Germany) using a fluorescence cartridge (λ_{ex} = 360/35 nm; λ_{em} = 465/35 nm; integration time 140 ms). All experiments were performed in triplicates.

Enzyme-linked immunosorbent assay (ELISA)

ELISA experiments were performed in black 384-well microtitre plates (high-binding, Greiner Bio-one, Frickenhausen, Germany) with an assay volume of 100 μL. All experiments were performed in triplicates.

Initial testing of the conjugate yield derived from the coupling with 1.5 equivalents hEPl wt or hEPl scC112S was determined using an affinity test. 100 µL of an in-house produced monoclonal antibody (HPT-104 [13], 2.25 µg/mL) were incubated at 4°C for 16 hours. After a threefold washing step using 200 µL washing buffer (phosphate buffered saline, 10 mmol/L phosphate, 0.3 mol/L NaCl, pH 7.4, 0.05% (v/v) Tween®20), free binding sites were blocked with 120 µL blocking buffer (0.5% (w/v) casein in washing buffer). After a further washing step, conjugate solutions were added with a total protein concentration of 2 µg/mL and incubated. After removal of unbound components by washing, signals were generated by addition of 50 µmol/L GD_4K-na in assay buffer (0.1 mol/L Tris–HCl, pH 8.0, 10% DMSO) and the fluorescence was monitored at 465 nm after excitation at 360 nm.

The microtitre plate for the EGF sandwich ELISA was coated with monoclonal mouse-anti-EGF antibody by incubating the solution (100 µL, 1 mg/L) at 4°C for 16 hours. After washing, blocking was performed with 120 µL Roti®-Block and the plate was washed again before the addition of a four-fold serial dilution of EGF (10 ng/mL to 39.1 pg/mL) in *E. coli* cell lysate (protein concentration 6 mg/L). After incubation for 1 hour and washing, 100 µL of a polyclonal rabbit-anti-EGF antibody (250 ng/mL) was added. Subsequently, the plate was washed and 100 µL of the diluted (1:1000) antibody-enzyme conjugate was incubated for one hour. After washing, 50 µmol/L GD_4K-na (in 0.1 mol/L Tris–HCl, pH 8.0, 10% DMSO) was added for signal development. End-point measurement was performed 4 hours after incubation at 37°C with λ_{ex} = 360 nm and λ_{em} = 465 nm. All experiments were performed in triplicates.

Results

Low reaction volumes are favoured for antibody-enzyme coupling reactions to obtain high coupling yields. Hence, proteins with high solubility are favoured for coupling. We therefore investigated, if the method of protein supercharging can produce mutants that are more suitable for coupling reactions compared to their wild type enzymes using human enteropeptidase light chain (hEPl) as a model enzyme (Figure 1). The light chain of the enteropeptidase holoenzyme, a 26 kDa enzyme-fragment, contains the catalytic triade and is therefore sufficient for most enzymatic applications. In an earlier study, a rational design of a mutant with a high surface charge (N6D, G21D, G22D, C112S, N141D, K209E) resulted in a more than 100-fold increase in enzyme solubility [11], which should affect the coupling reaction positively.

As a first step for the characterization of the enzyme, the kinetic parameters of the wild type and mutated enzyme were determined (Table 1 and Additional file 1). GD_4K-na was chosen as a specific substrate, because its cleavage can be detected with high sensitivity (λ_{ex} = 360 nm, λ_{em} = 465 nm). The data indicate that the mutation of the enzyme resulted in a slightly decreased affinity towards the substrate, as the Michaelis constant K_m was increased from 0.11 mmol/L (hEPl wt) to 0.65 mmol/L for hEPl scC112S. The insertion of aspartic acid and glutamic acid in the mutant decreased the net charge of the enzyme by a difference of 6. This could result in electrostatic repulsion between enzyme surface and the tetra-aspartyl-motif of the substrate, and could be an explanation for the slightly lower substrate affinity of the supercharged enzyme hEPl scC112S. On the other hand, the mutant enzyme possesses an increased turnover number for GD_4K-na, which confirms the data obtained in earlier studies for another substrate, Z-Lys-SBzl [11]. This means that at high concentrations of substrate (>1 mmol/L), the mutated variant hEPl scC112S would result in a faster signal increase upon digestion than the wild type enzyme.

In order to analyze the coupling efficiencies of the two enzymes, both enzymes were coupled to a secondary antibody under the same coupling conditions and an ELISA experiment was carried out. Both enzyme conjugates led to signals that differed distinctly from the negative control. The signal increase for the supercharged conjugate (Ab-hEPl scC112S) was significantly higher than that for the wild type conjugate (Ab-hEPl wt) (Figure 2). While Ab-hEPl scC112S induced a signal increase of 15758 fluorescence intensity units per minute, application of the wild type conjugate resulted in a signal increase of only 2611 fluorescence intensity units per minute.

This indicates a higher yield of supercharged enzyme-antibody conjugate. Control experiments without antigen did not show any signal differing from the negative control (Additional file 2). As the used substrate concentration for the ELISA experiment in this study is much lower (50 µmol/L) than the concentration needed to reach v_{max} (1 mmol/L), the faster cleavage in the ELISA obtained here can only origin from a higher amount of enzyme-antibody conjugates.

This demonstrates that the application of enzyme supercharging generated a highly soluble mutant, which is more suitable for coupling processes than its wild type form. Unfortunately, we were not able to further confirm this result via native or denaturing SDS-PAGE or Western Blot. Both methods did not show clear bands that could be used for a quantification of the coupling product (data not shown).

To further investigate possible bioanalytical applications of the synthesized conjugate, a coupling reaction with a 10-fold molar excess of enzyme (hEPl scC112S)

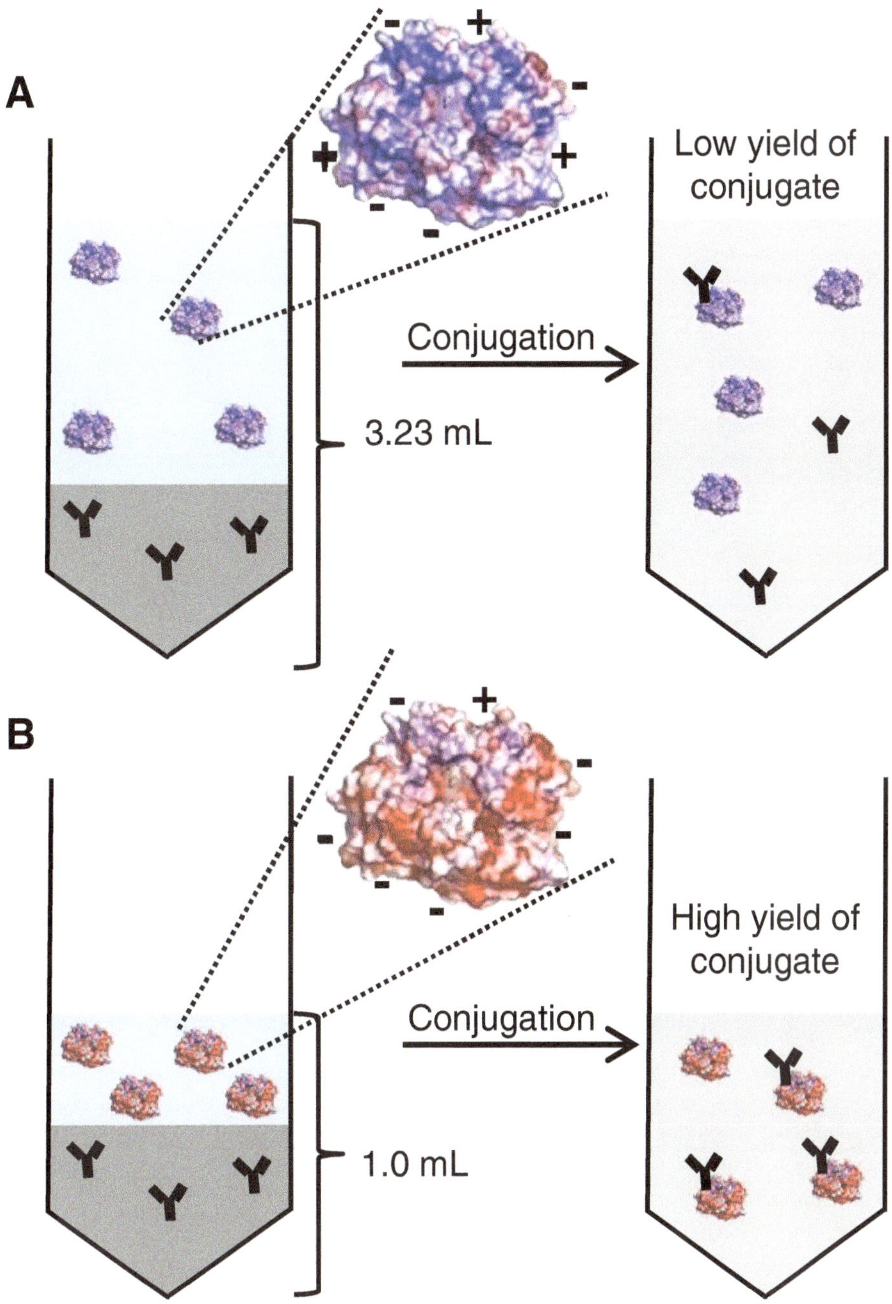

Figure 1 Schematic principle of the improvement of a protein-protein coupling reaction by enzyme surface supercharging. A high volume of wild type enzyme of the human enteropeptidase light chain is required due to its low solubility, resulting in a high total reaction volume and low yield of coupling product **(A)**. Highly soluble surface supercharged human enteropeptidase light chain permits an increase of the enzyme concentration prior to the coupling process **(B)**. Therefore the total reaction volume can be reduced to give a higher yield of desired product (antibody-enzyme conjugate).

Table 1 Kinetic parameters for the digestion of GD_4K-na by hEPI wt and hEPI scC112S

Enteropeptidase	K_m [mmol/L]	k_{cat} [s^{-1}]	k_{cat}/K_m [$(mmol/L)^{-1}$ s^{-1}]	Ref.
hEPI wt	0.11 ± 0.01	81.9 ± 9.8	745	This study
hEPI scC112S	0.65 ± 0.09	169.3 ± 24.5	260	This study
Human recombinant light chain	0.16 ± 0.01	115 ± 5	719	[14]

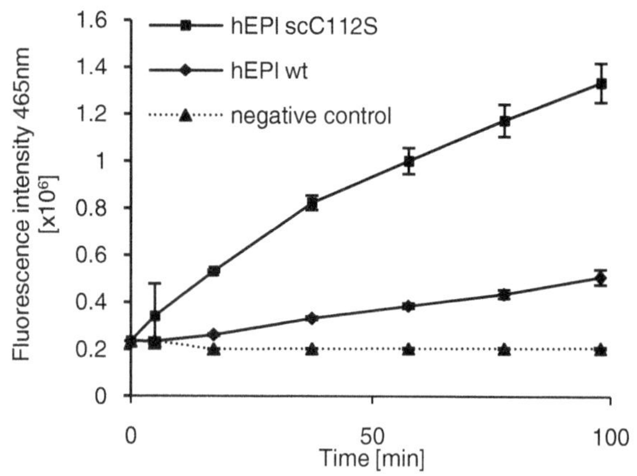

Figure 2 Assay performances of hEPI wt and hEPI scC112S conjugate solutions. Detection of an immobilized model antigen (HPT-104 antibody [13]) was performed using 2 μg/mL total protein of untreated conjugate solution. After incubation and washing, signal was generated by addition of 50 μmol/L GD_4K-na and monitored for 100 minutes. Errors = SD, n = 3.

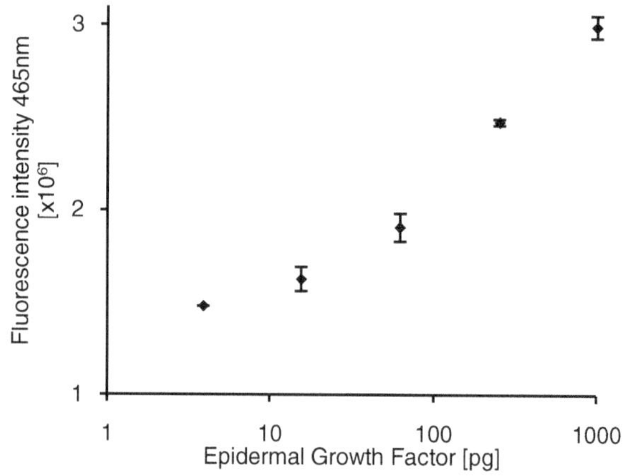

Figure 3 Use of a hEPI scC112S-antibody conjugate for the detection of EGF. Epidermal growth factor (in *E. coli* cell lysate as matrix) was detected using a sandwich ELISA and an in-house produced donkey-anti-rabbit/hEPI scC112S conjugate as secondary antibody. Signal was generated by addition of 50 μmol/L GD_4K-na and an end-point measurement was performed after 4 hours of incubation at 37°C. The limit of detection (S/B >3xSD) was 15.63 pg EGF. Errors = SD, n = 3.

was performed. The product was then utilized as a reporter molecule for the detection of epidermal growth factor (EGF). EGF is a low-molecular-weight polypeptide which affects many biochemical pathways by binding to its receptor and can be used as a cancer biomarker [15]. EGF was diluted in a complex mixture (*E. coli* cell lysate, 6 mg/L) and detected in a standard sandwich ELISA using two specific antibodies and the synthesized antibody conjugate (Ab-hEPI scC112S) was added for detection. The observed signals after addition of the peptide substrate GD_4K-na were dependent on the amount of analyte present and permitted a detection of EGF down to 15.6 pg (Figure 3). This indicates that the hEPI scC112S mutant can be used as a reporter molecule in ELISA, while the wild type enzyme is not suitable for this application. This originates from the low solubility of the wild type which does not allow a high-yielding coupling protocol. In contrast to this, the supercharging surface modification of the enzyme results in an increased protein solubility as well as heat stability, which subsequently lead to higher yields of the desired antibody-enzyme conjugate.

Discussion

Enzymes, like the serine protease enteropeptidase examined in this study, are widely used tools in biotechnology, industry and bioanalytics due to their catalytic function and their high specificity. However, many applications require the immobilization of these proteins on insoluble surfaces or on other biomolecules, such as other proteins. A very prominent example for enzyme-protein couplings are enzyme-antibody couplings and their subsequent use in enzyme immunoassays. As discussed earlier, various methods for enzyme-antibody couplings exist. However, the central problem of unsatisfactory protein solubility affects all of those coupling methods in a negative way.

One way to improve protein solubility is to optimize the buffer components of the coupling buffer by applying different ionic compounds or osmolytic stabilizers like sugars and polymers [16]. However, proteins are very sensitive to these conditions and slight modifications can cause protein precipitation, a loss in enzyme activity or antibody affinity. Other methods for the improvement of protein solubility exist, though they are not always suitable for the described problem. In 2009 Golovanov *et al.* showed, that protein solubility can be increased by a simple addition of charged amino acids. Addition of arginine and glutamic acid made several proteins accessible to NMR-studies by suppression of their aggregation tendencies [17]. Nevertheless, this method may not be suitable for protein-protein coupling reactions, since the additives possess the same reactive moieties as the proteins (e.g. amino groups). In case of protein modification with commonly used NHS-ester based linker molecules, linker reactivity would be quenched, as the reaction cannot occur specifically with the protein. This would decrease the protein modification degree and impact the overall coupling process negatively.

Also the introduction of a solubility-enhancement tag, as described in Lee *et al.* [18] and Zhou *et al.* [19] may have undesired effects, because it might alter the enzyme's activity and selectivity. In principle, a mutation of surface amino acids of proteins can of course also lead to undesired side-effects. However, a careful design of suitable amino acids can leave the enzyme unaffected. This has been shown by us before for the mutation of surface amino acids of the human enteropeptidase light chain by site-directed mutagenesis [11].

Site-directed mutagenesis was found not only to be a tool for tailoring enzyme activity and stability, but also to improve immobilization reactions. Methods were reviewed in detail before and are for example based on the production of mutants that were capable of site-directed immobilization via specific tags [20,21]. Such methods include the incorporation of particular amino acid sequences for affinity fixation (anti-tag antibody to tagged enzyme, avidin to biotinylated protein) or the addition of a Lys-containing tags for the covalent linkage via transglutaminase cross-coupling.

Coupling procedures for the synthesis of antibody-enzyme conjugates require particularly high yields, since any unlabeled antibody fraction decreases the assay sensitivity, especially because it is difficult to purify the resulting coupling product from unmodified antibody. In addition to that, the preparation of highly pure and active enzymes is enormously laborious, which makes them an expensive material. Therefore high coupling yields are desired not only in terms of assay performance, but also with regard to material costs.

In this study, we were able to demonstrate that site-directed mutagenesis can be used to optimize coupling reactions due to an enhancement of the protein solubility. For that purpose, we tested the potential of a supercharged version of human enteropeptidase light chain with respect to its antibody-coupling yield and could show that the method of supercharging can evolve highly soluble proteins that are suitable for coupling reactions. As proven by us earlier, the site-directed surface modification of the human enteropeptidase light chain resulted in a more than 100-fold increased solubility which allowed the use of a highly concentrated enzyme solution without causing protein aggregation. This allowed the covalent linkage to an antibody with significantly higher yields compared to the wild type enzyme. Lee and coworkers demonstrated earlier that the resulting enteropeptidase-antibody conjugates are attractive components for heterogeneous immunoassays [22]. With the new coupling strategy, we showed here that the medium-scale synthesis of this kind of enzyme-antibody conjugates is possible, if a highly soluble supercharged version of the enzyme is used.

In summary, we could show that the supercharged human enteropeptidase light chain can be effectively used for the production of an antibody-enzyme conjugate. This procedure allows the use of the mutant enzyme as a reporter molecule in a sensitive ELISA system, while the wild type enzyme failed for this application. We furthermore suggest that this strategy can in principle be used for the coupling of any desired enzyme to other proteins. Examples of such enzymes which may be used for establishing novel ELISA systems may include subtilisin [23], acetylcholinesterase [24] and lipase [25].

Conclusion

This study showed that surface supercharging of proteins is a suitable method for the improvement of protein-protein cross coupling reactions. Hence, the biotechnologically improved, highly soluble mutant hEPl scC112S could be utilized as a reporter molecule for a sensitive ELISA system. In contrast, the wild type enzyme could not be coupled to a satisfactory extend and is therefore not suitable for this application.

Additional files

Additional file 1: Characterization of enzyme kinetics for hEPl wt and hEPl scC112S. Measurements were performed in 0.1 mol/L Tris-buffer (pH 8.0, 10% DMSO) with a GD_4K-na substrate concentration range from 3.7 to 947 µmol/L. Enzymes were added to a final concentration of 10 nmol/L and fluorescence was recorded at 465 nm (λ_{ex} = 360 nm). K_m values were obtained from the fitted curves and were 0.11 mmol/L for hEPl wt and 0.65 mmol/L for hEPl scC112S. Errors = SD, n = 3.

Additional file 2: Negative controls in ELISA. In the absence of antigen, no unspecific binding by both conjugate mixtures could be detected. Conjugate solutions with a total protein concentration of 2 µg/mL were used; signal development was carried out by addition of 50 µmol/L GD_4K-na. Errors = SD, n = 3.

Abbreviations

EGF: Epidermal growth factor; ELISA: Enzyme-linked immunosorbent assay; FI: Fluorescence intensity; hEPl: Human enteropeptidase light chain; k_{cat}: Turnover number; K_m: Michaelis constant; scC112S: Surface charged mutant; SD: Standard deviation; SDS-PAGE: Sodium dodecyl sulphate polyacrylamide gel electrophoresis; v_{max}: Maximum rate; wt: Wild type.

Competing interests

The authors declare that they have no competing interests.

Authors' contributions

AAP designed and performed the experiments and wrote the manuscript. TZa designed the antibody-enzyme coupling strategy. KB provided all enzymes and helped with determination of enzyme kinetics. RH helped with the design of enzyme coupling and ELISA experiments. TZu helped with design of all experiments and wrote the manuscript. All authors corrected the manuscript. All authors read and approved the final manuscript.

Acknowledgment

We gratefully acknowledge the expert technical assistance from Stefanie Langanke. This work was supported by the Federal Ministry of Education and Research (BMBF) Innoprofile project no. 03IP604 to R.H. and the BMBF Go-Bio project no. 0315988 to T.Zu.

References

1. Leisola M, Jokela J, Pastinen O, Turunen O: **Industrial Use of Enzymes.** In *Physiology and Maintenance - Vol. II.* Oxford: EOLSS Publishers; 2001.
2. Lequin RM: **Enzyme Immunoassay (EIA)/Enzyme-Linked Immunosorbent Assay (ELISA).** *Clin Chem* 2005, **51**:2415–2418.
3. Datta S, Christena LR, Rajaram YRS: **Enzyme immobilization: an overview on techniques and support materials.** *3 Biotech* 2013, **3**:1–9.
4. Haugland RP: **Coupling of monoclonal antibodies with enzymes.** *Methods Mol Biol* 1995, **45**:235–243.

5. Avrameas S: **Coupling of enzymes to proteins with glutaraldehyde.** *Immunochemistry* 1969, **6**:43–52.
6. Engvall E, Perlmann P: **Enzyme-linked immunosorbent assay (ELISA) quantitative assay of immunoglobulin G.** *Immunochemistry* 1971, **8**:871–874.
7. Tresca JP, Ricoux R, Pontet M, Engler R: **Comparative activity of peroxidase-antibody conjugates with periodate and glutaraldehyde coupling according to an enzyme immunoassay.** *Ann Biol Clin (Paris)* 1995, **53**:227–231.
8. Lawrence MS, Phillips KJ, Liu DR: **Supercharging proteins can impart unusual resilience.** *J Am Chem Soc* 2007, **129**:10110–10112.
9. Thompson DB, Cronican JJ, Liu DR: **Protein Engineering for Therapeutics, Part B.** In *Protein Engineering for Therapeutics, Part B.* Elsevier; 2012.
10. Miklos AE, Kluwe C, Der BS, Pai S, Sircar A, Hughes RA, Berrondo M, Xu J, Codrea V, Buckley PE, Calm AM, Welsh HS, Warner CR, Zacharko MA, Carney JP, Gray JJ, Georgiou G, Kuhlman B, Ellington AD: **Structure-based design of supercharged, highly thermoresistant antibodies.** *Chem Biol* 2012, **19**:449–455.
11. Simeonov P, Berger-Hoffmann R, Hoffmann R, Sträter N, Zuchner T: **Surface supercharged human enteropeptidase light chain shows improved solubility and refolding yield.** *Protein Eng Des Sel* 2011, **24**:261–268.
12. Bradford MM: **A rapid and sensitive method for the quantitation of microgram quantities of protein utilizing the principle of protein-dye binding.** *Anal Biochem* 1976, **72**:248–254.
13. Singer D, Lehmann J, Hanisch K, Härtig W, Hoffmann R: **Neighbored phosphorylation sites as PHF-tau specific markers in Alzheimer's disease.** *Biochem Biophys Res Commun* 2006, **346**:819–828.
14. Gasparian ME, Ostapchenko VG, Schulga AA, Dolgikh DA, Kirpichnikov MP: **Expression, purification, and characterization of human enteropeptidase catalytic subunit in Escherichia coli.** *Protein Expr Purif* 2003, **31**:133–139.
15. Yarden Y: **The EGFR family and its ligands in human cancer.** *Eur J Cancer* 2001, **37**:3–8.
16. Schein CH: **Solubility as a function of protein structure and solvent components.** *Bio/technology (Nature Publishing Company)* 1990, **8**:308–317.
17. Golovanov AP, Hautbergue GM, Wilson SA, Lian L: **A simple method for improving protein solubility and long-term stability.** *J Am Chem Soc* 2004, **126**:8933–8939.
18. Lee EN, Kim YM, Lee HJ, Park SW, Jung HY, Lee JM, Ahn Y, Kim J: **Stabilizing peptide fusion for solving the stability and solubility problems of therapeutic proteins.** *Pharm Res* 2005, **22**:1735–1746.
19. Zhou P, Lugovskoy AA, Wagner G: **A solubility-enhancement tag (SET) for NMR studies of poorly behaving proteins.** *J Biomol NMR* 2001, **20**:11–14.
20. Hernandez K, Fernandez-Lafuente R: **Control of protein immobilization: coupling immobilization and site-directed mutagenesis to improve biocatalyst or biosensor performance.** *Enzym Microb Technol* 2011, **48**:107–122.
21. Steen Redeker E, Ta DT, Cortens D, Billen B, Guedens W, Adriaensens P: **Protein engineering for directed immobilization.** *Bioconjug Chem* 2013, **24**:1761–1777.
22. Lee Y, Jeong Y, Kang HJ, Chung SJ, Chung BH: **Cascade enzyme-linked immunosorbent assay (CELISA).** *Biosens Bioelectron* 2009, **25**:332–337.
23. Huang W, Wang J, Bhattacharyya D, Bachas LG: **Improving the activity of immobilized subtilisin by site-specific attachment to surfaces.** *Anal Chem* 1997, **69**:4601–4607.
24. Ivanov Y, Marinov I, Gabrovska K, Dimcheva N, Godjevargova T: **Amperometric biosensor based on a site-specific immobilization of acetylcholinesterase via affinity bonds on a nanostructured polymer membrane with integrated multiwall carbon nanotubes.** *J Mol Catal B Enzym* 2010, **63**:141–148.
25. Kimura Y, Tanaka A, Sonomoto K, Nihira T, Fukui S: **Application of immobilized lipase to hydrolysis of triacylglyceride.** *Eur J Appl Microbiol Biotechnol* 1983, **17**:107–112.

17

Direct analysis of mAb aggregates in mammalian cell culture supernatant

Albert J Paul*, Karen Schwab and Friedemann Hesse

Abstract

Background: Protein aggregation during monoclonal antibody (mAb) production can occur in upstream and downstream processing (DSP). Current methods to determine aggregate formation during cell culture include size exclusion chromatography (SEC) with a previous affinity chromatography step in order to remove disturbing cell culture components. The pre-purification step itself can already influence protein aggregation and therefore does not necessarily reflect the real aggregate content present in cell culture. To analyze mAb aggregate formation directly in the supernatant of Chinese hamster ovary (CHO) cell culture, we established a protocol, which allows aggregate quantification using SEC, without a falsifying pre-purification step.

Results: The use of a 3 μm silica SEC column or a SEC column tailored for mAb aggregate analysis allows the separation of mAb monomer and aggregates from disturbing cell culture components, which enables aggregate determination directly in the supernatant. Antibody aggregate analysis of a mAb-producing CHO DG44 cell line demonstrated the feasibility of the method. Astonishingly, the supernatant of the CHO cells consisted of over 75% mAb dimer and larger oligomers, representing a substantially higher aggregate content than reported in literature so far.

Conclusion: This study highlights that aggregate quantification directly in the cell culture supernatant using appropriate SEC columns with suitable mAb aggregate standards is feasible without falsification by previous affinity chromatography. Moreover, our results indicate that aggregate formation should be addressed directly in the cell culture and is not only a problem in DSP.

Keywords: Protein aggregation, Monoclonal antibodies, Mammalian cell culture, CHO cells

Background

Over the past ten years, the demand for monoclonal antibodies (mAbs) as biopharmaceutical drugs for the treatment of cancer and other diseases has increased [1-3]. Like other recombinant therapeutic proteins, mAbs are mainly produced in mammalian cells, usually in Chinese hamster ovary (CHO) cells [4]. Antibody manufacturing includes several steps, where environmental factors such as pH, temperature, ionic strength, protein concentration, oxygen and shear forces can lead to aggregate formation during upstream (USP) and downstream (DSP) processing [5,6]. Kramarcyk et al. reported up to 20-30% aggregate content of a partially purified mAb produced in CHO cells [7]. Self-association and formation of aggregates are a major concern for therapeutic applications, since aggregates influence drug performance and safety [8,9].

DSP offers the opportunity to remove aggregates, but this often leads to a reduction of protein yields. Another strategy involves reducing the formation of aggregates in the cell culture [10]. Jing and colleagues showed that proper control of culture conditions during USP successfully reduced the level of protein aggregation and improved the process yield [11]. To evaluate aggregate formation upstream, suitable analytical methods for aggregate detection and quantification are essential. This is challenging, since the size of aggregates may range from small oligomers to visible particles. Furthermore, host cell proteins (HCPs) and cell culture medium components may complicate aggregate detection. Nowadays, investigation of protein aggregation during cell cultivation is usually performed after a Protein A capture step [11-14]. Protein A affinity chromatography is a powerful tool for antibody purification, but it also exposes antibodies to pH-shifts [15]. This pH shift favors the formation of aggregates, since the aggregation rate of proteins

* Correspondence: paul@hochschule-bc.de
Institute of Applied Biotechnology (IAB), Biberach University of Applied Sciences, 88400 Biberach, Germany

is strongly influenced by pH [16,17]. Phillips et al. showed that acidic pH values of Protein A elution led to substantial protein aggregation and precipitation [18]. This implies that Protein A purification itself might influence the aggregation status, thus purified samples do not necessarily reflect the aggregation state of mAbs in cell culture.

Size exclusion chromatography (SEC) used in a high pressure liquid chromatography (HPLC) system is the most commonly applied analytical method for the analysis of soluble protein aggregates [19]. Typically, SEC analysis is performed after Protein A chromatography-based isolation of the mAbs from the cell culture supernatant, because cell culture components interfere with the direct analysis of mAb cell culture samples on SEC columns [10]. In the present study, we developed a procedure to analyze aggregate formation directly in the supernatant of CHO cells without a pre-purification step. Using a SEC column consisting of 3 μm silica particles or a SEC column especially tailored for mAb analysis due to selected pore size and column dimensions, we were able to separate mAb monomer and aggregates from interfering signals caused by DNA, host cell and culture medium components. The high resolution of these columns enabled quantification of mAb monomer and aggregate content directly from cell culture samples. In order to distinguish between the aggregates formed in cell culture, mAb aggregates differing in size and morphology were induced using different stress methods. The different high molecular weight (HMW) species were identified as dimers, tetramers, oligomers and large particles with hydrodynamic diameters greater than 100 nm. Additionally, stability and traceability of the induced mAb aggregates under conditions similar to the cell culture environment were shown. The induced aggregates served as standards for aggregate analysis in the culture supernatant. Finally, we analyzed aggregate formation of a mAb-producing CHO DG44 cell line to demonstrate that monomer and aggregates were detectable and quantifiable directly in the supernatant without a pre-purification step.

Methods

Monoclonal antibodies, cell line and media

Two aggregation-prone monoclonal antibodies (mAb1 and mAb2) were used as model proteins for this study. The mAbs were produced in CHO DG44 cells [20,21]. The cells were cultivated in defined protein-free medium (SFM4CHO, Thermo Scientific). A non-producing CHO DG44 cell line was used as control, which was supplemented with hypoxanthine and thymidine (HT). The antibodies were Protein A purified, filtered (0.2 μm) and stored in 20 mmol L^{-1} acetate at pH 3.5 for further studies. After stress induction the samples were filtered (0.2 μm) prior to SE-HPLC analysis.

Generation of different mAb aggregates

A pH-shift, high-salt concentration or freeze-thawing (FT) were used to induce different types of mAb aggregates. The antibodies were diluted to a final concentration of 1 mg mL^{-1} for the aggregation studies and analyzed immediately after stress-induction.

In order to generate pH-dependent aggregation, the mAbs were diluted in 100 mmol L^{-1} citric acid (2-hydroxypropane-1,2,3-tricarboxylic acid) and 200 mmol L^{-1} disodium hydrogen phosphate (Na_2HPO_4) with pH values ranging from pH 3–8. For exposure to high salt concentrations the mAbs were diluted to final sodium chloride (NaCl) concentrations ranging from 0.1 to 1.5 mol L^{-1}. For this purpose, a 5 mol L^{-1} NaCl stock solution was prepared using ultrapure water (Millipore). Moreover, both antibodies were subjected to multiple FT cycles. One FT cycle included storage at −80°C for 15 min, followed by thawing at 25°C for 15 min in a thermal mixer (HLC Biotech).

SE-HPLC analysis

Analysis of soluble aggregates was performed using an Agilent 1100 HPLC (Agilent Technologies) system and an UltiMate 3000 (Thermo Scientific) system. TSKgel G3000SWXL (Tosoh Bioscience), Yarra SEC-4000 (Phenomenex) and MAbPac SEC-1 (Thermo Scientific) were used as SEC columns. The column properties are listed in the Additional file 1: Table S1. The SEC separation was performed at ambient temperature isocratically using a mobile phase consisting of phosphate-buffered saline (PBS), which was adjusted to pH 7.2 and filtered 0.1 μm prior to use. Flow rates varied between 0.3, 0.5 and 0.8 mL min^{-1}. The respective amounts of mAb monomer, aggregates and fragments were quantified by calculation of the peak areas detected by the ultraviolet (UV) detectors. The UV signal of the Agilent 1100 system was displayed in mV, the signal of the Ultimate 3000 system in mAU. All samples were pre-filtered using 0.2 μm syringe filters (Phenomenex).

Dynamic light scattering (DLS)

Formation of large aggregates was determined using a Zetasizer 3000 HS instrument (Malvern Instruments) at 25°C. The average hydrodynamic diameter (Zave) of the HMW species was measured in disposable semi-micro UV cuvettes (Brand).

Molecular weight determination

The molecular weight of the aggregates was determined using SEC in combination with multi-angle light scattering (MALS). SEC-MALS was performed using an Äkta Explorer 100 instrument (GE Healthcare) equipped with a Yarra SEC-4000 column (Phenomenex), an Optilab T-Rex refractive index (RI) detector (Wyatt) and a DAWN

HELEOS 8+ MALS detector (Wyatt). The samples were analyzed at room temperature (RT) using 0.1 μm filtered PBS (pH 7.0).

Stability of the induced aggregates

To investigate aggregate stability, mAb2 was stressed using 500 mmol L^{-1} NaCl or three cycles of FT, respectively, in order to induce dimer and oligomer formation and measured after storage at RT for up to 72 h. To investigate the traceability under cell culture conditions, stressed antibody samples were spiked into SFM4CHO medium or CHO DG44 host cell supernatant and analyzed immediately afterwards.

Aggregate formation in mAb-producing CHO cell line

To investigate aggregate formation in cell culture, CHO DG44 cells producing mAb2 were analyzed directly after inoculation and after 144 h of cultivation, respectively. For this purpose, cells were seeded at an initial concentration of 4×10^5 mL^{-1} in 125 mL shake flasks (Corning) with a working volume of 25 mL. Cultivation was performed in SFM4CHO medium supplemented with 10 g L^{-1} glucose and 4 mmol L^{-1} glutamine at 37°C, 140 rpm, 80% humidity and 5% CO_2. For analysis of mAb aggregate content using SE-HPLC, cells were separated at 10.000 g for 5 min and the supernatant was filtered (0.2 μm).

Determination of DNA and host cell proteins

To ensure the absence of DNA and HCPs in the SEC fractions used for aggregate quantification, the corresponding fractions of the CHO mAb2 supernatant were collected (fraction volume: 98 μL) and analyzed. For DNA quantification, the fractions were analyzed using a NanoDrop 1000 spectrophotometer (Thermo Scientific). For the determination of HCP and culture medium components, the SEC fractions were separated using sodium dodecyl sulfate polyacrylamide gel electrophoresis (SDS-PAGE) and protein bands were detected using silver staining. To receive sufficient amounts for analysis the SEC fractions were pooled and concentrated. Concentration was performed using Vivaspin 500 with a 3 kDa molecular weight cut-off (Sartorius Stedim Biotech). SDS-PAGE was performed at 160 V for 70 min at RT under reducing conditions with an Amersham ECL Gel Box and a 4-20% Amersham ECL Gel (GE Healthcare). Purified mAb2 and supernatant of non-producing CHO cells served as controls. PageRuler Prestained Protein Ladder (Thermo Scientific) was used as size standard.

Results and discussion

SEC-based analysis of mAb aggregates in cell culture supernatant

MAbs produced in mammalian cells are secreted into the cell culture supernatant, where formation of aggregates can be detected using SEC analysis after a Protein A capture step [4,15]. This approach is time-consuming, laborious and can critically influence the aggregation status. Direct analysis of cell culture samples containing mAb using SEC is complicated, since components such as DNA, lipids and HCPs can interfere with the analysis of the product [10]. Accordingly, analysis of CHO mAb2 fermentation supernatant samples using a common SEC column (TSKgel G3000SWXL) was insufficient and therefore confirmed that separation of mAb monomer from other supernatant signals disturbed aggregate detection (Figure 1A). Since analysis of cell culture samples using SEC appeared to be a problem of separation capacity, columns promising higher separation efficiencies were tested. Using a SEC column specifically designed for mAb analysis (MAbPac SEC-1) showed that cell culture medium components eluted later than purified mAb monomer (Figure 1B). Serum-free media often contain growth factors, lipoproteins and other factors as growth-promoting supplements. It was confirmed that none of the media components interfered with the analysis of the mAb2 monomer as these components showed higher retention times. Furthermore, analysis of the supernatant from the CHO DG44 culture using a 3 μm SEC column (Yarra SEC-4000) revealed that also the cell culture components in the supernatant eluted later than the mAb monomer (Figure 1C). These host cell components, i.e. cellular proteins, DNA and other components usually complicate the detection of target proteins [22]. Our results indicated that cell culture and medium impurities eluted later than the mAb monomer using columns providing sufficiently high separation efficiency. Therefore, both columns were tested for their applicability for the analysis of mAb aggregates in cell culture samples.

Generation of different mAb aggregates

For aggregate detection during cultivation, one need to be capable of measuring and characterizing all types of aggregates potentially present in the culture broth. Therefore, different types of mAb aggregates were generated by applying different stress conditions. Since pH changes and increase of osmolality are both known phenomena occurring in cell culture, pH-shift and high salt concentration were used as induction methods. Furthermore, to generate high amounts of soluble aggregates, FT was additionally used as unnatural stress condition. To distinguish between the different HMW species, the molecular weight was determined using SEC-MALS. The various stress methods induced different mAb aggregates, which were used as standards for the analysis of cell culture supernatants.

pH-shift

As mentioned in the introduction, the pH-shift required for Protein A chromatography favors the formation of

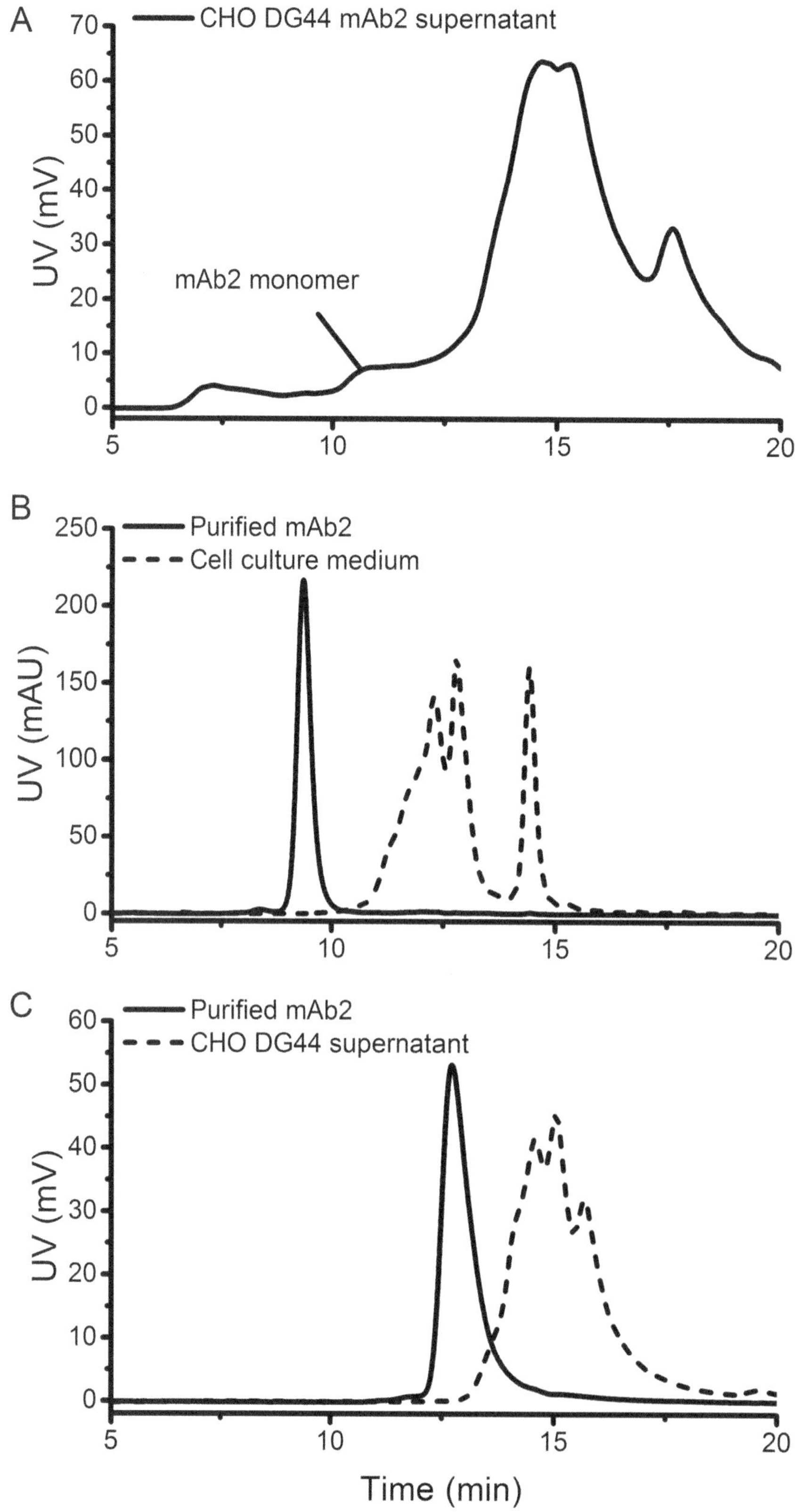

Figure 1 SEC analysis of mAb2 cell culture samples. CHO cell culture supernatant analyzed using a common SEC (TSKgel G3000SWXL) column **(A)**. Analysis of purified mAb2 compared to SFM4CHO medium using MAbPac SEC-1 **(B)** and CHO DG44 supernatant using Yarra SEC-4000 **(C)**.

aggregates and therefore may lead to reduced antibody yields [17]. Thus, a pH-shift can also be used to induce mAb aggregates. The different antibodies, mAb1 and mAb2, were exposed to different pH values ranging from pH 3–8, which resulted in the formation of small as well as large aggregates (Table 1). With increasing pH a loss of mAb2 monomer was observed accompanied by the formation of mAb1 and mAb2 dimer. At pH 3 mAb1 consisted of only 1.8% dimer and mAb2 of 1.7% dimer, whereas 3.4% and 6.3% dimer were visible at pH 8, respectively. Additionally, a pH shift induced the generation of mAb2 oligomer and a smaller fragment. Oligomer was elevated at pH 3, but less prominent for other pH values. The correlation between pH increase and dimer formation has already been previously described [23]. Ishikawa et al. showed a pH-dependent fragment formation of different antibodies [24]. The antibody used in their study showed most aggregate formation at neutral pH, which is consistent with our results. However, in their work the amount of lower molecular weight species was highest at pH 4, whereas mAb2 formed more fragments at higher pH values. Additionally, DLS revealed a pH-induced formation of large mAb aggregates. With an average diameter greater than 80 nm, the Zave of mAb1 at pH 5 and 6 was more than 2 times higher than at pH 3 (34.2 nm ±1.8 nm) and 4 (32.5 nm ±3.3 nm). MAb2 formed large aggregates at pH 5 with a diameter greater than 100 nm, whereas at other pH values the average diameter was smaller than 50 nm. Since both antibodies formed large aggregates and mAb2 was also degraded to fragments, pH-shift was not used for further stability studies and for the generation of aggregate standards.

NaCl

Increased salt concentration reduces the colloidal stability by suppressing electrostatic repulsion, leading to aggregation [25]. We used this mechanism to induce aggregation of the two model proteins. The use of different NaCl concentrations induced the formation of small aggregates as well as large aggregates (Table 1). In comparison to pH-shift induced aggregation, fewer aggregates were formed using NaCl, but it is worth mentioning that NaCl-induction specifically formed one smaller HMW species. Using SEC-MALS, the molecular weight of the NaCl-induced HMW species was determined as a 300 kDa mAb2 dimer (Figure 2A). In addition, the light scattering signal revealed the presence of a large aggregate population, which was not visible using UV detection. With increasing NaCl concentrations mAb1 monomer content decreased and dimer increased from 0.6% (w/o NaCl) to 2.0% (1.5 mol L^{-1} NaCl). Similar to mAb1 also mAb2 monomer decreased with increasing NaCl concentration, whereas the dimer content increased more than fourfold from 1.1% (w/o NaCl) to 4.6% (1.5 mol L^{-1} NaCl). In contrast to mAb2, mAb1 also formed some degradation products. Our results correlate with the results of Fesinmeyer et al., who reported that even in milli-molar concentrations, salts promote mAb aggregate formation [26]. They suggested anion binding as

Table 1 Formation of mAb dimer, oligomer, fragment and large aggregates using the different induction methods

	mAb1			mAb2			
	Dimer (%)	Oligomer (%)	Zave (nm)	Dimer (%)	Oligomer (%)	Fragment (%)	Zave (nm)
pH							
3	1.8	-	34.2 ± 1.8	1.7	1.8	-	39.6 ± 2.4
4	1.9	-	32.5 ± 3.3	2.1	0.9	-	49.5 ± 3.5
5	2.3	-	84.3 ± 15.0	2.0	0.9	3.5	144.9 ± 35.8
6	2.6	-	85.7 ± 1.9	3.5	0.6	5.7	47.9 ± 1.3
7	3.1	-	51.1 ± 5.3	4.0	0.5	4.5	45.7 ± 0.8
8	3.4	-	n. a.	6.3	0.6	6.3	n. a.
NaCl							
Control	0.6	-	33.7 ± 5.3	1.1	-	-	33.2 ± 0.2
500 mM	1.3	-	23.8 ± 0.2	3.0	-	-	41.8 ± 1.9
1 M	1.7	-	24.8 ± 0.6	4.2	-	-	56.6 ± 8.8
1.5 M	2.0	-	24.2 ± 1.6	4.6	-	-	78 ± 9.2
FT							
Control	0.6	-	19.7 ± 1.1	1.4	-	-	10.6 ± 1.1
1×	16.0	11.3	20.7 ± 1.8	17.1	3.6	-	14.4 ± 0.1
2×	20.2	16.3	26.6 ± 1.9	19.3	3.9	-	12.9 ± 0.5
3×	22.7	11.0	27.3 ± 2.8	21.0	6.6	-	12.1 ± 0.5

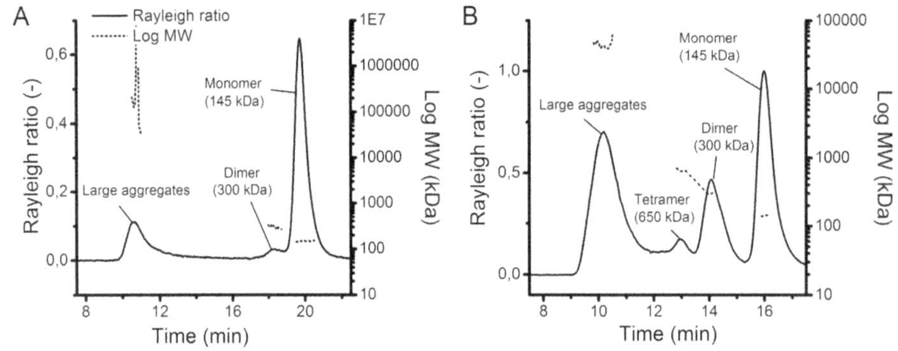

Figure 2 Molecular weight determination of mAb aggregate standards. MAb2 aggregate standards were induced using either NaCl **(A)** or FT **(B)** and analyzed using SEC-MALS.

possible reason for aggregation by lowering mAb conformational stability and reduced valence. Analysis with DLS revealed that mAb2 also formed large aggregates (Table 1). NaCl-induction increased the aggregate size up to an average diameter of around 75 nm (1.5 mol L^{-1} NaCl). However, mAb1 did not form large aggregates. Our work showed that increasing ionic strength by 100 mmol L^{-1} NaCl is sufficient to induce mAb dimers, and at concentrations higher than 500 mmol L^{-1} also to induce larger mAb aggregates. Cation concentration in the culture broth increases during the fermentation process due to pH control [27]. Consequently, osmolality in the bioreactor increases over time thus NaCl-induced aggregates are likely to be present in the cell culture broth. Since 500 mmol L^{-1} NaCl induced a significant amount of small aggregates (3.0% dimer) without forming large HMW species of mAb2, this concentration was used to induce mAb2 dimers for stability studies and the generation of aggregate standards.

Freeze-thawing (FT)

Protein aggregation by FT is attributed to partial unfolding caused by perturbing conditions like low temperature, buffer crystallization, exposure to the ice-liquid interface, adsorption to the container surface, increasing salt or protein concentration [28]. Using FT, both of our model proteins showed the highest aggregation propensity, which illustrated the impact of perturbation on protein stability (Table 1). The aggregates formed by FT were identified as dimer and tetramer (Figure 2B). Again, the light scattering signal showed the presence of a large aggregate population, which was already visible using NaCl-induction, but not measurable using UV detection. Increasing the number of FT cycles resulted in a loss of mAb1 monomer with a corresponding increase of mAb1 dimer (22.7%) and tetramer (11.0%) after three cycles of FT. Similarly, mAb2 comprised around 21.0% dimer and 6.6% tetramer after three cycles of FT. With more than 20% dimer formation, FT induced three times higher aggregate levels than pH-shift to pH 8 (6.3% dimer) and more than four times higher as the highest NaCl concentration (4.6% dimer). Furthermore, FT led to the formation of a significant amount of oligomers (11.0% mAb1, 6.6% mAb2), which were not formed to that extent using the other stress methods. FT also induced formation of large aggregates, which were detectable using DLS. With increasing FT cycle the hydrodynamic diameter of mAb1 aggregates increased from 20 nm to above 25 nm, corresponding to the monomer loss visible in the SEC chromatogram. Interestingly, mAb1 showed the highest transformation rate after three cycles of FT, whereas mAb2 formed most dimer, oligomer and large aggregates after one FT cycle. Our results coincide with Hawe et al., who observed an increase in Zave and the polydispersity index (PdI) caused by FT-induced large and heterogeneous aggregates [29]. Furthermore, our results are also in good accordance with other reports [30]. Since FT induced more aggregates than the other stress methods, this technique was chosen in addition to NaCl-induction for further stability studies and the generation of aggregate standards.

Stability of the induced aggregates

In order to ensure stability of the aggregate standards, mAb1 and mAb2 were stressed and analyzed after storage at RT. To ensure traceability of the standards in conditions similar to the cell culture environment, the induced aggregates were further analyzed after spiking into cell culture medium and the supernatant of non-producing CHO DG44 cells, respectively.

To evaluate stability of NaCl-induced aggregates, mAb1 and mAb2 aggregates were induced using 500 mmol L^{-1} NaCl and analyzed after storage at RT for up to 72 h (Figure 3A-D). MAb1 dimer content increased already after 2 h from 1.3% to 3.3% (Figure 3B) and mAb2 dimer increased within 72 h from 6.5% to 10.2% (Figure 3D), revealing that the aggregation reaction was not finished within this time period. These observations are in good accordance with other reports [26,31,32]. Stability analysis

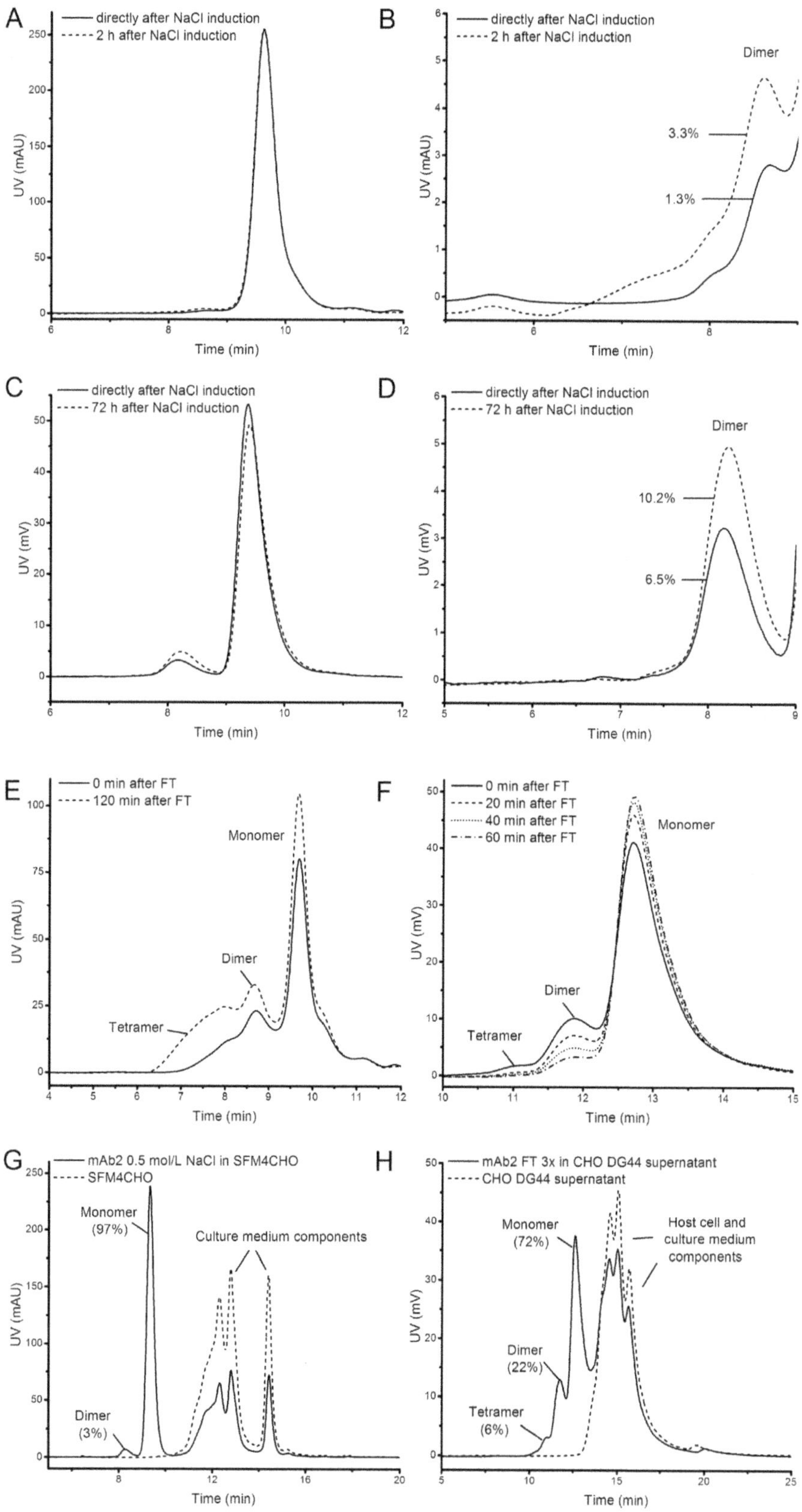

Figure 3 (See legend on next page.)

(See figure on previous page.)
Figure 3 Stability and traceability of mAb aggregate standards. The stability of NaCl-induced mAb1 **(A + B)** and mAb2 **(C + D)** aggregates was investigated. **B** is an enlargement of **A**, **D** is an enlargement of **C**. **(E)** Stability of FT-induced mAb1 using MAbPac SEC-1. **(F)** Stability of FT-induced mAb2 aggregates using Yarra SEC-4000. **(G)** Traceability of NaCl-induced mAb2 aggregates using MAbPac SEC-1. **(H)** Traceability of FT-induced mAb2 aggregates using Yarra SEC-4000 under cell culture conditions.

of FT-induced aggregates revealed that the aggregates formed were readily reversible (Figure 3E + F). MAb1 tetramer increased from 11% to 26% after 2 h storage at RT (Figure 3E). FT-induced aggregates of mAb2 seemed to be reversible. Increasing storage time at room temperature resulted in mAb2 monomer recovery with a corresponding decrease in dimer and oligomers (Figure 3F). After 1 h storage at RT, FT-induced mAb2 dimer decreased from 14% to 4% and the tetramer from 2.5% to 0.3%, whereas mAb2 monomer increased from 82.6% to 95.7%. This shift in aggregate distribution was also observed by Philo, who reported that incubation of Protein X at 29°C after FT led to a drop in dimer and larger aggregate content, with a corresponding increase in monomer [33]. Since Philo used Protein X, our results are the first report, to the best of our knowledge, which showed the reversibility of FT-induced aggregates for mAbs. Since the aggregates induced by NaCl and FT were unstable, all aggregate standards were prepared freshly for further experiments.

To study traceability under cell culture conditions, NaCl-induced mAb2 was spiked into SFM4CHO cell culture medium (Figure 3G). Monomer as well as mAb2 aggregates were still detectable in this matrix. Given that the medium components of the serum-free cell culture medium eluted later than mAb2 monomer, quantification of monomer and aggregate content was feasible. With 3% dimer and 97% monomer the values were consistent to NaCl-induced aggregate formation of mAb2 (Table 1). Furthermore, we showed that the dimers and tetramers induced by FT were also detectable and quantifiable in the supernatant of non-producing CHO DG44 cells (Figure 3H). HCPs and culture medium components eluted later than mAb2 monomer and aggregates without influencing aggregate quantification. In total 22% dimer, 6% oligomer and 72% remaining monomer was determined for mAb2, which correlated well to the results obtained from freeze-thawed mAb2 (Table 1). The spiking experiments demonstrated that the aggregates induced by the selected stress methods were stable during the experiment and still detectable under conditions expected in cell culture. Since the high resolution facilitated by the 3 μm SEC column enabled sufficient efficiency to separate mAb monomer and aggregates from CHO DG44 cell culture components, aggregate formation in the culture of a mAb-producing CHO cell line was investigated using this column.

Aggregate formation in a mAb-producing CHO cell line

To demonstrate the feasibility of the method, the cell-free supernatant derived from a culture of a mAb2-producing CHO cell line was analyzed for aggregate formation. In order to ensure typical growth, culture parameters such as viability and viable cell concentration as well as substrate, metabolite and product concentrations were measured during cultivation. Cell culture samples were taken every 24 h, and aggregation analysis was performed directly after inoculation and after 144 h of cultivation.

The cultivation showed typical growth, substrate and metabolite concentrations for mAb production (Additional file 2: Figure S1). Directly after inoculation neither monomer nor mAb aggregates were detected (Figure 4A + B). Chromatograms of the SE-HPLC analysis obtained from the supernatant of the three shake flasks showed identical results (Figure 4B). Retention times of the signals obtained immediately after the start of cultivation (14–16 min) corresponded to the elution times obtained in the analysis of the cell-free supernatant from a culture of a non-producing CHO DG44 host cell line (Figure 3H), indicating that the signals were caused by host cell impurities and medium components such as DNA, lipids, HCPs, secreted cellular metabolites and excess nutrients [10,34]. To ensure that these cell culture contaminants do not interfere with quantification of the mAb2 monomer and aggregates in the CHO supernatant, the corresponding chromatography fractions were collected and analyzed for DNA (Figure 4C) and HCPs (Figure 4D). DNA analysis of the collected SE-HPLC fractions revealed that the CHO mAb2 supernatant contained a significant amount of DNA (up to 15 ng μL^{-1}). The finally formulated recombinant mAb product has to contain less than 10 ng/dose DNA [35]. Therefore the impurities have to be removed from the cell culture supernatant after harvest using several DSP operations [22]. However, DNA was detected only in the fractions containing host cell and culture medium components and not observed in the SEC fractions used for mAb aggregate quantification. In order to determine the amount of HCPs, the corresponding SEC fractions were pooled, concentrated and separated using reducing SDS-PAGE with subsequent silver staining (Figure 4D). The supernatant of the non-producing CHO cells (DG44) and the supernatant of the mAb2-producing CHO cells (CHO mAb2) both showed a huge amount of signals, indicating presence of HCPs in the supernatant. The strongest bands in the mAb2-producing CHO

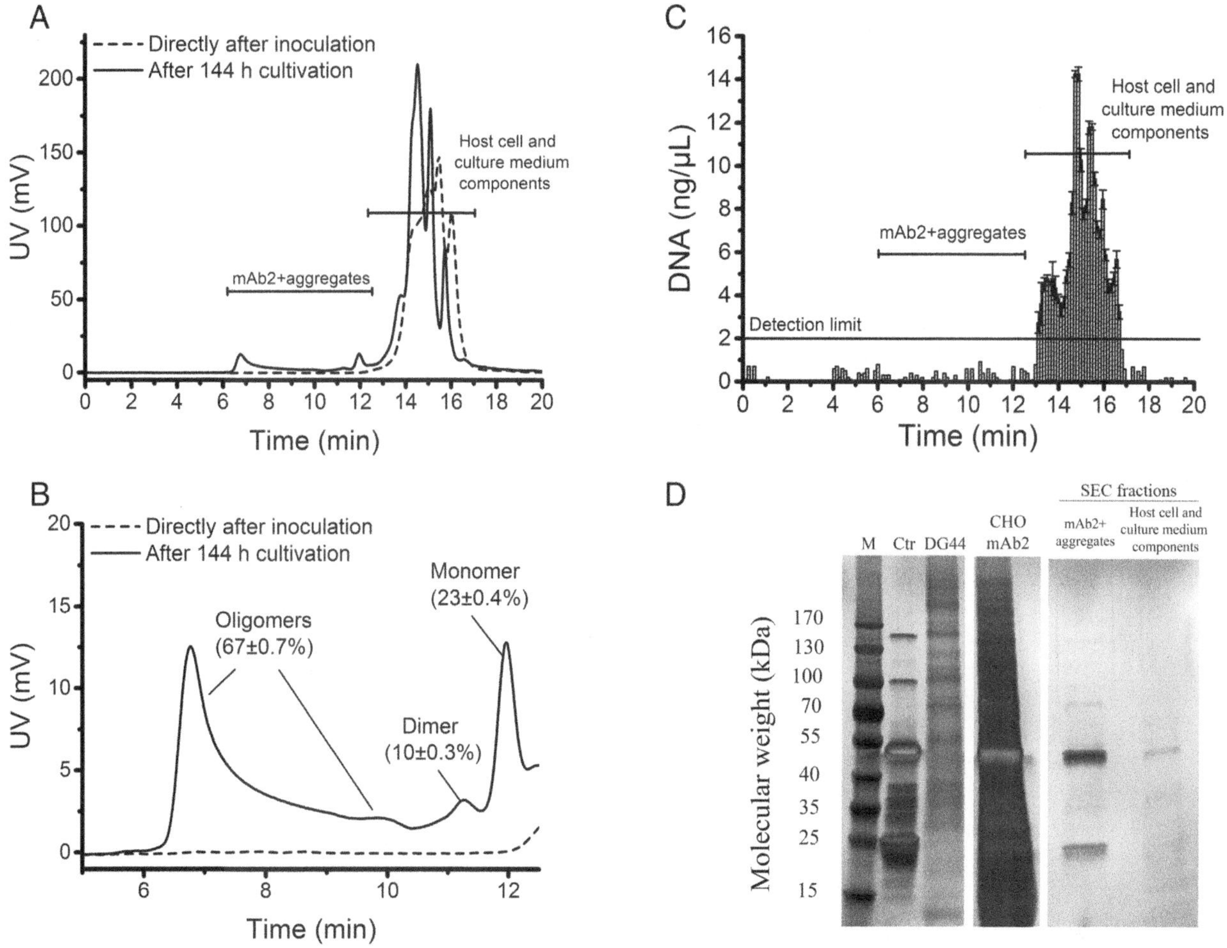

Figure 4 Analysis of mAb aggregate formation, DNA content, host cell and medium components in the supernatant of a mAb2-producing CHO cell line. Supernatant was analyzed for mAb aggregate formation directly after inoculation and after 144 h cultivation **(A + B)**. Amount of mAb2 monomer, dimer and oligomers was obtained from biological triplicates ± standard deviation. B is an enlargement of A. DNA content of collected SEC fractions was analyzed using NanoDrop 1000 spectrophotometer **(C)**. For determination of HCPs and culture medium components, SEC fractions were pooled, concentrated and analyzed using SDS-PAGE under reducing conditions with subsequent silver staining **(D)**.

supernatant were caused by the heavy chain (around 50 kDa) and the light chain (around 25 kDa) of reduced mAb2, also visible in the purified mAb2 control (Ctr). Analysis of the collected SEC fractions revealed that the mAb2 monomer and aggregate fraction contained mainly heavy and light chain, indicating that nearly no HCPs were present. The monomer and aggregate fraction showed a signal at 150 kDa caused by unreduced mAb2 as well as bands between heavy and light chain, which were also visible in the control, whereas the fractions collected from host cell and culture medium components showed additional signals. It is known that mammalian cells secrete HCPs together with the product [36]. Nevertheless, the mAb2 monomer and aggregate fractions collected from the SE-HPLC analysis contained no detectable amounts of DNA and were nearly free of HCP enabling mAb aggregate quantification directly in the supernatant of CHO cells using the SEC columns tested in this work. However, the results also revealed the presence of a large amount of DNA and cell culture components in the supernatant of CHO cell culture and emphasize the importance of the use of appropriate SEC columns for aggregate analysis.

These interfering components may be the reason why analysis of protein aggregation during cell culture was mainly performed after a Protein A capture step in other experiments [11-13,37]. Franco et al. used an analytical gel filtration column with 13 μm bead size, which had a lower resolution than the 3 μm SEC column used in our work. Ho et al. and Jing et al. performed their experiments with the same column type, which showed insufficient separation of mAb monomer from other cell culture signals in our experiments (Figure 1A). This could explain why in those experiments a pre-purification step was necessary to detect aggregates in cell culture samples.

Moreover, the amount of HCPs in the harvest product pool is dependent on the host organism, the protein of interest as well as culture and harvest conditions [38]. It is conceivable that the production cell line as well as the non-producing CHO DG44 cell line used in this work did not secrete HCPs, which interfere with mAb monomer and aggregates enabling a smooth quantification. Using our method, mAb monomer and aggregates were detectable directly in the supernatant of CHO mAb producer cells after 144 h of cultivation without interfering with host cell impurities (Figure 4A + B). Quantification of aggregate formation revealed that a huge amount of mAb consisted of aggregates. Comparison of the cell culture samples to retention times of the aggregated mAb standards indicated that the aggregates comprised of dimer and a huge amount of larger oligomers. Figure 4B shows the mean values of monomer, dimer and larger aggregates resulting from all shake flasks. With only 23% ±0.4% remaining monomer, the supernatant after 144 h comprised of 10% ±0.3% dimer and 67% ±0.7% larger oligomers (Figure 4B). This is the first report demonstrating such a high amount of aggregates during CHO cell culture. Franco et al. found about 10% aggregates in all harvest batches of hybridoma cells secreting a monoclonal anti-PSA antibody [13]. Other work using IgG-producing CHO cells reported aggregation rates from 20% up to 30% [7,39]. Ho et al. obtained an aggregate and fragment content of more than 50% by decreasing the light chain to heavy chain (LC:HC) ratio [37]. As mentioned before, most of these approaches measured mAb aggregate formation after a Protein A purification step. This step included loading of the mAbs to the Protein A column under neutral conditions, elution under acidic conditions and neutralization before applying SEC analysis [10,40]. During elution and neutralization the mAbs undergo critical pH values (pH 5 and 6), which induced very large aggregates in our pH-shift experiments (Table 1). With a diameter above 100 nm at pH 5, the mAb2 aggregates induced by pH-shift were by far larger than using the highest NaCl concentration and three cycles of FT. These large aggregates may be filtered during sample preparation or get stuck in the column matrix and could therefore be underrepresented in the subsequent SEC analysis. This would explain the varying amount of aggregates detected in DSP and emphasizes the importance of a more direct method for upstream aggregate analysis. Directly analyzing mAb cell culture samples using high resolution SEC columns does not include shifts to critical pH values. Although the mAbs used in this work were aggregation-prone, our results indicate how much protein may be lost due to formation of HMW species during cell culture. It can be assumed that the HMW species were product-related, since the supernatant of non-producing CHO cells showed no comparable signals (Figure 3H) and the SEC fractions used for mAb aggregate quantification contained neither DNA nor significant amounts of HCP and culture medium components (Figure 4C + D). Furthermore, the large aggregates detected in the supernatant of the mAb2-producing CHO cell line were also visible in the Rayleigh ratio after NaCl- and FT-induction of purified mAb2 (Figure 2). Our results demonstrate that SE-HPLC analysis using a high-resolution column in combination with suitable aggregate standards can be a valuable tool to determine aggregate formation in cell culture, as the method allows quantification of mAb aggregate content within 20 min directly in the cell culture supernatant without falsifying the results by labor-intensive pre-purification of the samples.

Conclusions

In this study, we have investigated SEC-based quantification of mAb aggregates directly in the supernatant of CHO cell cultures without a pre-purification step. High-resolution SEC columns enabled separation of mAb monomer and aggregates from interfering cell culture components such as DNA, host cell and culture medium contaminants. In order to distinguish between different HMW species formed in cell culture, mAb dimers, tetramers, oligomers and large aggregates with diameter above 100 nm were induced using different stress techniques to be used as aggregate standards. The induced aggregates were still detectable under conditions similar to the cell culture environment. Finally, we demonstrated that the method is applicable to quantify aggregates in the supernatant of a mAb2-producing CHO cell line. The results indicated that over 75% percent of mAb was aggregated proving that aggregate formation may already have its origin in the cell culture and is not only a problem in DSP.

Additional files

Additional file 1: Table S1. Specifications of SEC columns used in this study.

Additional file 2: Figure S1. Cell culture parameters of mAb2-producing CHO cell line. Viable cells and viability (A), glucose and lactate concentration (B), glutamine and ammonium concentration (C) and product formation (D) of a mAb-producing CHO culture. Samples were taken every 24 h over 144 h of cultivation. Viable cells and viability was measured using Cedex XS (Roche). Substrate, metabolite and product concentrations were measured using Konelab Arena 20 (Thermo Scientific).

Abbreviations

CHO: Chinese Hamster Ovary; DLS: Dynamic light scattering; DSP: Downstream processing; FT: Freeze-thawing; HCP: Host cell protein; HMW: High molecular weight; HPLC: High pressure liquid chromatography; mAbs: Monoclonal antibodies; MALS: Multi angle light scattering; PBS: Phosphate-buffered saline; RI: Refractive index; RT: Room temperature; SEC: Size exclusion chromatography; USP: Upstream processing; UV: Ultraviolet; Zave: Zeta average.

Competing interests

The authors declare that they have no competing interests.

Authors' contributions
AP performed all the experiments and drafted the manuscript. KS has been involved in drafting the manuscript and revising it critically for important intellectual content. FH conceived of the study, revised it critically and participated in its design and coordination and helped to draft the manuscript. All authors read and approved the final manuscript.

Acknowledgements
This research was supported by the German Federal Ministry of Education and Research (Grant No. 0315342A). The authors are grateful to Boehringer Ingelheim Pharma GmbH & Co for providing the mAb1-producing CHO cell line and Rentschler Biotechnologie GmbH for providing the mAb2-producing CHO cell line.

References
1. Sliwkowski MX, Mellman I: **Antibody therapeutics in cancer.** *Science* 2013, **341**:1192–1198.
2. Holliger P, Hudson PJ: **Engineered antibody fragments and the rise of single domains.** *Nat Biotechnol* 2005, **23**:1126–1136.
3. Leavy O: **Therapeutic antibodies: past, present and future.** *Nat Rev Immunol* 2010, **10**:297.
4. Wurm FM: **Production of recombinant protein therapeutics in cultivated mammalian cells.** *Nat Biotechnol* 2004, **22**:1393–1398.
5. Vázquez-Rey M, Lang DA: **Aggregates in monoclonal antibody manufacturing processes.** *Biotechnol Bioeng* 2011, **108**:1494–1508.
6. Dengl S, Wehmer M, Hesse F, Lipsmeier F, Popp O, Lang K: **Aggregation and chemical modification of monoclonal antibodies under upstream processing conditions.** *Pharm Res* 2013, **30**:1380–1399.
7. Kramarczyk JF, Kelley BD, Coffman JL: **High-throughput screening of chromatographic separations: II. Hydrophobic interaction.** *Biotechnol Bioeng* 2008, **100**:707–720.
8. Perchiacca JM, Tessier PM: **Engineering aggregation-resistant antibodies.** *Annu Rev Chem Biomol Eng* 2012, **3**:263–286.
9. Paul R, Graff-Meyer A, Stahlberg H, Lauer ME, Rufer AC, Beck H, Briguet A, Schnaible V, Buckel T, Boeckle S: **Structure and function of purified monoclonal antibody dimers induced by different stress conditions.** *Pharm Res* 2012, **29**:2047–2059.
10. Pan H, Chen K, Pulisic M, Apostol I, Huang G: **Quantitation of soluble aggregates in recombinant monoclonal antibody cell culture by pH-gradient protein A chromatography.** *Anal Biochem* 2009, **388**:273–278.
11. Jing Y, Borys M, Nayak S, Egan S, Qian Y, Pan S-H, Li ZJ: **Identification of cell culture conditions to control protein aggregation of IgG fusion proteins expressed in Chinese hamster ovary cells.** *Process Biochem* 2012, **47**:69–75.
12. Yang WC, Lu J, Nguyen NB, Zhang A, Healy NV, Kshirsagar R, Ryll T, Huang Y-M: **Addition of Valproic Acid to CHO Cell Fed-Batch Cultures Improves Monoclonal Antibody Titers.** *Mol Biotechnol* 2014, **56**(5):421–428.
13. Franco R, Daniela G, Fabrizio M, Ilaria G, Detlev H: **Influence of osmolarity and pH increase to achieve a reduction of monoclonal antibodies aggregates in a production process.** *Cytotechnology* 1999, **29**:11–25.
14. Rea JC, Moreno GT, Vampola L, Lou Y, van Haan B, Tremintin G, Simmons L, Nava A, Wang YJ, Farnan D: **Capillary size exclusion chromatography with picogram sensitivity for analysis of monoclonal antibodies purified from harvested cell culture fluid.** *J Chromatogr A* 2012, **1219**:140–146.
15. Cromwell MEM, Hilario E, Jacobson F: **Protein aggregation and bioprocessing.** *AAPS J* 2006, **8**:E572–E579.
16. Chi EY, Krishnan S, Randolph TW, Carpenter JF: **Physical stability of proteins in aqueous solution: mechanism and driving forces in nonnative protein aggregation.** *Pharm Res* 2003, **20**:1325–1336.
17. Arosio P, Rima S, Morbidelli M: **Aggregation mechanism of an IgG2 and two IgG1 monoclonal antibodies at low pH: from oligomers to larger aggregates.** *Pharm Res* 2013, **30**:641–654.
18. Phillips J, Drumm A, Harrison P, Bird P, Bhamra K, Berrie E, Hale G: **Manufacture and quality control of CAMPATH-1 antibodies for clinical trials.** *Cytotherapy* 2001, **3**:233–242.
19. Mahler H-C, Jiskoot W: *Analysis of Aggregates and Particles in Protein Pharmaceuticals.* Hoboken, New Jersey: John Wiley & Sons; 2012.
20. Urlaub G, Chasin LA: **Isolation of Chinese hamster cell mutants deficient in dihydrofolate reductase activity.** *Proc Natl Acad Sci U S A* 1980, **77**:4216–4220.
21. Urlaub G, Emmanuel K, Carothers AM, Chasin LA: **Deletion of the diploid dihydrofolate reductase locus from cultured mammalian cells.** *Cell* 1983, **33**(June):405–412.
22. Guiochon G, Beaver LA: **Separation science is the key to successful biopharmaceuticals.** *J Chromatogr A* 2011, **1218**:8836–8858.
23. Ahrer K, Buchacher A, Iberer G, Jungbauer A: **Detection of aggregate formation during production of human immunoglobulin G by means of light scattering.** *J Chromatogr A* 2004, **1043**:41–46.
24. Ishikawa T, Ito T, Endo R, Nakagawa K, Sawa E, Wakamatsu K: **Influence of pH on heat-induced aggregation and degradation of therapeutic monoclonal antibodies.** *Biol Pharm Bull* 2010, **33**:1413–1417.
25. Saito S, Hasegawa J, Kobayashi N, Tomitsuka T, Uchiyama S, Fukui K: **Effects of ionic strength and sugars on the aggregation propensity of monoclonal antibodies: influence of colloidal and conformational stabilities.** *Pharm Res* 2013, **30**:1263–1280.
26. Fesinmeyer RM, Hogan S, Saluja A, Brych SR, Kras E, Narhi LO, Brems DN, Gokarn YR: **Effect of ions on agitation- and temperature-induced aggregation reactions of antibodies.** *Pharm Res* 2009, **26**:903–913.
27. Schmelzer AE, DeZengotita VM, Miller WM: **Considerations for osmolality measurement under elevated pCO(2): comparison of vapor pressure and freezing point osmometry.** *Biotechnol Bioeng* 2000, **67**:189–196.
28. Kueltzo LA, Wang WEI, Randolph TW, Carpenter JF: **Effects of Solution Conditions, Processing Parameters, and Container Materials on Aggregation of a Monoclonal Antibody during Freeze-Thawing.** *J Pharm Sci* 2008, **97**:1801–1812.
29. Hawe A, Kasper JC, Friess W, Jiskoot W: **Structural properties of monoclonal antibody aggregates induced by freeze-thawing and thermal stress.** *Eur J Pharm Sci* 2009, **38**:79–87.
30. Printz M, Friess W: **Simultaneous Detection and Analysis of Protein Aggregation and Protein Unfolding by Size Exclusion Chromatography with Post Column Addition of the Fluorescent Dye BisANS.** *J Chromatogr A* 2012, **101**:826–837.
31. Arosio P: **On the role of salt type and concentration on the stability behavior of a monoclonal antibody solution.** *Biophys Chem* 2012, **16**:19–27.
32. Majumdar R, Manikwar P, Hickey JM, Samra HS, Sathish HA, Bishop SM, Middaugh CR, Volkin DB, Weis DD: **Effects of Salts from the Hofmeister Series on the Conformational Stability, Aggregation Propensity, and Local Flexibility of an IgG1 Monoclonal Antibody.** *Biochemistry* 2013, **52**:3376–3389.
33. Philo JS: **Is any measurement method optimal for all aggregate sizes and types?** *AAPS J* 2006, **8**:E564–E571.
34. Goh L-T, Yap MGS: **Determination of interferon-gamma in Chinese hamster ovary cell culture supernatant by coupled-column liquid chromatography.** *J Sep Sci* 2005, **28**:2104–2110.
35. Chon JH, Zarbis-Papastoitsis G: **Advances in the production and downstream processing of antibodies.** *N Biotechnol* 2011, **28**:458–463.
36. Hogwood CEM, Tait AS, Koloteva-levine N, Bracewell DG, Smales CM: **The Dynamics of the CHO Host Cell Protein Profile During Clarification and Protein A Capture in a Platform Antibody Purification Process.** *Biotechnol Bioeng* 2013, **110**:240–251.
37. Ho SCL, Koh EYC, van Beers M, Mueller M, Wan C, Teo G, Song Z, Tong YW, Bardor M, Yang Y: **Control of IgG LC:HC ratio in stably transfected CHO cells and study of the impact on expression, aggregation, glycosylation and conformational stability.** *J Biotechnol* 2013, **165**:157–166.
38. Jin M, Szapiel N, Zhang J, Hickey J, Ghose S: **Profiling of host cell proteins by two-dimensional difference gel electrophoresis (2D-DIGE): Implications for downstream process development.** *Biotechnol Bioeng* 2010, **105**:306–316.
39. Gomez N, Subramanian J, Ouyang J, Nguyen MDH, Hutchinson M, Sharma VK, Lin AA, Yuk IH: **Culture temperature modulates aggregation of recombinant antibody in CHO cells.** *Biotechnol Bioeng* 2012, **109**:125–136.
40. Koguma I, Yamashita S, Sato S, Okuyama K, Katakura Y: **Novel purification method of human immunoglobulin by using a thermo-responsive protein A.** *J Chromatogr A* 2013, **1305**:149–153.

Synthesis of Fructooligosaccharides by IslA4, a truncated inulosucrase from *Leuconostoc citreum*

Arlen Peña-Cardeña, María Elena Rodríguez-Alegría, Clarita Olvera and Agustín López Munguía*

Abstract

Background: IslA4 is a truncated single domain protein derived from the inulosucrase IslA, which is a multidomain fructosyltransferase produced by *Leuconostoc citreum*. IslA4 can synthesize high molecular weight inulin from sucrose, with a residual sucrose hydrolytic activity. IslA4 has been reported to retain the product specificity of the multidomain enzyme.

Results: Screening experiments to evaluate the influence of the reactions conditions, especially the sucrose and enzyme concentrations, on IslA4 product specificity revealed that high sucrose concentrations shifted the specificity of the reaction towards fructooligosaccharides (FOS) synthesis, which almost eliminated inulin synthesis and led to a considerable reduction in sucrose hydrolysis. Reactions with low IslA4 activity and a high sucrose activity allowed for high levels of FOS synthesis, where 70% sucrose was used for transfer reactions, with 65% corresponding to transfructosylation for the synthesis of FOS.

Conclusions: Domain truncation together with the selection of the appropriate reaction conditions resulted in the synthesis of various FOS, which were produced as the main transferase products of inulosucrase (IslA4). These results therefore demonstrate that bacterial fructosyltransferase could be used for the synthesis of inulin-type FOS.

Keywords: Fructosyltransferase, Fructooligosaccharides, Inulin, Inulosucrase, *Leuconostoc citreum*

Background

Fructooligosaccharides (β2-1, FOS) with a degree of polymerization (DP) in the range of 2–10, are one of the most important sources of carbon and energy for the beneficial microbiota (probiotics) that exist in the intestinal lumen of humans [1]. FOS are generally consumed directly as part of a healthy diet, including fruit and vegetables. FOS can also be consumed indirectly through functional foods, which can be enriched with FOS obtained by the hydrolysis of chicory inulin or produced industrially through enzymatic processes with commercial fructosyltransferases (FTFs) from fungal species like *Aspergillus* [2]. In the latter case, fungal enzymes are the preferred choice of enzyme because of their high specificity towards low molecular weight inulins, low hydrolytic activity and high stability [3,4]. In contrast, bacterial inulosucrases can be used to synthesize high molecular weight inulin with low specificity towards FOS production [5-7]. FTFs (EC 2.4.1.-) are enzymes that catalyze the transfer of the fructosyl moiety of sucrose to different acceptors, resulting in the synthesis of fructans with different molecular weights depending on the specificity of the enzyme. The fructosyl unit can also be transferred to water, resulting in the hydrolysis of sucrose. According to the classification system for carbohydrate-active enzymes, bacterial FTFs belong to family 68 of the glycoside hydrolases (GH68). Most FTFs are 45 to 64 kDa in length and consist of a single catalytic domain with a five-bladed β-propeller fold that encloses a funnel-like central cavity where the conserved catalytic residues are located. FTFs can be classified as inulosucrases (EC 2.4.1.9), which synthesize β2-1 linked fructans (inulin), or levansucrases (EC 2.4.1.10), which produce fructans with β2-6 linkages (levan) [8].

We previously reported the isolation of inulosucrase from *Leuconostoc citreum* CW28. This particular enzyme was designated IslA and was found to synthesize high-molecular-weight inulin. IslA is a multidomain enzyme

* Correspondence: agustin@ibt.unam.mx
Departamento de Ingeniería Celular y Biocatálisis, Instituto de Biotecnología, Universidad Nacional Autónoma de México, Cuernavaca, Morelos C.P. 62210, México

containing additional regions at both the amino- and carboxyl-terminals of its catalytic domain, which are similar to those found in glucosyltransferases [9]. In this context, IslA4 is a truncated form of IslA that retains only the five-bladed β-propeller catalytic domain. IslA4 is therefore similar to several other fructosyltransferases previously reported in the literature, including the levansucrases from *Bacillus subtilis* (SacB) and *Gluconacetobacter diazotrophicus*, and the InuJ inulosucrase from *Lactobacillus johnsonii* NCC 533. We previously studied the effects of the additional domains of IslA on its overall properties and found that IslA4 was similar to IslA, in the sense that it produced mainly high molecular weight inulin. However, IslA4 exhibits a much higher hydrolytic activity than IslA under the same reaction conditions [10]. Interestingly, IslA developed a high level of hydrolytic activity following the elimination of some of its additional domains, and achieved similar activity to a single domain fructosyltransferase, such as levansucrase SacB, which can hydrolyze as much as 80% of the sucrose substrate [11]. It has been demonstrated that reaction specificity (i.e., hydrolysis or transferase) as well as product specificity (i.e., type and size of fructan or FOS) in fructosyltransferases is strongly dependent on the reaction conditions, including the substrate concentration and temperature [12,13], the form in which the enzyme is applied, such as free or immobilized [14], the presence of organic solvents or co-solvents [15], and the source of the enzyme [16].

In this study, we have evaluated the effect of the reaction conditions on the specificity of IslA4 and the truncated form of inulosucrase IslA, in an attempt to identify efficient enzymes for the synthesis of inulin-type FOS.

Results and discussion

Influence of substrate and enzyme concentration on IslA4 reaction specificity

A common characteristic of fructosyltransferases is their ability to transfer the fructosyl moieties of a substrate to an acceptor molecule (the fructan growing chain) or water, leading to the hydrolysis of the substrate. The effects of the IslaA4 and sucrose concentrations on the transfer and hydrolysis reactions of the fructosyl moiety are shown in Figure 1. As in many other FTF cases, transferase activity is favored by high substrate concentrations, because of the higher amount of sucrose molecules available with respect to water for the initial transfer of the fructosyl residue [17]. Similar behavior has also been reported for levansucrase from *B. subtilis*, where almost 80% of the sucrose was used for transferase reactions at sucrose concentrations greater than 1,750 mM [18]. In contrast, SacB becomes predominantly hydrolytic in terms of its activity at low sucrose concentrations (12 mM) [19]. Furthermore, levansucrase from *G. diazotrophicus* displayed only hydrolytic activity at sucrose concentrations lower than 60 mM [20]. An inverse reaction specificity effect was observed for the enzyme concentration of IslA4, where an increase in the enzyme concentration from 1 to 10 U mL^{-1} led to an increase in the level of hydrolytic activity, regardless of the sucrose concentration. Reactions involving a low enzyme concentration and high substrate concentration (e.g., 1 U mL^{-1} and 2,046 M, respectively) were therefore determined to be suitable reaction conditions for high transferase efficiencies, despite the lengthy reaction times required. In contrast, a high IslA4 concentration (10 U mL^{-1}) coupled with a low sucrose concentration (292 mM) led to 90% of the sucrose being hydrolyzed.

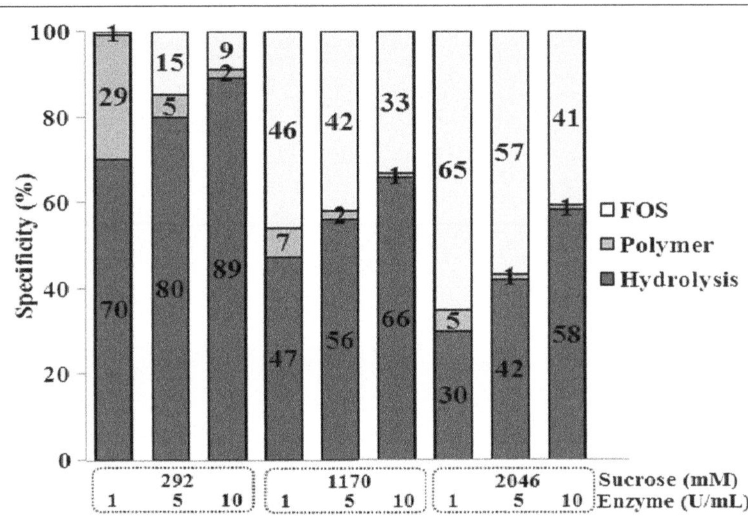

Figure 1 Transferase efficiency of IslA4 as a function of enzyme and substrate concentration. All reactions are carried out at 30°C in 50 mM phosphate (pH 6.0) containing 1 mM $CaCl_2$ and allowed to proceed until they reached a sucrose conversion of approximately 90%.

Similar behavior has also been reported for several other enzymes, such as SacB from *B. subtilis*, which transferred 78% of 1,750 mM sucrose at low enzyme concentrations [18].

With regard to other inulosucrases, it has been reported that the transfructosylation activity of *L. reuteri* 121 increased when the substrate and enzyme concentration were increased from 200 to 1800 mM and 40 to 130 U mg^{-1}, respectively, with FOS being identified as the main transferase product together with a relatively small amount of polymeric inulin material. In contrast, the same enzyme becomes predominantly hydrolytic in its activity at concentrations lower than 200 mM [21]. Similar results have also been reported for IslA, which became increasingly efficient in terms of ability to synthesize inulin when the initial concentration of sucrose in the reaction medium was increased [13]. Several other partially truncated forms of IslA have also exhibited similar reaction specificity behavior [10].

An extreme example of this behavior is levansucrase SacB, which shares 39% identity with inulosucrase IslA4. In this case, the enzyme concentration has a much stronger influence on the reaction specificity, with the hydrolysis of 300 mM sucrose increasing from 67 to 98% at 35°C when the enzyme activity is increased from 1 to 5 U mL^{-1} [18].

Product profile analysis

TLC analysis

The most interesting aspect of the product specificity exhibited by IslA4 is that although IslA is responsible for the synthesis of high molecular weight inulin, the combination of a truncated domain with the appropriate reaction conditions can results in the synthesis of a wide variety of FOS as the major transferase products. This effect can be readily visualized by TLC analysis, as shown in Figure 2. In this particular case, the FOS and inulin could both be clearly seen amongst the final reaction products when the products of reactions containing 1 U mL^{-1} of enzyme activity are analyzed at three different initial sucrose concentrations. A direct comparison with standard compounds suggested that 1-kestose (β-D-fructofuranosyl-(2 → 1)-β-D-fructofuranosyl-(2 → 1)-α-D-glucopyranoside) would be the first acceptor product, and that this product form as a consequence of a fructosyl transfer reaction to sucrose, followed by 1-nystose (β-D-fructofuranosyl-(2 → 1)-β-D-fructofuranosyl-(2 → 1)-β-D-fructofuranosyl-α-D-glucopyranoside) and f-nystose (β-D-fructofuranosyl-(2 → 1)-β-D-fructofuranosyl-(2→1)-β-D-fructofuranosyl-(2→1)-β-D-fructofuranosyl-α-D-glucopyranoside), which were only observed after an extended reaction time. However, an increase in the substrate concentration led to an increase in FOS synthesis, which was accompanied by an increase in transferase activity. Furthermore, there was a shift in the product profile with FOS with a low DP being observed at a sucrose concentration of 292 mM (mainly 1-kestose and nystose), while FOS with a DP greater than 10 fructose units were obtained at a sucrose concentration of 2,046 mM. Inulin was formed as the major synthetic product under the standard reaction conditions (i.e., 292 mM sucrose and

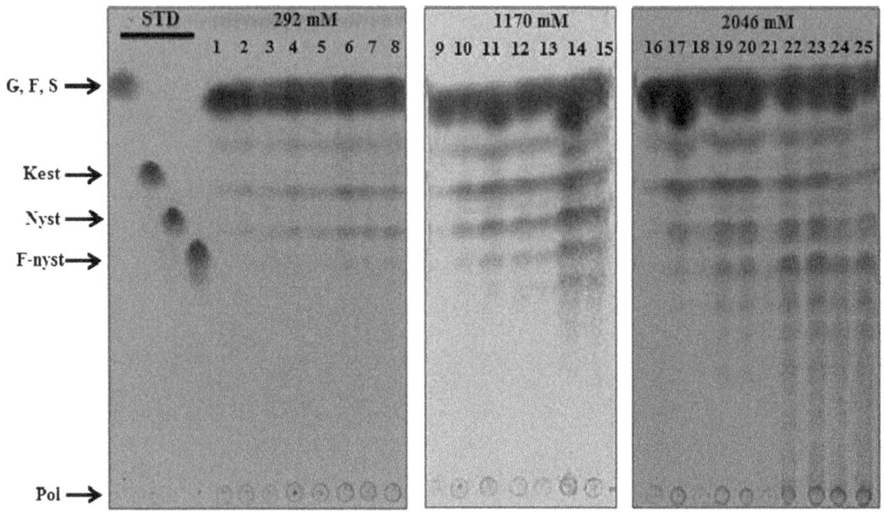

Figure 2 Influence of substrate concentration on the IslA4 product profile, as observed by TLC. All of the reactions are carried out with 1 U mL^{-1} of IslA4 at 30°C in 50 mM phosphate (pH 6.0) containing 1 mM $CaCl_2$, and were allowed to proceeded to a sucrose conversion of approximately 90%. Samples were taken at different time points. The numbers in the figures correspond to the following reaction times: *292 mM* (0, 2, 4, 6, 8, 10, 12 and 24 h); *1170 mM* (0, 3, 6, 9, 12, 25 and 34 h); and *2046 mM* (0, 3, 6, 9, 12, 24, 27, 30, 34 and 48 h). The STD lane contained the following standards: G, glucose; F, fructose; S, sucrose; Kest, 1-kestose (GF2); Nyst, 1,1-nystose (GF3); F-nyst, fructofuranosyl-nystose; and Pol, polymer.

1 U mL^{-1} of enzyme) using both IslA and its truncated form IslA4. Furthermore, the various FOS already described above were only observed as a small part of the product stream [10,22]. The reaction conditions can therefore be defined in detail to considerably reduce the inulin concentration in favor of forming FOS. In the case of *L. reuteri* 121 inulosucrase, 1-kestose and nystose were reported to be the only products obtained from 263 mM sucrose after 17 h of reaction [5]. However, an increase in the sucrose concentration to 840 mM led to the formation of additional FOS with a DP greater than five, regardless of the inulin concentration [23]. In the same context, InuJ and InuGA, which are inulosucrases from *L. johnsonii* NCC 533 [6] and *Lactobacillus gasseri* DSM 20143 [7], respectively, can synthesize inulin together with a wide range of FOS with DP values greater than five after long reaction times in reactions containing 600 mM of sucrose. Similar behavior has also been reported for *Zymomonas mobilis* levansucrase, where the product specificity shifted from levan to FOS when the sucrose concentration was increased to 800 mM [23]. Levansucrase from *Lactobacillus sanfranciscensis* TMW and a truncated version of the same protein also exhibited similar behavior [24].

One important thing to consider regarding the IslA4 reaction mechanism is that the low molecular weight FOS products appear to be synthesized by a non-processive mechanism. In this case, all of the FOS accumulating in the solution were derived from 1-kestose, which is the first reaction product formed when sucrose is used as an acceptor (Figure 2). It has been proposed previously that subsites +2 and +3, further away from the cleavage site (−1 and +1), could have a low affinity for the growing FOS chains, and that these sites would then release the product once the fructosyl unit had been transferred, representing the basis of a non-processive mechanism [21]. However, this mechanism does not appear to be applicable to the case of inulin, which is most likely synthesized through a processive mechanism because inulin is only observed in the reaction medium in this case as a high molecular weight product.

GPC and HPAEC-PAD analysis

The effect of the enzyme concentration and the substrate to enzyme ratio on the molecular weight distribution of the inulin synthesized by IslA4 was analyzed by GPC, and the results are shown in Figure 3. This result clearly shows that a product with an average molecular weight of 3,000 kDa was obtained following 14 min of elution. The distribution and size of the inulin produced was therefore independent of the enzyme concentration and of lower molecular weight than inulin produced by *L. reuteri* 121 [5] and *L. johnsonii* NCC 533 [6], which reached molecular weights as high as 10,000 and 40,000 kDa, respectively. This result also revealed that high enzyme concentrations favored the hydrolysis reaction, which resulted in the synthesis of small amounts of inulin (Figure 1). This behavior can therefore be explained in terms of the interactions between the enzyme molecules or enzyme-polymer molecules rather than the availability of active sites for a given sucrose concentration.

HPAEC-PAD analysis allowed for the identification of the synthesized FOS. An HPAEC-PAD chromatogram of the products obtained at three different enzyme concentrations is shown in Figure 4. Based on the retention times of the standards, the signals obtained at 4.5, 7.0, 8.5, 9.0 and 11.5 min were attributed to 1-kestose, 6-kestose (β-D-fructofuranosyl-(2→6)-β-D-fructofuranosyl-(2→6)-α-D-glucopyranoside), neokestose (β-D-Fructofuranosyl-(1→6)-β-D-fructofuranosyl-α-D-glucopyranoside), nystose and f-nystose,

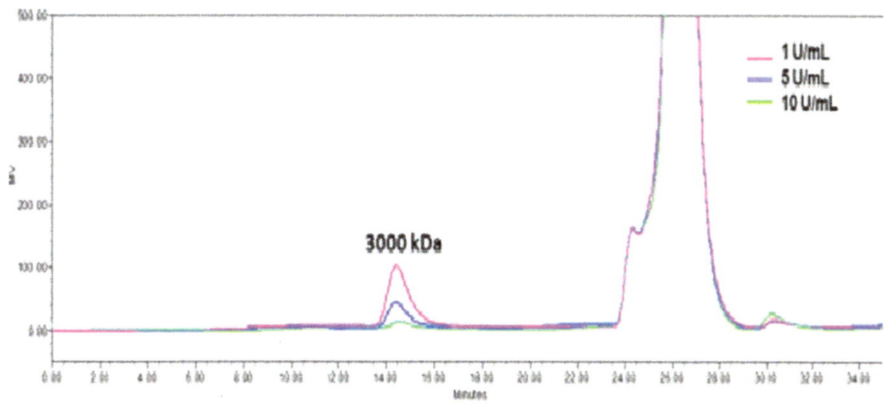

Figure 3 Influence of enzyme concentration on the molecular weight distribution of inulin synthesized by IslA4 as observed by GPC. Reactions were carried out at three different enzyme concentrations (i.e., 1, 5 and 10 U mL^{-1} of IslA4) with 292 mM sucrose at 30°C in 50 mM phosphate (pH 6.0) containing 1 mM $CaCl_2$, and were allowed to proceed until they reached a sucrose conversion of approximately 90%.

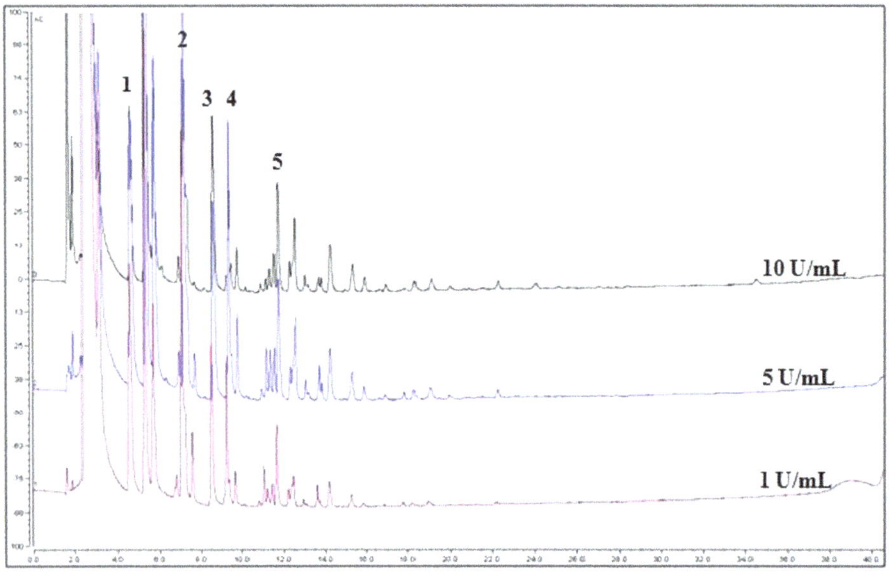

Figure 4 Influence of enzyme concentration on products synthesized by ISIA4 as observed by HPAEC-PAD analysis. Reactions were carried out at three different enzyme concentrations (i.e., 1, 5 and 10 U mL^{-1} of IslA4) with 292 mM of sucrose at 30°C in 50 mM phosphate (pH 6.0) containing 1 mM $CaCl_2$, and were allowed to proceed until they reached a sucrose conversion of approximately 90%. Identified oligosaccharides: 1) 1-kestose; 2) 6-kestose; 3) neokestose; 4) nystose; and 5) f-nystose.

respectively. Although the amperometric detection intensity is not always proportional to the amount of product, the amount of synthesized FOS decreased in the current system as the molecular weight of the product increased, in a similar manner to that reported for Inu from *L reuteri* 121 [5]. This effect can be observed quantitatively in Figure 2. Unidentified products were eluted between 1-kestose and 6-kestose (products 1 and 2), and between nystose and f-nystose (products 4 and 5). Although these products have not yet been identified, the results suggest that the smallest FOS obtained after an elution time of 5 min was *blastose* (β-D-fructofuranosyl-(6→2)-α,β-D-glucopyranoside), which is an isomer of sucrose. Blastose could be produced by the direct transfer of fructose to glucose in C6 or by the hydrolysis of the trisaccharide neokestose (product 3), which is a transfer product resulting from the transfer of fructose to the C6-position of the glucose unit in sucrose. Interestingly, these unidentified products were also reported in inulosucrase from *L. reuteri*121 [5] and *L. johnsonii* NCC 533 [6]. However, the formation of these products does not represent a general property of inulosucrases, because these products were not observed in InuGA and InuGB from *L. gasseri*, where linear β 2-1 FOS was the only FOS obtained from the direct transfer of fructose to the C1-position of the fructose acceptor [7].

The highest transferase yield for the reaction was obtained at the highest sucrose concentration and the resulting FOS profile is shown in Figure 5. Once again, the reaction led to the formation of large amounts of 1-kestose, 6-kestose, neokestose, nystose and f-nystose, as well as a large amount of FOS with a DP >5. Despite the limitations of the separation equipment, several products were observed with retention time in excess of 20 min. This behavior was similar to that reported for *L. reuteri* 121, which gave FOS with a higher DP than nystose when it was synthesized under high substrate concentrations [21].

Quantitative analysis of the inulin and FOS synthesized by IslA4

The amount of fructose transferred and incorporated into inulin (F_{POL}) was determined from a reference plot relating inulin GPC elution area to inulin concentration, which was constructed according to the procedure described above using inulin precipitated with ethanol and subsequently freeze-dried. With this methodology in hand, it was possible to demonstrate that although the inulin synthesis was independent of the substrate concentration, this factor is still an important function of enzyme concentration and activity. Overall, the level of inulin synthesis was found to be negligible at high enzyme concentrations, as shown in Table 1. These results therefore suggested that the increase in transferase activity that results from the rise in substrate concentration could be attributed to the exclusive synthesis of FOS, as already shown by TLC and HPAEC-PAD (Figures 2 and 4).

In quantitative terms, it was possible to reduce the amount of inulin from 16 to 4 or 1 g L^{-1} by increasing

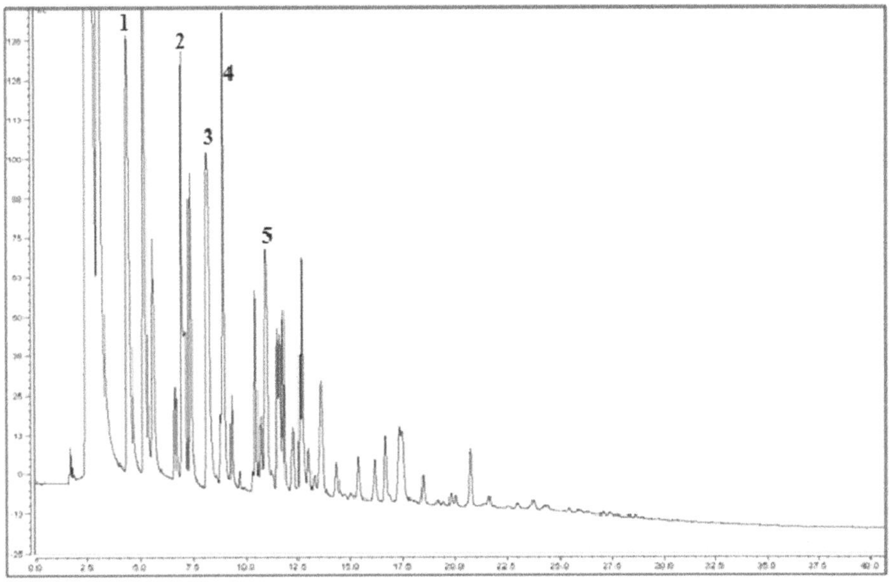

Figure 5 IslA4 product profile under high transferase conditions as observed by HPAEC-PAD. Reactions were carried out with 2,046 mM of sucrose and 1 U mL^{-1} of IslA4. The known oligosaccharides have been indicated as follows: 1) 1-kestose; 2) 6-kestose; 3) neokestose; 4) nystose; and 5) f-nystose.

the IslA4 concentration from 1 to 5 or 10 U mL^{-1} respectively, at any of the three sucrose concentrations studied. The highest amount of inulin produced under these reaction conditions was significantly higher than that reported for inulosucrase from *L. reuteri* 121, where only 0.8 g L^{-1} inulin was obtained from 90 g L^{-1} of sucrose. Although no explanation was offered for the result obtained in this particular case, the incubation times reported would only have allowed for an 18% conversion [5]. In the same context, the highest polymer yield (16 g L^{-1}) obtained with IslA4 was very similar to the yield of inulin calculated for the inulosucrases InuGA and InuGB from *L. gasseri*, which produced 11 and 13 g L^{-1} of inulin, respectively, from their reactions with 205 g L^{-1} of sucrose [7]. In contrast, InuJ from *L. johnsonii* NCC 533 allowed for the synthesis of 1.5 g L^{-1} of inulin from 200 g L^{-1} of sucrose [6]. Taken together, these results suggest that IslA4 is a much more suitable enzyme for the synthesis of inulin than any of the other inulosucrases currently available, even at low substrate concentrations (292 mM), where hydrolysis is the dominant reaction pathway. However, based on the FOS concentration, F_{FOS} (mM), which can be indirectly calculated according to the method described above, a low concentration of F_{FOS} (mM) was obtained when the reactions were conducted at 292 mM, which was the lowest sucrose concentration evaluated (Table 1). As shown in Figure 1, these reaction conditions also induced high levels of hydrolysis, as exemplified by the reaction conducted with 10 U mL^{-1} of enzyme activity, where up to 89% of the substrate was converted to glucose and fructose. Furthermore, increasing the sucrose concentration from 1,170 to 2,046 mM in reactions with 1 U mL^{-1} of enzyme activity, led to a two-fold increase in the amount of fructose incorporated in the FOS, despite the presence of a large amount of hydrolyzed sucrose (around 600 mM) in the mixture.

Table 1 Quantitative analysis of the products synthesized by IslA4 under different reaction conditions used

Sucrose (mM)	IslA4 (U mL^{-1})	Sucrose consumed (%)	F_{POL} (mM)	F_{FOS} (mM)
	1	92 ± 1.11	95 ± 4.3	3 ± 6.78
292	5	90 ± 1.37	18 ± 0.4	49 ± 2.56
	10	93 ± 0.96	6 ± 0.1	29 ± 1.50
	1	94 ± 0.45	87 ± 0.5	584 ± 0.12
1170	5	93 ± 1.59	27 ± 0.1	465 ± 0.21
	10	97 ± 1.67	11 ± 0.1	407 ± 0.10
	1	87 ± 3.40	89 ± 0.2	1251 ± 1.20
2046	5	90 ± 0.87	23 ± 0.2	1093 ± 0.18
	10	94 ± 2.81	5 ± 0.3	808 ± 0.05

Values shown are means ± SD of the results from three independent experiments.

Figure 1 provides a detailed overview of the fate of each transferred fructose, in terms of whether it goes to water, inulin or FOS. Based on this analysis, it can be concluded that the majority of the FOS synthesis was conducted from 2046 mM sucrose in reactions with 1 U mL^{-1} of enzyme, where 70% sucrose was used for transfer reactions, and 65% of this sucrose was directed towards a transfructosylation pathway for the synthesis of FOS. In contrast, the use of a low sucrose concentration (292 mM) led not only to a decrease in the transferase reaction but also favored polymer synthesis.

Conclusions

An efficient set of reaction conditions have been established for the synthesis of FOS and/or inulin using IslA4 inulosucrase. In particular, the results of the current study have shown that the enzyme specificity can be shifted almost exclusively from the synthesis of high molecular weight inulin to the synthesis of FOS. This shift was made possible in this truncated form of the multidomain inulosucrase IslA through the careful selection of the enzyme and substrate concentration conditions. The results of the current study have also shown that the use of a high sucrose concentration in conjunction with a low IslA4 concentration results in the highly efficient synthesis of low molecular weight FOS, including 1-kestose, 6-kestose, neokestose, nystose and f-nystose.

Methods

Preparation of *Escherichia coli* cell extracts and purification of IslA4

Escherichia coli BL21 cells transformed with the pET28a-*islA4* gene were grown in Luria-Bertani broth containing 50 μg mL^{-1} kanamycin at 37°C and 120 rpm. The culture was induced with 0.2 mM isopropyl β-D-1-thyogalactopyranoside (Gold Biotechnology, Inc., St Louis, MO USA). Once the cells reached an optical density of 0.6 (600 nm) they were incubated at 18°C for 6 additional hours. The cells were then harvested by centrifugation (2,500 × *g* for 10 min) and the resulting pellet washed twice with a 50 mM phosphate buffer (pH 6.0) containing 1 mM $CaCl_2$ before being re-suspended in 5 mL of the same buffer containing pepstatin (0.7 μg mL^{-1}) and Complete (one tablet to 50 mL of cell extract), which were added as protease inhibitors, and sonicated. Cell debris was removed by centrifugation (21,130 × *g* for 30 min) and the supernatant containing the enzyme was recovered for further processing. The protein concentration was determined according to the Bradford method [25] using the Bio-Rad reagent, with bovine serum albumin (BSA) (albumin fraction V; Sigma Aldrich, St. Louis, MO, USA) being used as a standard. The enzyme was purified by His-tag affinity chromatography. A bed volume of 600 μL of Ni-nitroacetic acid (Ni-NTA) resin (Qiagen) was used to bind the protein from 5 mL of cell extract. The resin was equilibrated with 500 μL of pH 7.0 binding buffer containing 50 mM NaH_2PO_4, 300 mM NaCl and 10 mM imidazole. The cell extract was diluted with an equivalent amount of the binding buffer and incubated overnight at 4°C under agitation with the equilibrated resin. The proteins from the extract were subsequently eluted from the resin by successive additions of buffer, with increasing amounts of imidazole (i.e., 30, 60 and 100 mM of imidazole). Finally, the enzyme was eluted with buffer containing 250 mM imidazole. The protein was dialyzed against 50 mM phosphate buffer (pH 6.0) containing 1 mM $CaCl_2$ and concentrated by ultrafiltration on an Amicon Ultra centrifugal Filter 10,000 NMWL (Merck Millipore, Billerica, MA, USA). The purity of IslA4 was verified by 10% SDS-PAGE (Figure 6); it shows an adequate level of enzyme purity.

Activity assay

In order to have a general idea of the enzyme activity, initial reaction rates were measured at 30°C in 50 mM phosphate buffer (pH 6.0) containing 1 mM $CaCl_2$ in the presence of 292 mM sucrose. The global activity was expressed as the reducing sugars released from sucrose using the 3,5-dinitrosalicylic acid method (DNS) [26]. One *global* activity unit (U) was defined as the amount of enzyme required to produce 1 μmol of reducing sugars per minute. The specific hydrolysis and transferase activities were

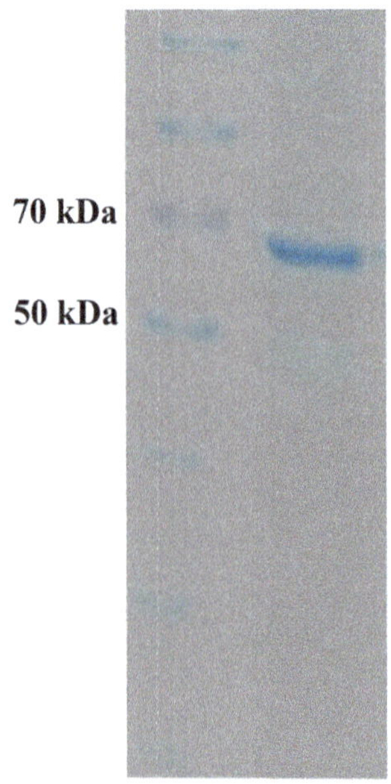

Figure 6 SDS-PAGE of the purified IslA4 inulosucrase form. The intracellular protein extracted from *E.coli*, was purified by affinity chromatography in Ni-nitroacetic acid (Ni-NTA) resin (Qiagen) and eluted with increasing amounts of imidazole. Afterwards, the enzyme was dialyzed against phosphate buffer and concentrated by ultrafiltration on an Amicon Ultra centrifugal Filter. 10,000 NMWL (Merck Millipore, Billerica, MA, USA).

measured by HPLC based on the amounts of fructose and glucose. The HPLC method used in the current study is described below.

Influence of reaction conditions on IslA4 product and reaction specificity

The enzymatic synthesis was carried out at 30°C in 50 mM phosphate (pH 6.0) containing 1 mM $CaCl_2$ at three different sucrose (292, 1,170 and 2,046 mM) and enzyme (1, 5 and 10 U mL^{-1}) concentrations to evaluate the influence of the reaction conditions on the product and reaction specificities of IslA4. All of the reactions were allowed to proceed until they reached a sucrose conversion of approximately 90%. The reaction specificity (i.e., hydrolysis to transferase ratio) was determined by measuring the amounts of glucose and fructose released during the reaction. Free glucose (Gf) was derived from hydrolyzed sucrose as well as the sucrose that was used as a fructose donor for the transferring reactions, whereas free fructose (Ff) was derived exclusively from the hydrolytic activity of the enzyme. The difference between the amounts of consumed and hydrolyzed sucrose could therefore be used to determine the amount of fructose used for transfructosylation (Ft), i.e., the transferase activity.

FOS, sugars and fructan analysis

TLC analysis of the oligosaccharide profile

The oligosaccharides produced by IslA4 were analyzed by TLC using silica gel 60 glass plates (Merck, Darmstadt, Germany). Samples were taken at different reaction times and spotted (1 μL) onto a TLC plate. Fructose, glucose and sucrose (Sigma Aldrich), and 1-kestose (GF2) nystose (GF3) and fructofuranosyl-nystose (Wako Pure Chemical Industries) were used as standards. The TLC plates were ran twice using a 15:9:6 (v/v/v) mixture of ethanol, butanol and water as the mobile phase. The plates were then air-dried and sprayed with an alcoholic solution of α-naphthol before being heated at 150°C.

Sugars analysis

Sugars were quantified by high-pressure liquid chromatography (HPLC) using a Waters 600E system controller (Waters Corp. Milford, MA, USA) equipped with a refraction index detector (Waters 410) using a Prevail Carbohydrate ES column (250 × 4.6 mm) at 30°C. The HPLC system was eluted with a mobile phase consisting of a 75:25 (v/v) mixture of acetonitrile and water at a flow rate of 1.0 mL min^{-1}. The molecular weight of the Inulin was determined by gel permeation chromatography (GPC) using a serial set of Ultrahydrogel (UG 500 and linear) columns at 30°C. The columns were eluted with a mobile phase sodium nitrate 100 mM at a flow rate of 0.8 mL min^{-1}. Oligosaccharides were separated and analyzed by high-performance anion exchange chromatography with pulsed amperometric detection (HPAEC-PAD, Dionex), using a CarboPac PAD-200 column (250 × 2 mm). The following gradient was used: eluent A at 100% (0 min), 99% (0.5 min), 80% (25 min), 20% (85 min) and 100% (95 min). Eluent A was 150 mM sodium hydroxide and eluent B was 150 mM sodium hydroxide in 500 mM sodium acetate.

Quantitative analysis of the products synthesized by IslA4

The polymer was precipitated from the supernatant by the addition of 2 volumes of cold 96% ethanol followed by centrifugation (2500 × *g* at 30 min). The material was then dissolved in MilliQ water and dialyzed against 50 mM phosphate buffer (pH 6.0) containing 1 mM $CaCl_2$, and the resulting polymers were freeze-dried and occupied to produce a standard curve, which was used to calculate the amount of fructose incorporated into the polymer (F_{POL}). For quantitative purposes, the level of fructose incorporated exclusively into FOS (F_{FOS}) was defined as the difference between Ft and F_{POL} (i.e., $F_{FOS} = Ft - F_{POL}$). Although this method meant that the amount of fructose incorporated into FOS was calculated in an indirect manner, the results still allowed for a comparative analysis.

Abbreviations

DP: Degree of polymerization; FTF: Fructosyltransferase; FOS: Fructooligosaccharide; GPC: Gel permeation chromatography; HPAEC-PAD: High performance anion exchange chromatography with pulsed amperometric detection; TLC: Thin-layer chromatography.

Competing interests

The authors declare that they have no competing interests.

Authors' contributions

APC: Conducted most of the experimental work as part of her MSc thesis. APC also wrote the preliminary manuscript. MERA: Provided technical support, conducted preliminary and complementary experiments, and was actively involved in general project discussions, as well as reviewing the manuscript. CO: Isolated the original enzyme and constructed the truncated versions. CO was also actively involved in general project discussions. ALM: Research director. Designed the project, conducted the research and wrote the final version of the manuscript. All authors read and approved the final manuscript.

Acknowledgments

This project was supported by PAPIIT-UNAM No. IN212311. Arlen Peña-Cardeña thanks CONACYT, México for her master scholarship (No. 254166). The authors would like to thank Fernando Gonzalez, Juan Manuel Hurtado and Roberto Rodriguez for technical support.

References

1. Biedrzycka E, Bielecka M. Prebiotic effectiveness of fructans of different degrees of polymerization. Trends Food Sci Tech. 2004;15:170–5.
2. Chi Z, Chi Z, Zhang T, Liu G, Yue L. Inulinase-expressing microorganism and applications of inulinases. Appl Microbiol Biotechnol. 2009;82:211–20.
3. Godshall MA. Future directions for the sugar industry. Int Sugar J. 2001;103:378–84.

4. Yun JW. Fructooligosaccharides—occurrence, preparation, and application. Enzyme Microb Technol. 1996;19:107–17.
5. van Hijum SAFT, van Geel-Schutten GH, Rahaoui H, van der Maarel MJEC, Dijkuizen L. Characterization of a novel fructosyltransferase from *Lactobacillus reuteri* that synthesizes high-molecular-weight inulin and inulin oligosaccharides. Appl Environ Microbiol. 2002;68:4390–8.
6. Anwar MA, Kralj S, van der Maarel MJEC, Dijkhuizen L. The probiotic *Lactobacillus johnsonii* NCC 533 produces high-molecular-mass inulin from sucrose by using an inulosucrase enzyme. Appl Environ Microbiol. 2008;74:3426–33.
7. Anwar MA, Kralj S, Villar Piqué A, Leemhuis H, van der Maarel MJEC, Dijkhuizen L. Inulin and levan synthesis by probiotic *Lactobacillus gasseri* strains: characterization of three novel fructansucrase enzymes and their fructan products. Microbiology. 2010;156:1264–74.
8. van Hijum SA, Kralj S, Ozimek LK, Dijkhuizen L, van Geel-Schutten IG. Structure-function relationships of glucansucrase and fructansucrase enzymes from lactic acid bacteria. Microbiol Mol Biol Rev. 2006;70:157–76.
9. Olivares-Illana V, López-Munguía A, Olvera C. Molecular characterization of inulosucrase from *Leuconostoc citreum*: a fructosyltransferase within glucosyltransferase. J Bacteriol. 2003;185:3606–12.
10. Del Moral S, Olvera C, Rodríguez ME, López-Munguía A. Functional role of the additional domains in inulosucrase (IslA) from *Leuconostoc citreum* CW28. BCM Biochem. 2008;9:6. doi:10.1186/1471-2091-9-6.
11. Ortiz-Soto ME, Rivera M, Rudiño-Piñera E, Olvera C, López-Munguía A. Selected mutations in *Bacillus subtilis* levansucrase semi-conserved regions affecting its biochemical properties. Protein Eng Des Sel. 2008;21:589–95.
12. Tanaka T, Oi S, Yamamoto T. Synthesis of Levan by Levansucrase. Some factors affecting the rate of synthesis and degree of polymerization of levan. J Biochem. 1979;85:287–93.
13. Ortiz-Soto ME, Olivares-Illana V, López-Munguía A. Biochemical properties of inulosucrase from *Leuconostoc citreum* CW28 used for inulin synthesis. Biocatal Biotrans. 2004;22:275–81.
14. Chambert R, Petit-Glatron MF. Immobilisation of levansucrase on calcium phosphate gel strongly increases its polymerase activity. Carbohydrate Res. 1993;244:129–36.
15. Castillo E, López-Munguía A. Synthesis of levan in water-miscible organic solvents. Biotechnol. 2004;114:209–17.
16. Falconer DJ, Rupendra M, Robyt JF. Biosynthesis of dextrans with different molecular weights by selecting the concentration of *Leuconostoc mesenteroides* B-512FMC dextransucrase, the sucrose concentration, and the temperature. Carbohydr Res. 2011;346:280–4.
17. Chuankhayan P, Hsieh CY, Huang YC, Hsieh YY, Guan HH, Hsieh YC, et al. Crystal structures of *Aspergillus japonicus* fructosyltransferase complex with donor/acceptor substrates reveal complete subsites in the active site for catalysis. J Biol Chem. 2010;285:23251–64.
18. Porras-Domínguez JR. Obtención de Fructooligosacáridos ridos a partir de levanas bacterianas mediante el uso de endo-levanasas. MSc Thesis. http://132.248.9.195/ptd2013/Presenciales/0693615/Index.html. Accessed 2012.
19. Chambert R, Gonzy-Tréboul G, Dedonder R. Kinetic studies of levansucrase of *Bacillus subtilis*. Eur J Biochem. 1974;41:285–300.
20. Song DD, Jacques NA. Purification and enzymatic properties of the fructosyltransferase of *Streptococcus salivarius* ATCC 25975. Biochem J. 1999;341:285–91.
21. Ozimek LK, Kralj S, van der Maarel MJEC, Dijkhuizen L. The levansucrase and inulosucrase enzymes of *Lactobacillus reuteri* 121 catalyse processive and non-processive transglycosylation reactions. Microbiology. 2006;152:1187–96.
22. Rodríguez-Alegría ME, Enciso-Rodríguez A, Ortiz-Soto ME, Cassani J, Olvera C, López-Munguía A. Fructooligosaccharide production by a truncated *Leuconostoc citreum* inulosucrase mutant. Biocatal Biotrans. 2010;28:51–9.
23. Crittenden RG, Doelle HW. Identification and characterization of the extracellular sucrases of *Zymomonas mobilis* ATCC 39676. Appl Microbiol Biotechnol. 1994;41:302–8.
24. Tieking M, Ehrmann MA, Vogel RF, Gänzle MG. Molecular and functional characterization of a levansucrase from the sourdough isolate *Lactobacillus sanfranciscensis* TMW1.392. Appl Microbiol Biotechnol. 2005;66:655–63.
25. Bradford MM. A rapid and sensitive method for the quantitation of microgram quantities of protein utilizing the principle of protein dye binding. Anal Biochem. 1976;72:248–54.
26. Miller GL. Use of dinitrosalicylic acid reagent for determination of reducing sugar. Anal Chem. 1959;31:426–8.

19

Optimization of production, purification and lyophilisation of cellobiose dehydrogenase by *Sclerotium rolfsii*

Christin Fischer*, Annett Krause and Thomas Kleinschmidt

Abstract

Background: The enzyme cellobiose dehydrogenase (CDH) can be used to oxidize lactose to lactobionic acid. As *Sclerotium rolfsii* is known to be a good producer of CDH, the aim of this paper was to simplify its production and secondly to systematically study its purification aiming for a high yield. Two preservation methods (freezing and freeze-drying) and the influence of several protectants were investigated.

Results: Production of cellobiose dehydrogenase was optimized leading to a more simplified medium composition. Purification of the enzyme was evaluated by determining breakthrough profiles on different ion exchange (IEX) and hydrophobic interaction (HIC) materials with regard to buffer composition. Highest purification with an acceptable loss during the capture step using IEX was obtained with a Q Sepharose XL medium and a 100 mM sodium acetate buffer at pH 4.5. Subsequent purification using hydrophobic interaction chromatography was done at 1.1 M ammonium sulfate concentration. Purification was moderate, yielding a specific activity of 11.9 U/mg (56% yield). However, as could be shown in a preliminary experiment, purity of the obtained enzyme solution was sufficient for its intended use to oxidize lactose to lactobionic acid. Various sugars and sugar alcohols were investigated to study their protective effect during lyophilisation and freezing at -20°C. Glucose and lactulose could be identified to have a high lyoprotective effect while loss of enzyme activity was high (77%) when using no additives.

Conclusion: By simplifying the cultivation medium of *Sclerotium rolfsii*, the costs of cellobiose dehydrogenase production could be reduced. Simultaneously, CDH production was increased by 21%. The production of lactobionic acid from lactose is possible using partially purified and unpurified enzyme. Storage at -20°C using 50% (w/v) glycerol was considered to be most suited for preservation of the enzyme.

Keywords: Cellobiose dehydrogenase, CDH, *Sclerotium rolfsii*, Enzyme purification, Lyophilisation, Cryoprotection, Lactobionic acid

Background

Cellobiose dehydrogenase (CDH, EC 1.1.99.18) is an extracellular enzyme produced by a number of wood-degrading fungi [1]. One of them is the plant pathogen *Sclerotium (Athelia) rolfsii* which is common in the tropics and subtropics and attacks mostly crops and vegetables. It is mainly producing cellulolytic enzymes in order to enter the host organism [2]. The biological function of CDH is controversial. For example, CDH increases the efficiency of cellulose degradation or reduces product inhibition of cellulases by oxidizing cellobiose to the corresponding lactone [1,3]. This theory was also confirmed several years later [4].

CDH is a monomeric protein consisting of a flavin and a heme domain. Both domains are connected by a protease-sensitive linker. When CDH is cleaved by proteases, it results in an active flavin domain and an inactive heme fragment. It is possible to distinguish between the holoenzyme and the flavin domain by using both one and two electron acceptors. Two electron acceptors like 2,6-dichlorophenol-indophenol (DCIP) can be reduced either directly at the flavin domain or via internal electron transfer at the heme domain. One electron acceptors

* Correspondence: c.fischer@bwp.hs-anhalt.de
Anhalt University of Applied Sciences, Department of Applied Biosciences and Process Engineering, Bernburger Str. 55, 06366 Köthen, Germany

like cytochrome c can only be used at the heme domain. Thus, they are used to detect only the intact enzyme.

As indicated by its name, cellobiose is the main substrate of the enzyme [5-9] indicated by a low Michaelis Menten constant. Other cellooligosaccharides are also favoured [10] but another common substrate is lactose. Figure 1 shows the reaction mechanism using DCIP as a redox mediator. The corresponding lactone is spontaneously oxidized to lactobionic acid which is rumored to have prebiotic effects [11]. Thus, an application in the food industry is a possible alternative to its current usage in biosensors for glucose ([12,13]) or lactose ([14-16]) and as a bleaching agent ([17-19]).

As *Sclerotium rolfsii* is known to be a good producer of CDH, the aim of this paper was to simplify its production and secondly to systematically study its purification aiming for a high yield. Two preservation methods (freezing and freeze-drying) and the influence of several protectants were investigated.

Methods

Organism and culture conditions

Sclerotium rolfsii strain CBS 191.62 was obtained from the Centraalbureau voor Schimmelcultures (Baarn, The Netherlands). The fungus was maintained on glucose-maltose Sabouraud agar plates, which were inoculated with a piece (diameter 1 cm) of overgrown agar and then incubated at 30°C for 5 to 7 days.

A medium propagated by Sachslehner et al. [2] was used in some studies. It contained 43 g/L α-Cellulose, 80 g/L peptone from meat, 2.5 g/L NH_4NO_3, 1.5 g/L $MgSO_4$ x $7H_2O$, 1.2 g/L KH_2PO_4, 0.6 g/L KCl , and 0.3 ml/L trace element solution (1.0 g/L $ZnSO_4$ x H_2O, 0.3 g/L $MnCl_2$ x $4H_2O$, 3.0 g/L H_3BO_3, 2.0 g/L $CoCl_2$ x $6H_2O$, 0.1 g/L $CuSO_4$ x $5H_2O$, 0.2 g/L $NiCl_2$x $6H_2O$ and 4.0 ml/L conc. H_2SO_4). The natural pH of the medium was 5.5. For optimizing the culture conditions of *Sclerotium rolfsii* with regard to maximum production of cellobiose dehydrogenase, start pH was varied (5.5, 5.0 and 4.0) by adding appropriate amounts of phosphoric acid. Additionally, a pH-stat method for a start pH value of 5.5 was carried out by measuring and readjusting the pH to 5.5 with 5 M NaOH on a daily basis.

Further, the composition of the medium was modified. The medium of Sachslehner et al. [2] was used as a reference. First, single salts were omitted from the medium described above to evaluate their influence on enzyme production. In a second series, a basal medium containing 43 g/L α-Cellulose, 80 g/L peptone from meat and 0.3 ml/L trace element solution was used. NH_4NO_3, $MgSO_4$, KH_2PO_4, KCl or none of them were added (amounts as in the reference medium).

All experiments were carried out in 300 ml unbaffled Erlenmeyer flasks containing 100 ml culture medium. Flasks were inoculated with 2 agar plugs (diameter 1 cm) of a freshly grown culture and incubated at 30°C and 150 rpm until enzyme activity remained constant. Samples of 1 ml were withdrawn starting after 7 days of incubation and analyzed for enzyme activity.

Enzyme purification

First, the mycelia were separated from the cultivation medium containing the CDH enzyme by filtration using a folded filter. This crude enzyme extract was desalted and concentrated to about one fifth of its initial volume using a Vivacell 250 filtration device (Sartorius AG, Germany) equipped with a PES membrane having a MWCO of 50 kDa. Within two dialysis steps and applying a pressure of 4 bar, conductivity was reduced from about 18 mS/cm to about 1 mS/cm indicating a successful salt removal. No enzyme loss was detected during this step.

Enzyme purification was carried out on an ÄKTA Prime Plus system (GE Healthcare) equipped with a 150 ml superloop for the application of large sample volumes. Different adsorbents for anion exchange (DEAE Sepharose FF as a weak and Q Sepharose XL as a strong ion exchanger, column volume: 5 ml, both by GE Healthcare) and hydrophobic interaction chromatography (Butyl-S FF, Butyl FF, Octyl FF, Phenyl HP, Phenyl FF (low sub), column volume: 1 ml, all by GE Healthcare) were tested to optimize purification procedure. With IEX, binding capacity was determined at

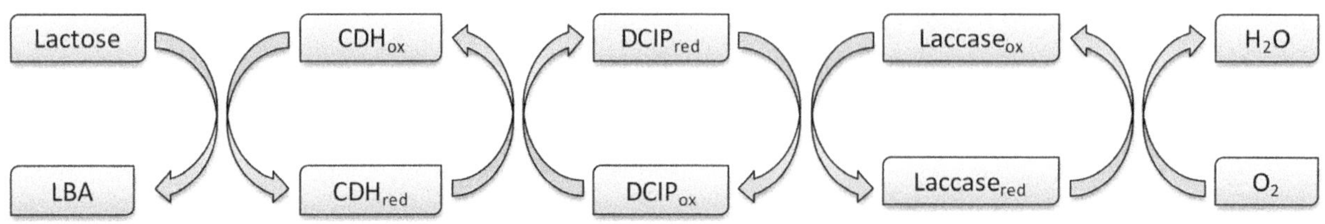

Figure 1 Reaction scheme of enzymatic oxidation of lactose to lactobionic acid (LBA). The used enzyme system consists of cellobiose dehydrogenase (CDH) and laccase with 2,6-dichlorophenol-indophenol (DCIP) as redox mediator.

various pH values (50 mM sodium acetate buffer) and buffer concentrations (20 – 200 mM sodium acetate, pH 4.5) by applying 110 ml of desalted and appropriately diluted crude enzyme extract (final concentration about 3 U/ml) onto the column. Fractions (2 ml) of the flow-through were collected and analyzed for CDH activity to determine a breakthrough profile. Also, elution conditions were optimized by analyzing the collected fractions for enzyme activity and protein content. Fractions having a higher specific activity than the applied sample were pooled. Ammonium sulfate precipitation was carried out by adding an appropriate amount of a 3.8 M solution to obtain the desired saturation before applying the sample to various HIC columns. Concentration and composition of the starting buffer were optimized as well.

Preservation of the enzyme

Seven sugars and two sugar alcohols were tested for their ability to serve as lyo- or cryoprotectants for CDH. Therefore 1 ml of purified enzyme solution (containing 11 U/ml) and 0.5 ml of lyoprotectant solution (to give a final concentration of 10 µmol per unit CDH) were mixed and either frozen at -20°C or freeze dried in 1.5 ml Eppendorf tubes. Freeze-drying was carried out in an Alpha 1-4 freeze-dryer (Martin Christ Gefriertrocknungsanlagen GmbH, Germany). Plate temperature was set to -50°C and within 3 hours a product temperature of -28°C was obtained. Vacuum (0.005 mbar) was applied and plate temperature was set to 30°C. The process was stopped after 4 to 5 days when product temperature was positive. Samples were rehydrated with 1 ml water immediately after drying and analyzed for residual enzyme activity. Frozen samples were thawed after 60 and 160 hours and also assayed for remaining CDH activity.

Enzyme activity assay

For measuring the enzyme concentration, the DCIP assay as propagated by Baminger et al. [20] was used. Therefore, 100 µl 300 mM lactose, 20 µl 200 mM NaF and 760 µl 100 mM sodium acetate buffer pH 4.0 were incubated at 30°C for at least 10 minutes before analysis. A solution of 3 mM DCIP (containing 10% v/v ethanol) was tempered separately. For analysis, 100 µl of DCIP solution were added to the lactose/NaF/buffer mixture and 20 µl appropriately diluted sample solution was added. After mixing the reduction of DCIP was measured at 520 nm every 5 seconds for 3 minutes. The extinction coefficient for DCIP at 520 nm and pH 4.0 was determined to be 6.9 mM^{-1} cm^{-1}. One unit was defined as the amount of enzyme that reduces 1 µmol DCIP per minute under the described assay conditions.

Protein determination

Protein content was determined according to the Bradford method [21]. Therefore 100 µl appropriately diluted sample solution were mixed with 1 ml of Bradford reagent (Roti®-Quant, Carl Roth GmbH, Germany) and incubated for 10 minutes at room temperature. Extinction was measured at 595 nm. Bovine serum albumin was used as a standard.

Application of CDH for lactobionic acid synthesis

The partial purified enzyme as well as the crude enzyme extract were used in a 4.8% lactose solution to synthesise lactobionic acid. Enzyme substrate ratio was set to 70 DCIP-Units per gram lactose. DCIP (1 µmol/unit CDH) was used as a redox mediator and laccase from *Trametes versicolor* (Sigma Aldrich, Germany) was added in a 5-fold excess over CDH activity. Reaction was carried out in a water bath at 35°C and 200 rpm. Samples were analyzed for glucose, galactose, lactose and lactobionic acid using HPLC. The column was a Hi-Plex Na column (300 mm x 7.7 mm from Agilent Technologies Deutschland GmbH) which was used at 0.3 $mLmin^{-1}$ (eluent 0.2% sodium azide in water) and 80°C.

Results and discussion

Influence of culture conditions on CDH yield

Influence of pH

Figure 2 shows the influence of the starting pH of the medium. At pH 5.0 CDH production begins noticeably at day 10, continues to rise straight and reaches a value of 7.3 U/ml after 17 days. With a slightly higher pH of 5.5 CDH production started a little earlier (day 7) and was significantly higher until day 10. Thenceforward the measured enzyme activity was a little higher compared to a starting pH of 5.0 but that difference was not significant. Also, after 17 days the produced enzyme amount was the same as with pH 5.0. Although cultivation at a starting pH of 5.5 seemed to be a little instable (as indicated by high standard deviations), this value was used for further studies as 5.5 is the natural pH of the medium and therefore no adjustment is necessary. There was only very little (0.1 U/ml) CDH production at a pH of 4.0 (data not shown). Therefore this pH is not suitable.

Since the pH of the culture medium shifts to lower levels during cultivation, it was assumed using a pH-stat method to cultivate at a constant pH value of 5.5 might further enhance CDH production. Therefore, an appropriate amount of 5 M NaOH was added on a daily basis. However, a maximum enzyme concentration of only 2.6 U/ml was reached after 13 days. It then slightly dropped down to 2.3 U/ml and remained on that level until the end of the experiment after 21 days. CDH production using cultivation without pH adjustment was 3

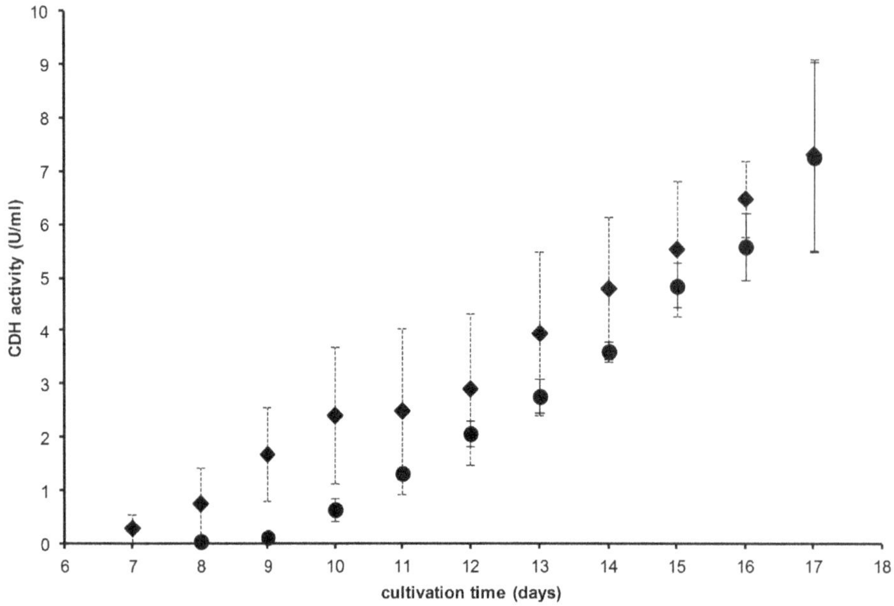

Figure 2 Influence of start pH on CDH production by *S. rolfsii*. pH 5.5 (♦, n = 18), pH 5.0 (●, n = 4).

times higher at that time. Consequently, control of pH is not favorable.

Influence of medium composition

Influence of medium composition was studied by omitting a particular salt in the medium proposed by Sachslehner et al. [2]. The latter was used as a reference medium. As can be seen from Figure 3, enzyme activity in the reference medium is constantly increasing from day 7 to 13 and then maintains at a nearly constant level from day 14 to 20.

KH_2PO_4 seems to have a negative impact on enzyme production as when omitting this salt, a significantly higher enzyme concentration could be reached as can be seen clearly from the data at days 14 to 17. Although, compared to the reference medium there seems to be a small decline in enzyme concentration with ongoing cultivation, so that at day 20 enzyme activity reached the

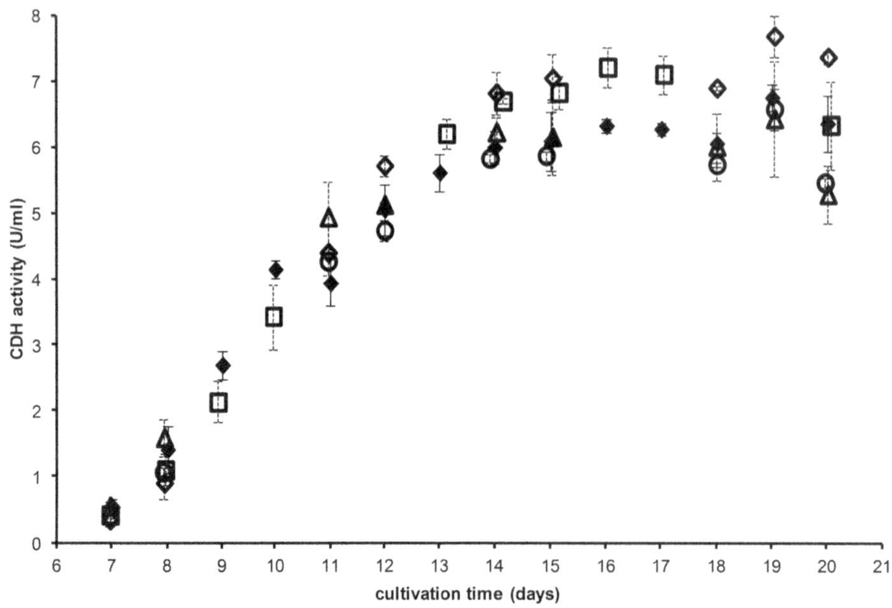

Figure 3 CDH production by *S. rolfsii*. reference medium (♦, n = 4), medium lacking individual salts (without NH_4NO_3 (◊)/KCl (Δ): n = 3, without KH_2PO_4 (□)/$MgSO_4$ (○): n = 2).

same value (6.3 U/ml). The same can be conclude for NH_4NO_3 as there is significantly higher enzyme production from day 12 to 20 when this salt is omitted. In contrast to KH_2PO_4, enzyme concentration is enhanced until the end of the experiment resulting in about 15% higher CDH concentration compared to the reference medium. On the other hand, KCl and $MgSO_4$ seem to have no impact on CDH production as when omitting these salts no significant difference in enzyme production was observed. However, there might be a tendency that CDH is depleted by the fungus if cultivation would be continued as indicated by considerably lower enzyme activity values at day 20.

A second approach was used to gain more insight in the influence of each salt. At that none or only one salt at a time was added to a basal medium containing only α-cellulose, peptone from meat and trace element solution (amounts as in the reference medium). In all cases enzyme production by *S. rolfsii* started at the same time and reached a maximum at day 19. There were no significant differences in CDH production in the first 15 days but the influence of all salts became more pronounced later (data not shown). As can be seen in Figure 4, no matter which salt was added, CDH production was higher compared to the reference medium. The effect was only marginal with KCl and NH_4NO_3 but more distinct when adding KH_2PO_4 or $MgSO_4$. This is in contrast to the results presented above which showed no effect of $MgSO_4$ and a negative impact of KH_2PO_4. However, with the presence of other salts, the conditions were different and therefore the impact of the individual salts could play a diverse role. As was expected, adding none of the tested salts to the basal medium also resulted in significantly higher enzyme concentration. On day 19, 21% more enzyme was produced in the medium containing α-cellulose, peptone from meat and trace element solution than in the reference medium. CDH concentration began to decline afterwards so one has to be careful to choose the right harvesting time. Also, when additionally omitting the trace element solution (i.e. only cultivating the fungus in α-cellulose and meat peptone solution), a similar CDH production was observed (data not shown).

Using the same cultivation method (flasks containing 100 ml medium), Sachslehner et al. [2] obtained an enzyme activity of 3.6 U/ml after 13 days of cultivation. Ludwig and Haltrich [22] achieved an enzyme concentration of 4.1 U/ml after 14 days. In 2003 they reported about 7 U/ml in the same medium after 13 days [23]. With reducing peptone from meat to 20 g/L and adding 30 g/L leucine instead, 11 U/ml were obtained after 13 days. This good result could not be confirmed in the present study as enzyme production using this medium composition was very low (1.2 ± 0.9 U/ml after 16 days ($n = 4$)). The difference might be due to the usage of meat peptone from different suppliers therefore resulting in a different composition.

Enzyme purification

Optimization of ion exchange chromatography

To choose an appropriate anion exchange medium, the binding capacity of two materials was determined by

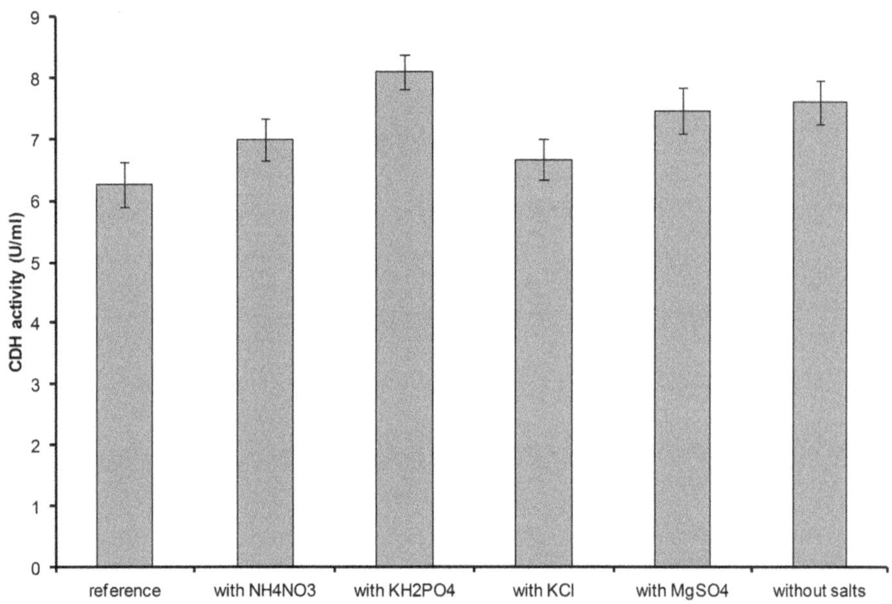

Figure 4 CDH production by *S. rolfsii* after 19 days of cultivation in different culture media. reference $n = 6$, NH_4No_3 $n = 3$, KH_2PO_4 $n = 2$, KCl $n = 4$, $MgSO_4$ $n = 5$, without salts $n = 5$.

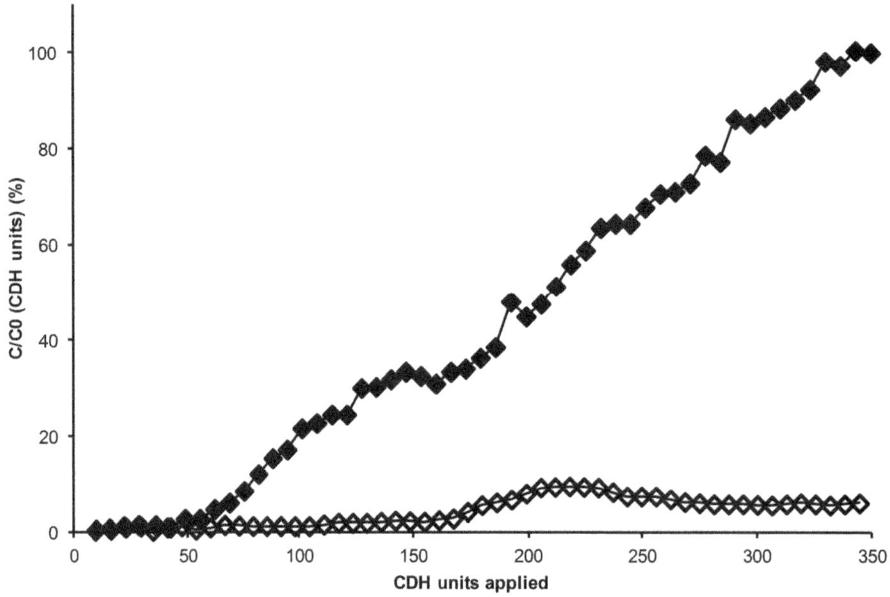

Figure 5 Breakthrough curves for CDH at pH 5.0. DEAE Sepharose FF (♦), Q Sepharose XL (◊).

applying 350 units onto the column and determining breakthrough curves. As can be seen from Figure 5, with the weak anion exchange medium (DEAE Sepharose FF) the breakthrough was negligible until about 50 units were applied. After that, the amount of unbound enzyme increases linearly thus reaching a value of c/c_0 of 100% (meaning the sample leaves the column as it is, no more enzyme binds onto the column) after 350 units were applied. In contrast, with the strong anion exchanger (Q Sepharose XL) at the same pH of 5.0, the breakthrough profile always stays below 10% therefore indicating to be more suitable for CDH binding in general.

Hence, the CDH breakthrough profiles for various pH values were studied using Q Sepharose XL. A pH of 4.5 is very close to the isoelectric point of the enzyme (as reported to be 4.2 [24]), however the breakthrough stays below 5% until 140 units are applied (Figure 6). After that, enzyme loss increases exponentially reaching 100% breakthrough after 215 units were applied. Even higher c/c_0 values were obtained meaning that not only

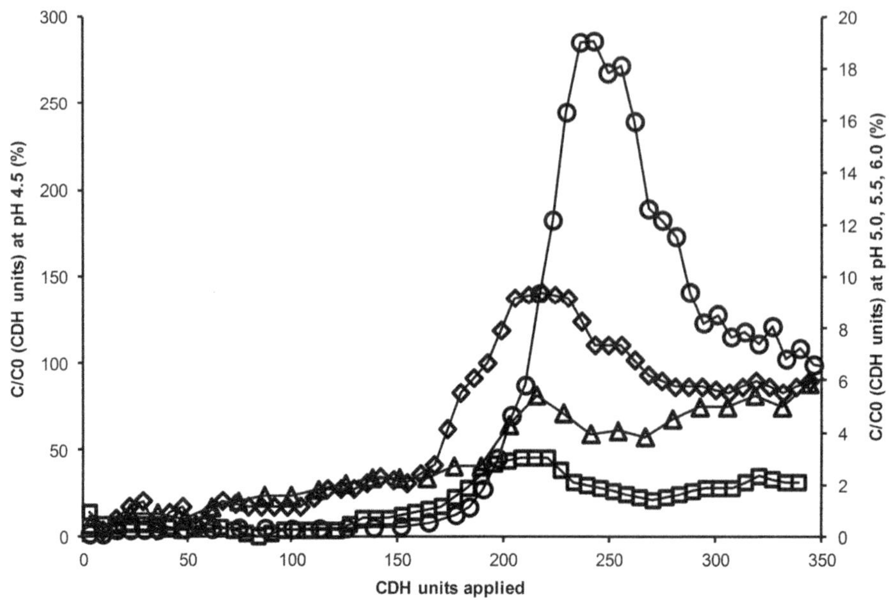

Figure 6 Breakthrough curves for CDH on Q Sepharose XL. pH 4.5 (○) on the left axis, pH 5.0 (◊), 5.5 (Δ) and 6.0 (□) on the right axis.

no enzyme from the feed solution binds but also that most of the already bound enzyme is washed from the column. When setting the start buffer to a pH of 5.5 or 6.0 the maximum breakthrough is 6% and 3% respectively. Table 1 shows the calculated dynamic binding capacity (defined at 5% breakthrough) for CDH for both columns at the studied pH values. For pH 6.0 dynamic binding capacity could not be determined properly as breakthrough stayed below the critical value of 5%. As can be seen, with the DEAE Sepharose FF column the dynamic binding capacity and therefore the amount of enzyme that should be applied to the column to keep loss at a minimum, is very low (about 60 units) compared to the dynamic binding capacity with the Q Sepharose XL column (about 180 units). As can be seen from Table 1, dynamic binding capacity is rising with increasing pH. This can be explained by a stronger net charge of the enzyme and therefore a better binding of CDH to the column material. In theory, when using pH values closer to the isoelectric point, the binding of other proteins should be less. However, as can be seen from Table 1, this effect is marginal as the specific enzyme activity determined in the elution peak is only a little higher at pH 4.5 (only 160 units were applied as otherwise all enzyme would be lost during sample application and wash) than at the other pH values studied. But, more other proteins compared to CDH bind on the DEAE Sepharose FF column as indicated by the low specific activity of 1 U/mg which is another reason for not choosing this material for purification. Nevertheless, Baminger et al. [24] used a DEAE Sepharose FF column as a capture step with a good purification effect. However, this result could not be obtained during this study (purification fold about 2.3 when applying 50 units).

Next, ionic strength of the starting buffer was varied from 20 mM to 200 mM sodium acetate (pH 4.5) to determine the maximum possible molarity where CDH still binds but most of the other proteins do not. About 130 units CDH were applied to stay below the previously determined dynamic binding capacity. Three samples were taken from each run and analyzed for CDH activity and protein content during sample application, washing, and elution. As expected, the amount of total bound protein is decreasing with increasing ionic strength (Figure 7). Concurrent, about 95% of the applied CDH is bound to the column at buffer concentrations from 20 to 50 mM. As a consequence, the purification factor is rising. At a sodium acetate concentration of 100 mM only 87.5% of the enzyme is bound. However, the purification is better because simultaneously to some enzyme loss, a comparatively larger amount of other proteins is also not able to bind. Indeed, when the start buffer has a concentration of 200 mM, nearly the entire enzyme is eluted during sample application and washing therefore resulting in a loss of 95%. A loss of about 12.5% as detected at 100 mM was decided to be acceptable. Therefore this ionic strength was used in further studies.

Table 1 dynamic binding capacity determined at 5% breakthrough and specific enzyme activity in the elute

Column	pH value	Dynamic binding capacity (Units)	Specific activity (U/mg)
Q Sepharose XL	4.5	140	1.6
Q Sepharose XL	5.0	180	1.5
Q Sepharose XL	5.5	210	1.5
Q Sepharose XL	6.0	> 350	1.5
DEAE Sepharose FF	5.0	62	1.0

Elution conditions were optimized by first using a linear gradient from 0 to 0.4 M NaCl in 100 mM sodium acetate buffer at various pH values (4.5, 6.0 and 8.0) in 20 column volumes (CV). Two not well resolved peaks were obtained when eluting with the same pH (4.5) as was used during sample application and wash. CDH eluted mainly in the first peak but had a distinct tailing into the second peak (Figure 8). When setting the pH to 6.0, peak separation was better because a number of other proteins were more strongly retained onto the column compared to CDH. Thus, purification factor was rising from 3.3 to 4.7. With further increasing of the pH to 8.0 CDH was also bound more strongly to the column material, i.e. some other proteins eluted first, followed by CDH and more proteins from this complex mixture. The purification factor remained the same (4.7). A pH value of 6.0 was considered to be most suited for elution.

Optimization of hydrophobic interaction chromatography

To choose an appropriate hydrophobic interaction medium for the second purification step, the binding properties of several materials was determined by applying 30 units onto the columns and determining breakthrough curves. From the various columns tested, only Phenyl HP was suited for CDH purification as with the other columns CDH binding was weak indicated by high c/c_0 values right from the beginning of sample application (data not shown). Using Butyl-S FF, Octyl FF and Phenyl FF low sub the entire enzyme was lost during sample application and washing. With the Butyl-FF column some of the enzyme was bound but loss was still high with 70%. Therefore all further experiments were carried out with a Phenyl HP column.

Next, ammonium sulfate concentration (0.5 M, 0.9 M, and 1.1 M) in the starting buffer was varied. Also, the effect of adding NaCl (0.2 M, 2.1 M) was studied. An ammonium sulfate concentration of 0.5 M is too low which could also be seen from the breakthrough

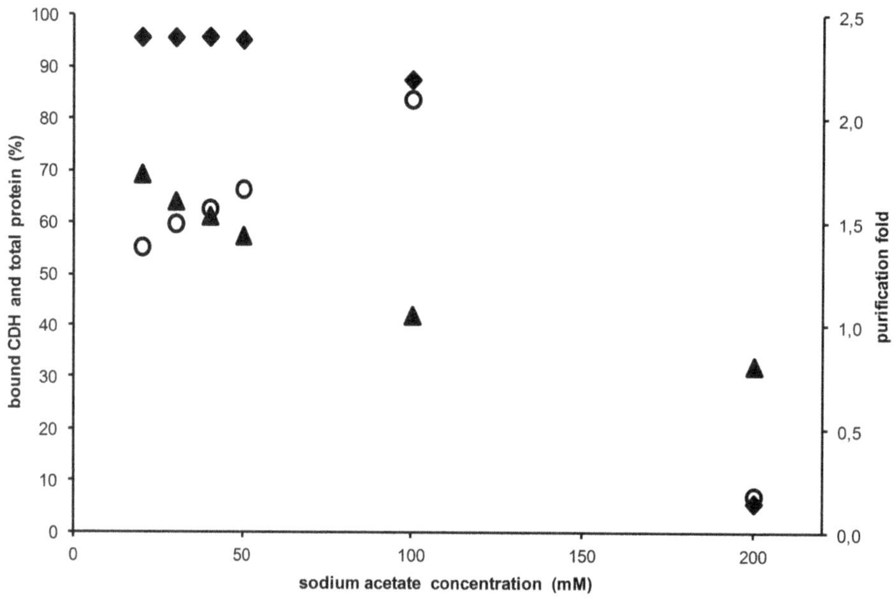

Figure 7 CDH purification as a function of buffer concentration on Q Sepharose XL. amount of bound CDH units (♦, left axis), total protein (▲, left axis), purification factor (○, right axis).

profile as most of the enzyme was not bound therefore resulting in a high loss of nearly 70%. Raising the concentration of $(NH_4)_2SO_4$ to 0.9 M (as was used by Baminger et al. [24] on a Phenyl Resource column) was also not sufficient as most of the target enzyme was eluting during the washing step (loss about 60%). Thus, 1.1 M ammonium sulfate was used as loss was only a third (19%). When adding 0.2 M NaCl to the starting buffer, enzyme loss could be minimized (about 10%). A further increase of NaCl to 2.1 M resulted in an enhanced binding of other proteins which in turn resulted in a worse separation during elution. To gain a certain specific activity more enzyme loss would have to be taken into account.

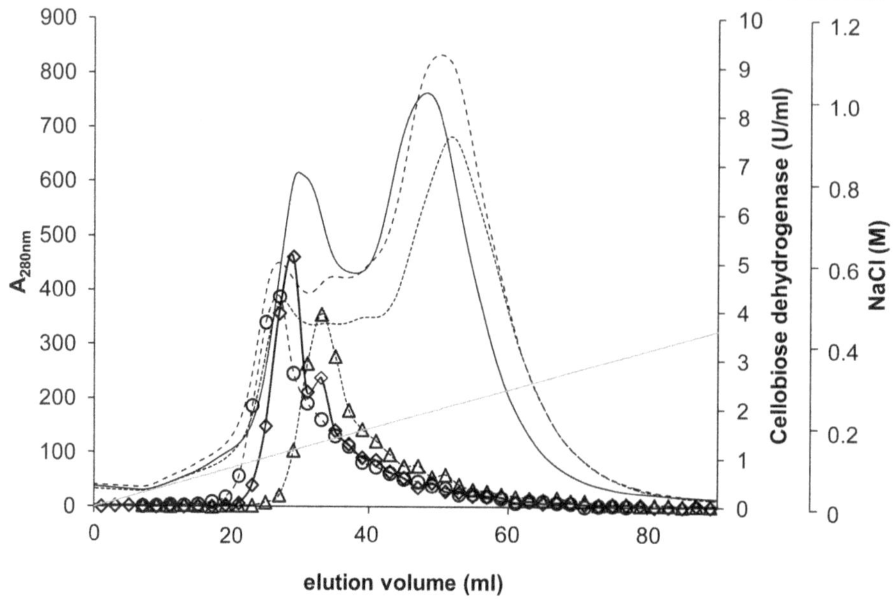

Figure 8 Elution profiles of CDH on Q Sepharose XL at various pH values. pH 4.5 (CDH ◊, A_{280nm} straight line), pH 6.0 (CDH ○, A_{280nm} dashed line), pH 8.0 (CDH Δ, A_{280nm} fine dashed line), elution buffer was 1 M NaCl in 100 mM sodium acetate, gradient from 0 to 0.4 M NaCl in 20 CV.

Table 2 Purification of cellobiose dehydrogenase from *Sclerotium rolfsii*

Purification step	CDH activity (Units)	Total protein (mg)	Specific activity (U/mg)	Purification fold	Yield (%)
Crude extract	1173	1112	1.1	1.0	100
IEX (Q Sepharose XL)	974	456	2.1	2.0	83
$(NH_4)_2SO_4$ precipitation	750	398	1.9	1.8	64
HIC (Phenyl HP)	661	56	11.9	11.3	56

Tandem purification protocol

A method was established that combined IEX and HIC to purify CDH further since enzyme purification using either IEX or HIC was not sufficient. As higher enzyme loading is possible for IEX on Q Sepharose XL, a pH of 5.0 was chosen for the starting buffer (100 mM sodium acetate). Sample solution was diluted 2-fold with starting buffer (resulting in about 20 U/ml) and applied to the column at a speed of 1 ml/min. All further steps were carried out at 5 ml/min. The optimized elution profile includes a linear increase from 0 to 0.15 M NaCl in 8 CV, a hold at 0.15 M NaCl for 4 CV and a stepwise increase to 1 M NaCl (hold 7 CV). After an ammonium

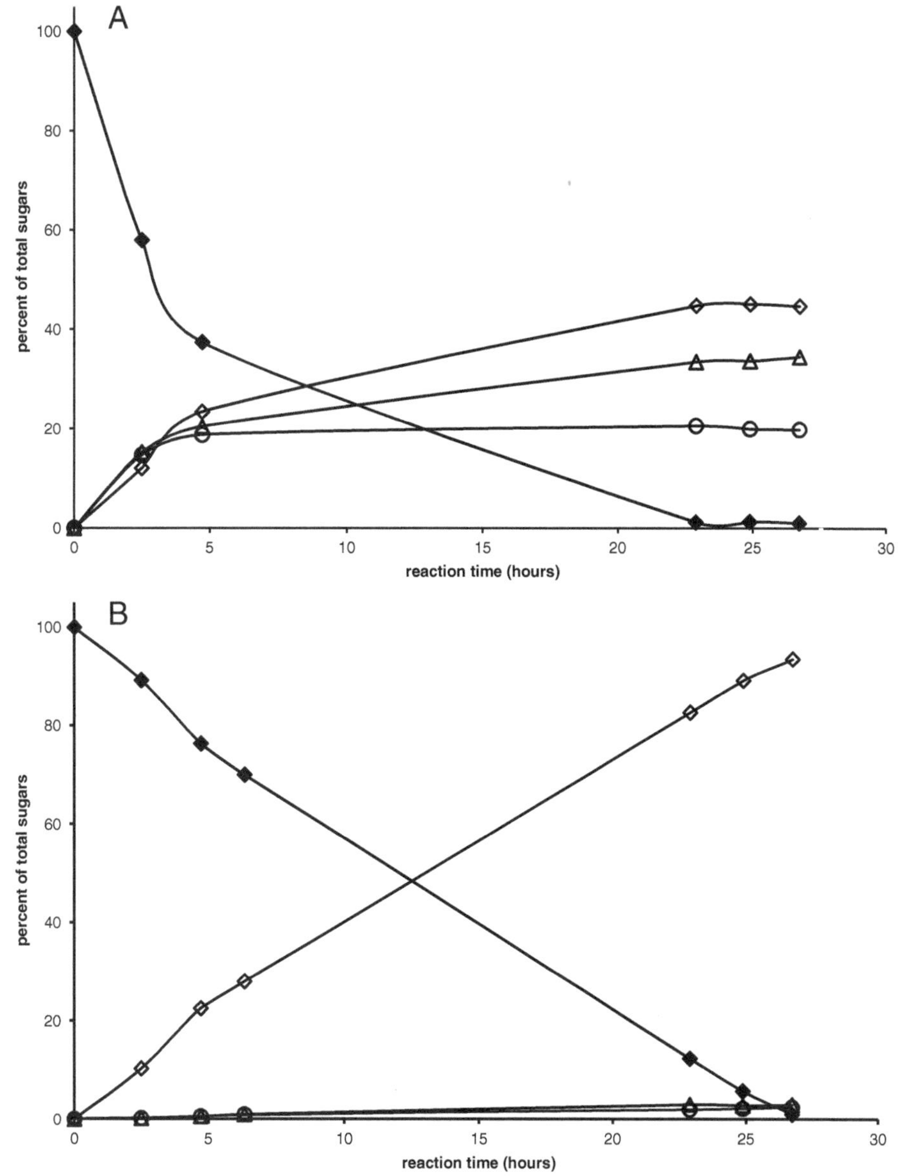

Figure 9 Oxidation of lactose using crude enzyme extract (A) and purified enzyme (B). lactobionic acid (◊), glucose (○), galactose (Δ), lactose (♦).

sulfate precipitation the supernatant was applied to HIC on a Phenyl HP column (volume: 5 ml). The equilibration buffer contained 1.1 M ammonium sulfate and 0.2 M NaCl in 100 mM sodium acetate buffer pH 5.0. Flow rate was set to 5 ml/min for the entire procedure. Elution was performed by increasing buffer B (50 mM sodium acetate pH 5.0) from 0 to 100% in 8 CV. Table 2 shows the overall purification protocol.

However, the obtained enzyme solution after hydrophobic interaction chromatography still contains some other proteins than CDH as could be detected by SDS-PAGE, i.e. the final enzyme solution is not 100% pure. To evaluate if the other proteins have a negative effect on the oxidation of lactose to lactobionic acid (i.e. hydrolysis into glucose and galactose catalyzed by enzymes like α-galactosidase or β-glucosidase which are also produced by the fungus [2]), the enzyme solution was used in a 4.8% lactose solution. For comparison, the crude enzyme extract was used in another experiment.

As can be seen clearly from Figure 9A, using unpurified extract leads to high amounts of glucose and galactose caused by α-galactosidases or β-glucosidases as described before. Thus, yield of lactobionic acid was only 45%. Using partial purified enzyme as described in this paper (Figure 9B), only minor amounts of monosaccharides (3% galactose, 2.5% glucose) could be detected and therefore the obtained purity is sufficient for the intended use of CDH. In 27 hours it was possible to obtain a lactobionic acid yield of 94% which is more than twice as much compared to the crude enzyme solution. However, one might also consider using the unpurified extract depending on the exact application, thus saving costs for enzyme purification. Lactobionic acid might further be tested regarding its potential prebiotic effect [11]. The tolerance in humans is already shown [25]. It can further be used as a calcium salt in food applications as a stabilizer. It is also commonly used in the Wisconsin transplantation solution as an organ preservative [26].

Preservation of purified CDH

Experiments for evaluating two possibilities (freezing and freeze-drying) to preserve CDH were carried out at high enzyme concentration (11,000 U/L) as residual activity is generally higher at higher enzyme concentration [27,28]. Sugars are known to be effective in stabilizing enzymes during freeze-drying therefore acting as lyoprotectants [27,29,30] but no studies on CDH recovery during freeze-drying are known so far.

Figure 10 shows the residual CDH activity compared to the sample before freezing or freeze-drying. As can be seen, without using a lyoprotectant most of the CDH is inactivated during freeze-drying leaving a residual activity of only 23%. Glycerol could stabilize the enzyme to some extent while mannitol could not. In general, it was more effective using sugars as stabilizing agents than sugar alcohols. With glucose and lactulose no loss in activity was observed. With the other disaccharides lactose, cellobiose, and maltose over 80% of CDH could be retained while galactose and raffinose are considered to be less effective. However, residual activity was a lot higher compared to the two sugar alcohols tested.

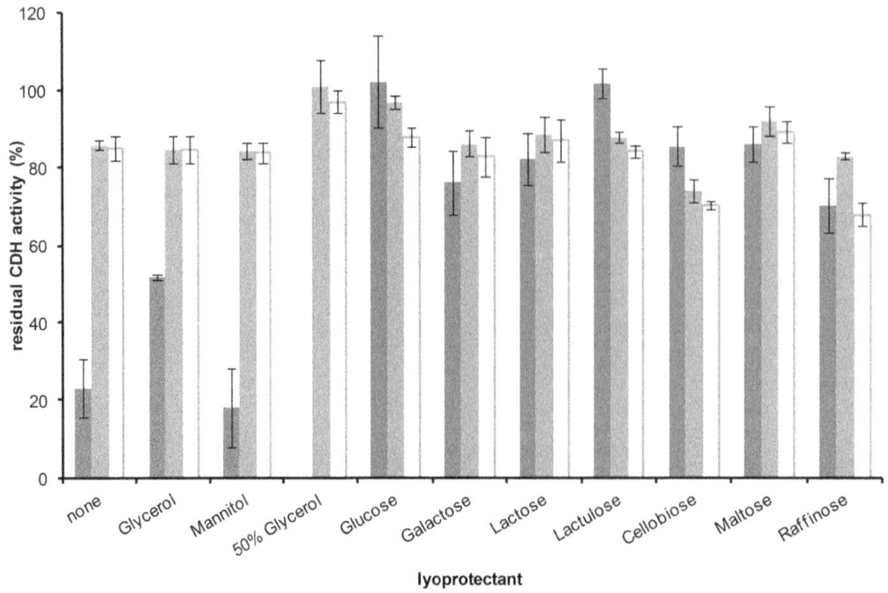

Figure 10 Preservation of CDH. The figure shows the residual CDH activity after lyophilisation (dark grey) or freezing after 60 hours (light grey) or 160 hours (white) in the absence (none) or presence of different lyoprotectants (10 μmol/unit), n = 3.

In another series, storage of the enzyme at -20°C was evaluated. In general, a solution of 50% (w/v) glycerol is often used when trying to make enzymes storable. Therefore, as expected, using high glycerol concentrations gave a residual activity of 100% after 60 hours of storage. However, after 160 hours a small decline is observed. Also, glucose seems to be very effective although enzyme stability might not be satisfying as indicated by the relatively high drop in CDH activity after 160 hours compared to stable activity values when using lactose, glycerol, mannitol or no lyoprotectant. On the other hand, cellobiose clearly has a negative effect in enzyme stability during freezing as residual activity values are below the ones when using no additives. Raffinose, similar to glucose, is stabilizing the enzyme at the beginning of the process but with prolonged storage it seems to be less effective and therefore being not suitable.

Generally, freezing at -20°C effects enzyme activity to a much lesser extent than freeze-drying as can be seen from the data when using no further agents. Although it is shown that with adding glucose or lactulose no activity is lost, the process of freeze-drying is obviously more cost and time consuming than simply freezing the solution at -20°C. Additionally, when adding the right amount of glycerol (no effect was observed when using only 10 µmol per unit) it is also possible to retain all of the enzyme activity. It has to be noted that the experiments carried out during this study are only considered to be preliminary. More research is necessary in studying enzyme behavior during storage, for example determining the half-life time.

Conclusions

To produce cellobiose dehydrogenase from *Sclerotium rolfsii*, it was possible to simplify a rather complex cultivation medium to a basic medium consisting only of α-cellulose, peptone from meat, and trace elements. Additionally, enzyme production was increased by 21%. Purification of the enzyme was studied systematically resulting in a yield of 56%. Although other proteins were detected in the resulting enzyme solution, its purity was considered to be sufficient for oxidizing lactose to lactobionic acid. With freeze-drying of CDH, glucose and lactulose could be identified to be good lyoprotectants. However, freezing at -20°C is preferred for storage as this method being much simpler and residual activity being 100% when using 50% (w/v) glycerol.

Competing interests

The authors declare that they have no competing interests.

Authors' contributions

CF designed the study, carried out the experimental work and drafted the manuscript. AK participated in the design of the study and commented on the manuscript. TK conceived the study, participated in its design and commented on the manuscript. All authors read and approved the final manuscript.

Acknowledgements

The authors would like to thank Julia Streithoff for her support during the experimental work.

References

1. Henriksson G, Johansson G, Pettersson G: **A critical review of cellobiose dehydrogenases.** *J Biotechnol* 2000, **78**(2):93–113.
2. Sachslehner A, Haltrich D, Nidetzky B, Kulbe KD: **Production of Hemicellulose- and Cellulose-Degrading Enzymes by Various Strains of *Sclerotium Rolfsii*.** *Appl Biochem Biotechnol* 1997, **63–65**(1–3):189–201.
3. Cameron MD, Aust SD: **Cellobiose dehydrogenase - an extracellular fungal flavocytochrome.** *Enzyme Microb Technol* 2001, **28**(2–3):129–138.
4. Kruså M, Lennholm H, Henriksson G: **Pre-treatment of cellulose by cellobiose dehydrogenase increases the degradation rate by hydrolytic cellulases.** *Cellul Chem Technol* 2007, **41**(2–3):105–111.
5. Karapetyan KN, Fedorova TV, Vasil'chenko LG, Ludwig R, Haltrich D, Rabinovich ML: **Properties of neutral cellobiose dehydrogenase from the ascomycete *Chaetomium sp.* INBI 2-26(-) and comparison with basidiomycetous cellobiose dehydrogenases.** *J Biotechnol* 2006, **121**(1):34–48.
6. Subramaniam SS, Nagalla SR, Renganathan V: **Cloning and Characterization of a Thermostable Cellobiose Dehydrogenase from *Sporotrichum thermophile*.** *Arch Biochem Biophys* 1999, **365**(2):223–230.
7. Fang J, Liu W, Gao PJ: **Cellobiose Dehydrogenase from *Schizophyllum commune*: Purification and Study of Some Catalytic, Inactivation, and Cellulose-Binding Properties.** *Arch Biochem Biophys* 1998, **353**(1):37–46.
8. Saha T, Ghosh D, Mukherjee S, Bose S, Mukherjee M: **Cellobiose dehydrogenase production by the mycelial culture of the mushroom *Termitomyces clypeatus*.** *Process Biochem* 2008, **43**(6):634–641.
9. Henriksson G, Sild V, Szabó IJ, Pettersson G, Johansson G: **Substrate specificity of cellobiose dehydrogenase from *Phanerochaete chrysosporium*.** *Biochim Biophys Acta Protein Struct Mol Enzymol* 1998, **1383**(1):48–54.
10. Ayers AR, Ayers SB, Eriksson KE: **Cellobiose Oxidase, Purification and Partial Characterization of a Hemoprotein from *Sporotrichum pulverulentum*.** *Eur J Biochem* 1978, **90**(1):171–181.
11. Schaafsma G: **Lactose and lactose derivatives as bioactive ingredients in human nutrition.** *Int Dairy J* 2008, **18**(5):458–465.
12. Tasca F, Zafar MN, Harreither W, Nöll G, Ludwig R, Gorton L: **A third generation glucose biosensor based on cellobiose dehydrogenase from Corynascus thermophilus and single-walled carbon nanotubes.** *Analyst* 2011, **136**(10):2033–2036.
13. Zafar MN, Safina G, Ludwig R, Gorton L: **Characteristics of third-generation glucose biosensors based on Corynascus thermophilus cellobiose dehydrogenase immobilized on commercially available screen-printed electrodes working under physiological conditions.** *Anal Biochem* 2012, **425**(1):36–42.
14. Yakovleva M, Buzas O, Matsumura H, Samejima M, Igarashi K, Larsson PO, Gorton L, Danielsson B: **A novel combined thermometric and amperometric biosensor for lactose determination based on immobilised cellobiose dehydrogenase.** *Biosensors Bioelectron* 2012, **31**(1):251–256.
15. Tasca F, Ludwig R, Gorton L, Antiochia R: **Determination of lactose by a novel third generation biosensor based on a cellobiose dehydrogenase and aryl diazonium modified single wall carbon nanotubes electrode.** *Sensors Actuators B Chem* 2013, **177**:64–69.
16. Knöös P, Schulz C, Piculell L, Ludwig R, Gorton L, Wahlgren M: **Quantifying the release of lactose from polymer matrix tablets with an amperometric biosensor utilizing cellobiose dehydrogenase.** *Int J Pharm* 2014, **468**(1–2):121–132.
17. Enayatzamir K, Tabandeh F, Yakhchali B, Alikhani HA, Rodríguez Couto S: **Assessment of the joint effect of laccase and cellobiose dehydrogenase on the decolouration of different synthetic dyes.** *J Hazard Mater* 2009, **169**(1–3):176–181.
18. Tilli S, Ciullini I, Scozzafava A, Briganti F: **Differential decolorization of textile dyes in mixtures and the joint effect of laccase and cellobiose dehydrogenase activities present in extracellular extracts from Funalia trogii.** *Enzyme Microb Technol* 2011, **49**(5):465–471.

19. Flitsch A, Prasetyo EN, Sygmund C, Ludwig R, Nyanhongo GS, Guebitz GM: **Cellulose oxidation and bleaching processes based on recombinant Myriococcum thermophilum cellobiose dehydrogenase.** *Enzyme Microb Technol* 2013, **52**(1):60–67.
20. Baminger U, Nidetzky B, Kulbe KD, Haltrich D: **A simple assay for measuring cellobiose dehydrogenase activity in the presence of laccase.** *J Microbiol Methods* 1999, **35**(3):253–259.
21. Bradford MM: **A rapid and sensitive method for the quantitation of microgram quantities of protein utilizing the principle of protein dye binding.** *Anal Biochem* 1976, **72**(1–2):248–254.
22. Ludwig R, Haltrich D: **Cellobiose dehydrogenase production by *Sclerotium* species pathogenic to plants.** *Lett Appl Microbiol* 2002, **35**(3):261–266.
23. Ludwig R, Haltrich D: **Optimisation of cellobiose dehydrogenase production by the fungus *Sclerotium (Athelia) rolfsii.*** *Appl Microbiol Biotechnol* 2003, **61**(1):32–39.
24. Baminger U, Subramaniam SS, Renganathan V, Haltrich D: **Purification and Characterization of Cellobiose Dehydrogenase from the Plant Pathogen *Sclerotium (Athelia) rolfsii.*** *Appl Environ Microbiol* 2001, **67**(4):1766–1774.
25. Van Dokkum W, Wezendonk LJW, van Aken-Schneider P, Kistemaker C: **The tolerance of lactobionic acid in man.** 1994. TNO Report, V94.115, TNO Nutrition and Food Research, Zeist, The Netherlands.
26. Playne MJ, Crittenden RG: **Galacto-oligosaccharides and Other Products Derived from Lactose.** In *Advanced Dairy Chemistry, Volume 3: Lactose, Water, Salts and Minor Constituents.* Heidelberg: Springer- Verlag; 2009:121–201.
27. Tanaka K, Takeda T, Miyajima K: **Cryoprotective Effect of Saccharides on Denaturation of Catalase by Freeze-Drying.** *Chem Pharm Bull (Tokyo)* 1991, **39**(5):1091–1094.
28. Capolongo A, Barresi AA, Rovero G: **Freeze-drying of lignin peroxidase: Influence of lyoprotectants on enzyme activity and stability.** *J Chem Technol Biotechnol* 2003, **78**(1):56–63.
29. D'Andrea G, Salucci ML, Avigliano L: **Effect of Lyoprotectants on Ascorbate Oxidase Activity after Freeze-Drying and Storage.** *Process Biochem* 1995, **31**(2):173–178.
30. Corveleyn S, Remon JP: **Maltodextrins as Lyoprotectants in the Lyophilization of a Model Protein.** *LDH Pharm Res* 1996, **13**(1):146–150.

20

Fast and economic immobilization methods described for non-commercial *Pseudomonas* lipases

Silvia Cesarini, Belén Infanzón, F I Javier Pastor and Pilar Diaz*

Abstract

Background: There is an increasing interest to seek new enzyme preparations for the development of new products derived from bioprocesses to obtain alternative bio-based materials. In this context, four non-commercial lipases from *Pseudomonas* species were prepared, immobilized on different low-cost supports, and examined for potential biotechnological applications.

Results: To reduce costs of eventual scaling-up, the new lipases were obtained directly from crude cell extracts or from growth culture supernatants, and immobilized by simple adsorption on Accurel EP100, Accurel MP1000 and Celite®545. The enzymes evaluated were LipA and LipC from *Pseudomonas* sp. 42A2, a thermostable mutant of LipC, and LipI.3 from *Pseudomonas* CR611, which were produced in either homologous or heterologous hosts. Best immobilization results were obtained on Accurel EP100 for LipA and on Accurel MP1000 for LipC and its thermostable variant. Lip I.3, requiring a refolding step, was poorly immobilized on all supports tested (best results for Accurel MP1000). To test the behavior of immobilized lipases, they were assayed in triolein transesterification, where the best results were observed for lipases immobilized on Accurel MP1000.

Conclusions: The suggested protocol does not require protein purification and uses crude enzymes immobilized by a fast adsorption technique on low-cost supports, which makes the method suitable for an eventual scaling up aimed at biotechnological applications. Therefore, a fast, simple and economic method for lipase preparation and immobilization has been set up. The low price of the supports tested and the simplicity of the procedure, skipping the tedious and expensive purification steps, will contribute to cost reduction in biotechnological lipase-catalyzed processes.

Keywords: *Pseudomonas*, Lipase, Immobilization, Accurel, Celite, FAMEs

Background

According to the Roadmap of the European Commission for Environmental and Sustainable Development (July 2013), *"it is now time to move towards an energy and resource efficient economy, based on the use of nature's toolbox"*. Bio-inspired processes or enzymatic reactions have a low environmental impact, reduce the amount of waste material and minimize costs, thus serving the requirements to integrate environmental sustainability with economic growth and welfare (Roadmap of the European Commission for Environmental and Sustainable Development, July 2013). Therefore, there is an increasing interest to seek new enzyme preparations and to develop new products derived from bioprocesses to obtain alternative bio-based materials [1,2]. In this context, biotechnological production of biodiesel with lipases has received great consideration in recent years and is undergoing a fast development [3]. Employment of lipases as biocatalysts in the transesterification of triglycerides looks attractive because of the mild reaction conditions required and the easy recovery of glycerol, without an elaborate purification procedure involving production of chemical wastes. Moreover, the enzymatic process tolerates the water content of oil and increases the biodiesel yield by avoiding the typical soap formation caused by alkaline transesterification [4].

Enzyme immobilization is a well reported technology that allows application of enzymes in many biocatalyzed

* Correspondence: pdiaz@ub.edu
Department of Microbiology, University of Barcelona, Av. Diagonal 643, 08028 Barcelona, Spain

processes like in lipase-mediated biodiesel production [5]. In general, immobilization allows reuse of the biocatalyst, makes the product recovery easier, and frequently enhances the enzyme resistance against inactivation by different denaturants, providing more stable catalysts [6]. Efficiency and recyclability of immobilized enzymes depends not only on the procedure and support utilized but also on the specific enzyme used and the type of process where it is applied [7]. Various immobilization procedures like adsorption, cross-linking, encapsulation or entrapment have been employed on lipases used for biodiesel production [8]. However, most immobilization procedures use sophisticated protocols for lipase entrapment on expensive supports [9], not suitable for a real scale-up, causing an increase in the costs of industrial processes. Therefore, adsorption, used in this work, is a convenient system for lipase immobilization that could be successfully applied to large volume industrial processes, as in biodiesel production, because of its simplicity and low cost [10,11]. This type of immobilization occurs via binding of the lipase onto the surface of the support by weak forces, such as Van der Waals or hydrophobic interactions or through dispersion forces [8]. Adsorbed lipases can be prepared and used under mild conditions without significant activity loss, and the associated process is very simple, as the carrier can be easily recovered for repeated immobilization rounds [5]. Lipases can be adsorbed onto many different supports but generally, porous materials are the best option. These supports allow the effective dispersion of enzyme molecules on a large surface, thus allowing a higher number of enzyme molecules to deliver their catalytic potentiality [12]. In order to obtain stable, efficient and economic enzyme preparations suitable for potential biotechnological applications, in this work three hydrophobic and porous materials were tested as supports for lipase immobilization: two polypropylene matrices (Accurel EP100 and MP1000) and a diatomaceous silica (Celite®545).

Accurel EP100 and MP1000 are commercially available hydrophobic, macroporous, low-density polypropylene powders that display a large surface area for adsorption because of their very small particle size [13,14]. These preparations are extremely resistant towards organic solvents [14-16], a feature that increases their interest for lipase immobilization when aimed at oil transesterification [12,17]. Celite®545 consists of highly porous diatomaceous beads composed of silica (SiO_2), also containing some other inorganic oxides. Because of its chemical inertness and special interconnected pore structure, Celite®545 constitutes a very suitable support for physical adsorption [18] which has been extensively used in immobilized-lipase biodiesel production [11,19,20].

Previous works aimed at lipase-mediated biodiesel synthesis used mainly commercial enzyme preparations. As an alternative, in the present work we assayed four new lipases from *Pseudomonas* sp. 42A2 and *Pseudomonas* sp. CR-611 [21-23], previously isolated and characterized in our laboratory [24]. Classical enzyme immobilization generally uses purified lipases, a step that requires laboratory materials and equipment that considerably increases the preparation time, thus raising the price of enzyme preparations and the total costs of the process. Therefore, we tried to avoid purification by considering the high hydrophobic nature of lipases and probably of their selective adsorption on the hydrophobic and porous supports used. Accordingly, we describe here the preparation of four non-commercial lipases for potential biotechnological applications. A simple immobilization protocol by fast adsorption of crude enzymes onto three different, inexpensive supports was developed. As a possible application, the behavior of the immobilized lipase preparations was assayed in triolein transesterification for fatty acid methyl esters (FAMEs) production.

Results

Lipase production

Pseudomonas sp. 42A2 extracellular lipases LipA and LipC were purified and fully characterized in a previous work [21]. LipA is a robust, mesophilic enzyme whereas LipC is more thermolabile and shows a cold-adapted behavior [21,22]. LipCmut, instead, is a LipC thermostable variant previously obtained in our laboratory, still keeping the wild type cold-adapted behavior, thus being a good candidate for biotechnological applications [22,25]. For lipase production, genes coding for LipA, LipC and LipCmut were ligated to pBBR1MCS vector together with their cognate foldase gene *lip*H [21,22]. These constructions were cloned and expressed in the homologous host *Pseudomonas* PABST7.1, where lipases are naturally folded by LipH and further secreted to the culture medium in their fully active conformation [21,22]. Supernatants from each strain were obtained and directly used for activity determination and immobilization. The use of *Pseudomonas aeruginosa* PAO1 mutant strain deficient for chromosomal lipase production (PABST7.1), guarantees that all extracellular lipolytic activity found in supernatants from LipA, LipC or LipCmut recombinant clones corresponds indeed to each individually cloned lipase. Therefore, use of the selected strains as hosts for multicopy plasmid-encoded lipases allowed us to skip the purification steps for each enzyme, thus reducing time and costs of the entire enzyme production process.

Pseudomonas sp. CR611 lipase LipI.3 was previously cloned in *E. coli* 5 K, where it is expressed in an inactive form as inclusion bodies [23]. Nevertheless, enzymatic preparations of LipI.3 adequate for evaluation were successfully obtained from inclusion bodies by refolding in urea, as previously described [23]. However, activity recovery from inclusion bodies makes the whole process longer and more expensive, indicating that further improvements of the expression system and enzyme preparation protocol are required to make LipI.3 a faster and more efficient enzyme system. Several attempts to increase Lip I.3 expression were conducted by means of using different vectors and hosts, but in all cases the enzyme was found in the form of inclusion bodies. Consequently, only low but sufficient amounts of active LipI.3 were obtained from the initial construction in order to test this lipase for immobilization on different supports with the same loading charge.

Effectiveness of lipase immobilization

Adsorption of lipases on hydrophobic polymeric supports has been reported previously [12,13]. However, previous works mainly assayed existing commercial lipases. In this work we wished to discover the behavior of four non-commercial lipases towards immobilization using different kinds of supports, for further biotechnological applications. Thus, direct supernatant samples of PABST7.1 bearing LipA, LipC or LipCmut constructions, and refolded crude cell extracts of *E. coli* LipI.3 were successfully immobilized on EP100, MP1000 and Celite®545. During adsorption on EP100 and MP1000, residual activity corresponding to unbound lipases was measured in supernatants of the immobilization batches as an indicator of the process yield. For this purpose, supernatant samples were obtained at different times, and the decrease of residual activity was plotted for immobilization assessment. As shown in Figure 1 (EP100) and Figure 2 (MP1000), all four lipases tested were efficiently adsorbed on the Accurel matrices used. A fast decrease of residual activity could be observed during the first 5 h incubation for all tested enzymes. For LipC and LipCmut, 90% immobilization was produced within the first hour of incubation, both on EP100 and MP100. Adsorption of LipA on polypropylene matrices proceeded slightly slower, but all lipolytic activity of the batch supernatant was lost after 24 h incubation, indicating a complete immobilization. Enzyme-loaded EP100 and MP1000 were recovered through vacuum filtration and lipase activity of the supports was measured. As shown in Table 1, in general, immobilization on MP1000 resulted in the most active enzyme preparations. Since immobilization on Celite®545 followed a different protocol, only the final activity adsorbed to the silica could be measured (Table 1). Celite®545 showed to be the best support for LipA, although this enzyme also displayed good activity when adsorbed to MP1000. LipC displayed the highest activity values after immobilization on EP100 and MP1000, whereas the catalytic potential of LipCmut and LipI.3 was better preserved on MP1000. Taking into consideration the great variation of activity values found among replica assays (Table 1), and trying to describe a common support for the four lipases tested, it could be concluded that Accurel MP1000 would be the support of choice for immobilization of LipC, LipCmut and LipI.3, whereas LipA would be better immobilized on Celite®545. In general, immobilized lipases displayed full activity over a long period, indicating a good stability of the biocatalysts [10].

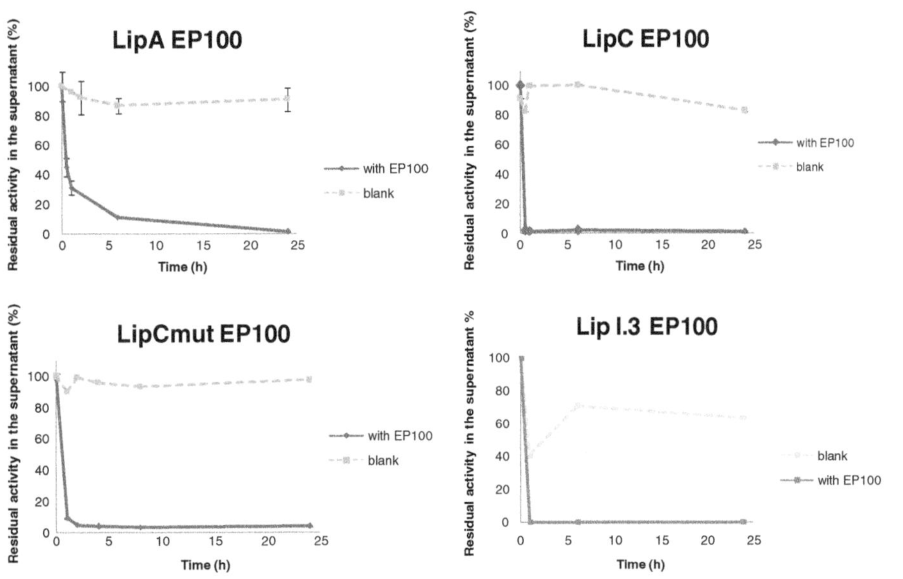

Figure 1 Immobilization on EP100. Continuous lines show the decrease of residual activity in immobilization batch supernatants when incubated with the support. As a control, dotted lines show the same supernatants incubated without any support: initial activity was maintained during the 24 h immobilization process, as shown by recovery of initial activity.

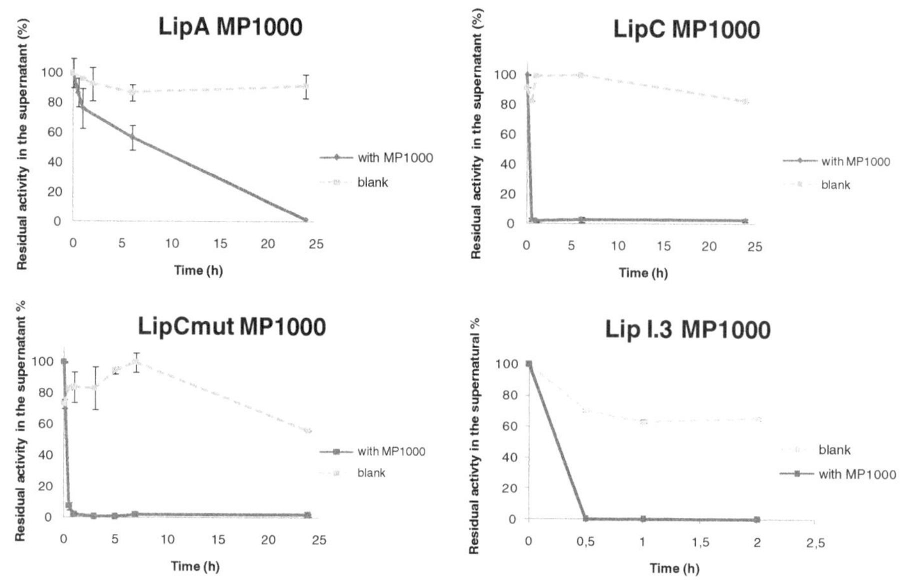

Figure 2 Immobilization on MP1000. Continuous lines show the decrease of residual activity in immobilization batch supernatants when incubated with the support. As a control, dotted lines show the same supernatants incubated without any support: initial activity was maintained during the 24 h immobilization process, as shown by almost full recovery of initial activity. Samples were taken mainly during the first 8 h of incubation to better define the rapid decrease of supernatant residual activity. The apparent activity decrease observed for LipCmut control sample was probably due to a different degree of substrate emulsion in samples assayed after 8 h incubation. High immobilization rates were obtained after 2 h incubation.

FAMEs production

As a possible biotechnological application, the four immobilized lipases were assayed in preliminary transesterification reactions for FAMEs production. The transesterification reaction conditions used were an adaptation of a previous protocol to the immobilized enzymes [26]. Impure triolein, (Sigma) containing also di- and monolein, was used as a substrate to simulate a natural vegetable oil, generally constituted by a mixture of glyceride forms. The reaction mixture contained the oil substrate, immobilized enzyme, and a defined amount of water and methanol, as described in materials and methods. The resulting products were analyzed through TLC (Figure 3), used as a fast and simple qualitative system for confirmation of FAMEs synthesis. A control sample showing the substrate impurity is visible in the first lane of Figure 3, where several spots appear, the most abundant corresponding to triolein. An industrially synthesized biodiesel sample, loaded in the second lane, was used as a control for FAMEs production. FAMEs appearance in course of the reaction with lipases LipA, LipC, LipCmut and LipI.3 can be identified in Figure 3 as spots migrating at the same position as the industrially synthesized biodiesel (lane 2). Although moderate, production of FAMEs by LipA, LipC and LipCmut preparations was detectable. Unfortunately, no FAMEs spots were revealed in the case of LipI.3-mediated transesterifications. Independently of the behavior of LipI.3, the highest yield in FAMEs production could be detected in reactions carried out with LipA-EP100, LipA-Celite®545 and with LipC and LipCmut immobilized on the two Accurel matrices, where discernible biodiesel spots appeared. Once demonstrated that new, non-commercial lipases can be efficiently immobilized on low-cost supports following a simple and fast protocol, and having shown that these lipases can perform transesterification reactions, further experiments are being conducted to increase the immobilized activity and to assess the re-usability of such preparations [10].

Table 1 Enzyme activity after immobilization on EP100, MP1000 and Celite®545

	LipA	LipC	LipCmut	LipI.3
EP100 ($U\ g^{-1}$ support)	6.8 ± 0.8	42.7 ± 1.4	28.6 ± 0.1	34.3 ± 1.1
MP1000 ($U\ g^{-1}$ support)	36.0 ± 1.7	39.4 ± 7.8	62.6 ± 0.1	86.2 ± 6.2
Celite®545 ($U\ g^{-1}$ support)	48.3 ± 1.0	7.2 ± 3.0	21.7 ± 0.2	13.8 ± 1.9

Discussion

Lipases LipA, LipC, LipCmut and LipI.3 were efficiently immobilized on all assayed supports with no need for time-consuming and expensive purification protocols. Use of the selected strains as hosts for multicopy plasmid-encoded lipases allowed high enzyme production and permitted to skip the purification steps for each enzyme [27], thus reducing time and costs of the entire

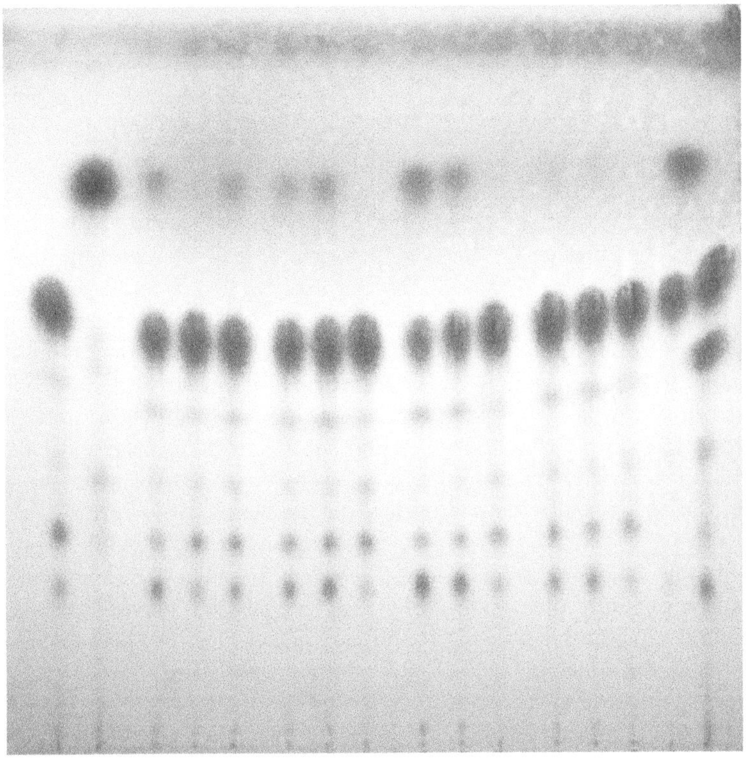

Figure 3 TLC showing the products resulting from transesterification reactions. T: triolein 65% (substrate); BD: industrial biodiesel (product control); C: transesterification reaction performed with Novozym 435, shown as a control; C-: negative control of a transesterification reaction run without alcohol. Enzymes immobilized on 1) EP100; 2) MP1000; 3) Celite®545. Reaction products of a) LipA; b) LipC; c) LipCmut; d) LipI.3. FAMEs production was clearly detected in lanes 1a, 3a, 1b, 2b, 1c, 2c and C, where spots at the same height of industrial biodiesel appeared.

enzyme isolation process. Moreover, supernatants from PABST7.1 cultures generally produce a very low amount of total protein (ca. 0.033 mg mL^{-1}), where each cloned lipase is the most abundant enzyme, as they are produced in higher copy number than any other secreted, chromosome-born protein. Therefore, the corresponding culture supernatants obtained can be considered as semi-purified enzyme preparations [10]. Cost and time savings in lipase isolation would be a great advantage in future scaling up of enzyme production for immobilization or biotechnological application. Although several attempts were performed to obtain higher expression yields of Lip I.3, low expression rates were obtained, indicating that new strategies should be developed for further lipase production improvement. Nevertheless, the amount of enzyme obtained by inclusion body refolding was sufficient for the immobilization assays performed.

In contrast to other immobilization procedures using sophisticated protocols and expensive supports [7,9], economic matrices were tested here to make the process industrially feasible. All four lipases tested were efficiently adsorbed on the Accurel matrices used, where a fast decrease of residual activity could be observed during the first 5 h incubation. For these supports, lipolytic activity of the batch supernatant was lost after 24 h incubation, indicating a complete immobilization. Immobilization on Celite®545 provided the best results for LipA. However, although the protocol for immobilization on Celite®545 is extremely fast, cheap and easy [9], the use of cold acetone in high proportions for enzyme precipitation could become a disadvantage during an eventual scaling up of the immobilization process. Therefore, we suggest polypropylene supports like Accurel MP1000 as the more appropriate carriers for lipases LipA, LipC, LipCmut and LipI.3, considering their possible role in biocatalysis.

It has been reported that adsorption occurs via hydrophobic interactions and is generally more effective for lipases than for other proteins [28]. Considering the high hydrophobicity shown by lipases we can assume that the highest proportion of proteins adsorbed on the supports assayed here were the lipases tested, with only minor adsorption of other proteins derived from supernatants or urea-dissolved cell extracts. It has also been reported that the success of hydrophobic materials in lipase adsorption depends on the structural features of "true" lipases, that is, those bearing a structural domain called "lid" [29]. The "lid" contributes to provide an open and a closed configuration to the enzyme. In the closed conformation, the "lid" exposes the hydrophobic residues towards the catalytic site, thus preventing substrate binding and activity. But in the presence of a hydrophobic interface, the "lid" changes its conformation and exposes the catalytic site to the substrate (triglyceride), in a phenomenon known as "interfacial activation" [29]. LipA, LipC and LipCmut possess a "lid" that covers the active site and is related to the affinity of lipases towards hydrophobic matrices [30]. Although LipA bears a "lid", it does not show interfacial activation [21] a fact that could justify its different immobilization behavior with respect to LipC and LipCmut, which display interfacial activation and resulted here better immobilized on hydrophobic Accurel supports. Interestingly, LipI.3, belonging to a different lipase family, does not have a canonical "lid" but displayed the highest activity when immobilized on the hydrophobic support Accurel MP1000, probably due to a major exposure of its catalytic site. In the case of LipC and LipCmut, it is likely that during adsorption on polypropylene matrices, these lipases change their conformation in the presence of the hydrophobic support surface, being thus adsorbed in the open configuration and therefore being more active than when immobilized on Celite®545. To understand why immobilization of LipA on Accurel EP100 produced the lowest activity recovery, another important factor should be considered: the orientation of the adsorbed enzyme. Indeed, due to the hydrophobic nature of the support, it is possible that some lipase molecules interact with the support through the very hydrophobic surface surrounding the active site. This might hinder the substrate reaching of the active site [12], producing low activity values. It has also been shown that enzyme loading is directly correlated with the activity of the immobilized biocatalyst [31]. At this respect, it was reported that when *Thermomyces lanuginosus* lipase was immobilized on EP100 at low loading, part of the enzyme molecules were completely inactivated: with a loading of 6.6 mg of protein g^{-1} support, only 70% of the active sites were titrable compared to the 93% titrable active sites obtained at a loading of 55 mg g^{-1} support [15]. Moreover, immobilization might also lead to the inactivation of some lipase molecules by the distortion of the tertiary structure of the enzyme caused by the enzyme–support interaction. This would lead to structural rearrangements that could cause a decrease in lipase activity. All these effects may affect to a different extent each immobilized enzyme, thus making necessary to assay each individual lipase for immobilization efficiency on the desired support prior to application and scaling up for industrial purposes.

Immobilized lipases were evaluated for FAMEs production using triolein as a model substrate. LipA, LipC and LipCmut produced a visible FAMEs spot when analyzed by TLC, indicating that they are candidates for further study and process improvement. Enzyme or reaction engineering could contribute to increase the FAMEs yield obtained here. Lack of FAMEs formation with Lip I.3 could be explained in terms of substrate specificity, knowing that triolein is not a good substrate for this enzyme when used in hydrolysis reactions [23]. Alternatively, lack of transesterification activity could be due to a negative effect of immobilization on Lip I.3 when used in synthesis reactions, to a low enzyme concentration, or to inhibition by methanol, reasons that deserve further investigation.

The results obtained demonstrate that non-commercial lipases LipA, LipC, LipCmut and LipI.3 can efficiently be immobilized on all assayed supports, being polypropylene matrices (Accurel EP100, for LipA, and Accurel MP1000, for LipC and its variant) the best supports to be employed, especially in transesterification reactions, since they do not involve elaborate protocols, they are easy to recover and sufficiently cheap to be used in an eventual scaling up of the process. Immobilized LipA, LipC and LipCmut were capable to produce FAMEs from triolein under the conditions used here; therefore, they deserve further investigation.

Conclusions

From the results obtained it can be concluded that it is possible to expand the existing enzyme toolbox by testing new, non-commercial enzymes produced by easy and fast procedures, which can further be immobilized by simple adsorption on economic supports. These methodologies can contribute to introduce new enzymatic tools for biotechnological applications such as biodiesel production, as they do not involve any time or solvent consumption, they are easy to conduct and sufficiently economic to be used in an eventual scaling up.

Methods

Materials and reagents

All chemicals and solvents were obtained from Sigma Aldrich and Serviquimia (Spain) and, except when stated, had quality for analysis. Accurel EP100 (particle size 200–500 μm) was from Azko Nobel (Obernburg, Germany; kindly provided by Dr. F. Valero). Accurel MP1000 (particle size under 1500 μm) was purchased from Membrana GmbH (Wuppertal, Germany). Celite®545 (diatomaceous earth, 30–80 mesh) was obtained from Sigma-Aldrich.

Bacterial strains and enzyme production

Pseudomonas strains and the plasmids used in this work are listed in Table 2. *Escherichia coli* 5 K, used as recipient strain for recombinant plasmids, was grown in Luria–Bertani medium (LB, Panreac, Spain) at 37°C and

Table 2 Strains used

	Strain	Features	Reference
	Pseudomonas sp. 42A2	Wild type	[33]
	Pseudomonas sp. CR-611	Wild type	[24]
	P. aeruginosa PABST7.1	Δ*lipA* Δ*lip*H miniD-180 (*tet*A *tet*R *lac*lq P*lac*UV5-T7 gene1)	[32] Kindly provided by Dr. Rosenau.
LipA	*P. aeruginosa* PABST7.1 lipAHpBB	Contains *lip*A and *lip*H 42A2 gene	[21]
LipC	*P. aeruginosa* PABST7.1 lipCHpBB	Contains *lip*C and *lip*H 42A2 gene	[22]
LipCmut	*P. aeruginosa* PABST7.1 lipCHpBB (variant D2_H8)	Contains mutated *lip*C and wild type *lip*H 42A2 gene	[22]
LipI.3	E. coli 5 K (pGEM-T Lip I.3)	Contains *lip*I.3 CR-611gene	[23]

supplemented with 50 μg chloramphenicol mL^{-1} when necessary. *Pseudomonas* strains were routinely grown in LB medium (supplemented with 400 μg mL^{-1} chloramphenicol and 50 μg mL^{-1} tetracycline, when applied) at 30°C on a reciprocal rotary shaker (180 rev min^{-1}). *Pseudomonas* PABST7.1 is a PAO1 mutant deficient for *lip*A and *lip*H genes, where LipC is inactive due to the lack of the specific foldase LipH [32].

Pseudomonas sp. 42A2 lipases, LipA and LipC [21,33], and the thermostable variant LipCmut [22] were previously cloned with their specific foldase (LipH) in pBBR1MCS vector [34] and transformed into homologous *Pseudomonas* PABST7.1. All of them are naturally secreted lipases [21] and the enzyme recovery was easily carried out by centrifugation of the culture broth (20 min at 8000 rpm).

LipI.3, isolated from *Pseudomonas* sp. CR-611 [35], was previously cloned in pGEMT and transformed into *E. coli* 5 k [23] where it is expressed as inclusion bodies in an inactive form. Inclusion bodies were obtained by French press cell-disruption and solubilised according to Kojima's method [36]. Using a refolding protocol with urea 8 M, active LipI.3 was obtained in phosphate buffer 50 mM pH 7, after urea removal by dialysis [23].

Enzyme immobilization

Direct culture supernatants were used for LipA, LipC and LipCmut immobilization, whereas LipI.3 was immobilized from a solubilised inclusion body solution, thus skipping the high costs and time-consuming steps of protein purification.

Adsorption on EP100 and MP1000

Supports were pretreated following an adaptation from Guillen and coworkers [37] where the polypropylene matrices were activated with ethanol (3 mL per gram of MP1000 or EP100), followed by alcohol removal by vacuum filtration. Immobilization was carried out by mixing the pretreated supports with 5 mL cell extract or supernatant samples, following a ratio of 1 mg protein per g of EP100 and 2 mg protein per g of MP1000. Orbital shaking of the mixture was carried out for 24 h at each enzyme's optimum temperature, being 4°C for LipC and LipCmut, and 30°C for LipA and LipI.3. Control samples were run without any support under the same conditions. Supernatant samples from immobilization mixtures were taken at different times and their residual lipase activity was assayed as described [38]. The enzyme-loaded supports, containing the immobilized lipases, were recovered by vacuum filtration, dried in a speed vacuum and stored at 4°C after lipase activity determination.

Adsorption on celite®545

Immobilization on Celite®545 did not require any pretreatment and was performed by mixing 5 mL of cell extracts or supernatants with the support, using a ratio of 0.5 mg protein per g Celite®545. The mixture was incubated for 30 min at 4°C and precipitation of the enzyme onto the support was carried out with 5 mL cold acetone. Immobilized enzymes were then collected by vacuum filtration and dried with a speed vacuum concentrator prior to enzyme activity determination and storage at 4°C [9].

Activity assays

Activity of free enzyme in cell extract and supernatant fractions, or in supernatants from the immobilization mixture was analyzed by measuring the release of *para*-nitrophenol (*p*NP) from *p*NP-derivative fatty acid substrates, as a result of enzymatic hydrolysis, measured at $\lambda = 405$ nm as previously described [21,38]. One unit of activity was defined as the amount of enzyme that released 1 mol of *p*NP per minute under the assay conditions used. Lipase activity was routinely assayed at the optimum temperature for each enzyme (30°C for LipA and LipI.3; 4°C for LipC and LipCmut), using the standard conditions: 20 mM Tris HCl buffer at optimum pH of each enzyme (pH 8 for LipA, LipC and LipCmut; pH 5.5 for LipI.3), and 0.6% Triton X-100®. $CaCl_2$, at a final concentration of 20 mM, was added to the reaction mixture for LipI.3 activity assays [23].

Activity of immobilized lipases was measured using 2 mg support carrying the immobilized enzyme adsorbed on EP100 and MP1000, and 10 mg enzyme-loaded support for Celite®545. The reaction mixture was incubated at each enzyme's optimum temperature and was stopped by 2 min centrifugation at 14,000 rpm. Activity determination was carried out in 400 μL Tris 100 mM plus 100 μL *p*NP-buffer, as described above. One unit of immobilized lipase activity was defined as the amount of enzyme that released 1 mol of *p*NP per minute and per g of support, under the assay conditions used [21,38]. Protein concentration of free and immobilized samples was determined as previously described [39]. When required, dilution of lipase solutions was done using buffer Tris 20 mM pH 8 for LipA, LipC and LipCmut, and pH 7 for LipI.3 [21,23].

Transesterification assays

FAMEs synthesis reactions were carried out in 2 ml glass vials for 24 h with maximum agitation in a horizontal vortex. LipC and LipCmut reactions were carried out at room temperature, according to the cold-adapted behavior of these enzymes [21,22]. A temperature of 30°C was used for LipA and LipI.3 [21,23]. The reaction mixtures contained 1 g triolein of 65% purity to simulate the composition of mixed vegetable oils. Methanol 15% w/w of oil was used as the second substrate, and was added in two steps (7.5% each) at 0 h and 7 h, to prevent possible enzyme inhibition. Water (10% w/w of oil) was added to the reaction mixture to increase enzyme efficiency [26] and, for LipI.3, 20 mM $CaCl_2$ was also added [23]. In order to compare the effectiveness of immobilized lipases in transesterification assays, all enzymatic preparations were loaded of 2U lipase, calculated from their hydrolytic activity when immobilized. As positive control, the same reaction conditions were applied to a commercial immobilized lipase (Novozym 435, Novozymes DK), whereas in negative controls, the same reaction was performed without methanol for detection of possible hydrolysis processes. After 24 h reaction, the products were evaporated in a speed vacuum concentrator at 60°C for 1½ h to remove excess methanol and the upper layer, obtained by centrifugation during the same evaporation process, was prepared for TLC analysis.

Thin layer chromatography

AluGram UV254 silica gel plates (Mecherey Nagel) were used as stationary phase for TLC analysis, and a mixture of hexane-diethyl ether-acetic acid (80:20:1.5 v/v/v) was used as the mobile phase. Commercial triolein (Sigma) and a sample of industrially synthesized biodiesel (BE- Biodiesel, Spain) were used as standards for glycerides and FAMEs, respectively. Enzymatic reaction products and standards were diluted 1:50 in hexane, and 8 μL samples were then applied to a 20 × 20 cm plate. 15% phosphomolibdic acid in ethanol was used as color developer through a fine spray. Spot visualization was completed by mildly heating the plate with an air dryer.

Competing interests

No interest or financial relationships that could influence the author's objectivity exist regarding the information provided in it. Neither patent or stock ownership, membership of a company board of directors, membership of an advisory board or committee for a company, and consultancy for or receipt of speaker's fees from a company apply to the manuscript authors who are a Post-doctoral fellow, a PhD student, and two public University professors. Therefore, the authors declare that they have no competing interests.

Authors' contribution

SC participated in the design of the study, carried out most of the experiments, organized and interpreted the data, and drafted the manuscript. BI made some of the immobilization experiments, performed the TLC assays and contributed to the manuscript draft. FIJ P contributed to critical discussion, and revised the manuscript. PD coordinated the project and the design of the study, critically interpreted the data and corrected the manuscript. All authors read and approved the manuscript.

Acknowledgements

This work was financed by the Scientific and Technological Research Council (MINECO, Spain), grant CTQ2010-21183-C02-02/PPQ, by the IV Pla de Recerca de Catalunya, grant 2009SGR-819, by PCI-AECID project A203563511, and by the Generalitat de Catalunya to the "Xarxa de Referència en Biotecnologia" (XRB). S. Cesarini and B. Infanzón acknowledge doctoral fellowships from the Spanish Ministry of Science and Education (AP2008-04579) and from MAEC-AECID, respectively. We thank Dr. F. Valero for kindly providing Accurel EP100.

References

1. Drepper T, Eggert T, Hummel W, Leggewie C, Pohl M, Frank R, Wilhelm S, Jaeger KE: **Novel biocatalysts for white biotechnology.** *Biotechnol J* 2006, 1:777–786.
2. Casas Godoy L, Duquesne S, Bordes F, Sandoval G, Marty A: **Lipases: an Overview.** In *Lipases and Phospholipases: Methods and Protocols*, Volume 861. Edited by Sandoval G. Totowa, Nj USA: Humana Press Inc, 999 Riverview Dr, Ste 208; 2012:07512-1165. 3–30.
3. Ghaly AE, Dave D, Brooks MS, Budge S: **Production of biodiesel by enzymatic transesterification: review.** *Am J Biochem Biotechnol* 2010, **6**(2):54–76.
4. Dizge N, Aydiner C, Imer DY, Bayramoglu M, Tanriseven A, Keskinlera B: **Biodiesel production from sunflower, soybean, and waste cooking oils by transesterification using lipase immobilized onto a novel microporous polymer.** *Bioresour Technol* 2009, **100**(6):1983–1991.
5. Tan TW, Lu JK, Nie KL, Deng L, Wang F: **Biodiesel production with immobilized lipase: a review.** *Biotechnol Adv* 2010, **28**(6):937–937.
6. Tischer W, Kasche V: **Immobilized enzymes: crystals or carriers?** *Trends Biotechnol* 1999, **17**(8):326–335.
7. Li X, Huang SS, Xu L, Yan YJ: **Improving activity and enantioselectivity of lipase via immobilization on macroporous resin for resolution of racemic 1-phenylethanol in non-aqueous medium.** *BMC Biotechnol* 2013, **13**:9.
8. Jegannathan KR, Abang S, Poncelet D, Chan ES, Ravindra P: **Production of biodiesel using immobilized lipase - A critical review.** *Crit Rev Biotechnol* 2008, **28**(4):253–264.
9. Ferrer M, Plou FJ, Fuentes G, Cruces MA, Andersen L, Kirk O, Christensen M, Ballesteros A: **Effect of the immobilization method of lipase from *Thermomyces lanuginosus* on sucrose acylation.** *Biocatal Biotransform* 2002, **20**(1):63–71.
10. Cesarini S, Pastor FIJ, Diaz P: **Improvement of P. aeruginosa 42A2 lipase preparations for FAMEsproduction, both in immobilized and soluble form.** *J Mol Catal B Enzym* 2014, **99**:1–7.

11. Lumor SE, Akoh CC: **Esterification and Hydrolytic Activities of *Candida rugosa* Lipase Isoform 1 (LIP1) Immobilized on Celite 545, Duolite A7, and Sephadex G-25.** *J Agric Food Chem* 2008, **56**(21):10396–10398.
12. Salis A, Pinna M, Monduzzi M, Solinas V: **Comparison among immobilised lipases on macroporous polypropylene toward biodiesel synthesis.** *J Mol Catal B Enzym* 2008, **54**(1–2):19–26.
13. Gangoiti J, Santos M, Llama MJ, Serra JL: **Production of Chiral (R)-3-Hydroxyoctanoic Acid Monomers, Catalyzed by *Pseudomonas fluorescens* GK13 Poly(3-Hydroxyoctanoic Acid) Depolymerase.** *Appl Environ Microbiol* 2010, **76**(11):3554–3560.
14. Al-Duri B, Goddard R, Bosley J: **Characterisation of a novel support for biocatalysis in supercritical carbon dioxide.** *J Mol Cat B-Enzym* 2001, **11**(4–6):825–834.
15. Persson M, Wehtje E, Adlercreutz P: **Factors governing the activity of lyophilised and immobilised lipase preparations in organic solvents.** *Chembiochem* 2002, **3**(6):566–571.
16. Persson M, Mladenoska I, Wehtje E, Adlercreutz P: **Preparation of lipases for use in organic solvents.** *Enzym Microb Technol* 2002, **31**(6):833–841.
17. Severac E, Galy O, Turon F, Pantel CA, Condoret JS, Monsan P, Marty A: **Selection of CalB immobilization method to be used in continuous oil transesterification: analysis of the economical impact.** *Enzym Microb Technol* 2011, **48**(1):61–70.
18. Chang SF, Chang SW, Yen YH, Shieh CJ: **Optimum immobilization of *Candida rugosa* lipase on Celite by RSM.** *Appl Clay Sci* 2007, **37**(1–2):67–73.
19. Sagiroglu A: **Conversion of sunflower oil to biodiesel by alcoholysis using immobilized lipase.** *Artif Cells Blood Substitutes Biotechnol* 2008, **36**(2):138–149.
20. Shah S, Gupta MN: **Lipase catalyzed preparation of biodiesel from Jatropha oil in a solvent free system.** *Process Biochem* 2007, **42**(3):409–414.
21. Bofill C, Prim N, Mormeneo M, Manresa A, Pastor FIJ, Diaz P: **Differential behaviour of *Pseudomonas* sp 42A2 LipC, a lipase showing greater versatility than its counterpart LipA.** *Biochimie* 2010, **92**(3):307–316.
22. Cesarini S, Bofill C, Pastor FIJ, Reetz MT, Diaz P: **A thermostable variant of *P. aeruginosa* cold-adapted LipC obtained by rational design and saturation mutagenesis.** *Process Biochem* 2012, **47**(12):2064–2071.
23. Panizza P, Syfantou N, Pastor FIJ, Rodriguez S, Diaz P: **Acidic lipase Lip I.3 from a *Pseudomonas fluorescens*-like strain displays unusual properties and shows activity on secondary alcohols.** *J Appl Microbiol* 2013, **114**(3):722–732.
24. Ruiz C, Pastor FIJ, Diaz P: **Isolation of lipid- and polysaccharide-degrading micro-organisms from subtropical forest soil, and analysis of lipolytic strain *Bacillus* sp CR-179.** *Lett Appl Microbiol* 2005, **40**(3):218–227.
25. Bassegoda A, Cesarini S, Diaz P: **Lipase Improvement: Goals and Strategies. Minireview.** *Comput Struct Biotechnol J* 2012, **2**(3):1–8.
26. Cesarini S, Diaz P, Nielsen P: **Exploring a new, soluble lipase for FAMEs production in water-containing systems using crude soybean oil as a feedstock.** *Process Biochem* 2013, **48**(3):484–487.
27. Cesarini S, Pastor FIJ, Diaz P: **Improvement of *P. aeruginosa* 42A2 lipase preparations for FAMEs production, both in immobilized and soluble form.** *J Mol Catal B-Enzym* 2013, **99**:1–7.
28. Gitlesen T, Bauer M, Adlercreutz P: **Adsorption of lipase on polypropylene powder.** *Biochim Et Biophys Acta-Lipids Lipid Metab* 1997, **1345**(2):188–196.
29. Jaeger KE, Dijkstra BW, Reetz MT: **Bacterial biocatalysts: molecular biology, three-dimensional structures, and biotechnological applications of lipases.** *Annu Rev Microbiol* 1999, **53**:315.
30. Nardini M, Lang DA, Liebeton K, Jaeger KE, Dijkstra BM: **Crystal structure of *Pseudomonas aeruginosa* lipase in the open conformation - The prototype for family I.1 of bacterial lipases.** *J Biol Chem* 2000, **275**(40):31219–31225.
31. Salis A, Svensson I, Monduzzi M, Solinas V, Adlercreutz P: **The atypical lipase B from *Candida antarctica* is better adapted for organic media than the typical lipase from *Thermomyces lanuginosa*.** *BBA-Proteins Proteomics* 2003, **1646**(1–2):145–151.
32. Jaeger KE, Schneidinger B, Rosenau F, Werner M, Lang D, Dijkstra BW, Schimossek K, Zonta A, Reetz MT: **Bacterial lipases for biotechnological applications.** *J Mol Catal B-Enzym* 1997, **3**(1–4):3–12.
33. Mercade E, Robert M, Espuny MJ, Bosch MP, Manresa MA, Parra JL: **New surfactant isolated from *Pseudomonas* 42A2.** *J Am Oil Chem Soc* 1988, **65**:1915–1916.
34. Kovach ME, Phillips RW, Elzer PH, Roop RM, Peterson KM: **pBBR1MCS: a broad-host-range cloning vector.** *BioTechniques* 1994, **16**:800–802.
35. Prim N, Bofill C, Pastor FIJ, Diaz P: **Esterase EstA6 from *Pseudomonas* sp CR-611 is a novel member in the utmost conserved cluster of family VI bacterial lipolytic enzymes.** *Biochimie* 2006, **88**(7):859–867.
36. Kojima Y, Kobayashi M, Shimizu S: **A novel lipase from *Pseudomonas fluorescens* HU380: gene cloning, overproduction, renaturation-activation, two-step purification, and characterization.** *J Biosci Bioeng* 2003, **96**(3):242–249.
37. Guillen M, Benaiges MD, Valero F: **Immobilization and Stability of a *Rhizopus oryzae* Lipase Expressed in Pichia pastoris: comparison Between Native and Recombinant Variants.** *Biotechnol Prog* 2011, **27**(5):1232–1241.
38. Ruiz C, Falcocchio S, Xoxi E, Pastor FIJ, Diaz P, Saso L: **Activation and inhibition of *Candida rugosa* and *Bacillus*-related lipases by saturated fatty acids, evaluated by a new colorimetric microassay.** *Biochim Et Biophys Acta-Gen Subj* 2004, **1672**(3):184–191.
39. Bradford MM: **Rapid and sensitive method for quantitation of microgram quantities of protein utilizing principle of protein dye binding.** *Anal Biochem* 1976, **72**(1–2):248–254.

Preparation of cell-permeable Cre recombinase by expressed protein ligation

Soo Kyung Lyu and Hyockman Kwon*

Abstract

Background: Protein transduction is safer than viral vector-mediated transduction for the delivery of a therapeutic protein into a cell. Fusion proteins with an arginine-rich cell-penetrating peptide have been produced in *E. coli*, but the low solubility of the fusion protein expressed in *E. coli* impedes the large-scale production of fusion proteins from *E. coli*.

Results: Expressed protein ligation is a semisynthetic method to ligate a bacterially expressed protein with a chemically synthesized peptide. In this study, we developed expressed protein ligation-based techniques to conjugate synthetic polyarginine peptides to Cre recombinase. The conjugation efficiency of this technique was higher than 80%. Using this method, we prepared semisynthetic Cre with poly-L-arginine (ssCre-R9), poly-D-arginine (ssCre-dR9) and biotin (ssCre-dR9-biotin). We found that ssCre-R9 was delivered to the cell to a comparable level or more efficiently compared with Cre-R11 and TAT-Cre expressed as recombinant fusion proteins in *E. coli*. We also found that the poly-D-arginine cell-penetrating peptide was more effective than the poly-L-arginine cell-penetrating peptide for the delivery of Cre into cell. We visualized the cell transduced with ssCre-dR9-biotin using avidin-FITC.

Conclusions: Collectively, the results demonstrate that expressed protein ligation is an excellent technique for the production of cell-permeable Cre recombinase with polyarginine cell-penetrating peptides. In addition, this approach will extend the use of cell-permeable proteins to more sophisticated applications, such as cell imaging.

Keywords: Expressed protein ligation, Cre recombinase, Polyarginine, Cell-penetrating peptides, Intracellular delivery

Background

Protein delivery is regarded as a safer and useful alternative to gene delivery for the induction of genome editing and alterations in the gene expression profile of cells. Gene delivery through the transfection of plasmid DNAs or viral transductions inevitably imposes potential complications to gene alteration and is subject to safety problems. In contrast, there is a slight possibility that a genome would be altered by proteins delivered directly into cells. Furthermore, proteins become degraded and are eventually removed from cells. This property makes protein delivery more suitable for cell therapy applications in which a transient, rather than a sustained, action of transcription factors is required.

Due to the impermeability of proteins through a non-polar cell membrane, protein delivery typically requires a protein delivery vector. Cell-penetrating peptides (CPPs) have emerged as the most promising protein delivery vector and are distinct from liposomes and other encapsulation technologies. CPPs are defined as short peptides (10–30 amino acids) that facilitate the cellular uptake of large macromolecules with a low membrane permeability [1]. Since the discovery that Tat48–60 enables the HIV-1 Tat protein to penetrate cells and activates HIV-1-specific target genes [2-4], several CPPs, including penetratin and transportan, have been reported. CPPs are either arginine-rich, amphipathic, or hydrophobic [5]. Recently, a cell-type-specific CPP, Xentry, is found from the X-protein of the hepatitis B virus [6]. The arginine-rich CPPs are short basic peptides highly enriched with arginine and lysine residues. The guanidinium groups of arginines play a central role in the cell penetrating activity of the arginine-rich CPPs, and in fact, homooligomers of arginine (7–12 arginines) are significantly more effective than native arginine-rich CPPs, such as Tat_{48-60} [7,8].

* Correspondence: hmkwon@hufs.ac.kr
Department of Bioscience and Biotechnology and Protein Research Center for Bio-Industry, Hankuk University of Foreign Studies, Yongin 449-791, Republic of Korea

The different characteristics in the peptide sequences of CPPs suggest that their translocation mechanisms are not identical. Most importantly, the arginine-rich CPPs are less toxic than the amphipathic CPPs, and the arginine-rich CPPs do not cause membrane leakage at low micromolar concentrations where the arginine-rich CPPs are highly effective for cellular uptake [9-11]. Considering their low toxicity and high efficiency, the arginine-rich CPPs have been preferentially selected as protein delivery vectors. For example, the polyarginine fusion proteins of Oct4, Sox2, Klf4, and c-Myc have been delivered into somatic cells to reprogram them into iPSCs [12,13], and TAT-Pdx1, TAT-Ngn3, and TAT-MafA have been used to stimulate pancreatic β-cell differentiation [14-16]. Moreover, R7-CARM1 has been delivered to human mesenchymal stem cells to modulate their differentiation potential [17].

The arginine-rich CPP proteins have been produced in *E. coli* as recombinant fusion proteins. However, the fusion of the arginine-rich CPPs to proteins markedly decreases the solubility of the proteins, and when these are overexpressed in *E. coli*, they generally form the inclusion bodies. To obtain the soluble and active form of the proteins, the inclusion bodies have to be solubilized by denaturation agents, such as urea, and then the proteins have to be refolded [12,14,16,18]. Alternatively, the temperature used to induce protein expression has to be lowered to maintain the solubility of the arginine-rich CPP proteins when expressed in *E. coli* [15,17]. However, either method severely decreases the efficiency of the recovery of the active proteins, which impedes the large-scale production of the arginine-rich CPP proteins from *E. coli*.

Expressed protein ligation (EPL) is a semisynthetic technique that conjugates a synthetic peptide containing an N-terminal cysteine to a C-terminal thioester through a native chemical ligation reaction (Figure 1). A C-terminal thioester of a protein is generated by the thiolysis of a recombinant intein fusion protein expressed in *E. coli*. The orthogonal reactivity, the production of a native peptide bond, and the high yield in an aqueous buffer at neutral pH and low temperature make EPL a powerful ligation method that introduces a synthetic peptide into recombinant proteins. EPL has been used to incorporate unnatural amino acids, fluorophores, and post-translational modifications to proteins [19]. In this study, we employed EPL to conjugate a synthetic polyarginine CPP polypeptide to a recombinant protein expressed in a soluble form in *E. coli*. The major advantage of this EPL-based technique over the preparation of recombinant polyarginine CPP

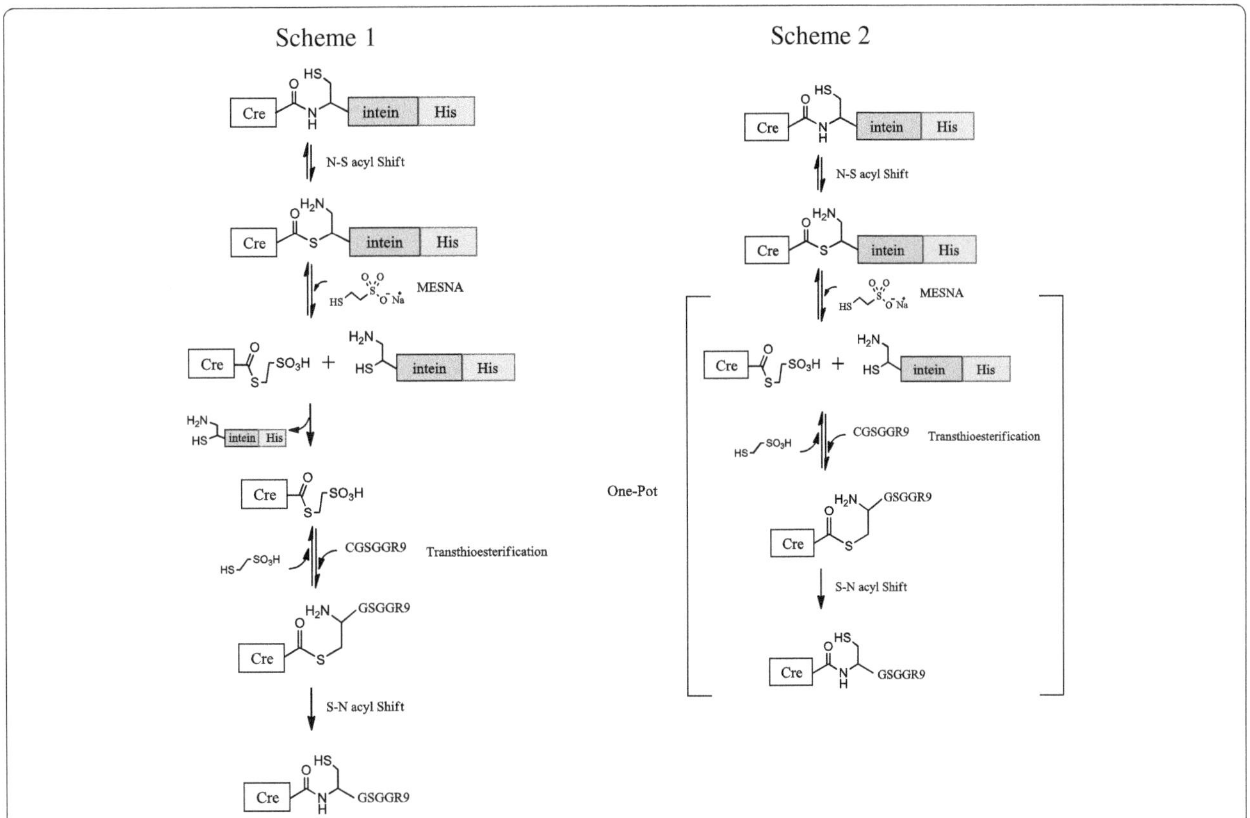

Figure 1 General scheme for expressed protein ligation. In scheme 1, the Cre-thioester is isolated from the intein-His for further reaction with the thiol group of an N-terminal cysteine of the peptides. In scheme 2, the thiolysis by MESNA, the transthioesterification reaction, and the S-N acyl Shift proceed in one-pot.

fusion proteins from *E. coli* is that the EPL-based technique is not subject to the solubility problem as long as protein can be produced as soluble form in *E. coli.* In addition, the EPL-based technique allows the preparation of cell-permeable proteins with novel arginine-rich CPPs composed of unnatural amino acids, guanidinium-rich peptoids, or biotin.

The applicability of the EPL-based technique for the preparation of arginine-rich CPP proteins was previously tested using the enhanced green fluorescent protein (EGFP) as a model protein [20]. The flow cytometry study showed that EGFP conjugated with arginine 12-mer (R12) prepared by the EPL-based technique was delivered to cells with an efficiency comparable to that obtained with EGFP-R12 expressed as a fusion protein in *E. coli.* However, the EGFP-R12 molecules taken up by cells appeared to accumulate at the cell surface or be entrapped in endosomes, and it was uncertain whether these were delivered to the cytoplasm. The bio-activity of EGFP-12R in the cells was not examined in the study.

The Cre recombinase from bacteriophage P1 is a DNA sequence-specific recombinase that catalyzes the recombination of DNA between two loxP sites. Cre has been widely used to generate a tissue-specific or inducible knockout in transgenic mice or conditional gene expression in cells. Cre carries the nuclear localization sequence (NLS)-like elements that direct the efficient entry of Cre protein into the nucleus of mammalian cells [21]. In addition, cells harboring a loxP-STOP-loxP reporter gene, such as β-galactosidase or EGFP, allow us to monitor Cre-mediated recombination by Cre protein [22,23]. In this study, we prepared a cell-permeable Cre protein through the conjugation of synthetic polyarginine CPPs to recombinant Cre expressed in *E. coli* using EPL. The semisynthetic Cre-R9 (ssCre-R9) prepared by EPL had the similar enzymatic activity as Cre. We found that ssCre-R9 efficiently led to recombination events in living cells. It was previously reported that *d-isomer* polyarginine CPPs are more efficient than *l-isomer* polyarginine CPPs for the delivery of fluorescently labeled peptides [7]. Biotin (vitamin H or B-7) is a water-soluble vitamin, and the strong biotin-avidin interaction allows imaging of the cells that uptake a biotin-containing protein. Because EPL allows the incorporation of *d-isomer* amino acids and peptide mimetics into a protein, we prepared Cre proteins with the *d-isomer* of R9 (ssCre-dR9) and biotin (ssCre-dR9-biotin) by EPL and explored whether these elaborations in the resulting CPP enhance the effectiveness of the intracellular delivery or allow cell imaging.

Results and discussion

The Cre-intein-His protein was expressed in *E. coli* and purified in its native form by Ni^{2+}-agarose column chromatography. We generated a Cre-thioester by intein-mediated cleavage using MESNA and then ligated the resultant Cre-thioester with CGSGG-R9 or CGSGG-dR9 by native chemical ligation. To minimize any potential adverse effect of MESNA on the activity of Cre, we first performed the intein-mediated cleavage reaction with 50 mM MESNA at 4°C for 24 h. Under this condition, more than 95% of the Cre-intein-His proteins were cleaved into Cre and the intein-His (lane 4 and lane 5 of Figure 2A).

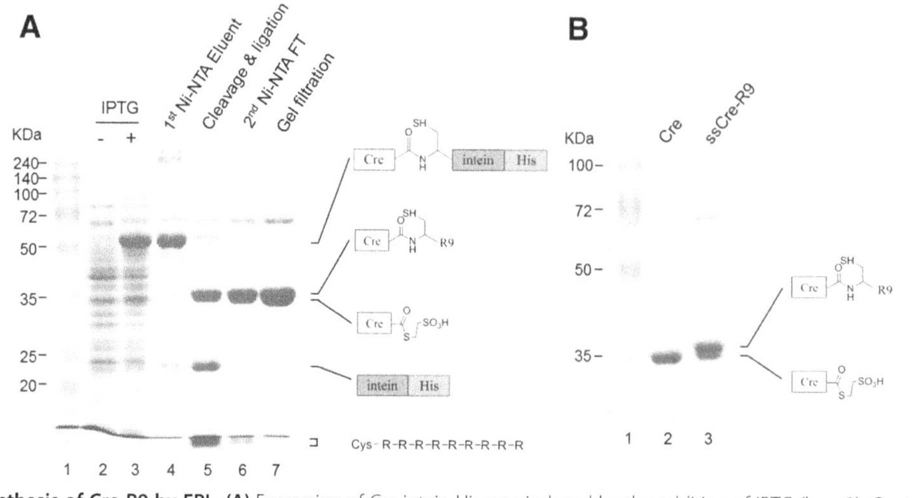

Figure 2 Semi-synthesis of Cre-R9 by EPL. (A) Expression of Cre-intein-His was induced by the addition of IPTG (lane 3). Cre-intein-His was purified by Ni-NTA chromatography (lane 4). After dialysis, Cre-intein-His was cleaved by MESNA and ligated with CGSGG-R9 peptide at 4°C for 24 h (lane 5). Cre-R9 was isolated from intein-His and CGSGG-R9 peptide by Ni-NTA chromatography (lane 6) and gel filtration (lane 7). Lane 1, size marker. The 12% SDS-PAGE gel was stained with Coomassie Brilliant Blue R-250. **(B)** The 12.5% SDS-PAGE gel was run for a longer running time, and it separated Cre-R9 from Cre-thioester as the distinct bands (lane 3). Cre was prepared by the DTT-mediated cleavage of Cre-intein-His (lane 2). Lane 1, size marker. The gel was stained with Coomassie Brilliant Blue R–250. The ratio of Cre-R9:Cre-thioester, which was estimated from the integrated peak areas of densitometry analysis of lane 3, was 84:16.

Second, we explored the direct conversion of the Cre-intein-His protein into the ligation product of Cre-R9 through a two-step one-pot reaction (Scheme 2 of Figure 1). EPL proceeds through the following three reactions; the intein-mediated cleavage reaction by thiol reagents, the transthioesterification reaction between the thiol group of an N-terminal cysteine of the peptides and the Cre-thioester, and the spontaneous intramolecular $S \rightarrow N$ acyl shift that generates an amide bond at the ligation junction. The first two reactions are reversible, but the last reaction is irreversible. The irreversibility of the last amide-forming reaction allows the two-step one-pot reaction, in which the intein-mediated cleavage and the native chemical ligation proceed in one pot, resulting in a high yield of the final ligation product. We conducted the two-step one-pot EPL reaction with 2 mg/ml Cre-intein-His and a 10-fold excess of CGSGG-R9 in 50 mM MESNA, 50 mM NaH_2PO_4, and 250 mM NaCl, pH 8.0 at 4°C for 24 h. The intein-His proteins, MESNA, and CGSGG-R9 were then removed from the final ligation product by Ni^{2+}-agarose column chromatography and subsequent gel filtration chromatography (Figure 2A). Cre-R9 migrates slowly compared to Cre and forms the distinct up-shifted band on SDS-PAGE (Figure 2B). Conjugation efficiency was calculated from integrated peak areas of densitometry analysis of the Coomassie-stained SDS-PAGE gels. Typically, 84% of Cre was conjugated into Cre-R9, and the final yield of Cre-R9 was more than 2 mg from 20 mg of the Cre-intein-His protein. We named the final Cre-R9 product produced by EPL as the semisynthetic Cre-R9 (ssCre-R9).

We next determined whether Cre proteins retain their activity after the conjugation of R9 peptides by expressed protein ligation. Control Cre was generated from the Cre-intein-His protein by intein-mediated cleavage with DTT instead of MESNA. Assays of the Cre enzyme activity measured the release of a circular Km^r from the linearized pGEM-T with floxed-Km^r by the site-specific recombination of Cre (Figure 3). The maximum activity was observed at 80 ng/reaction for both the control Cre and ssCre-R9. This finding demonstrates that the conjugation of CGSGG-R9 to Cre by EPL does not impair the enzymatic activity of Cre.

One major issue in developing a semisynthetic method to produce cell-permeable Cre proteins is whether the functionalities of cell-permeable Cre proteins are maintained during the semisynthetic process. Human TE671 (LoxP-LacZ) cells harbor a loxP-STOP-loxP LacZ reporter gene, which allows us to assess the cellular uptake, subcellular localization and enzymatic activity in the nucleus of Cre proteins (Figure 4A). In order for the transduced Cre proteins to elicit recombination, the exogenous Cre protein is required to enter the cell, translocate to the nucleus, and remove the STOP sequence by recombination at two loxP sites. It results in the expression of *lacZ*, and the cells expressing β-galactosidase turns blue in X-gal staining assay. The transfection of pBS185CMV-Cre plasmid DNA using Lipofectamine induced recombination in 40% of cells (Figure 4). It indicates that Cre proteins without any additional NLS are able to enter the nucleus and elicit recombination in human TE671 cells.

When treated with the control Cre (R9-CPP minus), few *lacZ*-expressing cells were observed (Figure 4). In contrast, treatment with ssCre-R9 showed a dose-dependent increase in recombination. Treatment with 80 μg/ml ssCre-R9 induced recombination in 94% of cells. The $EC50_{app}$ value was measured to be 42.7 μg/ml (1.07 μM) (Table 1). Because 84% of Cre was conjugated to Cre-R9 in the

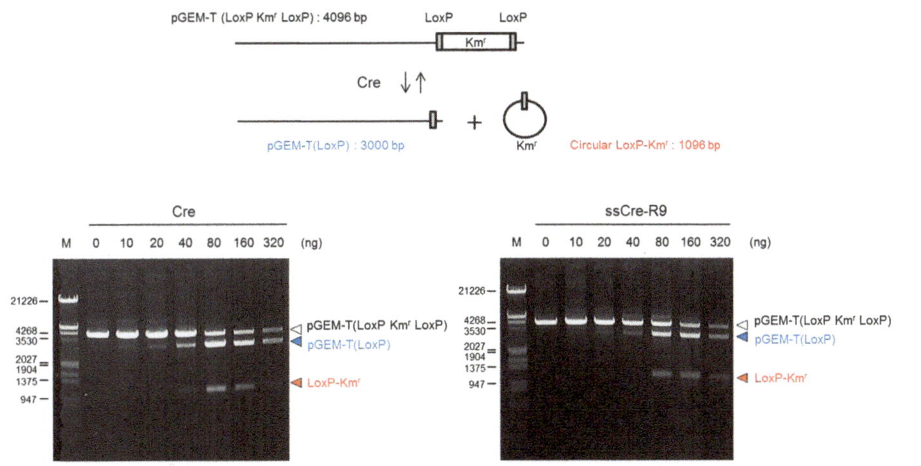

Figure 3 Activity of Cre-R9 remains intact after EPL. Linearized pGEM-T (LoxP Km^r LoxP) is 4,096 bp in length and contains a LoxP Km^r LoxP at one end. Cre-mediated recombination between two LoxP sites results in a linear (3,000 bp) pGEM-T (LoxP) and a circular (1,096 bp) LoxP-Km^r. Cre and the semisynthetic Cre-R9 (ssCre-R9) were incubated with the linearized pGEM-T (LoxP Km^r LoxP) at 37°C for 1 h. After isolated from proteins by phenol extraction, DNA was analyzed by agarose gel electrophoresis. M, Lambda DNA digested with EcoRI and HindIII.

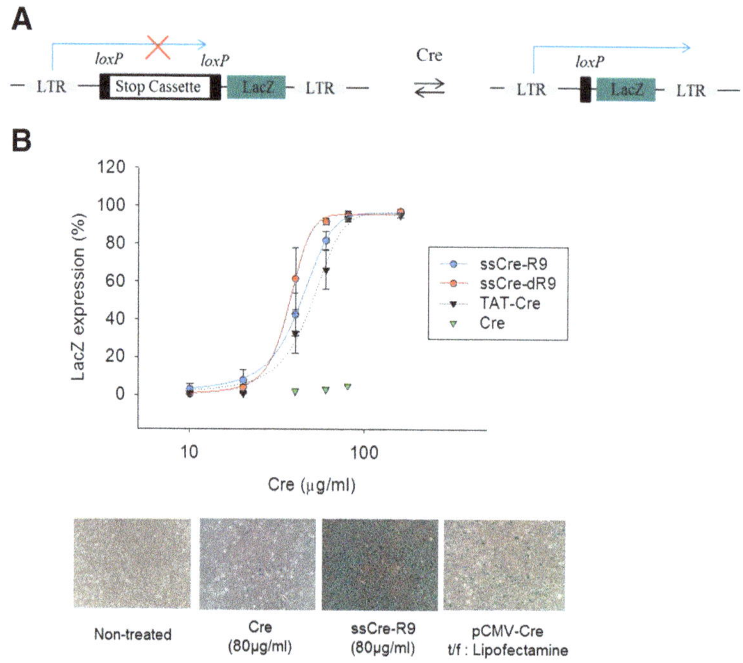

Figure 4 Protein transduction of the semisynthetic Cre-R9 and Cre-dR9. (A) Cre-mediated recombination of the floxed STOP cassette leads to LacZ expression in human TE671 (LoxP-LacZ) cells. **(B)** The cells were exposed to varying doses of Cre, ssCre-R9, ssCre-dR9, and TAT-Cre in OPTI-MEM for 5 h. The cells were further incubated with the serum-containing media for a total of 48 h, and then the percentage of LacZ-expressing cells was determined by X-gal staining (n = 4; mean ± SD). The pCMV-Cre expression DNA plasmid was transfected using Lipofectamine.

ssCre-R9 preparation (Figure 2B), the $EC50_{real}$ of ssCre-R9 was estimated to be 35.9 μg/ml (0.9 μM). TAT-Cre and Cre-R11 purified as recombinant fusion proteins from *E. coli* are commercially available, and the EC50 of TAT-Cre was measured to be 46.6 μg/ml (1.14 μM) in a parallel experiment (Table 1). Transduction activity of Cre-R11 appeared to be very low; the number of *lacZ*-expressing cells obtained through the treatment of TE671 cells with Cre-R11 was too low to calculate the EC50 value. This result demonstrates that ssCre-R9 is more or at least equally effective for intracellular delivery than the recombinant TAT-Cre or Cre-R11 fusion protein. As reported for the recombinant TAT-Cre protein, the delivery of ssCre-R9 into the cells was also rapid, and treatment with 60 μg/ml ssCre-R9 for 1 h induced recombination in 65% of the cells (Additional file 1: Figure S1).

Table 1 EC50 values of Cre recombinase

Cre recombinase	$EC50^{a}_{app}$	$EC50^{b}_{real}$
ssCre-R9	42.7 ± 3.8 μg/ml (1.07 μM)	35.9 ± 3.2 μg/ml (0.9 μM)
ssCre-dR9	37.2 ± 4.1 μg/ml (0.93 μM)	31.2 ± 3.4 μg/ml (0.78 μM)
TAT-Cre (A company)	-	46.6 ± 6.0 μg/ml (1.14 μM)
Cre-R11 (B company)	-	NA[c]

[a]Four independent experiments were performed (n = 4; mean ± SD). [b]$EC50_{real}$ is defined as $EC50_{real} = EC50_{app}$ x the conjugation efficiency. [c]NA, not available. The transduction activity of Cre-R11 was too low for $EC50_{real}$ to be measured.

ssCre-R9 was non-toxic to the cells at concentrations up to 160 μg/ml (Additional file 2: Figure S2).

It has been shown that the *d-isomer* R9 peptide labeled with a fluorescent probe had a five-fold lower Km value than the *l-isomer* R9 peptide for cellular uptake [7]. However, the cellular uptake mechanisms of CPPs vary depending on the size and nature of the cargos [24], and it has not been rigorously examined yet whether the *d-isomer* polyarginine CPPs works better than the *l-isomer* polyarginine CPPs for the delivery of large-size protein cargos [20]. In this study, we conjugated CGSGG-dR9 to Cre by EPL and produced ssCre-dR9 with a similar conjugation efficiency, purity, and overall yield as ssCre-R9. To compare the abilities of ssCre-R9 and ssCre-dR9 to enter cells, we treated human TE671 (loxP-LacZ) cells with ssCre-R9 and ssCre-dR9 and scored the recombination events based on the transduced Cre proteins (Figure 4). At concentrations less than 20 μg/ml, ssCre-R9 and ssCre-dR9 were equally non-effective. However, the dose-dependent increase in recombination obtained with ssCre-dR9 was steeper than that obtained with ssCre-R9, and the $EC50_{app}$ values for ssCre-dR9 and ssCre-R9 were 37.2 μg/ml and 42.7 μg/ml, respectively (Table 1). We consistently observed that the EC50 of ssCre-dR9 was at least 10% lower than that of ssCre-R9, although the difference in the EC50 values between ssCre-dR9 and ssCre-R9 was not as marked as the difference in the Km values between the *d-isomer* R9 peptide and the *l-isomer*

R9 peptide. Wender et al. measured the initial rates of cellular uptake of the peptides at 3°C for 240 s by FACS after trypsin treatment [7], and this method cannot differentiate the peptides delivered to the cytosol from those incorporated into the plasma membrane or trapped in endosomes. These differences in the assay methods and in the size and nature of the cargos may explain the discrepancy in the extents of the effects of the *d-isomer* on the efficiency of cellular uptake. Nonetheless, these results clearly demonstrate that the *d-isomer* polyarginine is more efficient than the *l-isomer* polyarginine even for the delivery of large-size protein cargos.

The guanidinium group of arginine-rich CPPs forms a bidentate bond with the negative-charged lipid components of a membrane and generates a negative Gaussian membrane curvature, which induces the temporal and dynamic topological transformation of the membrane structure to the *mesh phase* [10,25]. CPPs have been proposed to translocate into the inside surface of a membrane via an adaptive translocation during this phase transition. Therefore, the engineering of the membrane curvature may be a strategy for the promotion of the cellular uptake of CPPs. Indeed, it was found that epsin N-terminal peptide 1–18 (EpN18), a peptide that induces a positive curvature, accelerates the translocation of R8 peptides into cells [26]. However, the effect of membrane curvature on the cellular uptake of CPPs with large-size protein cargos has not yet been examined. In this study, we treated human TE671 (LoxP-LacZ) cells with ssCre-R9 and ssCre-dR9 in the presence of EpN18 and scored the recombination events (Figure 5). EpN18 promotes recombination in a dose-dependent manner when used at a concentration near the EC50 of ssCre-R9 or ssCre-dR9, 40 μg/ml. The acceleration effect of EpN18 started to be observed at 10 μM, and recombination was observed in 91% of cells exposed to 80 μM EpN18 and 40 μg/ml ssCre-dR9. The acceleration effect of EpN18 was also observed in the cells treated with a suboptimal concentration of ssCre-R9 and ssCre-dR9, 20 μg/ml. In this case, however, the acceleration effect of EpN18 became obvious at higher concentrations, such as 40 and 80 μM. Taken together, these results demonstrate that EpN18 is able to promote the cellular uptake of polyarginine CPPs with large-size protein cargos. Because EpN18 is not toxic to cells [26], EpN18 may be useful as a booster for the protein transduction of ssCre-R9 or ssCre-dR9 to cell types that are not amenable to protein transduction.

The visualization of the cells transduced with therapeutic cell-permeable proteins will be highly advantageous for post-treatment evaluation. As the first step to explore this possibility, we synthesized dR9-biotin [NH2-Cys-Gly-Ser-Gly-Gly-D (Arg)9-Gly-Gly-Lys (biotin)-amide] and conjugated dR9-biotin to Cre by EPL. ssCre-dR9-biotin was produced with a similar purity and overall yield as ssCre-dR9. We found little difference in the transduction efficiency between ssCre-dR9 and ssCre-dR9-biotin. To visualize the ssCre-dR9-biotin delivered into cells, the cells were treated with ssCre-dR9-biotin for 5 h, washed with PBS, fixed with paraformaldehyde, and then stained with avidin-FITC. Images taken using a confocal laser scanning microscope showed both cytoplasmic and nuclear staining with a distinct pattern (Figure 6). A previous study raised the possibility that cell fixation leads to the artifactual redistribution of CPP [27]. Therefore, it is unlikely that ssCre-dR9-biotin has the same subcellular localization in living cells demonstrated by avidin-FITC staining in this study. Nonetheless, these results clearly demonstrated the applicability of EPL for the preparation of cell-permeable proteins with biotin for cell imaging. For *in vivo* imaging of whole animals or cells, it will be more advantageous to use CPPs with fluorophores instead of biotin. It will be interesting to devise a near-infrared (NIR) fluorophore that can be located near CPP without hampering the delivery and activity of proteins in cells.

Conclusions

In this study, we devised an EPL-based technique to conjugate synthetic polyarginine CPPs to Cre protein expressed in *E. coli* without impairing the enzymatic activity of Cre. The ligation product, ssCre-R9, exhibited comparable or

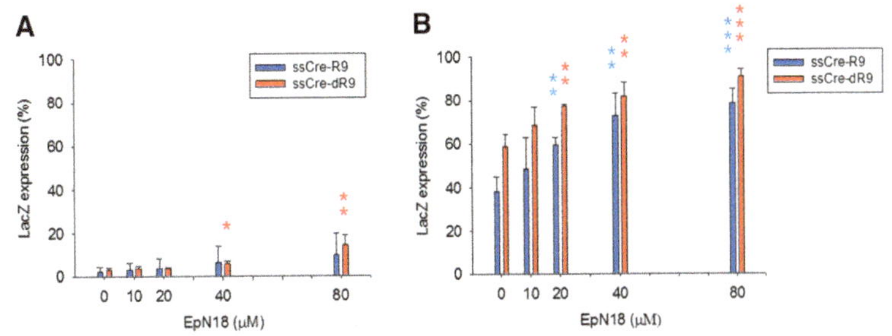

Figure 5 Enhancement of the transduction of ssCre-R9 and ssCre-dR9 by EpN18. Human TE671 (LoxP-LacZ) cells were exposed to either 20 μg/m **(A)** or 40 μg/ml **(B)** Cre in the presence of varying amounts of EpN18 for 5 h. The cells were further incubated with the serum-containing media for a total of 48 h, and the expression of LacZ was then examined by X-gal staining (*t*-test; * $P < 0.05$, ** $P < 0.01$, *** $P < 0.001$). Error bars, mean ± SD (n = 3).

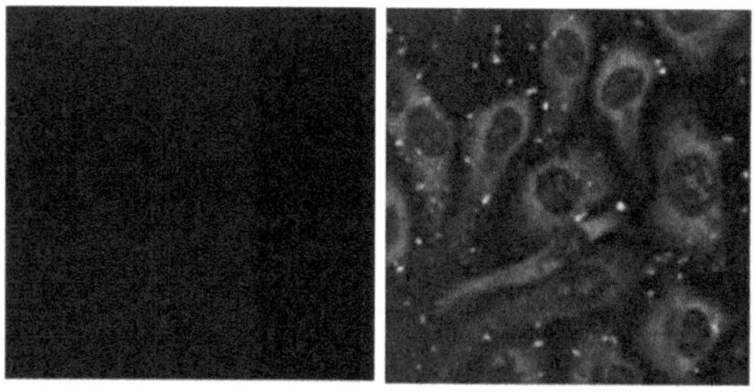

Figure 6 Cellular imaging of the internalized ssCre-dR9-biotin. Human TE671 cells were treated with 80 μg/ml ssCre-dR9-biotin for 5 h (ssCre-dR9-biotin) or left untreated (Control). The cells were fixed and then sequentially incubated with avidin-FITC, anti-avidin-biotin antibody, and avidin-FITC. The fluorescence images were taken under a confocal laser scanning microscope.

even better transduction activity compared with the recombinant TAT-Cre and Cre-R11 fusion proteins directly expressed in *E. coli*. Because all of the steps of EPL are designed to proceed in the presence of high salt concentrations, this approach avoids the technical hurdle caused by the decrease in the solubility of polyarginine fusion proteins when expressed in *E. coli*. In addition, this EPL-based technique allowed us to prepare ssCre-dR9 and ssCre-dR9-biotin containing poly-*D*-arginine and biotin as part of the CPPs. ssCre-dR9 resulted in improved protein transduction compared with Cre-R9, and ssCre-dR9-biotin allowed the visualization of the cells transduced with ssCre-dR9-biotin. The findings demonstrated the feasibility of this EPL-based technique for the preparation of cell-permeable proteins with higher transduction efficiency and more elaborate properties for use in more sophisticated applications, such as cell imaging. EPL should find broad use for the introduction of cell permeability to therapeutic proteins, such as transcription factors or antibodies.

Methods

Peptides

CGSGG-R9 [NH2-Cys-Gly-Ser-Gly-Gly-(Arg) 9-amide], CGSGG-dR9 [NH2-Cys-Gly-Ser-Gly-Gly-D (Arg) 9-amide], EpN18 [26], and dR9-biotin [NH2-Cys-Gly-Ser-Gly-Gly-D (Arg) 9-Gly-Gly-Lys (biotin)-amide] were synthesized and provided by Peptron Inc. The peptides were purified by high-performance liquid chromatography to a purity >90% and confirmed by mass analysis.

Preparation of the Cre-intein fusion protein

Cre recombinase DNA was obtained from pBS185CMV-Cre (Addgene plasmid 11916) through two amplification steps, first using the forward primer 5′-GAG CAT ATG TCC AAT TTA CTG ACC GTA CAC-3′ and the reverse primer 5′-GTA CTC GAG ATC GCC ATC TTC CAG GAC-3′ and then using the forward primer 5′-GAG CAT ATG TCC AAT TTA CTG ACC GTA CAC-3′ and the reverse primer 5′-GCG TGC TCT TCC GCA GTA CTC GAG ATC GCC ATC-3′ [28]. The amplified DNA was inserted into the Nde I and Sap I sites on the pTWIN_His vector (Jena Bioscience) to construct an expression plasmid for a recombinant fusion protein of Cre and mini-intein (derived from Mycobacterium Xenopi gyrase A, Mxe GyrA) combined with a 6× His-tag. Leu-Glu-Tyr was inserted between Cre and intein to improve the intein-mediated cleavage. The pTWIN_Cre plasmid encoding the Cre-intein-His fusion protein was transformed into *E. coli* BL21 (DE3) cells. Then, 50-ml LB medium overnight culture were inoculated into 1 L of LB medium with ampicillin (50 μg/ml) and grown at 37°C. When OD_{600nm} reached to 0.6, yeast extract was added to 1% and then the expression of Cre-intein-His was induced by 1 mM isopropyl-β-D-thiogalactopyranoside (IPTG) at 37°C for 3 h. The *E. coli* cells were harvested by centrifugation and resuspended with 5× packed cell volume of lysis buffer (20 mM NaH_2PO_4, 300 mM NaCl, and 10 mM imidazole, pH 8.0), treated with 1 mg/ml lysozyme on ice for 30 min, and then lysed via sonication. The soluble supernatant was collected by centrifugation at 10,000 × g and mixed with 10 ml of the 50% slurry of Ni-NTA resin (Qiagen) at 4°C for 1 h. The resin slurry was packed into a column and washed with five column volumes of washing buffer (50 mM NaH_2PO_4, 300 mM NaCl, and 50 mM imidazole, pH 8.0). The Cre-intein-His protein was eluted from the column with elution buffer (50 mM NaH_2PO_4, 300 mM NaCl, and 250 mM imidazole, pH 8.0). The fractions containing the Cre-intein-His protein were combined and dialyzed against cleavage buffer (50 mM NaH_2PO_4 and 250 mM NaCl, pH 8.0) at 4°C overnight.

Intein-mediated cleavage and ligation of Cre-thioester with polyarginine polypeptides

The intein-mediated cleavage to generate the Cre-thioester and the ligation of the Cre-thioester with polyarginine polypeptides were performed at the same time. The Cre-intein-His protein in cleavage buffer (50 mM NaH_2PO_4 and 250 mM NaCl, pH 8.0) was incubated with 50 mM 2-mercaptoethanesulfonate (MESNA) and a 10-fold molar excess of CGSGG-R9 or CGSGG-dR9 at 4°C for 24 h. To separate Cre-R9 from the Cre-intein-His or intein-His proteins, the reaction mixture was loaded onto a Ni-NTA column equilibrated with lysis buffer, and the flow-through fraction was collected. Cre-R9 was treated with dithiothreitol (DTT) to a final concentration of 10 mM, incubated on ice for 30 min, and then concentrated using a Pierce concentrator 9K (Thermo) at 4°C. Any leftover MESNA and peptides in Cre-R9 were removed by a spin column packed with Sephadex G25 Fine resin

GE Healthcare) equilibrated with storage buffer (50 mM NaH_2PO_4, 300 mM NaCl, and 1 mM DTT, pH 8.0).

Cell culture

Human TE671 (LoxP-LacZ) cells were purchased from Allele Biotech. The cells were cultured at 37°C in a 5% humidified CO_2 incubator in Dulbecco's Modified Eagle's medium (DMEM) with 10% fetal bovine serum, 100 U/ml penicillin G, and 100 μg/ml streptomycin.

Protein transduction assay

TAT-Cre and Cre-R11 were purchased commercially. To measure the rate of Cre-R9 internalization and subsequent recombination, human TE671 (LoxP-LacZ) cells were plated onto 24-well plate at 7×10^3 cells/well. After one day, the cells were washed once with OPTI-MEM (Gibco) and incubated with 300 μl of OPTI-MEM containing Cre proteins for 5 h. The medium was changed to DMEM containing 10% FBS, and the cells were then further incubated for 43 h. The recombination at LoxP sites by Cre proteins in the cells was measured by a LacZ enzyme assay. The cells were washed twice with cold PBS and fixed with 300 μl of a fixation solution (1% formaldehyde and 0.2% glutaraldehyde in PBS) at room temperature for 5 min. The cells were washed three times with PBS and incubated with 300 μl of an X-gal staining solution (4 mM K-ferrocyanide, 4 mM K-ferricyanide, 2 mM $MgCl_2$, and 0.4 mg/ml X-gal in PBS) at 37°C overnight.

In vitro assay of Cre activity

Cre protein was added to 50 μl of reaction buffer (50 mM Tris–HCl, 33 mM NaCl, and 10 mM $MgCl_2$, pH 7.5) containing 0.3 μg of linearized pGEM-T (floxed-Km^r) and incubated at 37°C for 1 h. Cre was then heat-inactivated by incubation at 70°C for 10 min. After isolation from proteins by phenol extraction and EtOH precipitation, DNA was analyzed by agarose gel electrophoresis.

Fluorescence imaging

Human TE671 cells grown on gelatin-coated glass cover-slips were incubated with 80 μg/ml ssCre-dR9-biotin in OPTI-MEM for 5 h. The cells were washed three times with PBS and fixed in 3.7% (v/v) paraformaldehyde for 15 min at room temperature. The cells were sequentially incubated with avidin-FITC, anti-avidin-biotin antibody, and avidin-FITC each for 30 min at room temperature. After mounting with 90% glycerol and 0.1% *o*-phenylenediamine in PBS, cellular images were obtained using a confocal laser scanning microscope.

Additional files

Additional file 1: Figure S1. One hour of incubation with exogenous ssCre-R9 is sufficient for Cre-mediated recombination in human TE671 (LoxP-LacZ) cells. Cells were exposed to 60 μg/ml of ssCre-R9 in OPTI-MEM for the indicated time. Cells were further incubated with the serum-containing media for the total 48 h, and then the expression of LacZ was examined by X-gal staining (n = 4; mean ± SD).

Additional file 2: Figure S2. ssCre-R9 does not affect cell viability. Human TE671 (LoxP-LacZ) cells were incubated with ssCre-R9 in OPTI-MEM for 5 h and further incubated with the serum-containing media for a total of 48 h. Cell viability was measured by MTT assay (n = 3; mean ± SD).

Abbreviations

R12: Arginine 12-mer; CPP: Cell-penetrating peptide; EGFP: Enhanced green fluorescent protein; EPL: Expressed protein ligation; EpN18: Epsin N-terminal peptide 1–18; NLS: The nuclear localization sequence; ssCre-R9: Semisynthetic Cre-R9.

Competing interests

The authors are affiliated with Hankuk University of Foreign Studies (HUFS). The following patents have been applied for by HUFS: Patent Application Publication No. 10-2014-0118121. PCT/KR2014/008796. The authors declare that they have no other competing interests.

Authors' contributions

SL performed all of the experiments and wrote/co-edited the manuscript. HK conceived, organized, and directed the research study and wrote/co-edited the manuscript. All of the authors read and approved the final manuscript.

Acknowledgments

We gratefully acknowledge the gift of pBS185 CMV-Cre from Brian Sauer and of pGEM-T (floxed-Km^r) from Hyune Hwan Lee. This work was supported by the Gyeonggi Regional Research Center (GRRC) Program of Gyeonggi Province (GRRC-HUFS-2014-B04).

References

1. Heitz F, Morris MC, Divita G. Twenty years of cell-penetrating peptides: from molecular mechanisms to therapeutics. Br J Pharmacol. 2009;157:195–206.
2. Green M, Loewenstein PM. Autonomous functional domains of chemically synthesized human immunodeficiency virus Tat trans-activator protein. Cell. 1988;55:1179–88.
3. Frankel AD, Pabo CO. Cellular uptake of the tat protein from human immunodeficiency virus. Cell. 1988;55:1189–93.
4. Vives E, Brodin P, Lebleu B. A truncated HIV-1 Tat protein basic domain rapidly translocates through the plasma membrane and accumulates in the cell nucleus. J Biol Chem. 1997;272:16010–7.
5. Gautam A, Singh H, Tyagi A, Chaudhary K, Kumar R, Kapoor P, et al. CPPsite: a curated database of cell penetrating peptides. Database (Oxford). 2012;2012:bas015.
6. Montrose K, Yang Y, Sun X, Wiles S, Krissansen GW. Xentry, a new class of cell-penetrating peptide uniquely equipped for delivery of drugs. Sci Rep. 2013;3:1661.
7. Wender PA, Mitchell DJ, Pattabiraman K, Pelkey ET, Steinman L, Rothbard JB. The design, synthesis, and evaluation of molecules that enable or enhance cellular uptake: peptoid molecular transporters. Proc Natl Acad Sci U S A. 2000;97:13003–8.
8. Futaki S, Suzuki T, Ohashi W, Yagami T, Tanaka S, Ueda K, et al. Arginine-rich peptides. An abundant source of membrane-permeable peptides having potential as carriers for intracellular protein delivery. J Biol Chem. 2001;276:5836–40.
9. Madani F, Lindberg S, Langel U, Futaki S, Gräslund A. Mechanisms of cellular uptake of cell-penetrating peptides. J Biophys. 2011;2011:414729.
10. Sakamoto K, Aburai K, Morishita T, Sakai K, Sakai H, Abe M, et al. Bioinspired mechanism for the translocation of peptide through the cell membrane. Chem Lett. 2012;41:1078–80.
11. Takayama K, Hirose H, Tanaka G, Pujals S, Katayama S, Nakase I, et al. Effect of the attachment of a penetration accelerating sequence and the influence

of hydrophobicity on octaarginine-mediated intracellular delivery. Mol Pharm. 2012;9:1222–30.
12. Zhou H, Wu S, Joo JY, Zhu S, Han DW, Lin T, et al. Generation of induced pluripotent stem cells using recombinant proteins. Cell Stem Cell. 2009;4:381–4.
13. Kim D, Kim CH, Moon JI, Chung YG, Chang MY, Han BS, et al. Generation of human induced pluripotent stem cells by direct delivery of reprogramming proteins. Cell Stem Cell. 2009;4:472–6.
14. Dominguez-Bendala J, Klein D, Ribeiro M, Ricordi C, Inverardi L, Pastori R, et al. TAT-mediated neurogenin 3 protein transduction stimulates pancreatic endocrine differentiation *in vitro*. Diabetes. 2005;54:720–6.
15. Kwon YD, Oh SK, Kim HS, Ku SY, Kim SH, Choi YM, et al. Cellular manipulation of human embryonic stem cells by TAT-PDX1 protein transduction. Mol Ther. 2005;12:28–32.
16. Vargas N, Alvarez-Cubela S, Giraldo JA, Nieto M, Fort NM, Cechin S, et al. TAT-mediated transduction of MafA protein in utero results in enhanced pancreatic insulin expression and changes in islet morphology. PLoS One. 2011;6:e22364.
17. Jo J, Song H, Park SG, Lee SH, Ko JJ, Park JH, et al. Regulation of differentiation potential of human mesenchymal stem cells by intracytoplasmic delivery of coactivator-associated arginine methyltransferase 1 protein using cell-penetrating peptide. Stem Cells. 2012;30:1703–13.
18. Yang WC, Welsh JP, Lee J, Cooke JP, Swartz JR. Solubility partner IF2 Domain I enables high yield synthesis of transducible transcription factors in Escherichia coli. Protein Expr Purif. 2011;80:145–51.
19. Flavell RR, Muir TW. Expressed protein ligation (EPL) in the study of signal transduction, ion conduction, and chromatin biology. Acc Chem Res. 2009;42:107–16.
20. Yu HH, Nakase I, Pujals S, Hirose H, Tanaka G, Katayama S, et al. Expressed protein ligation for the preparation of fusion proteins with cell penetrating peptides for endotoxin removal and intracellular delivery. Biochim Biophys Acta. 2010;1798:2249–57.
21. Le Y, Gagneten S, Tombaccini D, Bethke B, Sauer B. Nuclear targeting determinants of the phage P1 cre DNA recombinase. Nucleic Acids Res. 1999;27:4703–9.
22. Jo D, Nashabi A, Doxsee C, Lin Q, Unutmaz D, Chen J, et al. Epigenetic regulation of gene structure and function with a cell-permeable Cre recombinase. Nat Biotechnol. 2001;19:929–33.
23. Wadia JS, Stan RV, Dowdy SF. Transducible TAT-HA fusogenic peptide enhances escape of TAT-fusion proteins after lipid raft macropinocytosis. Nat Med. 2004;10:310–5.
24. Mishra A, Lai GH, Schmidt NW, Sun VZ, Rodriguez AR, Tong R, et al. Translocation of HIV TAT peptide and analogues induced by multiplexed membrane and cytoskeletal interactions. Proc Natl Acad Sci U S A. 2011;108:16883–8.
25. Mishra A, Gordon VD, Yang L, Coridan R, Wong GC. HIV TAT forms pores in membranes by inducing saddle-splay curvature: potential role of bidentate hydrogen bonding. Angew Chem Int Ed Engl. 2008;47:2986–9.
26. Pujals S, Miyamae H, Afonin S, Murayama T, Hirose H, Nakase I, et al. Curvature engineering: positive membrane curvature induced by epsin N-terminal peptide boosts internalization of octaarginine. ACS Chem Biol. 2013;8:1894–9.
27. Richard JP, Melikov K, Vives E, Ramos C, Verbeure B, Gait MJ, et al. Cell-penetrating peptides. a reevaluation of the mechanism of cellular uptake. J Biol Chem. 2003;278:585–90.
28. Sauer B, Henderson N. Targeted insertion of exogenous DNA into the eukaryotic genome by the Cre recombinase. New Biol. 1990;2:441–9.

22

Does size matter? Separations on guard columns for fast sample analysis applied to bioenergy research

Stefan Bauer* and Ana B. Ibáñez

Abstract

Background: Increasing sample throughput is needed when large numbers of samples have to be processed. In chromatography, one strategy is to reduce column length for decreased analysis time. Therefore, the feasibility of analyzing samples simply on a guard column was explored using refractive index and ultraviolet detection. Results from the guard columns were compared to the analyses using the standard 300 mm Aminex HPX-87H column which is widely applied to the analysis of samples from many biotechnology- and bioenergy-related experiments such as biomass conversions or fermentations.

Results: The 50 mm Rezex RFQ Fast Acid H^+ guard column was able to separate the most common fermentation products (ethanol, acetone, iso- and n-butanol) and promising precursors (furfural and 5-hydroxymethylfurfural) of biofuels and value-added chemicals. Compound profiles in fermentation samples were analyzed with similar accuracy compared to results using the 300 mm column. However, separation of glucose and xylose was not achieved. Nevertheless, it was possible to monitor the consumption of one of the two sugars during fermentation if the other one was absent or remained constant over the course of the experiment. If correct peak integration and interference subtraction was applied, concentration profiles from enzymatic digestibility experiments and even more complex samples (e.g. acetone-butanol-ethanol (ABE) fermentation) were reliably obtained. With the 50 mm guard column, samples were analyzed up to ten-times faster compared to the 300 mm column. A further decrease in analysis time was achieved by using the 30 mm Micro Guard Cation H guard column. This column is especially suitable for the rapid analysis of compounds with long elution times on the standard 300 mm column, such as biofuel-related alcohols (e.g., n-butanol, n-hexanol) and furan- and tetrahydrofuran-type molecules.

Conclusion: Applied to a suitable set of samples, separations on a guard column can give rapid and sufficiently accurate information on compound changes over the course of an experiment. Therefore, it is an inexpensive and ideal tool for processing a large amount of samples, such as in screening or discovery experiments, where detecting relative changes is often sufficient to identify promising candidates for further analysis.

Keywords: Ion exclusion chromatography, Aminex HPX-87H, Rezex RFQ, Micro Guard Cation H, Enzymatic digestibility assay, Glucose, Ethanol, Butanol, Furans, Fermentation

* Correspondence: stefan.bauer@berkeley.edu
Energy Biosciences Institute, University of California, Berkeley, CA 94720, USA

Background

A popular liquid chromatography column used for the analysis of biomass conversion and fermentation products is based on a polymer matrix of polystyrene-divinylbenzene (e.g. Bio-Rad Aminex HPX-87H, Phenomenex Rezex ROA and RFQ, Shodex SH1821, Sigma-Aldrich Supelcogel C610H and others) [1–6]. It is operated in isocratic elution mode enabling the connection to a refractive index detector (RID) which provides universal compound detection. This set up allows for analysis of a wide range of compounds. With a standard 300 mm length, this type of column provides baseline separation of the main cell wall sugars (glucose, xylose and arabinose) of many lignocellulosic biomass feedstocks. This column is also used for the analysis of organic acids, alcohols (e.g., ethanol, n-butanol) and sugar degradation products (e.g., 5-hydroxymethylfurfural, furfural) [7–10]. It only requires acidified water as mobile phase and only minimal sample preparation.

Typical HPLC methods aim to separate complex mixtures in order to individually detect isolated compounds – particularly when a non-specific detector like RID is used. In our experience from supporting researchers in an organization comprising more than 500 scientists dedicated to bioenergy research, many analytical questions can be reduced to identifying the change in a few compounds over the course of the experiment. Furthermore, reducing analysis time is often more important than obtaining precise absolute quantification data. This is especially true for the results from screening experiments involving a larger amount of samples. The detection of relative changes is often sufficient to identify promising candidates for further analysis.

Fast methods targeting individual compounds in complex mixtures have previously been established and include, for example, gas chromatography [11] spectroscopy [12], or enzymatic assays, such as for sugars, sugar alcohols, organic acids, and ethanol [13–17]. Several of these target analytes can even be measured in an automated mode using a biochemistry analyzer (YSI) and results are obtained in minutes [18]. However, all such methods either require costly equipment or individual analysis kits for every single analyte. Therefore, it is tempting to use existing analytical instrumentation (HPLC) and reduce analysis time for higher sample throughput. In this respect, Scarlata and Hyman [6] have successfully shown that reducing the column length to 100 mm (plus a 30 mm guard column) can reduce the analysis time by a factor of five compared to the 300 mm column, but still provide adequate accurate results from a biomass compositional analysis (glucose, xylose, arabinose, acetate, 5-hydroxymethylfurfural, furfural). In our studies, we explored a further reduction of the analysis time by using only a guard column as the simplest and most basic column available. For this purpose, samples generated by various research groups in our institute were analyzed on the standard 300 mm Aminex HPX-87H column and compared to results obtained from an analysis on a guard column (50 mm Rezex RFQ Fast Acid H^+ as well as 30 mm Micro Guard Cation H).

Results and discussion

A short column length is often used for faster separation especially in conjunction with mass spectrometric detection. These shorter columns have usually smaller particle sizes (3.5 μm and lower) which greatly improve resolution. However, columns like the Aminex HPX-87H operate in a mixture of size- and ion-exchange/exclusion. They consist of a resin with a rather large particle size (8–9 μm). As a result, a reduction in column length significantly lowers column efficiency. For example, a reduction of the column length to 50 mm results in practically unresolved glucose and xylose (see peak 1 and 2 in Fig. 6a). These types of columns can therefore not be directly compared to other columns. Despite the lower separation efficiency, the shorter columns can still provide adequate chromatographic resolution when used with appropriate samples in which compounds co-eluting with the target analytes are either not present or their concentration is sufficiently low so they do not significantly interfere with quantification.

Figure 1 shows the elution profile of some of the most common fermentation products (ethanol, acetone, iso-butanol, n-butanol) and sugar degradation products (5-hydroxymethylfurfural (5-HMF), furfural) on the 50 mm Rezex RFQ Fast Acid H^+ guard column ("50 mm guard column") operated at different temperatures at a flow rate of 1.0 mL/min. The degradation products 5-HMF and furfural are promising precursors for the generation of biofuels and value-added chemicals [3, 19, 20]. Good separation of all these compounds is achieved at a column temperature of 30 °C (Fig. 1a). At this temperature, ethanol (peak 1) and acetone (peak 2) are almost baseline separated. By increasing the column temperature the alcohols elute later and 5-HMF (peak 5) and furfural (peak 6) elute earlier. The retention time of acetone (peak 2) is practically not affected by temperature (Fig. 1a-c). Ethanol/acetone (peak 1/peak 2) separation is deteriorated at a higher temperature and 5-HMF (peak 5) is co-eluting with n-butanol (peak 4) at 55 °C (Fig. 1b) and with iso-butanol (peak 3) at 80 °C (Fig. 1c). The column temperature must therefore be chosen based on the sample composition so that target analytes do not overlap. For example, as mentioned above, a temperature of 30 °C has to be applied for successful separation of ethanol and acetone. This column temperature is also beneficial for the faster elution of iso-and n-butanol (Fig. 1a). In contrast, 5-HMF and furfural are eluted faster at a higher temperature. However, depending on which butanol isomer

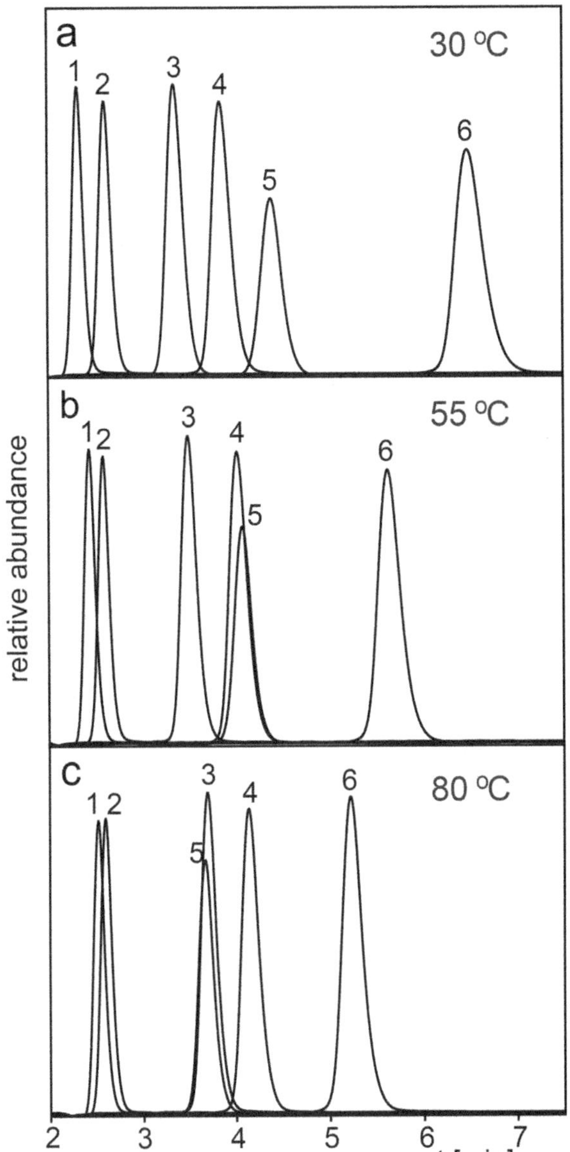

Fig. 1 Separation of ethanol (peak 1), acetone (peak 2), iso-butanol (peak 3), n-butanol (peak 4), 5-HMF (peak 5) and furfural (peak 6) on the 50 mm guard column at a column temperature of (**a**) 30 °C, (**b**) 55 °C and (**c**) 80 °C.HPLC parameters: 50 mm × 7.8 mm RFQ Fast Acid H^+ column, mobile phase 5 mM sulfuric acid, flow rate 1.0 mL/min, refractive index detection

is present, the column temperature needs to be correctly chosen since 5-HMF coelutes either with n-butanol at 55 °C (Fig. 1b) or iso-butanol at 80 °C (Fig. 1c).

Using a 20 μL injection volume, the determined linear calibrations for the compounds under study ranged from 0.005-0.02 mg/mL to 50 mg/mL (20 mg/mL for glucose and xylose) with RI and UV detection (Table 1). UV detection (280 nm) resulted in an about 100-fold increased sensitivity for 5-HMF and furfural compared to RI detection (calibration range 50 ng/mL – 0.25 mg/mL).

Table 1 Calibration results for each compound

Compound	Detection	Linear quantification range	Regression coefficient
Ethanol	RID	0.01 – 50 mg/mL	0.9995
Acetone	RID	0.01 – 50 mg/mL	0.9996
	UV 265 nm	0.01 – 50 mg/mL	0.9999
	UV 285 nm	0.02 – 50 mg/mL	0.9995
iso-Butanol	RID	0.005 – 50 mg/mL	0.9999
n-Butanol	RID	0.005 – 50 mg/mL	0.9997
5-HMF	RID	0.005 – 50 mg/mL	0.9999
	UV 280 nm	0.00005 – 0.25 mg/mL	0.9999
Furfural	RID	0.005 – 50 mg/mL	0.9995
	UV 280 nm	0.00005 – 0.25 mg/mL	0.9999
Glucose	RID	0.005 – 20 mg/mL	0.9999
Xylose	RID	0.005 – 20 mg/mL	0.9999

Chromatography conditions: 50 mm × 7.8 mm RFQ Fast Acid H^+ kept at 30 °C, flow rate: 1.0 ml/min of 5 mM sulfuric acid, 20 μL injection volume; Lower limit of the calibration range was determined as the concentration level with a relative standard deviation (RSD) <10 % (n = 5)

Since 80 °C is close to the maximum operating temperature of the column (85 °C), we chose a column temperature of 55 °C for most of the following experiments. This is good balance between fast analysis time and column life time.

Application to analysis of n-butanol and iso-butanol production during fermentations

These two alcohols elute later than most other components present in the fermentation media. Fig. 2a shows a representative chromatogram from an *E.coli* fermentation producing n-butanol (peak 1) which eluted at 3.7 min. Fig. 2b shows the determined n-butanol production of three *E. coli* strains after incubation for 24 h, 48 h and 72 h. The n-butanol profile obtained with the 50 mm guard column was very similar compared to the 300 mm Bio-Rad HPX-87H column ("300 mm column"). Values reached 90.9-98.0 % of the values obtained with the longer column (except for the BW25113 values which reached only 80.7-84.1 % due to the low concentration affecting accuracy). The DH1 strain is able to produce more n-butanol than BW25113, which almost showed no significant n-butanol levels. Usually, BW25113 produces about half the concentration of DH1 (Niwen Kong, personal communication). Unexpectedly, the BW25113 strain used in this experiment did not perform as expected. The deletion of AcrB efflux pumps in DH1Δ*acrB* lead to even higher n-butanol levels. AcrB pumps are large proteins and are used to actively secret n-butanol from the cell. However, they can be considered "energy

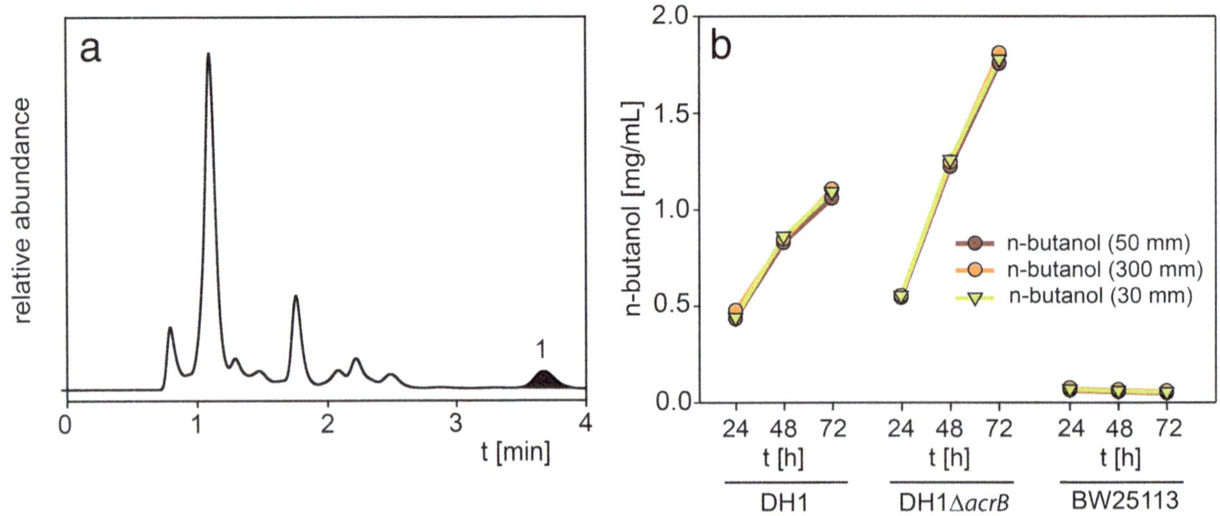

Fig. 2 Representative chromatogram of the supernatant from *Escherichia coli* strain DH1 fermentation analyzed with (**a**) the 50 mm guard column showing n-butanol (peak 1) elution, and (**b**) n-butanol concentration of the supernatant from *E. coli* strains DH1, DH1Δ*acrB* and BW25113 fermentations after 24 h, 48 h and 72 h analyzed with the 50 mm guard column, 300 mm column and 30 mm guard column. HPLC parameters: column temperature 55 °C, mobile phase 5 mM sulfuric acid, flow rate 1.0 mL/min (0.6 mL/min for 300 mm column), refractive index detection

drainers" since they use a big portion of the cellular resources needed for n-butanol synthesis. At the levels measured here, n-butanol is not toxic for the *E. coli* cells and therefore the AcrB pumps present in the DH1 strain are not a "de-toxification"-advantage over the DH1Δ*acrB* strain which has more energy resources available for n-butanol production.

A similar chromatogram was obtained for iso-butanol production from *E. coli* strains sFAB5441 and sFAB5692 (Fig. 3a). Iso-butanol (peak 1) eluted at 3.4 min in a region where no interference from other media components were expected. The values measured with the 50 mm guard column were again almost identical to the ones obtained with the 300 mm column (in the range of 96.7-100.4 %, Fig. 3b). In this experiment, the decrease of iso-butanol after 48 h for the sFAB 5441 strain was explained by evaporation due to a leak in the cap of the flask.

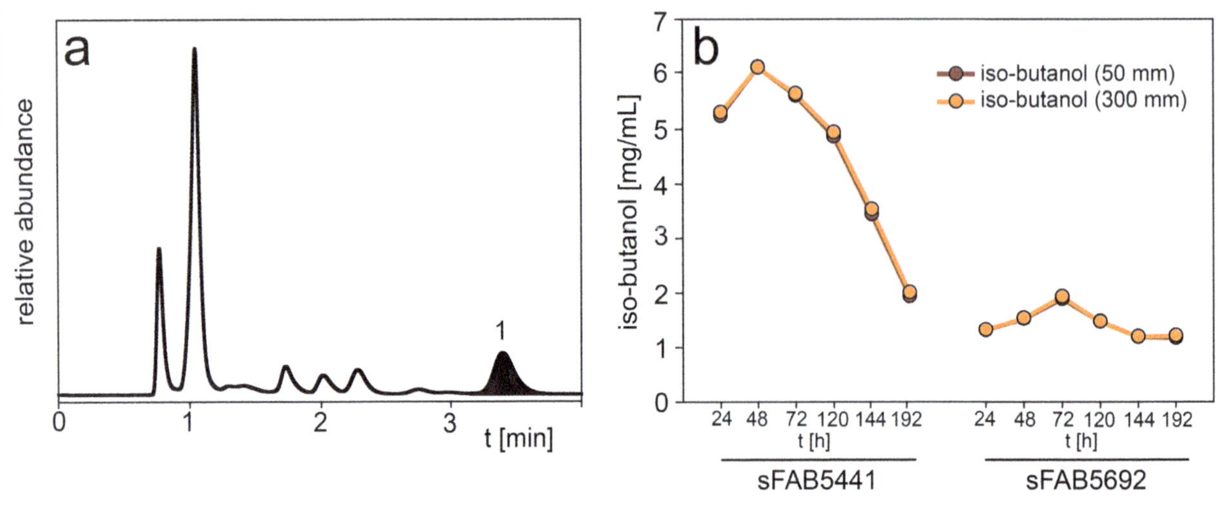

Fig. 3 Representative chromatogram of the supernatant from *Escherichia coli* strain sFAB5692 fermentation analyzed with (**a**) the 50 mm guard column showing iso-butanol (peak 1) elution, and (**b**) iso-butanol concentration profile of the supernatant from *E. coli* strains sFAB5441 and sFAB5692 fermentation between 24 h and 192 h analyzed with the 50 mm guard column and 300 mm column. HPLC parameters: column temperature 55 °C, mobile phase 5 mM sulfuric acid, flow rate 1.0 mL/min (0.6 mL/min for 300 mm column), refractive index detection

The analysis of n- and iso-butanol was performed with a 4 min total run time and this is about 9–10 times faster compared to the standard 300 mm column.

Application to analysis of 5-hydroxymethylfurfural (5-HMF) and furfural

5-HMF and especially furfural are compounds with long elution times on the standard 300 mm HPX-87H column (at a flow rate of 0.6 mL/min and 50 °C about 34 and 52 min, respectively). Experiments requiring the analysis of these compounds often include determination of the degree of their removal or reduction during membrane detoxification of hydrolysate [21] or the rate of sugar degradation during pretreatment or conversion steps [9, 10]. For the purpose of this study, three biomass hydrolysates prepared under different severity conditions were analyzed. Fig. 4a depicts the elution profiles for the two compounds on the 50 mm guard column. Hydrolysate 1 was prepared with a lower sulfuric acid concentration (0.5 %) and incubated at a lower temperature (158 °C) causing less sugar degradation. Therefore, it shows a lower 5-HMF (peak 1) and a lower furfural (peak 2) concentration compared to the other two hydrolysates which were prepared with a higher acid concentration (1 % and 1.5 %, respectively) and a higher temperature (180 °C and 190 °C, respectively). As illustrated in Fig. 4b, the 5-HMF analysis results with the 50 mm guard column compared to the 300 mm column were lower for hydrolysate 1 (88.2 %, 0.15 vs. 0.17 mg/mL), lower for hydrolysate 2 (93.6 %, 1.31 vs 1.40 mg/mL) and lower for hydrolysate 3 (84.1 %, 0.74 vs. 0.88 mg/mL). However, the furfural amounts determined with the 50 mm guard column were in the range of 96.8 %-103.8 % of the 300 mm column results (1.49 vs. 1.54 mg/mL, 1.81 vs. 1.87 mg/mL, and 2.46 vs. 2.37 mg/mL, respectively) and can be considered very comparable. In conclusion, the different pretreatment severity profiles were adequately reflected by analysis with the 50 mm guard column.

Application to analysis of ethanol production and xylose and glucose consumption during fermentations

When glucose and xylose are present in a mixture, their accurate determination using the 50 mm guard column is challenging since the separation of these two sugars is practically not possible. However, the analysis can be performed when either one of the two is present in high abundance or one of the two monosaccharides concentrations is constant over the course of the experiment. Ethanol eluted at 2.2 min which was close to other fermentation media compounds but was still sufficiently isolated for selective measurement in these experiments (Fig. 5a, Fig. 6a-c).

The consumption of xylose and production of ethanol of a modified *Saccharomyces cerevisiae* strain analyzed with the 50 mm guard column and 300 mm column is shown in Fig. 5b and a representative chromatogram is shown in Fig. 5a. The xylose amounts measured with the 50 mm guard column were in the range of 100.5-105.4 % and the ethanol amounts were in the range of 97.7-101.9 % (except for the two lower concentrations) of the values obtained with the 300 mm column. The two lower concentrations of ethanol showed 255.6 % (0.23 vs. 0.09 mg/mL) and 115.6 % (0.89 vs. 0.77 mg/mL) deviation most likely due to interfering co-eluting compounds at these lower concentrations. Overall, the consumption/production profile measured with the 50 mm guard column is very similar to the one obtained with the 300 mm column. However, the total run time using the 50 mm guard

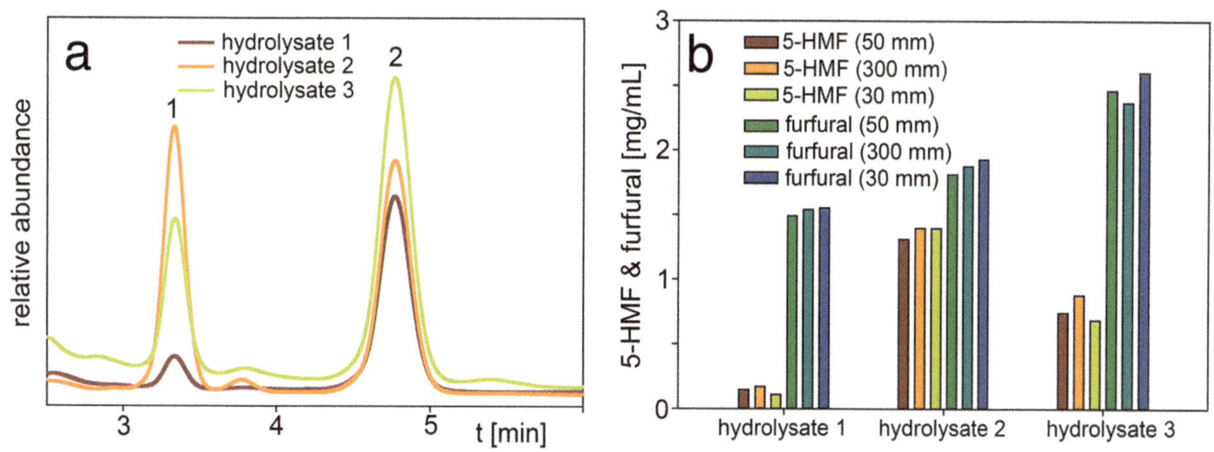

Fig. 4 Chromatogram of hydrolysate 1, hydrolysate 2 and hydrolysate 3 analyzed with (**a**) the 50 mm guard column showing 5-HMF (peak 1) and furfural (peak 2) elution, and (**b**) 5-HMF and furfural concentrations of hydrolysate 1, hydrolysate 2 and hydrolysate 3 analyzed with the 50 mm guard column, 300 mm column and 30 mm guard column. HPLC parameters: column temperature 55 °C, mobile phase 5 mM sulfuric acid, flow rate 1.0 mL/min (0.6 mL/min for 300 mm column), refractive index detection

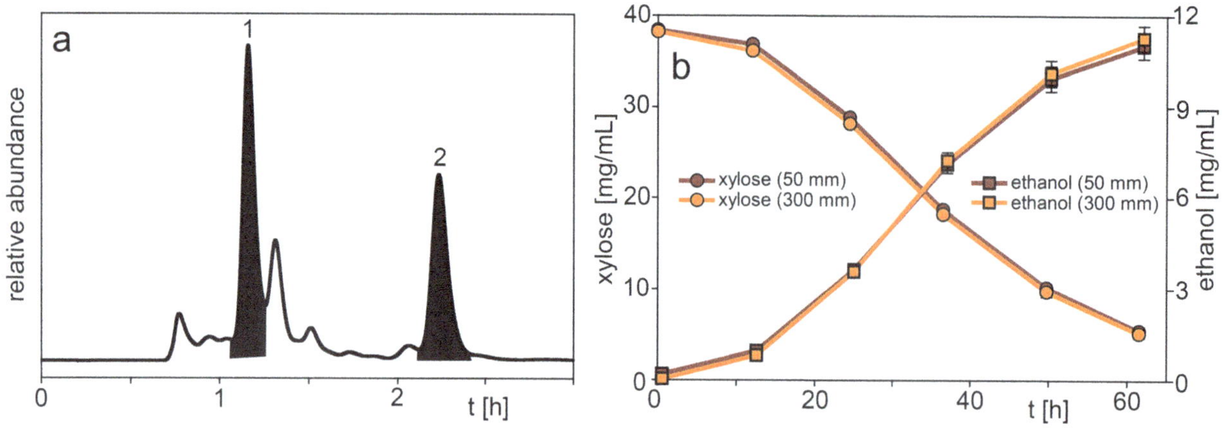

Fig. 5 Representative chromatogram of the supernatant from a *Saccharomyces cerevisiae* SA-1-X123 strain fermentation analyzed with (**a**) the 50 mm guard column showing xylose (peak 1) and ethanol (peak 2) elution, and (**b**) xylose and ethanol concentration profile of the supernatant from a modified *S. cerevisiae* strain fermentation between 0 h and 62 h analyzed with the 50 mm guard column and 300 mm column. HPLC parameters: column temperature 55 °C, mobile phase 5 mM sulfuric acid, flow rate 1.0 mL/min (0.6 mL/min for 300 mm column), refractive index detection

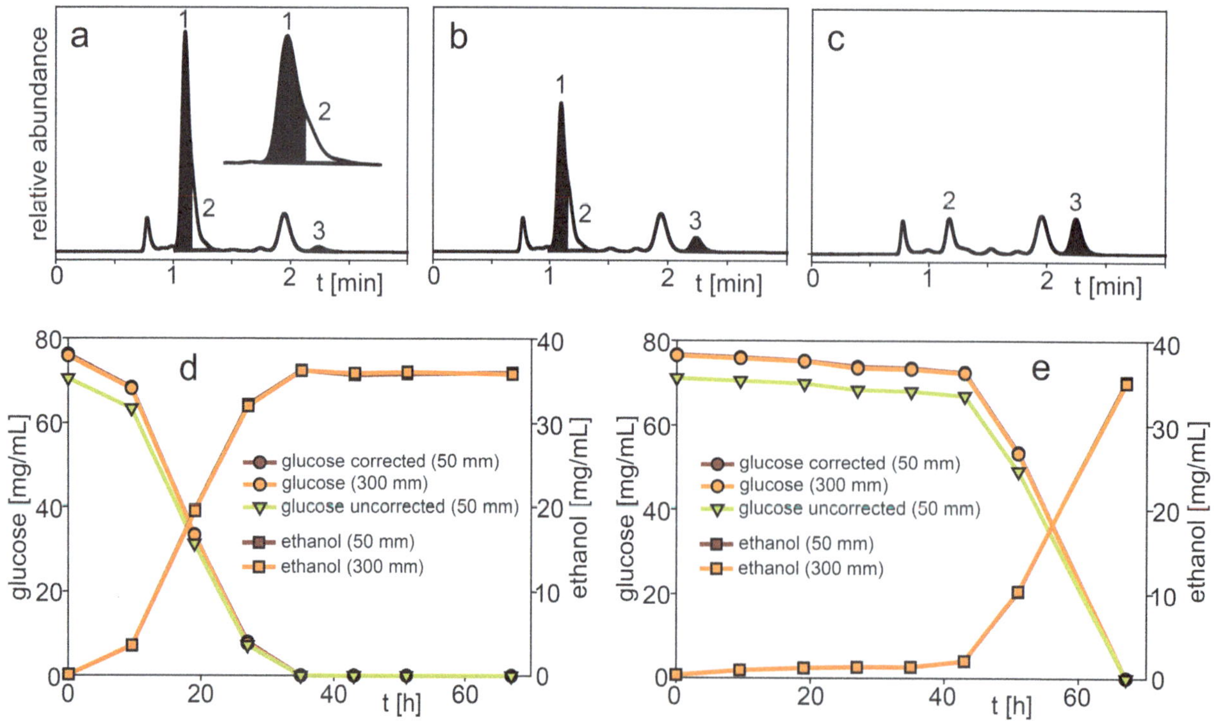

Fig. 6 Chromatogram of the supernatant from *Saccharomyces cerevisiae* SA-1 strain fermentation supplemented with 25 % of pH adjusted hydrolysate 3 analyzed with the 50 mm guard column showing glucose (peak 1), xylose (peak 2) and ethanol (peak 3) elution after (**a**) 42 h, (**b**) 50 h and (**c**) 67 h, and xylose and ethanol concentration profile of the supernatant between 0 h and 62 h from a *S. cerevisiae* SA-1 strain fermentation supplemented with (**d**) 25 % of pervaporation detoxified hydrolysate 3 and (**e**) 25 % of pH adjusted hydrolysate 3, analyzed with the 50 mm guard column and 300 mm column. HPLC parameters: column temperature 55 °C, mobile phase 5 mM sulfuric acid, flow rate 1.0 mL/min (0.6 mL/min for 300 mm column), refractive index detection

column was only 3 min and ethanol elution was completed at 2.5 min. This is about 10 times faster compared to the 300 mm column.

In another experiment, *S. cerevisiae* SA-1 was grown in a medium supplemented with 25 % hydrolysate 3 that was either pH adjusted or detoxified by membrane pervaporation which reduced the concentration of toxic compounds such as furfural, formic and acetic acid [21]. The hydrolysate was obtained by a dilute acid and heat pretreatment of *Miscanthus* biomass in order to release hemicellulosic sugars and improve cellulose digestibility. During this process, the harsh conditions lead to sugar and lignin degradation resulting in the production of a fleet of fermentation inhibitors like furans, phenolics and organic acids which could also interfere with the analysis [22, 23]. Representative chromatograms of the sample with pH-adjusted hydrolysate at time points 42 h, 50 h and 67 h analyzed with the 50 mm guard column are shown in Fig. 6a-c. A decrease of glucose (peak 1) and increase of ethanol (peak 3) can be observed over the time course whereas the remaining chromatogram is mostly the same. The inset in Fig. 6a shows the close elution of glucose (peak 1) and xylose (peak 2) resulting in a peak with a shoulder making accurate integration difficult. Fig. 6d and 6e show the measured glucose and ethanol concentrations over the course of the experiments with the membrane-detoxified hydrolysate and the pH-adjusted hydrolysate, respectively. The values for "glucose uncorrected" were obtained when the attempt was made to integrate glucose separately from the closely-eluting xylose peak (Fig. 6a inset). The determined "glucose uncorrected" concentrations for experiments with the pH-adjusted and with the detoxified hydrolysate were 91.0-93.6 % and 91.6-93.0 %, respectively, of the concentrations obtained when analyzed with the 300 mm column. This indicates that xylose interfered with a correct glucose area determination. The ethanol concentrations were almost identical with 99.4-100.6 % and 99.5-100.2 %, respectively, compared to the results of the 300 mm column. When glucose and xylose were integrated together as one peak and the area of the xylose peak (peak 2, Fig. 6c) of the 67 h time point was subtracted (where glucose concentration was close to 0), "glucose corrected" values were obtained which were closer to the results of the 300 mm column (94.8-100.6 % and 99.5-99.9 %, respectively). Even without glucose correction, the effect of the toxic hydrolysate on delaying the glucose fermentation and production of ethanol until around 42 h into the fermentation was clearly observed with the 50 mm column (Fig. 6e).

Application to analysis of glucose release from enzymatic digestibility assays of pretreated biomass

The enzymatic release of glucose from cellulose after biomass pretreatment is often performed not only to determine of the effectiveness of the pretreatment itself but also for the evaluation of the hydrolytic potential of new enzymes. A commonly used method is the pretreatment of biomass with dilute acid (e.g. sulfuric acid) at higher temperatures (140–200 °C) which leaves behind a solid cake consisting of mainly lignin and decrystallized cellulose [19, 24, 25]. After washing and/or neutralization this cake is incubated with cellulose-degrading enzymes in either citrate or acetate buffer at pH 5 in order to release glucose [26]. As shown in the chromatogram of the supernatant after enzymatic digestion of "pretreated *Miscanthus* biomass" (Fig. 7a), citrate/citric acid (peak 1) of the citrate buffer is interfering with glucose (peak 2) at lower and higher concentrations on the 50 mm guard column. Furthermore, sorbitol (peak 3) derived from the commercial enzyme mixture (peak 3) also interferes, but to a much lower extent. Using the 300 mm column, citric acid (peak 1), glucose (peak 2) and sorbitol (peak 3) can be separated (Fig. 7b). As observed earlier with the glucose/xylose integration (Fig. 6), the attempt to integrate the glucose peak separately (now without the citric acid peak contribution) lead to glucan conversions ("glucan uncorrected") that deviated by 2.7-7.1 % from the ones obtained using the 300 mm column (Fig. 7d). For example, glucan conversion after 72 h was 82.4 % (50 mm guard column) vs 77.8 % (300 mm column). At time point 0 h (Fig. 7a), the observed glucose peak (peak 2) was derived from the enzyme mixture used in the experiment. Therefore, this area had to be subtracted as a blank from the glucose area of following time points. Similar as before (Fig. 6e), if the citric acid and glucose peaks at timepoint 0 h were integrated as one peak and subtracted from following peaks (citric acid + glucose) the values obtained with the 50 mm guard column ("glucan corrected") were in the range of 98.7-99.7 % of the values with the 300 mm column (Fig. 7d). The glucan conversion after 72 h was 77.6 % (50 mm guard column) vs 77.8 % (300 mm column). The elution on the 50 mm guard column is completed after 1.4 min and is therefore about 9 times faster compared to 12.5 min with the 300 mm column.

Using acetate buffer, only sorbitol (peak 3) derived from the commercial enzyme mixture interfered with glucose on the 50 mm guard column (Fig. 7c) because acetate (acetic acid, peak 4) eluted later at 1.75 min. A "split-peak" integration of glucose (peak 2) and sorbitol (peak 3) is sufficient to obtain glucan conversions within 96.2-99.8 % of the ones resulting from using the 300 mm column. The 72 h conversion was 77.5 % (50 mm guard column) vs 78.5 % (300 mm column) and the run time was again about 9 times faster (1.8 min vs 16.5 min) (Fig. 7e).

It is noteworthy to mention that the acetate buffered samples can also be run with an analysis time of only

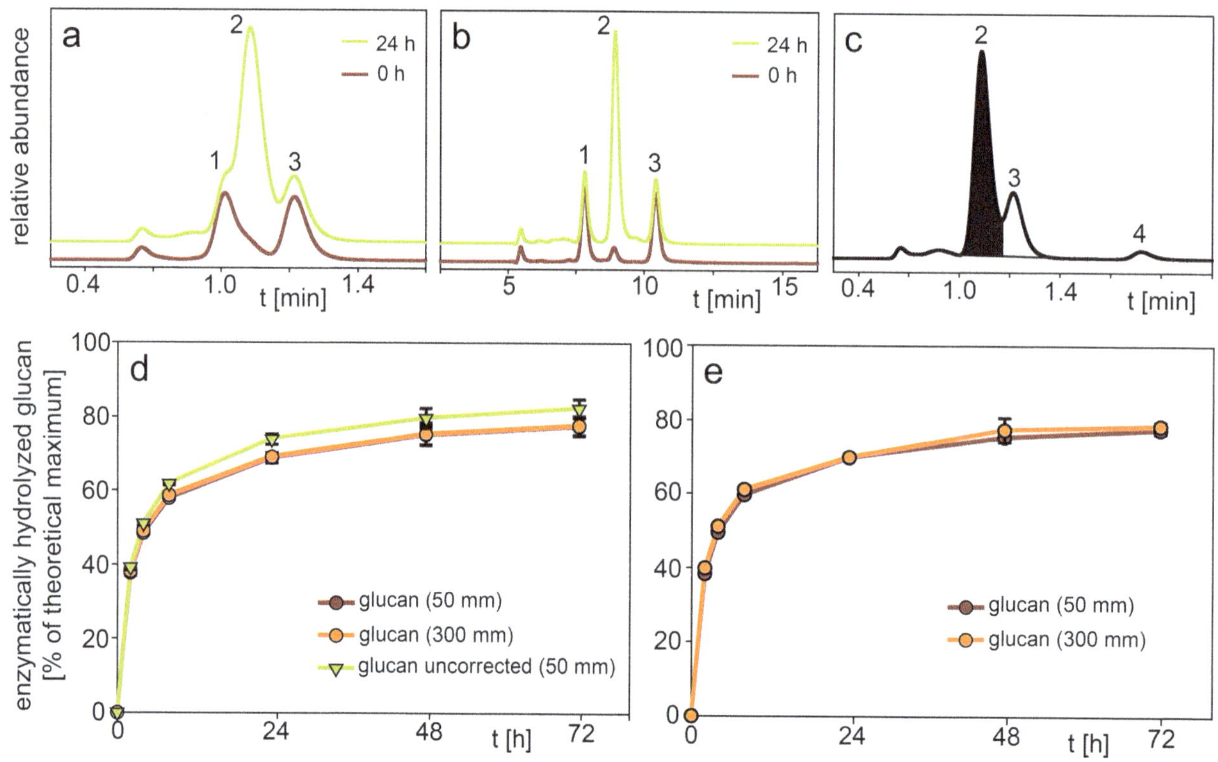

Fig. 7 Chromatogram of the supernatant from the enzymatic digestion of "pretreated *Miscanthus* biomass" showing citric acid (peak 1), glucose (peak 2), sorbitol (peak 3) and acetic acid (peak 4) elution after (**a**) 0 h and 24 h digestion in citrate buffer analyzed with the 50 mm guard column, (**b**) 0 h and 24 h digestion in citrate buffer analyzed with the 300 mm column, (**c**) 24 h digestion in acetate buffer analyzed with the 50 mm guard column; and enzymatically released glucan concentration profile of the supernatant from the enzymatic digestion of "pretreated *Miscanthus* biomass"between 0 h and 72 h using (**d**) citrate buffer (**e**) acetate buffer analyzed with the 50 mm guard column and 300 mm column. HPLC parameters: column temperature 55 °C, mobile phase 5 mM sulfuric acid, flow rate 1.0 mL/min (0.6 mL/min for 300 mm column), refractive index detection. Error bars represent +/− 1 standard deviation (n = 3)

1.4 min as the citrate buffered samples. The acetate peak (peak 4) will then be eluted within the first 0.4 min of the next sample (not interfering with the glucose peak).

Furthermore, the sample analysis time of 1.4 min is fast enough to compete with other higher-throughput glucose methods like using a YSI bioanalyzer for glucose determination. With the "overlap injection" function of the autosampler, which allows rapid injection of the next sample after a run is completed, about 1000 samples (> two 384-well plates) can be processed per day.

Application to analysis of ethanol, acetone and n-butanol production during acetone-butanol-ethanol (ABE) fermentation

The analysis of the products from acetone-butanol-ethanol (ABE) fermentation with a short column is challenging since *Clostridia* secret a fleet of other compounds (e.g., butyric acid, lactic acid) which can co-elute with the target analytes [8]. The use of an additional UV detector is beneficial for more accurate compound detection and quantification. Fig. 8a shows the chromatogram of a representative sample with RI detection. Ethanol (peak 1) and n-butanol (peak 2) have to be analyzed using the unspecific RI detector. For n-butanol, the concentration profile obtained with the 50 mm guard column is very similar to the 300 mm column (Fig. 8d), since interfering compounds were absent or in low concentration at this retention time. The measured n-butanol concentrations were on average 97.0 % +/− 3.8 % (range 98.1-100.3 %) of the values obtained with the 300 mm column, except for early time points (<16 h) due to lower concentrations affecting accuracy, and the 90 h and 112 h time points with 86.8 % and 105.3 % deviation.

In the 210 nm UV trace of the same sample (Fig. 8b) a compound (peak 3) is visible that is interfering with the ethanol determination using RI detection (note that in the UV trace a compound had a slightly earlier retention time since the RI detector is in line after the UV detector). However, with the 300 mm column, peak 3 and ethanol can be separated (data not shown). If ethanol was determined only by RI detection, concentrations were overestimated and the values ranged 117.5-149.3 % (average 126.9 % +/− 8.4 %) of the values obtained with

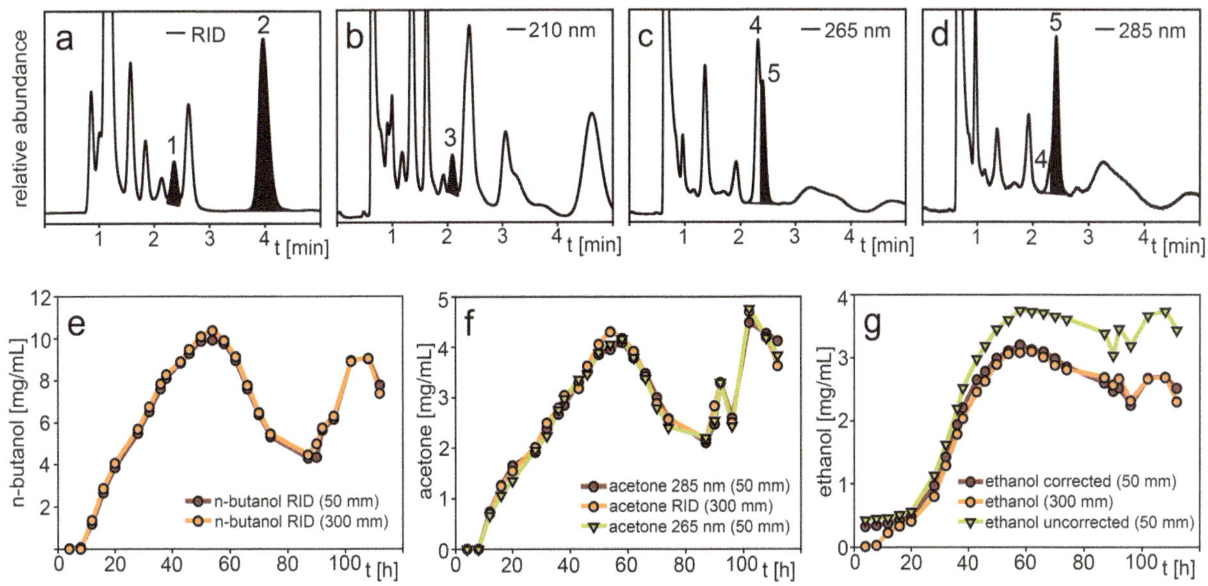

Fig. 8 Chromatogram of the supernatant from a *Clostridium acetobutylicum* ABE fermentation after 23 h analyzed with the 50 mm guard column showing ethanol (peak 1), n-butanol (peak 2), ethanol interference (peak 3), acetone interference (peak 4) and acetone (peak 5) analyzed with (**a**) refractive index detection (**b**) UV 210 nm detection (**c**) UV 265 nm detection, (**d**) UV 285 nm detection, and concentration profile of the supernatant from a *Chlostridium acetobutylicum* ABE fermentation of (**e**) n-butanol, (**f**) acetone, (**g**) ethanol, between 0 h and 112 h analyzed with the 50 mm guard column and 300 mm column. HPLC parameters: column temperature 30 °C, mobile phase 5 mM sulfuric acid, flow rate 1.0 mL/min (0.6 mL/min for 300 mm column), refractive index and UV (210 nm, 265 nm, 285 nm) detection

the 300 mm column (Fig. 8g). The interfering compound was present early in the fermentation process when ethanol production was zero or very minimal. Therefore, it should be possible to measure the area of peak 3 with 210 nm UV detection and correlate it to the area obtained with RI detection. In this way, the interference of peak 3 in RID mode should be reduced by measuring the sum of the area (ethanol + peak 3) with RID and subtracting the theoretical "RID area contribution" of peak 3 calculated from its 210 nm UV area. However, we noticed that the low intensity of peak 3 in the earlier time points made correlation of UV and RID signal less accurate. We therefore used an empirical method to determine the UV/RID correlation by adjusting the correlation factor so the corrected ethanol concentration from the 50 mm guard column best matched the 300 mm column results. This correction factor was applied to the UV-to-RID conversion of peak 3 at all time-points of the fermentation. "Ethanol corrected" values of average 102.8 % +/− 4.8 % (range 94.6-110.8 %) of the 300 mm values were obtained (Fig. 8g), excluding the initial 4–20 h timepoints.

Acetone (peak 5) is more selectively analyzed at a wavelength of 265 nm (absorbance maximum) but in the ABE fermentation sample an interfering peak (peak 4) was present (Fig. 8c). By applying a higher wavelength (285 nm) the area of the interfering peak was greatly reduced (Fig. 8d). Interestingly, despite the interference, even at 265 nm good analysis data (average 96.7 % +/− 5.1 %, range 85.5-105.8 % of the values using the 300 mm column, Fig. 8f) was obtained. The results from the 285 nm detection were similar (average 99.2 % +/− 5.6 %, range 87.3-106.5 % of the value using the 300 mm column, Fig. 8f), excluding the 112 h time point (113.6 %).

The increase and decrease of the profiles of ethanol, acetone and n-butanol during the course of the fermentation was caused by an applied membrane pervaporation which removed these components from the fermentation medium. During the first 50 h, the microbial production outpaced the membrane removal until the accumulated solvents became toxic to the microorganism and stopped further secretion. At this point, a decline in concentration was observed due to removal by membrane pervaporation (50–90 h). Since the concentration of the solvents was eventually reduced to a sub-toxic level, the cells resumed production of acetone, n-butanol and ethanol later in the experiment (90–112 h). Overall, analysis on the 50 mm guard column was able to monitor these changes with sufficient accuracy and with a sample analysis time of only 5 min compared to about 45 min on the 300 mm column.

Application of the Bio-Rad Micro Guard Cation H guard column

An even shorter guard column is the 30 mm x 4.6 mm i.d. Bio-Rad Micro Guard Cation H ("30 mm guard column"). It can be especially well used for the

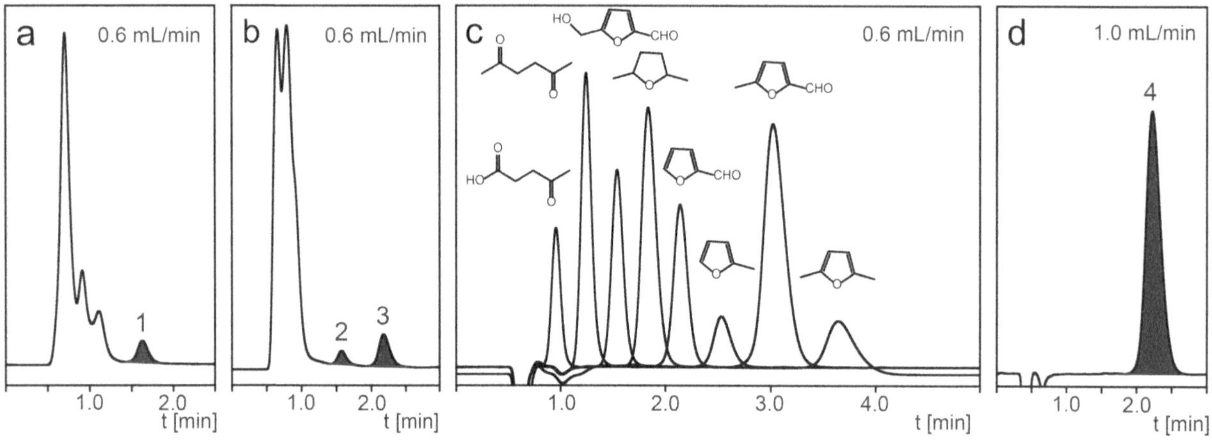

Fig. 9 Chromatogram of (**a**) the supernatant of *Escherichia coli* strain DH1 fermentation showing n-butanol (peak 1) elution at 0.6 ml/min flow rate, (**b**) the hydrolysate 3 showing 5-HMF (peak 2) and furfural (peak 3) elution at 0.6 ml/min flow rate, (**c**) a mixture of levulinic acid, 2,5-hexanedione, 5-HMF, 2,5-dimethyltetrahydrofuran, furfural, methylfuran, 5-methylfurfural and 2,5-dimethylfuran at 0.6 ml/min flow rate, (**d**) a standard of n-hexanol (peak 4) at 1.0 ml/min flow rate, using the 30 mm guard column. HPLC parameters: column temperature 55 °C, mobile phase 5 mM sulfuric acid, flow rate 0.6 mL/min (1.0 mL/min for n-hexanol), refractive index detection

analysis of compounds that elute late on the 300 mm column. For example, Fig. 9a shows its application to the n-butanol analysis mentioned earlier. The analysis results were very similar to the ones obtained with the 50 mm guard column with results ranging 92.0-101.8 % for DH1, 98.2-101.0 % for DH1Δ*acrB* and 88.8-91.6 % for BW25113 profile analysis compared to the 300 mm column (Fig. 2b). However, by using a standard 0.6 mL/min flow rate, the analysis time was cut in half (2 min) compared to the 50 mm column (4 min).

Applied to the analysis of 5-HMF and furfural (Fig. 9b), the analysis time (2.5-3 min) was further reduced by more than half compared to the 50 mm guard column (6–6.5 min, Fig. 4a). Furfural concentrations were slightly higher compared to the 300 mm column: 100.6 % (1.55 vs 1.54 mg/mL), 102.7 % (1.93 vs 1.88 mg/mL) and 109.7 % (2.60 vs 2.37 mg/mL) for hydrolysate 1–3, respectively (Fig. 4b). The concentration for 5-HMF was lower for hydrolysate 1 (64.7 %, 0.11 vs 0.17 mg/mL) and hydrolysate 3 (78.4 %, 0.69 vs 0.88 mg/mL) but identical for hydrolysate 2 (1.40 mg/mL).

An interesting application of this guard column is the analysis of some promising liquid fuel molecules and their precursors [3, 27]. Most of these types of molecules are usually analyzed by gas chromatography. If analyzed by liquid chromatography on the 300 mm column, the mobile phase has to be modified with a higher concentration of an organic solvent, e.g. acetonitrile, in order to reduce their long retention time [27]. Instead of changing the mobile phase composition, we have explored the reduction of the retention time by using the 30 mm guard column. Fig. 9c shows the nearly baseline separation of a standard mixture of levulinic acid, 2,5-hexanedione, 5-HMF, 2,5-dimethyltetrahydrofuran, furfural, methylfuran, 5-methylfurfural and 2,5-dimethylfuran, eluting within only 4 min. Since the latter three compounds elute after furfural their elution times are in the range of 1–2 h on the 300 mm column (data not shown). Interestingly, despite the structural similarity, especially of the furans, the 30 mm guard column provided sufficient separation efficiency for these molecules. When the flow rate on the 30 mm guard column was increased to 1 mL/min a further reduction of the analysis time was achieved. At this flow rate, other compounds eluting within the first 25 min on the 300 mm column now eluted in less than 1 min on the 30 mm guard column (data not shown). For example, the elution of n-hexanol was at 2.3 min at a flow rate of 1 mL/min (Fig. 9d) compared to 96 min on the standard 300 mm column (data not shown). In general, the 30 mm guard column operated at 1.0 ml/min elutes compounds about 4 times faster than the 50 mm guard column (data not shown). Examples are, e.g., 2,5-dimethylfuran (2.3 vs 10.2 min), n-hexanol (2.2 vs 9.3 min) and 2,5-dimethyltetrahydrofuran (1.1 vs 4.3 min). The 30 mm guard column can therefore be an ideal tool for the rapid analysis of selected late-eluting compounds. We therefore encourage researchers to explore this guard column for the analysis of compounds that seem to be "invisible" to the RID when injected on the 300 mm column. This invisibility is most likely due to the effect that compounds remain on the column for a long time and are eluted as broad peaks appearing more like a "baseline-drift", which makes peak detection and compound quantification difficult or impossible.

Conclusion

We have used examples from various experiments to show that a 50 mm guard column can be successfully used for the analysis of samples from many bioenergy- and biotechnology-relevant experimental setups. With this column, samples can be analyzed up to ten-times faster compared to the standard 300 mm column with very comparable results. Therefore, it is an ideal tool for processing a large amount of samples, such as in screening or discovery experiments. By applying correct peak integration and interference subtraction, concentration profiles from even more complex samples are reliably obtained. A further decrease of the analysis time was achieved by using a 30 mm guard column that has been shown to be especially suitable for the rapid analysis of compounds with long elution times on the standard 300 mm column, including biofuel-related alcohols (e.g. n-butanol, n-hexanol) and furan- and tetrahydrofuran-type molecules. Owing to the universal applicability and the ease of use as well the numerous citations in the literature, we suspect that almost all labs involved in the fields of biotechnology or biomass conversion technology are equipped with an HPLC-UV/RID instrument and should therefore be able to easily apply the methods presented here. Besides the faster analysis time, guard columns only cost a fraction of their counterparts they were designed to protect. We therefore encourage researchers to explore the feasibility of applying the separation on a guard column to their experimental setup in order to significantly reduce analysis time.

Methods

HPLC analysis

Samples were either filtered (0.45 μm) or centrifuged (10 min, 14,000 g) and analyzed using a 1200 series high-pressure liquid chromatography system (Agilent Technologies, Santa Clara, CA, USA) consisting of an autosampler with tray cooling, binary pump, degasser, thermostated column compartment, diode array detector (DAD) and refractive index detector (RI) connected in series. The supernatant was injected onto either a 300 mm × 7.8 mm (length × inner diameter) Aminex HPX-87H (Bio-Rad, Richmond, CA, USA) column with 9 μm particle size, 8 % cross-linkage equipped with a 30 × 4.6 mm micro-guard Cation H guard column cartridge (Bio-Rad, Richmond, CA) or onto a 50 × 7.8 mm Rezex™ RFQ-Fast Acid H^+ guard column (Phenomenex, Torrance, CA, USA) with 8 μm particle size, 8 % cross-linkage or onto a 30 × 4.6 mm micro-guard Cation H guard column cartridge (Bio Rad, Richmond, CA, USA). Compounds were eluted either at 30, 55 or 80 °C at a flow rate of either 0.6 or 1.0 mL using a mobile phase of 5 mM sulfuric acid.

n-butanol fermentation

DH1, DH1Δ*acrB* and BW25113 *E. coli* strains were inoculated at an OD600 of 0.35-0.45 in 250 mL screw-capped baffled Erlenmeyer flasks in TB (terrific broth) medium (1 L contains 12 g casein peptone, 24 g yeast extract, 4 mL glycerol, 2.2 g potassium phosphate monobasic, 9.4 g potassium phosphate dibasic, Fisher Scientific, Pittburgh, PA, USA) and grown anaerobically at 30 °C and 215 rpm for up to 72 h. DH1Δ*acrB* strain was constructed by removing the efflux pump subunit acrB from the chromosome according to established procedures [28].

iso-butanol fermentation

sFAB5441 is *E. coli* Dh5αZ1 strain (with LacI and TetR on chromosome) expressing iso-butanol pathway genes. The plasmids for these pathways (pSA55 and pSA69) were obtained from James Liao [29]. Strain sFAB5692 has the same 2 plasmid as sFAB5441 but the genetic background of the strain is AL329 with several deletions (adhE; frdB/frdC; fnr; IdhA; ptA; pflB, adhP, eutG, yiaY, yjgB, betA, fucO, eutE) to improve pyruvate production and iso-butanol yield [30].

All strains were grown aerobically in shake flasks in LB medium and then inoculated in M9 media containing 5 g/L yeast extract, ~50 g/L glucose, and 1000-fold dilution of A5 trace metal mix (2.86 g H_3BO_3, 1.81 g $MnCl_2x4H_2O$, 0.222 g $ZnSO_4x7H_2O$, 0.39 g $Na_2MoO_4x2H_2O$, 0.079 g $CuSO_4x5H_2O$, 49.4 mg $Co(NO_3)2x6H_2O$ per liter water) at 37 °C and grown to an OD600 of 0.6-0.8. Cells were then induced with isopropyl-β-D-thiogalactopyranoside (IPTG) and then grown at 30 °C [29].

Pretreated biomass and hydrolysates

A two-step dilute acid pilot plant pretreatment of *Miscanthus* X *giganteus* (around 1 inch size) was performed in a pilot plant at Andritz, Glens Falls, NY. In the first step, the biomass was heat-pretreated with 0.5 % (w/w) sulfuric acid at 12.5 % (w/w) solid loading applying a rapid steam-driven heating ramp to 158 °C and a holding time of 20 min. The liquid was removed from the solids/pretreated biomass by squeezing out in a hydraulic press and was referred to as "hydrolysate 1". In the second step, the pretreated biomass was again heat-pretreated with 1 % (w/w) sulfuric acid at 12.5 % (w/w) solid loading applying a rapid steam-driven heating ramp to 180 °C and a holding time of 4 min. The pressed out liquid from this second stage was referred to as "hydrolysate 2".

Another pretreatment was performed in a pilot plant of the National Renewable Energy Lab (NREL), Golden, CO. *Miscanthus* X *giganteus* (around 1 inch size) was heat-pretreated with 1.5 % (w/w) sulfuric acid at 25 % (w/w) solid loading applying a rapid steam-driven heating ramp to 190 °C and a holding time of 1 min and subsequent rapid pressure release. The pretreated biomass was separated from the mixture by centrifugation and referred to as "pretreated *Miscanthus* biomass". The obtained liquid phase was referred to as "hydrolysate 3".

Xylose to ethanol fermentation

Saccharomyces cerevisiae SA-1-X123 (Brazilian industrial strain) including the X123 cassette for xylose fermentation was grown from an OD600 of 1 in yeast-peptone-xylose (YPX) medium containing 40 g/L xylose aerobically at 30 °C and 100 rpm for 62 h [31, 32].

Glucose to ethanol fermentation

"Hydrolysate 3" was either adjusted to pH 5.5 with KOH, then centrifuged and filter sterilized or detoxified by 24 h pervaporation as described previously [21]. *Saccharomyces cerevisiae* SA-1 was provided by the Yeast Biochemistry and Technology Laboratory, Biological Science Department, Luiz de Queiroz College of Agriculture, University of Sao Paulo, Brazil and was grown at 30 °C at 200 rpm in 10 mL of synthetic complete media (SC-80). SC-80 contains 80 g/L glucose, 2 g/L dropout mix (US Biological, Salem, MA, USA), 6.7 g/L yeast nitrogen base (Becton Dickinson, Franklin Lakes, NJ, USA), 19.5 g/L 2-ethanesulfonic acid (MES) buffer, and a small amount of KOH to adjust the pH to 5.5. After overnight growth, cells were harvested by centrifugation. Fermentation was performed in 25 mL Hungate bottles under anaerobic conditions. The fermentation broth contained either 25 % (v/v) of "hydrolysate 3" or pervaporation-detoxified "hydrolysate 3" (with added water to match the amount removed by pervaporation), with the addition of the components of SC-80 to match SC-80 levels, and harvested SA-1 yeast cells to obtain an initial OD600 of 0.3. The fermentation was performed at 34 °C and 200 rpm for 67 h.

Enzymatic digestibility assay

"Pretreated *Miscanthus* biomass" was extensively washed with de-ionized water followed by consecutive centrifugation steps until pH of decanted water reached pH 5. Excess water was then removed from the biomass by manually squeezing the biomass between paper towels. An equivalent of biomass containing 1 g of cellulose was used in 25 mL liquid volume digestion reactions (considering water from biomass) in 50 mL Falcon screw cap tubes. Final digestion reactions contained either 0.05 M sodium citrate buffer (pH 4.8) or 0.05 M sodium acetate buffer (pH 4.8) and 30 FPU cellulase from *Trichoderma reesei* ATC 26291 (Sigma-Aldrich, St. Louis, MO, USA), 20 U Novozymes 188 (Sigma Aldrich, St. Louis, MO, USA) and 0.02 % (w/v) sodium azide. The digestion reactions were incubated at 50 °C and 200 rpm for 72 h.

ABE fermentation and pervaporation

Fermentations were carried out at 37 °C and 200 rpm in 3-L bioreactors (Bioengineering AG, Switzerland) with a 2 L working volume [33, 34]. Seed culture of *Clostridium acetobutylicum* ATCC824 (purchased from the American Type Culture Collection, Manassas, VA, USA) in clostridia growth medium (CGM, 100 mL) was prepared in a 150 mL anaerobic serum bottle at 37 °C until OD600 reached 2.0 and 60 mL of the seed culture was used to inoculate the bioreactor. The bioreactor automatically kept the pH ≥ 5.0 during the fermentation, using a 5 M KOH solution. Nitrogen gas was inserted into the bioreactor at a rate of 200 mL/min to maintain an anaerobic environment. Losses of volatiles, through the gas exhaust port, were minimized by using a cooling condenser attached to a RTE7 water bath (Thermo Fisher-Scientific, Sunnyvale, CA, USA) kept at 4 °C. The system was attached to pervaporation laboratory bench test unit built by Sulzer Chemtech, Neunkirchen, Germany.

After 18 h, pervaporation started with a polystyrene-block-polydimethylsiloxane-block-polystyrene (SDS) block copolymer membrane (37 cm^2, 2 μm thickness (on support)) [21]. The fermentation broth of the bioreactor was continuously passed over the membrane and back into the bioreactor by using a peristaltic pump (model # 7553–70, Cole-Parmer, Vernon, IL, USA,). ABE fermentation by *C. acetobutylicum* occurs in two steps: an acidogenesis phase wherein the microbes mainly produce acetic acid and butyric acid, followed by a solventogenesis phase wherein the microbes mainly produce ABE.

Competing interests

The authors declare that they have no competing interests.

Authors' contributions

SB is the academic responsible for funding and supervising the research in addition to coordinating the experimental design, experimental setup, data analysis and drafting the manuscript. ABI performed the experimental setup and data analysis. Both authors read and approved the final manuscript.

Acknowledgements

This work was funded by the Energy Biosciences Institute. "Hydrolysate 3" and "pretreated *Miscanthus* biomass" was provided by the National Renewable Energy Laboratory, 1617 Cole Blvd., Golden, CO 80401, a national laboratory of the U.S. Department of Energy managed by the Alliance for Sustainable Energy, LLC for the U.S. Department of Energy under Contract Number DE-AC36-08GO28308. The authors thank Niwen Kong, Vivek Mutalik, Thalita Peixoto Basso, Jeffrey Skerker, Zach Baer, Jon Kuchenreuther, and Eric Sacia for providing samples as well as information on sample generation and on the experimental setup. The authors also thank Mara Bryan for valuable comments and suggestions improving the manuscript. Publication made possible in part by support from the Berkeley Research Impact Initative (BRII) sponsored by the UC Berkeley Library.

References

1. Erdei B, Barta Z, Sipos B, Réczey K, Galbe M, Zacchi G. Ethanol production from mixtures of wheat straw and wheat meal. Biotechnol Biofuels. 2010;3:16.
2. Saha BC, Iten LB, Cotta MA, Wu YW. Dilute acid pretreatment, enzymatic saccharification and fermentation of wheat straw to ethanol. Process Biochem. 2005;40:3693–700.
3. Román-Leshkov Y, Barrett CJ, Liu ZY, Dumesic JA. Production of dimethylfuran for liquid fuels from biomass-derived carbohydrates. Nature. 2007;447:982–6.

4. Brandt A, Ray MJ, To TQ, Leak DJ, Murphy RJ, Welton T. Ionic liquid pretreatment of lignocellulosic biomass with ionic liquid-water mixtures. Green Chem. 2012;13:2489–99.
5. Sluiter A, Hames B, Ruiz R, Scarlata C, Sluiter J, Templeton D, Crocker D: Determination of Structural Carbohydrates and Lignin in Biomass. In Laboratory Analytical Procedure (LAP). National Renewable Energy Laboratory (NREL), Golden, CO. Revised Version 2012. http://www.nrel.gov/biomass/analytical_procedures.html. Accessed November 2014.
6. Scarlata CJ, Hyman DA. Development and validation of a fast high pressure liquid chromatography method for the analysis of lignocellulosic biomass hydrolysis and fermentation products. J Chromatogr A. 2010;1217:2082–7.
7. Buday Z, Linden JC, Karim MN. Improved acetone-butanol fermentation analysis using subambient HPLC column temperature. Enzyme Microb Technol. 1990;12:24–7.
8. Kumar M, Saini S, Gayen K. Acetone-butanol-ethanol fermentation analysis using only high performance liquid chromatography. Anal Methods. 2014;6:774–81.
9. Binder JB, Blank JJ, Cefali AV, Raines RT. Synthesis of furfural from xylose and xylan. Chem Sus Chem. 2010;3:1268–72.
10. Cai CM, Nagane N, Kumar R, Wyman CE. Coupling metal halides with a co-solvent to produce furfural and 5-HMF at high yields directly from lignocellulosic biomass as an integrated biofuels strategy. Green Chem. 2014;16:3819–29.
11. Li H, Chai X-S, Deng Y, Zhan H, Fu Z. Rapid determination of ethanol in fermentation liquor by full evaporation headspace gas chromatography. J Chromatogr A. 2009;1216:169–72.
12. Mendes LS, Oliveira FCC, Suarez PAZ, Rubim JC. Determination of ethanol in fuel ethanol and beverages by Fourier transform (FT)-near infrared and FT-Raman spectrometries. Anal Chim Acta. 2003;493:219–31.
13. Vermeir S, Nicolaï BM, Jans K, Maes G, Lammertyn J. High-throughput microplate enzymatic assay for fast sugar and acid quantification in apple and tomato. J Agric Food Chem. 2007;55:3240–8.
14. Birnberg PR, Brenner ML. A one-step assay for sucrose with sucrose phosphorylase. Anal Biochem. 1984;142:556–61.
15. Kristoffersen L, Skuterud B, Larssen BR, Skurtveit S, Smith-Kielland A. Fast quantification of ethanol in whole blood specimens by the enzymatic alcohol dehydrogenase method. Optimization by experimental design. J Anal Toxicol. 2005;29:66–70.
16. Velterop JS, Vos F. A rapid an inexpensive microplate assay for the enzymatic determination of glucose, fructose, sucrose, L-malate and citrate in tomato (Lycopersicon esculentum) extracts and in orange juice. Phytochem Anal. 2011;12:299–304.
17. Raba J, Mottola HA. Glucose oxidase as an analytical reagent. Critic Rev Anal Chem. 1995;25:1–42.
18. YSI 2950 biochemistry analyzer [http://www.ysilifesciences.com/index.php?page=ysi-2950-biochemistry-analyzer]
19. Wyman CE. Aqueous Pretreatment of Plant Biomass for Biological and Chemical Conversion to Fuels and Chemicals. New York: Wiley; 2013.
20. Ebert J: Furfural: Future feedstock for fuels and chemicals. Biomass Magazine [http://biomassmagazine.com/articles/1950/furfural-future-feedstock-for-fuels-and-chemicals]
21. Greer DR, Basso TP, Ibanez AB, Bauer S, Skerker JM, Ozcam AE, et al. Fermentation of hydrolysate detoxified by pervaporation through block copolymer membranes. Green Chem. 2014;16:4206–13.
22. Mitchell VD, Taylor CM, Bauer S. Comprehensive analysis of monomeric phenolics in dilute acid plant hydrolysates. Bioenerg Res. 2014;7:654–69.
23. Ibanez A, Bauer S. Analytical method for the determination of organic acids in dilute acid pretreated biomass hydrolysate by liquid chromatography time-of-flight mass spectrometry. Biotechnol Biofuels. 2014;7:145.
24. Humbird D, Davis R, Tao L, Kinchin C, Hsu D, Aden A, Schoen P, Lukas J, Olthof B, Worley M, Sexton D, Dudgeon D: Process Design and Economics for Biochemical Conversion of Lignocellulosic Biomass to Ethanol: Dilute-Acid Pretreatment and Enzymatic Hydrolysis of Corn Stover. Technical Report NREL/TP-5100-47764. United States National Renewable Energy Laboratory, US Department of Energy. May 2011.
25. Kumar P, Barrett DM, Delwiche MJ, Stroeve P. Methods for pretreatment of lignocellulosic biomass for efficient hydrolysis and biofuel production. Ind Eng Chem Rev. 2009;48:3713–29.
26. Selig M, Weiss N, Ji Y: Enzymatic Saccharification of Lignocellulosic Biomass. In Laboratory Analytical Procedure (LAP). National Renewable Energy Laboratory (NREL), Golden, CO. 2008. http://www.nrel.gov/biomass/analytical_procedures.html. Accessed November 2014.
27. Yuan J-P, Chen F. Simultaneous separation and determination of sugars, ascorbic acid and furanic compounds by HPLC-dual detection. Food Chem. 1999;64:423–7.
28. Fisher MA, Boyarskiy S, Yamada MR, Kong N, Bauer S, Tullman E. Enhancing tolerance to short-chain alcohols by engineering the *Escherichia coli* AcrB efflux pump to secrete the non-native substrate n-butanol. ACS Synth Biol. 2014;3:30–40.
29. Atsumi S, Hanai T, Liao JC. Non-fermentative pathways for synthesis of branched-chain higher alcohols as biofuels. Nature. 2008;451:86–90.
30. Rodriguez GM, Atsumi S. Isobutyraldehyde production from *Escherichia coli* by removing aldehyde reductase activity. Microb Cell Fact. 2012;11:90.
31. Kim SR, Park Y-C, Jin Y-S, Seo J-H. Strain engineering of *Saccharomyces cerevisae* for enhanced xylose metabolism. Biotechnol Adv. 2013;31:851–61.
32. Kim SR, Skerker JM, Kang W, Lesmana A, Wei N, Arkin AP, et al. Rational and evolutionary engineering approaches uncover a small set of genetic changes efficient for rapid xylose fermentation in Saccharomyces cerevisiae. PLoS One. 2013;8, e57048.
33. Anbarasan P, Baer ZC, Sreekumar S, Gross E, Binder JB, Blanch HW, et al. Integration of chemical catalysis with extractive fermentation to produce fuels. Nature. 2012;491:235–9.
34. Alsaker KV, Paredes C, Papoutsakis ET. Metabolite stress and tolerance in the production of biofuels and chemicals: gene-expression-based systemsanalysis of butanol, butyrate, and acetate stresses in the anaerobe Clostridium acetobutylicum. Biotechnol Bioeng. 2010;105:1131–47.

In planta production of ELPylated spidroin-based proteins results in non-cytotoxic biopolymers

Valeska Hauptmann[1], Matthias Menzel[2], Nicola Weichert[1], Kerstin Reimers[3], Uwe Spohn[2] and Udo Conrad[1*]

Abstract

Background: Spider silk is a tear-resistant and elastic biopolymer that has outstanding mechanical properties. Additionally, exiguous immunogenicity is anticipated for spider silks. Therefore, spider silk represents a potential ideal biomaterial for medical applications. All known spider silk proteins, so-called spidroins, reveal a composite nature of silk-specific units, allowing the recombinant production of individual and combined segments.

Results: In this report, a miniaturized spidroin gene, named *VSO1* that contains repetitive motifs of *MaSp1* has been synthesized and combined to form multimers of distinct lengths, which were heterologously expressed as elastin-like peptide (ELP) fusion proteins in tobacco. The elastic penetration moduli of layered proteins were analyzed for different spidroin-based biopolymers. Moreover, we present the first immunological analysis of synthetic spidroin-based biopolymers. Characterization of the binding behavior of the sera after immunization by competitive ELISA suggested that the humoral immune response is mainly directed against the fusion partner ELP. In addition, cytocompatibility studies with murine embryonic fibroblasts indicated that recombinant spidroin-based biopolymers, in solution or as coated proteins, are well tolerated.

Conclusion: The results show that spidroin-based biopolymers can induce humoral immune responses that are dependent on the fusion partner and the overall protein structure. Furthermore, cytocompatibility assays gave no indication of spidroin-derived cytotoxicity, suggesting that recombinant produced biopolymers composed of spider silk-like repetitive elements are suitable for biomedical applications.

Keywords: Spider silk, Immunogenicity, Synthetic spidroin, Biomaterial, Biocompatibility, Atomic force microscopy

Background

Recent biomedical developments in the field of tissue engineering require protein-based biomaterials as scaffolds that have been optimized for substantial extensibility, long-term stability, self-assembly and low energy loss [1]. Further important properties are biocompatibility and economical production, as well as purification systems, which could be provided by plant-based production of elastin-like peptide (ELP) fusion proteins, referred to as ELPylated proteins [2,3]. Since ancient times, spider silk has been known for its extraordinary properties. Bygone cultures used the secretory product of the spider spinning glands. For example, Ancient Greeks were aware of the boosting effect for wound healing and used spider webs to cover bleeding lesions [4]. Today, spider silk is known to promote the regeneration of nerves [5], and supports the proliferation of fibroblasts and keratinocyte cell lines [6]. In addition to these fascinating features, the lightweight and flexible biopolymer spider silk joins mechanical properties such as high toughness, tensile strength and stiffness that competes with man-made polymers [7].

Spider silks produced by orb-web-weaving spiders are composed of proteins that are commonly termed spidroins. An intensively investigated silk is the major ampullate silk, which is used by spiders as a structural element to build the web frame and the safety line; therefore, it is also called dragline silk. The well-known structure of major ampullate silk consists of two proteins called major ampullate spidroin 1 (MaSp1) and major ampullate spidroin 2 (MaSp2) [8]. The first partial sequence information from MaSp1 of *Nephila clavipes* was published in 1990 and revealed a high repetitive primary structure [9]. The protein consists of poly(A) blocks alternating with GGX (X = Y, L, Q) and $(GA)_n$ sequence motifs

* Correspondence: conradu@ipk-gatersleben.de
[1]Leibniz Institute of Plant Genetics and Crop Plant Research, Corrensstrasse 3, 06466 Stadt Seeland, OT Gatersleben, Germany
Full list of author information is available at the end of the article

[10]. Currently, the high number of repetitive peptide motifs in the core sequence is known as the key feature of spidroins. Clearly there is a strong relationship between the secondary structure, based on the unique motifs in the primary protein structure, and the outstanding properties of spider silks. The alanine-rich peptide regions form a β-sheet that provides remarkable strength to major ampullate silk [11]. Conversely, the high toughness arises from the glycine-rich peptide motifs, which likely induce the formation of β-turns and 3_{10} helices [7,12]. For the spidroin MaSp1, a molecular weight of up to 320 kDa is reported [13]. It is assumed that the size of spider silk proteins is a key factor in defining their mechanical properties, because all characterized native spider silks consist of proteins with high molecular weights [13].

The usage of native spider silk on a larger scale is not economically profitable. Currently, heterologous spidroin production is the method of choice to satisfy the demand of recombinant spider silk for research. For this purpose, the most widely used host system is the gram-negative bacterium *Escherichia coli.* Here, spider silk protein modules of mainly a low molecular weight, approximately 50 kDa, were produced [14-17]. However, heterologous expression of recombinant spider silk proteins in *E. coli* was found to be rather inefficient owing to the low production rate and instability of the spider silk gene [18]. Because of the highly repetitive nature of the proteins, DNA deletion in the spider silk gene, as well as transcription and translation errors were often observed during the reproduction of recombinant *E. coli* harboring the gene [19]. Furthermore, translational errors of proteins were caused by a depletion of the t-RNA pool owing to the high alanine and glycine content. Recently, expression of high molecular weight spider silk derivatives up to 285 kDa has been achieved by optimizing the glycyl-tRNA amount and glycine synthesis [20]. Other expression systems used for heterologous spidroin production are yeast [21], plants [22,23], insects [24] or mammalians [25].

The knowledge of the molecular structure of spider silk has inspired researchers to use the repeated modules of silks to develop synthetic spidroins. In addition, an approach using a synthetic gene can avoid the abovementioned difficulties during heterologous expression of spider silk proteins. The adaption of the codon usage to the t-RNA pool of the intended host system is a considerable advantage of synthetic genes. Furthermore, restriction sites necessary for cloning into expression vectors can be attached during gene synthesis and, therefore, prevent additional PCR reactions that often cause errors with highly repetitive genes. In previous studies, the synthetic spider silk protein SO1, which shows a 94% homology to MaSp1 of the golden silk spider, was successfully expressed in plants *(Nicotiana tabacum, Solanum tuberosum)* [22]. For further investigations, this synthetic spidroin was fused to 100 repeats of ELP to facilitate the purification of plant-produced spider silk-like proteins [26]. In addition, for other spider silk proteins ELPylation is a powerful technology for easy and economical purification [27].

In general, low immunogenicity is anticipated for highly repetitive proteins such as spider silk derivatives and ELP [28]. Here, we designed various synthetic spidroin-based fusion proteins consisting of several repetitive motifs of SO1 and accordingly MaSp1 termed $(VSO1)_n$-100xELP. We have designed these artificial fusion proteins to produce material for tissue engineering with suitable mechanical properties, low immunogenicity and cytocompatibility. After heterologous expression in plants, the spidroin-based fusion proteins were purified by a scalable and economical downstream processing procedure. Furthermore, these synthetic spidroins were used to analyze their mechanical properties and immunogenicity. By producing recombinant spider silks in different formats, the main requirements for biomedical applications, including biocompatibility, sufficient mechanical properties in terms of elasticity, hardness and stiffness, and low or even no immunogenicity, are met. Biocompatibility is defined as the quality of the biomaterial as not being toxic or having injurious effects on biological systems [19]. In the present paper we have measured the cytocompatibility with murine embryonic fibroblasts as an approach to estimate biocompatibility. Finally, we discuss the suitability of the different recombinant synthetic spider silk proteins for biomedical applications.

Results

Recombinant production of synthetic spidroin-based fusion proteins

Four different plant expression vectors coding for spidroin-based fusion proteins varying in the spidroin content have been designed to assess the mechanical properties, immunogenicity and cytotoxicity in relation to the protein size. Therefore, we created the synthetic spidroin gene *VSO1* containing characteristic repetitive motifs of *MaSp1* and *SO1*, respectively, which were used as model proteins for the blueprint. The synthesized gene was flanked by compatible but non-regenerative restriction sites (*Bam*HI and *Bgl*II) to enable the insertion of further synthetic genes for extension of the spidroin content. The resulting recombinant proteins contained the typical sequence motifs GGX, $(GA)_n$ and polyA blocks of MaSp1 in the primary sequence (Figure 1A). Additionally, the synthetic spidroin genes were combined with the gene encoding for the biopolymer 100xELP in an expression vector to facilitate ubiquitous expression in transgenic tobacco. For the retention of the recombinant proteins in the endoplasmic reticulum (ER), the expression cassette also contained the signal peptide (SP) LeB4 [29] and the ER retention signal KDEL [30] (Figure 1A). Thus, four different synthetic

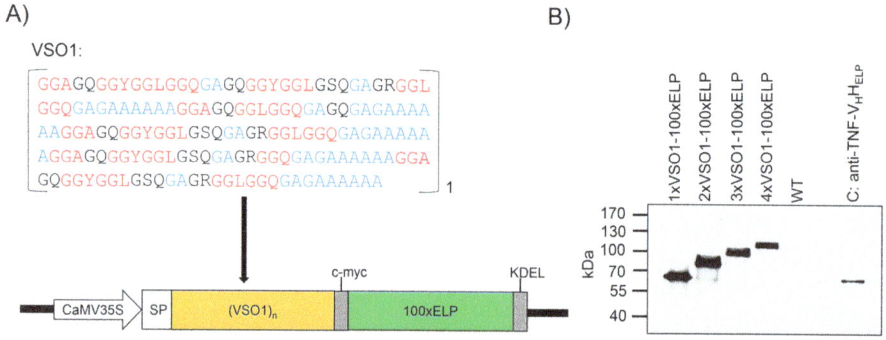

Figure 1 Expression of spidroin-based fusion proteins $(VSO1)_n$-100xELP in transgenic tobacco plants. A) Protein sequence of VSO1 and schematic representation of the plant expression cassette for $(VSO1)_n$-100 × ELP. In the protein sequence, repetitive motifs are highlighted in red and blue. Heterologous expression of synthetic spidroin $(VSO1)_n$ fused to 100 repeats of elastin-like peptide (100xELP) was controlled by the CaMV 35S promoter. SP: legumin B4 signal peptide; KDEL: ER retention signal; c-myc: detection tag. **B)** Immunoblot of leaf extracts from transgenic tobacco expressing spider silk fusion proteins of different sizes. Samples of 5 μg of total soluble protein were applied to the gel and separated by 8% SDS-PAGE, blotted and detected with an anti-c-myc antibody. WT: wild type *N. tabacum* cv. SNN. C+: 1 ng c-myc immunoblot standard anti-TNF-V_HH-100xELP.

spidroin-based biopolymers were designed with calculated molecular weights of 56 kDa (1xVSO1-100xELP), 69 kDa (2xVSO1-100xELP), 83 kDa (3xVSO1-100xELP) and 96 kDa (4xVSO1-100xELP). Initially, each expression vector was individually transformed into *Agrobacterium tumefaciens* (*A. tumefaciens*), followed by the generation of transgenic tobacco (*Nicotiana tabacum*) by leaf disc transformation. After selection of kanamycin-resistant tobacco plants, expression of fusion proteins was proven by immunoblotting. Analyzed leaf extracts showed the appropriate size increase from 1xVSO1-100xELP to 4xVSO1-100xELP (Figure 1B). The molecular weights were determined by SDS-PAGE and, accordingly, immunoblotting. Protein sizes appeared higher than the calculated masses, a phenomenon that was also observed for other ELP fusion proteins [31,32].

The principal reason for fusion of the synthetic spidroins to 100 repeats of ELP was to enable the chromatography-free purification by inverse transition cycling (ITC) [26,33], a low cost method for recovery of biopolymers. Here, synthetic spidroin-based fusion proteins were purified from tobacco leaves of transgenic plants *via* an advanced membrane based ITC [34], which was optimized for spider silks and performed according to Weichert *et al.* [27]. A heat incubation step at the beginning of the purification procedure leads to denaturation of the majority of the proteins. In the following cooling step (4°C), all fusion proteins became soluble and were separated by centrifugation and enriched by filtration after several temperature shifts. For further investigations purified proteins were analyzed as casted layers or in solution.

Mechanical investigation of synthetic spidroin-based biopolymer layers

Based on freshly prepared and smooth layers of the synthetic biopolymers 1xVSO1-100xELP, 2xVSO1-100xELP and 4xVSO1-100xELP, we performed Atomic Force Microscopy (AFM) based nanoindentation experiments to examine the relevance of the increasing spidroin content. Figure 2 shows the corresponding topographic images with the z-ranges of 9.6 nm (1xVSO1-100xELP), 7.4 nm (2xVSO1-100xELP) and 11.4 nm (4xVSO1-100xELP),

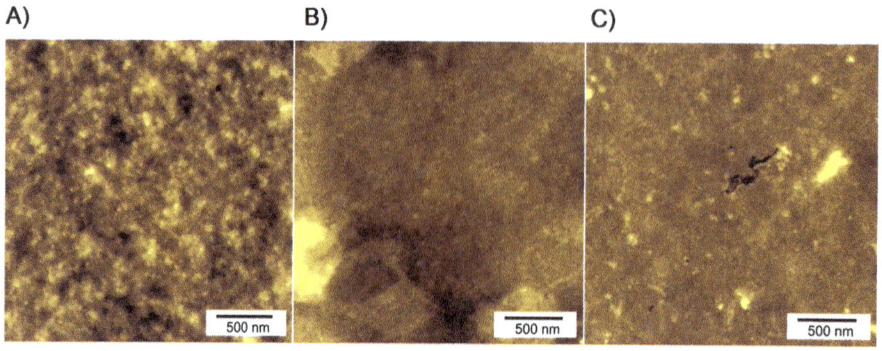

Figure 2 AFM topographic images of spidroin-based biopolymer layers. Surfaces have been taken in the intermitting tapping mode using 2.5 × 2.5 $μm^2$ scan sizes of **A)** 1xVSO1-100xELP, **B)** 2xVSO1-100xELP and **C)** 4xVSO1-100xELP.

respectively. All surfaces appeared homogenous with amorphous regions. It should be emphasized here that the roughness of the examined layers is very low, with roughness values < 2 nm (Table 1), fulfilling the presumption for AFM-based nanoindentation to examine the elastic penetration modulus E and therefore, the characterization of the stiffness of the biopolymer layers. In former studies we showed on the basis of layers prepared from a monomeric tagged spider silk protein (MaSp1) fused to 100 repeats of ELP in comparison to layers consisting of multimers of this protein that the elastic penetration modulus E increased by multimerization [27]. In this mentioned experiment, the mass relation between the spider silk part and ELP remains unchanged during the multimerization process. In the present study, the determination of the elastic penetration modulus E of biopolymer layers varying in the spidroin content enabled a more detailed characterization of the relation between the molecular weight of the whole biopolymer and the spidroin content. Experiments were performed at thin layers of each casted layer with a thickness of at least 1 μm and a mean surface roughness smaller than 2 nm for 2.5 × 2.5 μm^2 grids (Table 1). The E values presented in Table 2 are significantly different, as tested by the differential *t*-test [35]. It is important to notice that the E values of 2xVSO1-100xELP and 4xVSO1-100xELP were found to be considerably higher than this of 1xVSO1-100xELP. In comparison, AFM-based nanoindentation analyses of the fusion protein 100xELP revealed an E value of 2.74 GPa [27]. The differences between all mean values of the penetration modulus E are highly significant. It can be assumed that the increase of the spidroin content in the fusion proteins leads to an ascending E value, and therefore represents a higher stiffness of these materials.

To study in greater detail the influence of the increasing spidroin content in these ELPylated recombinant proteins, analyses of immunogenicity and cytotoxicity were performed and related to their potential use as biomaterials.

Immunogenicity of synthetic spidroin-based fusion proteins 1xVSO1-100xELP and 4xVSO1-100xELP

The use of spider silk variants for the production of protein scaffolds for tissue engineering or for engineering of drug delivery systems requires spidroin-based derivatives

Table 1 Surface roughness of various synthetic biopolymer layers

Substance	Mean surface roughness s_a (nm)	Squared surface roughness s_q (nm)
1xVSO1-100xELP	0.70	0.89
2xVSO1-100xELP	0.78	1.03
4xVSO1-100xELP	0.34	0.55

Defined grids of 2.5 × 2.5 μm^2 were analyzed.

Table 2 Elastic penetration moduli of recombinant spidroin-based fusion protein layers

Substance	Elastic penetration modulus E (GPa)
1xVSO1-100xELP	2.73 ± 0.02
2xVSO1-100xELP	3.85 ± 0.02
4xVSO1-100xELP	4.51 ± 0.03
LDPE Riblene®FL30	0.172 ± 0.001

GPa: Gigapascal; ELP: elastin-like peptide.
Values were obtained by AFM-based nanoindentation under the following conditions: temperature = 23°C, relative humidity = 35%, system sensitivity = 22.5 nm/V, spring constant = 59.3 N/m, tip radius = 112 nm, setpoint control = 500 nN, relative setpoint = 2500 nN, z-closed loop active, z-length = 1 μm, retract delay = 10 ms, grid size = 25 × 25 μm^2, data point 35 × 35 per run with an orthogonal and lateral inter-sampling point distance of at least 715 nm. Low density polyethylene (LDPE) Riblene® FL30 foil was used as reference.

that do not induce immune responses. Natural spider silks do not cause an inflammatory response [36]. A low immunogenicity for 100xELP is anticipated [28], and has been shown experimentally by studying the T-cell response [32]. In this report, we ask the question whether different-sized fusion proteins of synthetic spidroins and ELP can induce a humoral immune response and if the increasing spidroin content in addition to the higher molecular weight has an influence. For the determination of the immunogenicity elicited by heterologous synthetic spidroin-based biopolymers one group of C57BL/6 J mice (animals 1–4) was immunized with 1xVSO1-100xELP and another (animals 5–8) with 4xVSO1-100xELP. After the fourth immunization, the immune responses in mice were determined by an indirect ELISA against the injected antigen (Figure 3A and B). All sera showed a strong immune response against their antigen, but the antisera raised against the 4xVSO1-100xELP showed binding at higher dilutions, indicating a stronger immune response. Cross-reactivity against the other antigen with a different molecular weight was detected; however, this was very low (Figure 4). We next asked the question if this relative specificity is due to specific structural epitopes formed by the different fusion proteins and we also wanted to explain cross-reactivity. For further immunological analyses, serum 1, which was isolated from a mouse immunized with 1xVSO1-100xELP, and serum 5, which was from a mouse that had been immunized with 4xVSO1-100xELP, were selected. First we analyzed whether the immune response was directed against the spidroin region (VSO1), against the c-myc tag and/or against ELP. For this purpose 1xVSO1-100xELP, 4xVSO1-100xELP, 100xELP and the recombinant protein anti-TNF-V_HH [37], containing a c-myc tag, were included into the immunoblotting analyses. Additionally, the immune response to the fusion protein anti-TNF-V_HH-100xELP [37], which also contained the c-myc tag, was analyzed. All proteins were synthesized *in planta*. After developing immunoblots with the appropriate mouse serum, synthetic spidroins were detected, along with anti-

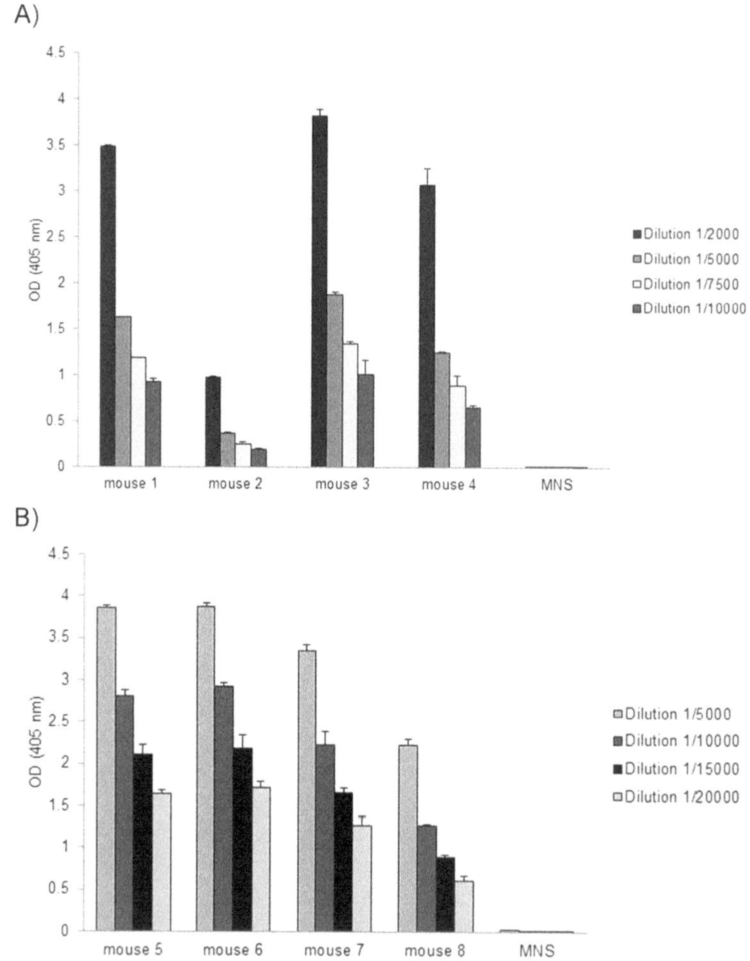

Figure 3 Specific immune response of mouse sera determined by indirect ELISA. Samples of 500 ng per well of **A)** 1xVSO1-100xELP and **B)** 4x VSO1-100xELP were coated to a microtiter plate and the reaction against mouse sera was analyzed for different dilutions. The antigen-antibody complex was detected using goat anti-mouse IgG conjugated with alkaline phosphatase. Extinction was measured at 405 nm (ordinate).

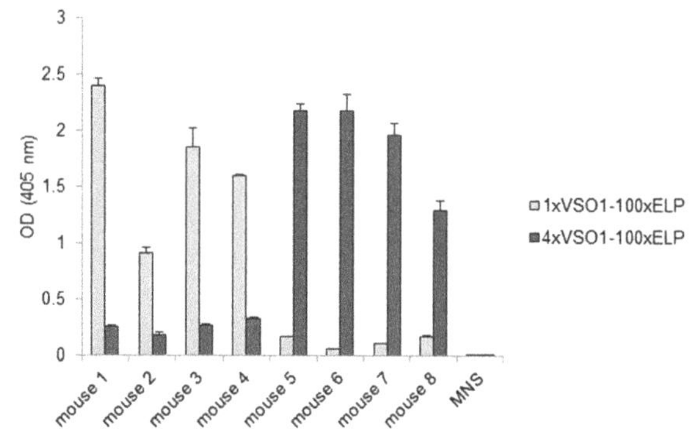

Figure 4 Determination of the cross-reaction of mouse sera by indirect ELISA. Antigens 1xVSO1-100xELP and 4xVSO1-100xELP were coated and the cross-reaction against the other antigen was analyzed. Mouse sera specific against 1xVSO1-100xELP were applied at a dilution of 1:2000 and mouse sera specific against 4xVSO1-100xELP were diluted at 1:5000. Extinction was measured at 405 nm (ordinate).

TNF-V_HH-100xELP and 100xELP proteins (Figure 5A and B). Anti-TNF-V_HH was not recognized by sera isolated from immunized mice, but it was detected by immunoblotting with the anti-c-myc antibody, implicating that the c-myc tag did not induce an immune response (Figure 5A-D). All other proteins mentioned here contained a c-myc tag and were detectable *via* immunoblotting using an anti-c-myc antibody (Figure 5C and D). After incubation with mouse normal serum, no nonspecific reaction was detectable (Figure 5E and F). The relative specificity for the proteins used for immunization could not be seen after immunoblotting. Serum 1 and serum 5 showed comparable binding behavior (Figures 5 and 6). With serum 1 against 1xVSO1-100xELP it was also possible to detect 4xVSO1-100xELP. Analogous antigen-antibody recognition occurred for serum 5 and 1xVSO1-100xELP. The epitopes responsible for relative specificity against the injected proteins in the indirect ELISA are possibly destroyed by denaturation during SDS-PAGE.

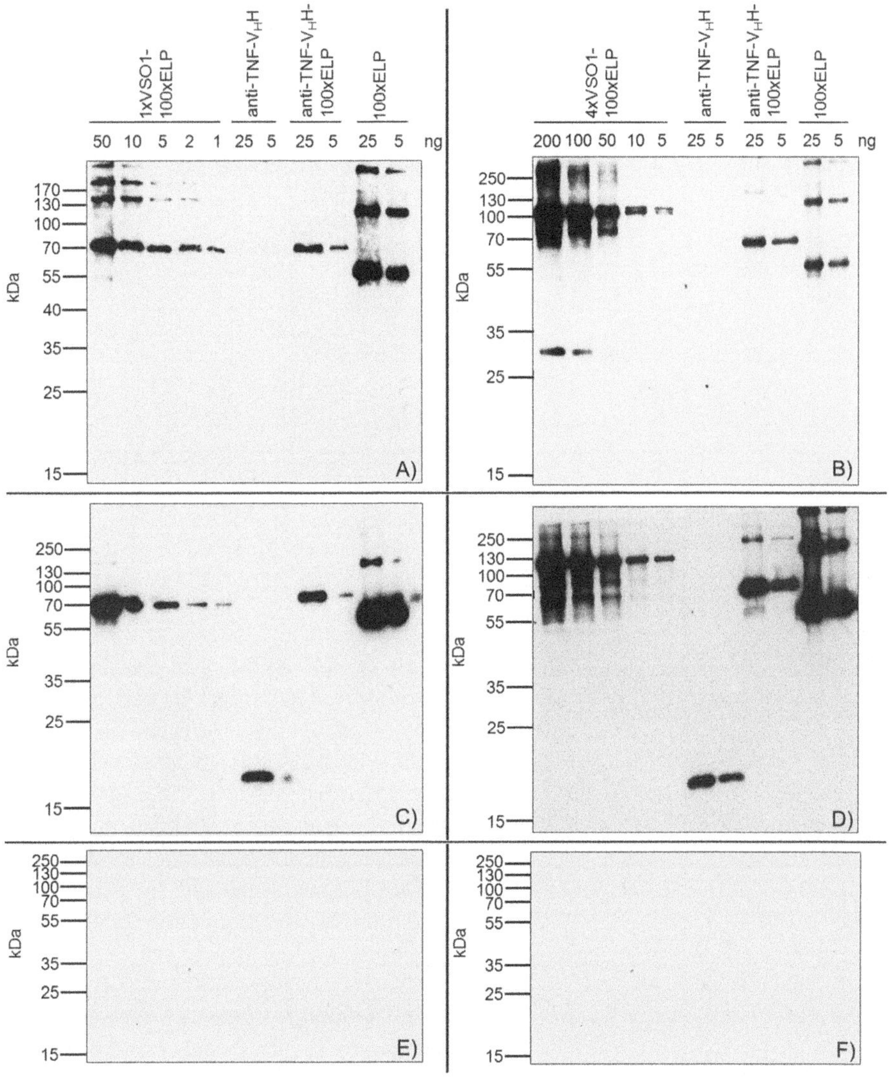

Figure 5 Comparative immunoblots of the immune response of plant-expressed biopolymers to different antibodies. Recombinant proteins were separated on 12% SDS-PAGE, electrotransferred and detected with **A)** mouse serum number 1, **B)** mouse serum number 5, **C)** and **D)** anti-c-myc monoclonal antibody, and **E)** and **F)** mouse normal serum.

Additionally, both sera analyzed in more detail showed significant binding to 100xELP in immunoblotting analysis and to anti-TNF-V_HH-100xELP (Figure 6A and B). Here, we analyzed the binding to ELP in more detail. Polyclonal antibodies are a mixture of different antibodies by nature, with their own respective kinetic characteristics and, in many cases, with differences in epitope specificity. Based on this heterogeneity, the rate constants derived and affinities calculated based on such measurements should be regarded as a mean of different subpopulations. In the case of ELP, the number of different epitopes is limited, because it is a highly repetitive protein consisting of pentamers of very similar if not equal amino acid sequences. Therefore, the complexity is also limited and calculation of a mean dissociation constant K_d is a useful tool to characterize the immune response against ELP-based epitopes. For the selected sera, optimal antigen concentrations were predetermined in dilution analyses and further evaluation of the immunogenicity of ELP was performed by a competitive ELISA. For this purpose, 1xVSO1-100xELP and 4xVSO1-100xELP, respectively, was adsorbed to the polystyrene surface and the binding of the relevant antisera was measured. This binding was inhibited by competition with 100xELP. The calculated mean dissociation constant K_d for the complex 1xVSO1-100xELP-anti 1xVSO1-100xELP from animal 1 and the competitor 100xELP was 25 nM (Figure 7A) and the mean K_d for 4xVSO1-100xELP-anti 4xVSO1-100xELP from animal 5 and 100xELP was 374 nM (Figure 7B). In the immunogenic 1xVSO1-100xELP, the relative ELP content is much

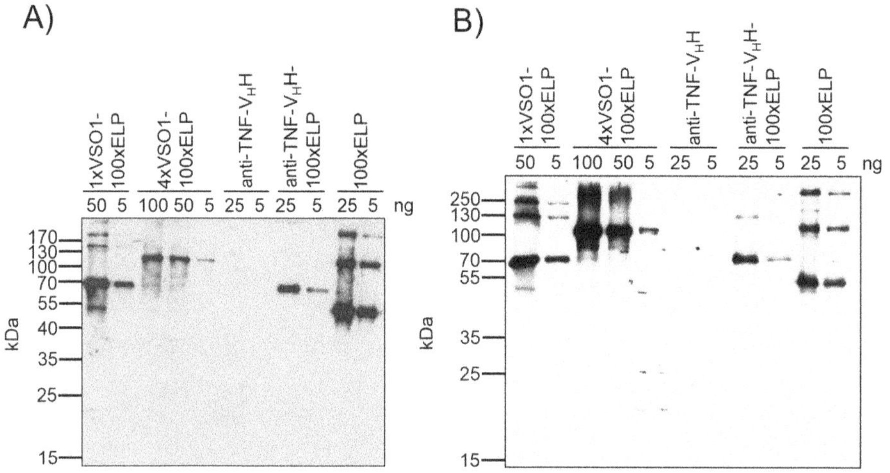

Figure 6 Immunoblots for the characterization of the capability for cross-reaction of antigens with serum from other spidroins. Both spidroin-based biopolymers, 1xVSO1-100xELP and 4xVSO1-100xELP, and the control proteins, anti-TNF-V_HH, anti-TNF-V_HH-100xELP and 100xELP, were separated using 12% SDS-PAGE, electrotransferred and detected with **A)** mouse serum 1 or **B)** mouse serum 5.

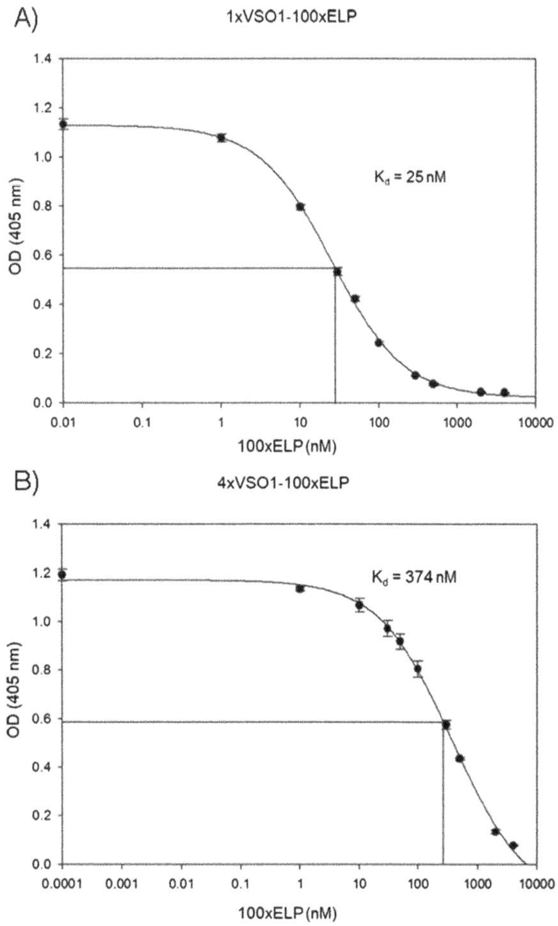

Figure 7 Characterization of the immune response to ELP by competitive ELISA. Analyses for a complex consisting of **A)** 1xVSO1-100xELP and mouse serum 1 competing to 100xELP and **B)** 4xVSO1-100xELP, and mouse serum 5 competing to 100xELP.

higher; thus, in the antiserum a stronger binding (higher affinity) to ELP was induced. However, we cannot exclude that the position of the ELP part in the resulting protein structure of the fusion protein is different in 1xVSO1-100xELP and 4xVSO1-100xELP. The competition curves showed that binding to 1xVSO1-100xELP and to 4xVSO1-100xELP could be completely inhibited (Figure 7). This implicates that the humoral immune response is mainly directed against 100xELP. An overview of all analyzed immune responses is given in Table 3.

In vitro cytotoxicity assays of the synthetic spidroin-based biopolymers 1xVSO1-100xELP and 4xVSO1-100xELP

At first, cytotoxicity of the synthetic spidroins fused to ELP was assayed with soluble and coated proteins. In general, cell metabolic activity is determined as an indirect measure for cell vitality and proliferation. This is done by adding a compound to the culture medium, which is metabolized by the living cells to a fluorescent end product. Here, the influence of a recombinant spider silk protein on the metabolism of murine embryonic fibroblasts was tested by adding the protein to the cell culture medium or by coating the culture surface. In both cases an influence could be observed. 1xVSO1-100xELP and 4xVSO1-100xELP significantly stimulated the cells when metabolic activity was determined 24 hours after the addition of the protein (Figure 8A). The same trend was observed on day three; however, now the only statistically significant differences were between 100xELP and 1xVSO1-100xELP and 4xVSO1-100xELP, respectively (Figure 8B). The results for cells grown on protein-coated surfaces were similar (Figure 9A and B), although a significant effect was detectable later (Figure 9B).

Table 3 Overview of analyzed immune responses

antigen (MW) \ antibody	serum 1	serum 5	anti-c-myc antibody	mouse normal serum
1xVSO1-100xELP (56 kDa)	+	+	+	-
4xVSO1-100xELP (96 kDa)	+	+	+	-
anti-TNF-V_HH (15 kDa)	-	-	+	-
anti-TNF-V_HH-100xELP (57 kDa)	+	+	+	-
100xELP (42 kDa)	+	+	+	-

A positive reaction is labeled with a plus sign and no reaction is marked with a bar.

In addition, a surface for hemocompatibility of the fusion protein 100xELP was evaluated by direct hemolysis testing. The possible destruction of the erythrocyte membrane was measured by determining the amount of free plasma hemoglobin (Hb). The results showed that 100xELP was not hemolytic when covered on glass coverslips. While 100% relative hemolysis could be achieved in the samples containing catheters as a positive control the relative hemolysis observed in the samples with protein-coated glass coverslips were in the same range as the samples obtained by testing blood on high density polyethylene films used as the negative reference material (Figure 10). In general, the results of these cytotoxicity analyses examining cytotoxicity and hemocompatibility provided a positive preliminary indication for the use of the investigated synthetic spidroin-based biopolymers in medical applications.

Discussion

The high molecular weight of spider silk proteins is assumed to be a key factor that underpins their outstanding mechanical properties, because all characterized native spider silks consist of proteins with high molecular weights [13]. In this study, we analyzed the relation of the spidroin content in recombinant spidroin-based biopolymers and the corresponding mechanical properties. The elastic penetration modulus E was found to range from 2.7 GPa for 1xVSO1-100xELP to 4.5 GPa for 4xVSO1-100xELP; the synthetic biopolymer with the highest molecular weight that was used in this study. The significant increase of the elastic penetration modulus E, which correlates in a first approximation with the Young's modulus, appears to be due to the increase in the spidroin content. E values were also comparable to ELPylated MaSp1 derivatives, which were previously investigated [27]. One must take into account that the recombinant spidroin-based biopolymers analyzed here has approximately one-third of the molecular weight of the native MaSp1 and that protein layers were examined. In comparison, for native dragline silk a Young's modulus of approximately 12 GPa was determined [20]. Additionally, the mechanical characterization of a fiber made from native-sized recombinant spider silk protein showed a Young's modulus of 21 ± 4 GPa; however, in the same study it was also noted that proteins of lower molecular weight did not yield similar material properties [20]. Therefore, for further intended projects the production of native-sized spidroin containing biopolymers is commendable and has already been successfully performed *in planta* with an intein-based post-translational protein multimerization technology [38].

Low immunogenicity is anticipated for natural spider silks; this feature was exemplarily proven for major ampullate dragline silk collected from *Nephila clavipes* [39].

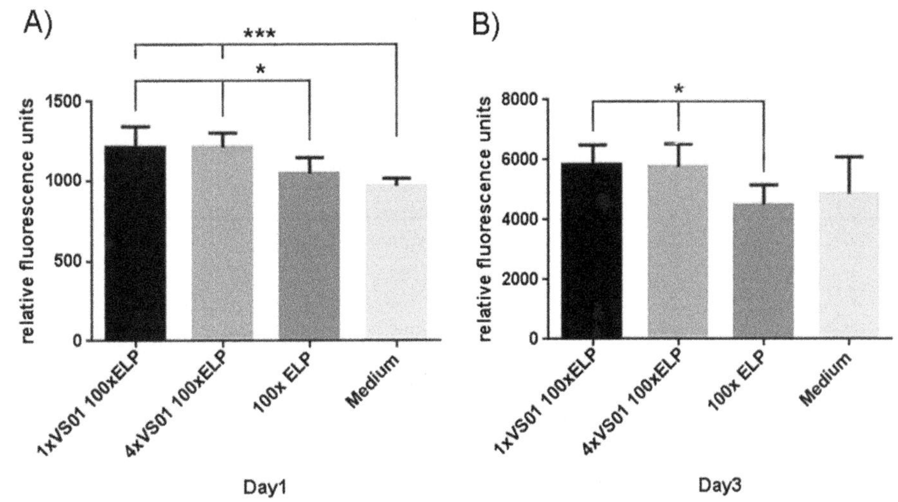

Figure 8 Metabolic activities of murine embryonic fibroblasts in contact with soluble synthetic biopolymers. The cells were cultured in medium supplemented with the indicated recombinant protein for 24 hours **(A)** or 72 hours **(B)**. The graphs depict a representative result of three independent repeats (n = 8, mean and SEM).

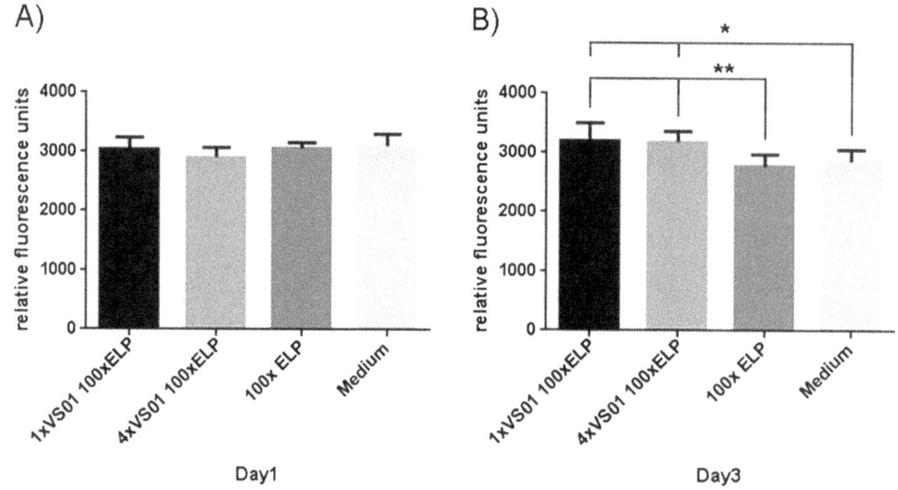

Figure 9 Metabolic activities of murine embryonic fibroblasts in contact with coated synthetic biopolymers. The cells were cultured on surfaces coated with the indicated recombinant protein for 24 hours **(A)** or 72 hours **(B)**. The graphs depict a representative result of three independent repeats (n = 8, mean and SEM).

Here, we presented the first results of a specific antibody response to a spidroin-based biopolymer, which was enabled by a collection of various mouse sera that were prepared after immunization of mice with soluble spidroin-ELP fusion proteins. The spidroin-based proteins additionally contain the c-myc tag. An immune response against the c myc tag could not be detected by immunoblotting. All the antigens used in the immunological analyses were produced in plants to allow comparisons in all experiments. The data found by indirect ELISA showed a relative specific reaction against the immunogen, either 1xVSO1-100xELP or 4xVSO1-100xELP. This data provides no insight into whether these specific antibodies in the sera bind to ELP- or spidroin-based epitopes or new overall structures in the different fusion proteins. After denaturation of the antigens and SDS-PAGE, the specific reactions were not observed in the immunoblotting experiments. Competitive ELISA data showed clearly that binding is completely inhibited by 100xELP. We conclude that the humoral immune response against the spidroin-based polymers is directed against epitopes involving 100xELP. The sera raised against 1xVSO1-100xELP bind to 100xELP with a > 15-fold higher mean dissociation constant. These differences in the affinity against ELP support the view (see above) that 1xVSO1-100xELP and 4xVSO1-100xELP induced antibodies against different epitopes. These antibodies with higher mean affinity could be induced either by the higher ELP proportion in the fusion protein compared with 4xSO1-100xELP or by specific structures occurring in the 1xVSO1-100xELP fusion protein. We conclude that, in general, spidroin-based biopolymer variants have to be tested according their immunogenicity in each single case. The structure of fusion proteins could lead to the induction of several new immune responses even if standard basic sequences are used.

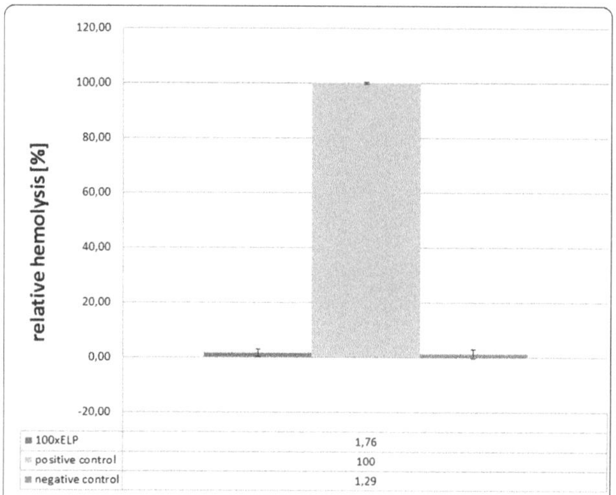

Figure 10 Hemolysis of human erythrocytes incubated with 100xELP. The biopolymer was coated on glass coverslips. Free Hb was determined and expressed as percent of the positive control (n = 3, mean and SEM).

Cytocompatibility is a prerequisite for the biomedical application of a material. Therefore, *in vitro* cytotoxicity assays are typically one of the first assessments carried out in the biological evaluation of biopolymers. Recombinant spider silk proteins have been tested in a number of different settings. In previous studies, it was shown that a plant-produced synthetic spidroin derived from MaSp1 and fused with hundred repeats of elastin-like peptides (ELP) resulted in a non-cytotoxic biopolymer that supported the proliferation of mammalian cells [26]. An example of a recombinant spider silk protein consisting of five glycine-rich segments alternating with four poly-alanine stretches connected to a non-repetitive globular C-terminal domain by a serine- and alanine-rich linker produced in *E. coli* was tested with primary human fibroblasts. The cells attached to the material and grew on different matrices such as meshes and foams [40]. 3T3 fibroblasts adhered to porous scaffolds of recombinant protein based on spidroin 1, even filling the deeper layers after 14 days [41]. However, a recombinant protein based on the silk of the European garden spider prevented adhesion and cell proliferation of BALB/3 T3 fibroblasts when coated on silicone surfaces [42]. Here, we found a positive stimulation of cellular metabolism, which indicates higher growth rates. This effect was dependent on the composition of the protein. Interestingly, we could even enhance the positive effect when the proteins were freely available in the medium. Although, we cannot rule out that this effect is based on the nutritive value of the proteins, this is rather unlikely owing to the negligible concentration of the protein present.

Biomaterials in contact with blood must show good hemocompatibility, which is often improved by coating. Sulfated silkworm fibroin has been used for this purpose in a study to enhance the hemocompatibility of poly(lactic-co-glycolic) acid vascular grafts [43]. Nevertheless, it has been discussed that larvae from two pyralid moths express silk proteins in their guts, their fat body and their hemocytes. It is assumed that these proteins also take part in immunity and coagulation [44]. We observed no hemolytic effect in the examined samples, indicating that they may serve as biocompatible coatings of blood-exposed implants.

Synthetic spidroins as biomaterials that are produced in biotechnology processes have potential use in a wide range of biomedical material applications. Prominent examples are scaffolds for tissue engineering (films, sponges, hydrogels) and drug delivery systems that trigger an effective immune response after vaccination with particle bound antigens [45,46]. In this study, we partially worked with casted proteins. Since the surface of the resulting films showed a very low roughness, applications as a coating or a wound dressing device is conceivable. Finally, the molecular weight of spidroin-based biopolymers did not influence the cytocompatibility of the casted films.

Conclusion

The main goal of the present study was the assessment of the immunological properties and cytotoxic effects of synthetic spidroin-based fusion proteins expressed *in planta*. Considering the rising elastic penetration modulus determined by AFM-based nanoindentation with increasing spidroin content, we assume additionally a first relationship between spidroin size and mechanical properties. All available antibody detection systems were used to determine epitope regions, including detection of the c-myc tag, characterization of mouse sera after immunization with the synthetic spidroin-based biopolymers and performance of competitive ELISA with the competitor ELP. Furthermore, analyses of cross-reactivity experiments gave no hint of an immunogenic region in the synthetic spidroin part of the fusion constructs. In the end, cytocompatibility studies provided no indication of spidroin-derived cytotoxicity. This implies that these plant-derived synthetic biopolymers are suitable for use as biomaterials.

Methods

Design of plasmids

Synthetic *1xVSO1* was produced by Geneart (Life Technologies, CA, USA) and contained repetitive gene motives from *Nephila clavipes* cDNA [GenBank: M37137.2]. In the course of the synthesis, the codon usage of synthetic *1xVSO1* was adapted to *N. tabacum*. After restriction digest with *Bam*HI and *Bgl*II the gene fragment was ligated into the vector 100xELP-pRTRA [47]. Further insertion of synthetic genes was facilitated with an additional restriction digest of the vector $(VSO1)_n$-100xELP-pRTRA with *Bam*HI and ligation of a *Bam*HI/*Bgl*II digested *1xVSO1* gene fragment. After ligation, a functional *Bam*HI restriction site was retained at the 5′-prime end of the synthetic gene. The resulting plasmid $(VSO1)_n$-100xELP-pRTRA contained a plant expression cassette consisting of the Cauliflower Mosaic Virus (CaMV) 35S promoter [48], the legumin B4 signal peptide (LeB4) [29], the synthetic gene $(\mathit{VSO1})_n$, a c-myc tag [49], 100 repeats of the fusion protein ELP [26] and the ER retention signal KDEL [50]. The expression cassettes were excised with the restriction enzyme *Hin*dIII and inserted individually into the binary vector pCB301-Kan [51], resulting in the plant expression vectors $(VSO1)_n$-100xELP-pCB301-Kan.

Production of plant-expressed synthetic spidroin-based fusion proteins

The binary vectors were transformed into the *A. tumefaciens* strain C58C1 (pGV2260) [52] by electroporation. For stable transformation of tobacco (*N. tabacum* cv. SNN), the leaf disc transformation method reported by Horsch *et al.* [53] was performed. The transgenic plants were cultured on Murashige-Skoog agar containing 50 mg/L kanamycin and analyzed by immunoblotting using an anti-c-myc antibody [54]. High expressing plants were cropped into soil and grown in a greenhouse for 4 to 6 weeks prior to harvesting the leaves. The fusion of the synthetic spider silks to 100xELP enabled protein purification *via* membrane-based inverse transition cycling (mITC) [34]. Therefore, frozen leaf material (–80°C) was crushed, added to preheated (85°C) 50 mM Tris–HCl (pH 8.0) and cooked for 1 hour. Further purification and desalting was performed as described [27]. For determining protein weight and storage, the purified proteins $(VSO1)_n$-100xELP were lyophilized (ALPHA2-4LSD; Christ, Osterode, Germany).

SDS-PAGE and immunoblotting analysis

For analysis of transgenic plants, leaf material was ground in liquid nitrogen. Sample buffer (72 mM Tris, 10% v/v glycerol, 2% w/v SDS, 5% w/v 2-mercaptoethanol and 0.0025 mM bromphenol blue, pH 6.8) was added and the homogenate was incubated for 10 min at 95°C. After centrifugation (30 min, 4°C, 12,000 rpm), the extract (supernatant) was kept and the protein concentration was determined by Bradford assay (Bio-Rad, Germany). Plant extracts or purified proteins, which were also analyzed by immunoblotting, were separated on reducing SDS-PAGE and electroblotted to a nitrocellulose membrane (Whatman GmbH, GE Healthcare, Germany) using 25 mM Tris, 0.1% w/v SDS, 192 mM glycine and 20% v/v methanol. For detection of the transgenic product, the membranes were blocked for 2 hours in 5% w/v fat-free dry milk dissolved in 180 mM NaCl and 20 mM Tris, pH 7.8. The primary antibody was either an anti-c-myc (9E10) supernatant [54] or an anti-$(VSO1)_n$-100xELP mouse serum. Therefore, two groups of four mice each (C57BL/6 J) were immunized with 1xVSO1-100xELP or 4xVSO1-100xELP. For the first immunization, 50 μg antigen and complete Freund's adjuvants (Difco, USA) were used. In the following three immunizations animals were boosted with 20 μg antigen and incomplete Freund's adjuvants (Difco, USA). Titers from blood samples were evaluated by ELISA and sera were collected one week after the fourth immunization. For immunoblotting analysis the secondary antibody used was a horseradish peroxidase-(HRP-) conjugated anti-mouse IgG from sheep (GE Healthcare UK Ltd., UK). Synthetic spidroin-based fusion proteins were detected by ECL (Amersham ECL Plus TM, GE Healthcare UK Ltd., UK).

Mechanical Testing of $(VSO1)_n$-100xELP

Protein layers for AFM imaging and AFM-based nanoindentation were casted by the drop to drop technique. Therefore, proteins were solubilized in water to a concentration of 1 mg/mL, successively dropped in 20 μL droplets onto glass slides and dried in a vacuum (Vacuum Concentrator 5301; Eppendorf, Germany) at room temperature until achieving layers of required thickness [27]. Measurements of protein layer thicknesses and topographical imaging were performed by an atomic force microscope Nanowizard®II (JPK Instruments, Germany) using either the Contact Mode with an MLCT silicon nitride cantilever (Bruker Cooperation, USA) or silicon cantilevers PPP-NCHR (NANOSENSORS™, NanoWorld AG, Switzerland) with tip radii below 7 nm for the Intermittent Contact Mode. Based on topographical information, roughness data were evaluated. AFM-based nanoindentation was performed to assess the elastic penetration modulus E. For that reason, the same AFM instrument was used to record and evaluate load penetration curves according to an advanced Hertzian model for spherical indenter geometry [27,55] as a course of the load dependent on a penetration depth between 10 to 15 nm. E-values were calculated from a large series of 1225 indentations, which were performed for each protein layer with an orthogonal and lateral inter-sampling point distance of 715 nm and this enabled the statistically evident calculation of E for each material (Additional file 1). Elastic penetration moduli E were calculated from the recorded load penetration curves. Here, a diamond-coated cantilever DT-NCHR #1 (NanoWorld AG, Switzerland) calibrated by the Thermal Noise Method was used. The exact geometry of the diamond-coated cantilever tips was checked by scanning electron microscopy (SEM) before and after indentation measurements.

Indirect and competitive ELISA

For evaluation of the antibody titer against the specific antigen and to examine the cross-reaction, an indirect ELISA was performed. Ninety-six-well plates (MaxiSorp™ Surface, Thermo Scientific Nunc A/S, Denmark) were coated overnight at room temperature with 500 ng of synthetic spidroin dissolved in 100 μL phosphate-buffered saline for phages (PPBS; 32 mM $Na_2HPO_4 \times 2\ H_2O$, 17 mM $NaH_2PO_4 \times H_2O$, 100 mM NaCl, pH 7.2). Bovine serum albumin (BSA, 3% w/v) in phosphate-buffered saline (PBS; 8 mM $Na_2HPO_4 \times 2\ H_2O$, 2 mM KH_2PO_4, 150 mM NaCl) supplemented with 0.05% (v/v) Tween-20 (PBS-T) was used as the negative control. Blocking was done with 130 μL/well using 3% (w/v) BSA in PBS-T (pH 7.6) for 2 hours at room temperature. Mouse sera were diluted with 3% BSA in PBS-T as indicated in the results section and applied in triplicate using a volume of 100 μL/well on the coated plates for 1.5 hours at 25°C. After five washing steps with PBS-T, a goat anti-mouse IgG conjugated with alkaline phosphatase (Sigma, USA) was added in a dilution of 1:2,000 in 1% (w/v) BSA-PBS-T. Plates were incubated for 1 hour at 25°C followed by five washing cycles with PBS-T. Bound antibodies were detected after the addition of the substrate p-nitrophenyl phosphate (1 mg/mL in 0.1 M diethanolamine-HCl, pH 9.8). The reaction was incubated at 37°C and the absorbance was measured at 405 nm within of 1 hour of the reaction being initiated.

For the competitive ELISA, antigens were dissolved in PPBS and 50 ng/well (1xVSO1-100xELP) or 100 ng/well (4xVSO1-100xELP), respectively, were coated to the microtiter plates. Blocking and washing were performed as mentioned above. Various concentrations of the competitor

100xELP (1 nmol to 4 μmol and without ELP) were premixed for 30 min at room temperature with either a 1:7,500 dilution of mouse serum 1 (against 1xVSO1-100xELP) or a 1:4,000 dilution of mouse serum 5 (against 4xVSO1-100xELP). Both competition partners were diluted in 3% (w/v) BSA in PBS-T. This premix was added in quintuplicates to the coated plates followed by incubation at 25°C for 1.5 hours; 3% BSA in PBS-T was used as negative control. Further processing of the assay was performed as described above.

Coating of the cell culture material for cytotoxicity tests

The synthetic spidroin-based biopolymers were diluted to a concentration of 50 μg/mL with phosphate-buffered saline without calcium and magnesium (PBS, Life Technologies). Glass coverslips for direct hemolysis assay, coated on both sides, and black microtiter plates for the determination of cell metabolism were coated with the protein by applying the protein solution at 4°C overnight. The liquid was then removed and cell culture plates were dried at room temperature. The coated materials were kept at 4°C or used immediately. PBS without the biopolymer was used as a control.

Cell metabolism assay

Murine embryonic fibroblasts were obtained from the ATCC (ATCC® SCRC-1045™) and cultured in Dulbecco's Modified Eagle's Medium (DMEM) high-glucose (Biochrom, Berlin Germany) with 1% penicillin/streptomycin (Biochrom, Berlin Germany), 1% sodium pyruvate (Biochrom, Berlin, Germany) and 10% fetal bovine serum (Biochrom, Berlin, Germany). The cultures were split thrice weekly and kept in a humidified atmosphere at 37°C and 5% CO_2. For the measurements, the cells were diluted to 5×10^4 cells/mL and seeded onto microtiter plates. All test samples were done either on surface-coated wells as indicated above or the protein was added to a final concentration of 50 μg/mL to the culture medium. After the indicated time points, 20 μL of CellTiter-Blue® solution (Promega, USA) was added to each well and incubated at 37°C for 2 hours. Fluorescent resorufin was measured by using 560 nm excitation and 590 nm emission filters. All tests were repeated at two independent times and performed at octuplicates. The data were analyzed by one-way ANOVA followed by Tukey's multiple comparisons test.

Preparation of blood

The blood was obtained from the Institut für Transfusionsmedizin, Medizinische Hochschule Hannover. The blood donors subscribed a declaration, that they agree that small amounts of their blood could be used for research proposals. Human blood was collected in S-Monovette tubes (Sarstedt, Germany) with citrate and used within 4 hours after collection by pooling equal amounts of blood. The Hb value of the pooled blood was measured and the blood was diluted with PBS to a total blood Hb concentration of 10 ± 1 mg/mL.

Direct hemolysis assay

The samples were covered with PBS resulting in a surface-to-PBS ratio of 3 cm^2/mL in screw-cap polypropylene tubes. Subsequently, pooled blood was added at a ratio of 1:7 and incubated at 37°C for 3 hours. During this time, the tubes were inverted carefully twice every 30 minutes. A sample of 1.8 ml was removed from the test tubes and centrifuged at 700–800 g for 15 minutes. Drabkin's solution (Sigma, Germany) was added to the samples at equal volumes and incubated at room temperature for 15 minutes. Two aliquots of each sample were transferred to a microtiter plate and the absorbance was measured at 540 nm. All measurements were done in triplicates and catheters (ARROWg + ard Blue, Arrow international) and high density polyethylene films (RM-C, Hatano Research Institute) were used as hemolytic and non-hemolytic controls, respectively.

Additional file

Additional file 1: Statistical distribution of elastic penetration moduli E for layers of synthetic biopolymers. Load penetration curves (n = 1225) were determined per protein sample layer with a thickness of at least 1 μm and a mean surface roughness smaller than 2 nm for a 2.5 × 2.5 μm^2 grid.

Competing interests

The authors declare that they have no competing interests.

Authors' contributions

The manuscript was written through contributions of all authors. VH and UC designed the study, performed most of the experiments, analyzed data, drafted and wrote the manuscript. MM performed AFM experiments and analyzed data. NW performed the protein purification and provided material. KR performed the cell metabolism assay, analyzed data and wrote the manuscript. US provided ideas and feedback and wrote the manuscript. All authors read and approved the final manuscript.

Acknowledgements

We kindly acknowledge the skillful technical support of Isolde Tillack, Christine Helmold and Ulrike Gresch. The authors are also grateful to Christina Reufsteck (BioMedImplant) for her support in the hemocompatibility tests and to Frank Rabenstein (JKI Quedlinburg) for his support on the immunization experiments. Further, we acknowledge all members of the International Society for Plant Molecular Farming for inspiring us and the exciting discussions about the production of spider silks. AFM analysis was financed by grant FKZ 22037511 of the Fachagentur Nachwachsende Rohstoffe e. V., supported by the Federal Ministry of Food and Agriculture, Germany.

Author details

[1]Leibniz Institute of Plant Genetics and Crop Plant Research, Corrensstrasse 3, 06466 Stadt Seeland, OT Gatersleben, Germany. [2]Fraunhofer Institute for Mechanics of Materials, Walter-Hülse-Strasse 1, 06120 Halle/Saale, Germany. [3]Department of Plastic, Hand and Reconstructive Surgery, Hannover Medical School, Podbielskistr. 380, 30659 Hannover, Germany.

References

1. Annabi N, Mithieux SM, Camci-Unal G, Dokmeci MR, Weiss AS, Khademhosseini A. Elastomeric Recombinant Protein-based Biomaterials. Biochem Eng J. 2013;77:110–8.
2. Floss DM, Schallau K, Rose-John S, Conrad U, Scheller J. Elastin-like polypeptides revolutionize recombinant protein expression and their biomedical application. Trends Biotechnol. 2010;28(1):37–45.
3. Floss DM, Conrad U, Rose-John S, Scheller J. ELP fusion technology for biopharmaceuticals. In: Schmidt SR, editor. Fusion protein technologies for biopharmaceuticals: Applications and challenges. Hoboken, NJ: Wiley Blackwell; 2013. p. 211–26.
4. Gerritsen VB. The tiptoe of an airbus. In: Protein Spotlight. Swiss Prot. 2002. http://web.expasy.org/spotlight/back_issues/024. Accessed 8 Aug 2013.
5. Radtke C, Allmeling C, Waldmann KH, Reimers K, Thies K, Schenk HC, et al. Spider silk constructs enhance axonal regeneration and remyelination in long nerve defects in sheep. PLoS One. 2011;6(2):e16990.
6. Wendt H, Hillmer A, Reimers K, Kuhbier JW, Schafer-Nolte F, Allmeling C, et al. Artificial skin–culturing of different skin cell lines for generating an artificial skin substitute on cross-weaved spider silk fibres. PLoS One. 2011;6(7):e21833.
7. Hinman MB, Jones JA, Lewis RV. Synthetic spider silk: a modular fiber. Trends Biotechnol. 2000;18(9):374–9.
8. Sponner A, Schlott B, Vollrath F, Unger E, Grosse F, Weisshart K. Characterization of the protein components of Nephila clavipes dragline silk. Biochemistry. 2005;44(12):4727–36.
9. Xu M, Lewis RV. Structure of a protein superfiber: spider dragline silk. Proc Natl Acad Sci U S A. 1990;87(18):7120–4.
10. Lewis RV. Spider silk: ancient ideas for new biomaterials. Chem Rev. 2006;106(9):3762–74.
11. Simmons AH, Michal CA, Jelinski LW. Molecular orientation and two-component nature of the crystalline fraction of spider dragline silk. Science. 1996;271(5245):84–7.
12. Hayashi CY, Shipley NH, Lewis RV. Hypotheses that correlate the sequence, structure, and mechanical properties of spider silk proteins. Int J Biol Macromol. 1999;24(2–3):271–5.
13. Ayoub NA, Garb JE, Tinghitella RM, Collin MA, Hayashi CY. Blueprint for a high-performance biomaterial: full-length spider dragline silk genes. PLoS One. 2007;2(6):e514.
14. Arcidiacono S, Mello C, Kaplan D, Cheley S, Bayley H. Purification and characterization of recombinant spider silk expressed in Escherichia coli. Appl Microbiol Biotechnol. 1998;49(1):31–8.
15. Huemmerich D, Helsen CW, Quedzuweit S, Oschmann J, Rudolph R, Scheibel T. Primary structure elements of spider dragline silks and their contribution to protein solubility. Biochemistry. 2004;43(42):13604–12.
16. Dams-Kozlowska H, Majer A, Tomasiewicz P, Lozinska J, Kaplan DL, Mackiewicz A. Purification and cytotoxicity of tag-free bioengineered spider silk proteins. J Biomed Mater Res Part A. 2012;101(2):456–64.
17. Winkler S, Szela S, Avtges P, Valluzzi R, Kirschner DA, Kaplan D. Designing recombinant spider silk proteins to control assembly. Int J Biol Macromol. 1999;24(2–3):265–70.
18. Fahnestock SR, Irwin SL. Synthetic spider dragline silk proteins and their production in Escherichia coli. Appl Microbiol Biotechnol. 1997;47(1):23–32.
19. Chung H, Kim TY, Lee SY. Recent advances in production of recombinant spider silk proteins. Curr Opin Biotech. 2012;23(6):957–64.
20. Xia XX, Qian ZG, Ki CS, Park YH, Kaplan DL, Lee SY. Native-sized recombinant spider silk protein produced in metabolically engineered Escherichia coli results in a strong fiber. Proc Natl Acad Sci U S A. 2010;107(32):14059–63.
21. Bogush VG, Sokolova OS, Davydova LI, Klinov DV, Sidoruk KV, Esipova NG, et al. A novel model system for design of biomaterials based on recombinant analogs of spider silk proteins. J Neuroimmune Pharmacol. 2009;4(1):17–27.
22. Scheller J, Guhrs KH, Grosse F, Conrad U. Production of spider silk proteins in tobacco and potato. Nat Biotechnol. 2001;19(6):573–7.
23. Menassa R, Zhu H, Karatzas CN, Lazaris A, Richman A, Brandle J. Spider dragline silk proteins in transgenic tobacco leaves: accumulation and field production. Plant Biotechnol J. 2004;2(5):431–8.
24. Wen H, Lan X, Zhang Y, Zhao T, Wang Y, Kajiura Z, et al. Transgenic silkworms (Bombyx mori) produce recombinant spider dragline silk in cocoons. Mol Biol Rep. 2010;37(4):1815–21.
25. Teule F, Miao YG, Sohn BH, Kim YS, Hull JJ, Fraser Jr MJ, et al. Silkworms transformed with chimeric silkworm/spider silk genes spin composite silk fibers with improved mechanical properties. Proc Natl Acad Sci U S A. 2012;109(3):923–8.
26. Scheller J, Henggeler D, Viviani A, Conrad U. Purification of spider silk-elastin from transgenic plants and application for human chondrocyte proliferation. Transgenic Res. 2004;13(1):51–7.
27. Weichert N, Hauptmann V, Menzel M, Schallau K, Gunkel P, Hertel TC, et al. Transglutamination allows production and characterization of native-sized ELPylated spider silk proteins from transgenic plants. Plant Biotechnol J. 2014;12(2):265–75.
28. Mackay JA, Chilkoti A. Temperature sensitive peptides: engineering hyperthermia-directed therapeutics. Int J Hyperthermia. 2008;24(6):483–95.
29. Baumlein H, Wobus U, Pustell J, Kafatos FC. The legumin gene family - structure of a B type gene of Vicia faba and a possible legumin gene specific regulatory element. Nucleic Acids Res. 1986;14(6):2707–20.
30. Fiedler U, Phillips J, Artsaenko O, Conrad U. Optimization of scFv antibody production in transgenic plants. Immunotechnology. 1997;3(3):205–16.
31. Ge X, Trabbic-Carlson K, Chilkoti A, Filipe CD. Purification of an elastin-like fusion protein by microfiltration. Biotechnol Bioeng. 2006;95(3):424–32.
32. Floss DM, Mockey M, Zanello G, Brosson D, Diogon M, Frutos R, et al. Expression and immunogenicity of the mycobacterial Ag85B/ESAT-6 antigens produced in transgenic plants by elastin-like peptide fusion strategy. J Biomed Biotechnol. 2010;2010:274346.
33. Meyer DE, Chilkoti A. Purification of recombinant proteins by fusion with thermally-responsive polypeptides. Nat Biotechnol. 1999;17(11):1112–5.
34. Phan HT, Conrad U. Membrane-based inverse transition cycling: an improved means for purifying plant-derived recombinant protein-elastin-like polypeptide fusions. Int J Mol Sci. 2011;12(5):2808–21.
35. Noack S, Schulze G. Statistical evaluation of analytical measurements - approximation of the integral limits of the R-distribution, T-distribution, and F-distribution. Fresen Z Anal Chem. 1980;304:250–4.
36. Allmeling C, Jokuszies A, Reimers K, Kall S, Choi CY, Brandes G, et al. Spider silk fibres in artificial nerve constructs promote peripheral nerve regeneration. Cell Prolif. 2008;41(3):408–20.
37. Conrad U, Plagmann I, Malchow S, Sack M, Floss DM, Kruglov AA, et al. ELPylated anti-human TNF therapeutic single-domain antibodies for prevention of lethal septic shock. Plant Biotechnol J. 2011;9(1):22–31.
38. Hauptmann V, Weichert N, Menzel M, Knoch D, Paege N, Scheller J, et al. Native-sized spider silk proteins synthesized in planta via intein-based multimerization. Transgenic Res. 2013;22(2):369–77.
39. Vollrath F, Barth P, Basedow A, Engstrom W, List H. Local tolerance to spider silks and protein polymers in vivo. In Vivo. 2002;16(4):229–34.
40. Widhe M, Bysell H, Nystedt S, Schenning I, Malmsten M, Johansson J, et al. Recombinant spider silk as matrices for cell culture. Biomaterials. 2010;31(36):9575–85.
41. Agapov II, Pustovalova OL, Moisenovich MM, Bogush VG, Sokolova OS, Sevastyanov VI, et al. Three-dimensional scaffold made from recombinant spider Silk protein for tissue engineering. Dokl Biochem Biophys. 2009;426:127–30.
42. Zeplin PH, Maksimovikj NC, Jordan MC, Nickel J, Lang G, Leimer AH, et al. Spider Silk Coatings as a Bioshield to reduce Periprosthetic Fibrous Capsule Formation. Adv Funct Mater. 2014;24:2658–66.
43. Liu H, Li X, Niu X, Zhou G, Li P, Fan Y. Improved hemocompatibility and endothelialization of vascular grafts by covalent immobilization of sulfated silk fibroin on poly(lactic-co-glycolic acid) scaffolds. Biomacromolecules. 2011;12(8):2914–24.
44. Korayem AM, Hauling T, Lesch C, Fabbri M, Lindgren M, Loseva O, et al. Evidence for an immune function of lepidopteran silk proteins. Biochem Biophys Res Commun. 2007;352(2):317–22.
45. Hofer M, Winter G, Myschik J. Recombinant spider silk particles for controlled delivery of protein drugs. Biomaterials. 2012;33(5):1554–62.
46. Spiess K, Lammel A, Scheibel T. Recombinant spider silk proteins for applications in biomaterials. Macromol Biosci. 2010;10(9):998–1007.
47. Floss DM, Sack M, Stadlmann J, Rademacher T, Scheller J, Stoger E, et al. Biochemical and functional characterization of anti-HIV antibody-ELP fusion proteins from transgenic plants. Plant Biotechnol J. 2008;6(4):379–91.
48. Franck A, Guilley H, Jonard G, Richards K, Hirth L. Nucleotide sequence of cauliflower mosaic virus DNA. Cell. 1980;21(1):285–94.
49. Munro S, Pelham HR. A C-terminal signal prevents secretion of luminal ER proteins. Cell. 1987;48(5):899–907.
50. Wandelt CI, Khan MR, Craig S, Schroeder HE, Spencer D, Higgins TJ. Vicilin with carboxy-terminal KDEL is retained in the endoplasmic reticulum and

accumulates to high levels in the leaves of transgenic plants. Plant J. 1992;2(2):181–92.

51. Gahrtz M, Conrad U. Immunomodulation of plant function by in vitro selected single-chain Fv intrabodies. Methods Mol Biol. 2009;483:289–312.
52. Deblaere R, Bytebier B, De Greve H, Deboeck F, Schell J, Van Montagu M, et al. Efficient octopine Ti plasmid-derived vectors for Agrobacterium-mediated gene transfer to plants. Nucleic Acids Res. 1985;13(13):4777–88.
53. Horsch RB, Fry JE, Hoffmann NL, Eichholtz D, Rogers SG, Fraley RT. A simple and general method for transferring genes into plants. Science. 1985;227(4691):1229–31.
54. Evan GI, Lewis GK, Ramsay G, Bishop JM. Isolation of monoclonal antibodies specific for human c-myc proto-oncogene product. Macromol Biosci. 1985;5(12):3610–6.
55. Lin DC, Dimitriadis EK, Horkay F. Robust strategies for automated AFM force curve analysis–I. Non-adhesive indentation of soft, inhomogeneous materials. J Biomech Eng. 2007;129(3):430–40.

24

Protein surface charge of trypsinogen changes its activation pattern

Karin Buettner[1], Thomas Kreisig[1], Norbert Sträter[1] and Thole Zuchner[1,2*]

Abstract

Background: Trypsinogen is the inactive precursor of trypsin, a serine protease that cleaves proteins and peptides after arginine and lysine residues. In this study, human trypsinogen was used as a model protein to study the influence of electrostatic forces on protein–protein interactions. Trypsinogen is active only after its eight-amino-acid-long activation peptide has been cleaved off by another protease, enteropeptidase. Trypsinogen can also be autoactivated without the involvement of enteropeptidase. This autoactivation process can occur if a trypsinogen molecule is activated by another trypsin molecule and therefore is based on a protein–protein interaction.

Results: Based on a rational protein design based on autoactivation-defective guinea pig trypsinogen, several amino acid residues, all located far away from the active site, were changed to modify the surface charge of human trypsinogen. The influence of the surface charge on the activation pattern of trypsinogen was investigated. The autoactivation properties of mutant trypsinogen were characterized in comparison to the recombinant wild-type enzyme. Surface-charged trypsinogen showed practically no autoactivation compared to the wild-type but could still be activated by enteropeptidase to the fully active trypsin. The kinetic parameters of surface-charged trypsinogen were comparable to the recombinant wild-type enzyme.

Conclusion: The variant with a modified surface charge compared to the wild-type enzyme showed a complete different activation pattern. Our study provides an example how directed modification of the protein surface charge can be utilized for the regulation of functional protein–protein interactions, as shown here for human trypsinogen.

Keywords: Protein design, Protein expression, Protein-interaction, Protein engineering, trypsin

Background

Protein–protein interactions are based on non-covalent interactions of amino acids on protein surfaces via van der Waals forces, hydrogen bridges, electrostatic interactions and hydrophobic effects. Other processes that can be influenced by electrostatic interactions include for example protein folding and protein stability, protein denaturation and solubilization or precipitation of proteins [1]. In addition, the binding of substrates by enzymes, the subsequent catalytic processes and also the formation of protein–protein complexes can be based on electrostatic interactions.

The human enzyme trypsinogen shows a distinct distribution of protein surface charges and may be an attractive model protein to study the influence of electrostatic forces on protein–protein interactions. The activation of trypsinogen by its natural activation enzyme, the serine protease enteropeptidase, is determined by a close interaction between the two proteases. Trypsinogen is the inactive precursor of trypsin (PRSS1, human cationic trypsinogen), which can cleave proteins and peptides after lysine and arginine residues and is activated by the membrane-bound enteropeptidase via cleavage of the N-terminal activation peptide located on the trypsinogen surface. The resulting trypsin itself activates other pancreatic zymogens of the digestive pathway [2]. Besides the specific activation of trypsinogen by enteropeptidase it is well known that trypsinogen can show autoactivation [3].

This trypsinogen autoactivation process is mostly based on the fact that active trypsin can activate trypsinogen by cleaving off its activation peptide in a similar way as occurs with the activation enzyme enteropeptidase. Subsequently, the newly formed trypsin can then accelerate the

* Correspondence: Thole.Zuechner@octapharma.com
[1]Institute of Bioanalytical Chemistry, Center for Biotechnology and Biomedicine, Universität Leipzig, 04103 Leipzig, Germany
[2]Current address: Octapharma Biopharmaceuticals GmbH, Im Neuenheimer Feld 590, Heidelberg 69120, Germany

activation of other trypsinogen molecules in a cascade reaction.

Several mutations that influence the autoactivation of human trypsinogen have been published and are discussed in the context of the development of pancreatitis [4,5]. A number of studies have investigated the influence of different mutations on the autoactivation process [6-8]. In these studies, autoactivation was initiated by addition of a small amount of trypsin. As the same amount and type of trypsin was used throughout these studies, this ensures that autoactivation was characterized in a comparable way.

Some studies [3,9] have also hypothesized that trypsinogen autoactivation may not require the presence of either active trypsin or active enteropeptidase, but that trypsinogen itself could have an inherent, though minimal, catalytic function and could therefore activate itself. However, physiological conditions of course include the possibility that trypsinogen activation could also occur to a minor extent through the non-specific cleavage by proteases other than trypsin or enteropeptidase. In addition, it may be experimentally challenging to exclude the presence of even the smallest amount of active protease in order to investigate this problem. We therefore decided to study the autoactivation process by adding a small defined amount of active trypsin to ensure comparability with published studies and congruent results.

Trypsinogen autoactivation as observed for the human enzyme differs from that observed in other organisms [10]. For example, trypsinogen from guinea pig is defective in autoactivation under conditions where human cationic and anionic trypsinogen rapidly autoactivate [8]. The reason for this different autoactivation behavior has not been clear until now. The guinea pig variant has a sequence identity of 75% with human cationic trypsinogen. However, when we compared the number of charged amino acids between human and guinea pig trypsinogen, we found a striking difference in the number of charged amino acids (26 vs. 42 in human cationic trypsinogen). More than this, when we looked at the position of these charged amino acids within the protein model, it turned out that a striking difference in the surface charge distribution between human and guinea pig wild-type trypsinogen can be observed (Figure 1a and b). Based on this observation, our aim was to study the influence of a modified protein surface charge of human trypsinogen on specific enzyme properties, and in particular on the autoactivation. Our hypothesis was, that the distinct distribution of surface charges could potentially influence the activation pattern of the enzyme. To study this in detail, we also had to consider that guinea pig trypsinogen contains a lysine at position 79 instead of the typical glutamine residue found in other mammalian trypsinogens. The functional role of this side chain is unclear, but the Glu79Lys mutation (E79K) in human cationic trypsinogen seems to be associated with chronic pancreatitis and also leads to a lower autoactivation compared to the wild type [11].

The rational optimization of surface charge–charge interactions has been described before, for example as a strategy to enhance protein stability [12]. The advantages of this method are that mutations on the protein surface are less likely to significantly affect the core structure and therefore the function of the protein [13]. One example is the generation of highly thermoresistant antibodies by increasing their surface charge [14]. The mutation of surface residues to charged amino acids is also known to decrease protein aggregation. Previously, the method of supercharging was demonstrated to increase the resistance of

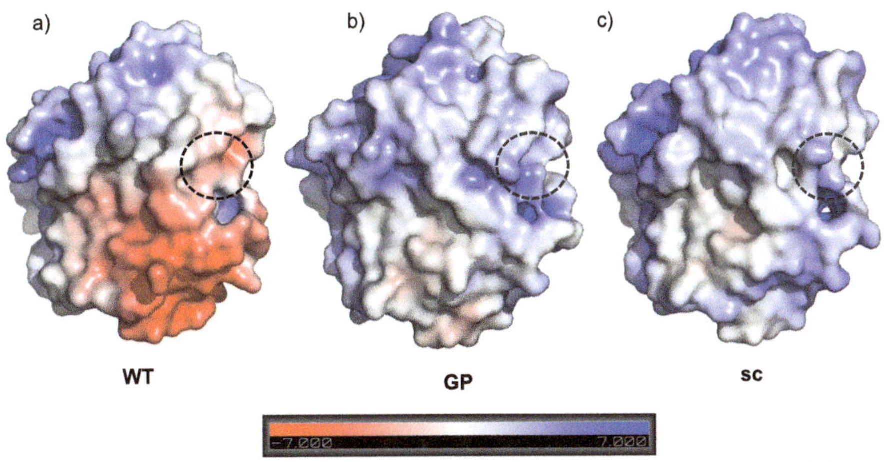

Figure 1 Electrostatic surface potential of (a) human cationic trypsinogen ("WT") in comparison to (b) modeled guinea pig trypsinogen ("GP") and (c) modified human trypsinogen ("sc"). The range of electrostatic surface potential is shown from -7 kT/e (red color) to +7 kT/e (blue color). N-terminal amino acids (V25 based on structure of trypsinogen 1TGN) are marked with a black circle. The substrate-binding cleft is located on the rear side of the shown model structures.

green fluorescent protein (GFP) to temperature-induced aggregation [15,16] to increase the solubility of a human enteropeptidase by more than 100-fold [17]. These studies reported the influence of protein surface charge on rather general protein properties such as solubility, aggregation properties and thermoresistance.

In this paper, we describe the modification of human trypsinogen by the exchange of several amino acid residues on the protein surface and its effect on specific properties of the protein such as the protein–protein interaction behavior of trypsinogen with enteropeptidase and active trypsin.

Results

Rational protein design

In this work, we examined the influence of the protein surface charge on protein–protein interactions of the serine protease human trypsinogen. Trypsinogen is an interesting model protein because it can be activated both by trypsin and by enteropeptidase. The resulting active trypsin can then contribute to the further activation of other trypsinogen molecules. As a result of the increasing amounts of active trypsin, a cascade reaction of autoactivation is observed. One interesting trypsinogen variant that shows nearly no autoactivation is the trypsinogen from guinea pig [8]. After we had done a sequence comparison between guinea pig and human cationic trypsinogen, several residues could be identified that are different in charge (Figure 2).

Interestingly, when we looked at the spatial distribution of these charged amino acids in the 3D structure of the protein, these charged amino acids are mostly located on the enzyme surface and distributed in a very distinct way (Figure 1). Inspired by this finding, we created a human trypsinogen mutant in which we tried to include a similar surface charge distribution as found by us for the guinea pig version. Although this kind of influence of a proteins surface charge on its activation pattern has never been described before, we aimed to examine its influence on the activation pattern of human trypsinogen.

After the amino acid sequence alignment of both variants, ten residues were identified that could lead to a different protein surface potential for trypsinogen. To prevent the undesirable loss of enzyme activity or specificity

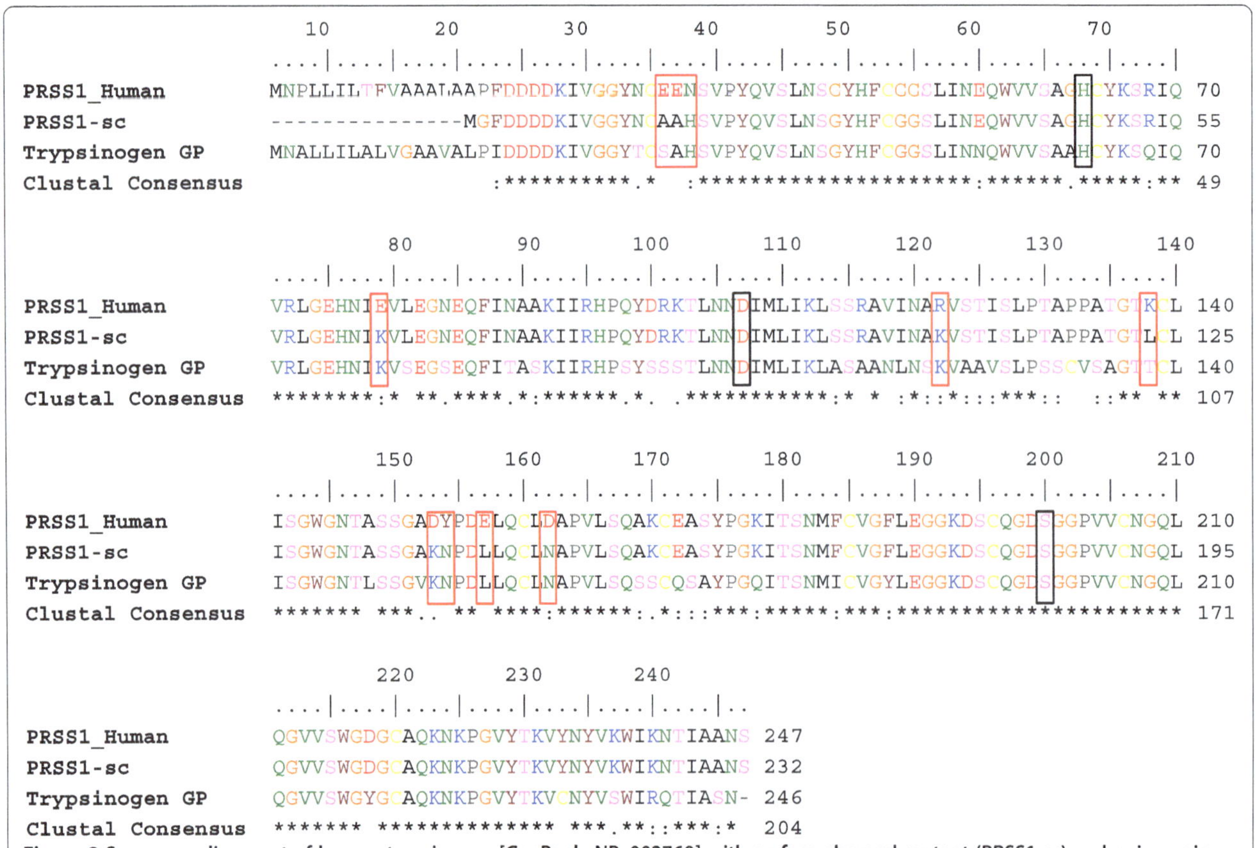

Figure 2 Sequence alignment of human trypsinogen [GenBank: NP_002760] with surface-charged mutant (PRSS1-sc) and guinea pig trypsinogen (Trypsinogen GP, [GenBank: NP_001166359]). Mutated positions of PRSS1 are E31A, E32A, N33H, E79K, R122K, K138L, D153K, Y154N, E157L, and D162N. The catalytic triad of trypsinogen is marked with black boxes. Exchanged amino acids in the mutant are indicated in red boxes. ClustalW [http://embnet.vital-it.ch/software/ClustalW.html] was used for the sequence alignment. Alignment was edited by Bioedit [http://www.mbio.ncsu.edu/bioedit/bioedit.html].

by mutations in the generated human trypsinogen variant, it was necessary to engineer the protein in such a way as to minimize the influence on the active center of the enzyme. Thus, mutated positions were checked to be on the surface of the protein and, in particular, to be distant from the activation sequence of the enzyme (distance >10 Å). The mutations were then performed in a way that the predicted electrostatic potential of the variant matched that of guinea pig trypsinogen.

In detail, the following ten mutations were introduced: E31A, E32A, N33H, E79K, R122K, K138L, D153K, Y154N, E157L, and D162N. We decided not to mutate position 138 like in the sequence of guinea pig trypsinogen, which has a threonine at this position to prevent posttranslational modification of OH-groups. The same consideration applied to position 31 and 154 [18,19].

In total, the trypsinogen sc variant has 16 negatively charged residues compared to the 22 in the wild-type enzyme and 21 positively charged residues compared to the 20 in the wild type. Thus, the net charge was increased from −2 to +5. Several mutations are known to potentially influence the autoactivation and autolysis of human trypsinogen like E79K and R122H [11,20]. Furthermore, the position R122 has the potential to influence the autolysis of the protein. It was therefore necessary to examine the influence of these positions on the human trypsinogen so as to allow a clear conclusion about the influence of the protein surface charge as compared to the effect of these point mutations. The effect of these mutations will be discussed later in this study.

Autoactivation of trypsinogen at pH 8

The autoactivation of human wild-type trypsinogen and the surface-charged variant (PRSS1-sc) was investigated at pH 8 and 37°C for several hours. To initiate the autoactivation and in accordance with several other studies [6-8], 10 nmol/L trypsin was added to 2 μmol/L trypsinogen. The time-course of the autoactivation was quantified by determination of the trypsin activity (Figure 3).

In agreement with previous studies, the human wild-type enzyme showed a high autoactivation [7]. This high autoactivation rate results in the complete activation of trypsinogen within 80 min. PRSS1-WT exhibits an activation rate constant of 30.9 nmol/L/min for activation of trypsinogen by trypsin. In contrast to this, the surface-charged PRSS1-sc variant showed nearly no detectable activity during the whole incubation time under the same conditions. This means, that the surface-charged trypsinogen variant shows a dramatically reduced activation rate when using trypsin as the activating enzyme. An activation rate constant of 0.00545 nmol/L/min, which is more than 5000 times slower than for the wild-type enzyme, was determined for the PRSS1-sc mutant. These results were also confirmed by SDS-PAGE analysis (Figure 4).

In SDS-PAGE, the band corresponding to human wild-type trypsinogen is detectable only in the 20 and 40 min samples. Even after 40 min the lower molecular weight band corresponding to the active trypsin molecule is the dominant species. In contrast, the surface-charged

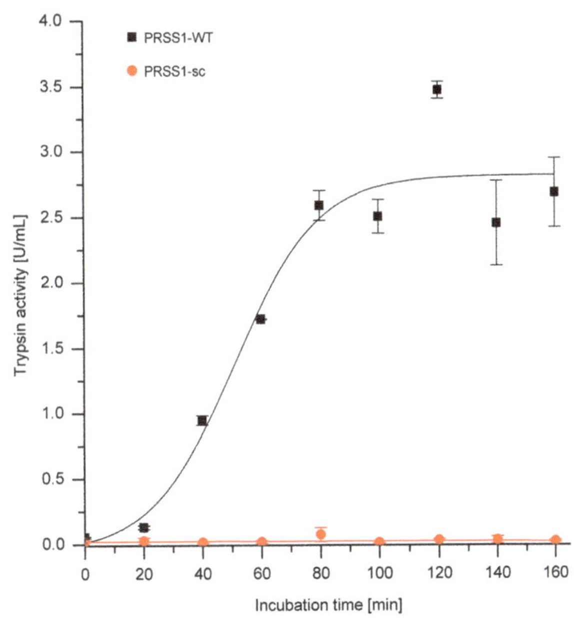

Figure 3 Autoactivation of trypsinogen at pH 8. Trypsinogen autoactivation was measured in 100 mmol/L Tris-HCl pH 8, 1 mmol/L $CaCl_2$ at 37°C with 2 μmol/L trypsinogen and 10 nmol/L trypsin as an initial starting concentration. Trypsin activity was determined with Cbz-GPR-pNA as substrate.

trypsinogen showed no detectable shift in the molecular weight and also no degradation of the enzyme. This confirms the finding of the activity measurement (Figure 3), i.e., that nearly no autoactivation occurs in the surface-charged variant.

The autoactivation of wild-type and surface-charged trypsinogen was also investigated at higher ionic strength to further confirm our hypothesis that autoactivation is influenced by the surface charge of the trypsinogen molecule and therefore the binding between the molecules. We could observe a decreased autoactivation of the wild-type trypsinogen compared to the experiment without the addition of salt (Figure 5).

The PRSS1-sc mutant contains the mutation E79K, besides the other mutations described above. The E79K mutation has previously been described to cause a decreased autoactivation [11]. Therefore, we also generated the point mutant PRSS1-E79K and investigated its autoactivation properties in the same way as for wild-type trypsinogen. Under the conditions used in this study, the E79K mutant showed an autoactivation that was comparable to that of the wild-type trypsinogen (Additional file 1: Figure S1). This finding seemingly contradicts the previously published result [11] where this mutant did not show autoactivation. However, in the study published by Teich et al., no trypsin was added to initiate the autoactivation and the activation behavior was not measured for a prolonged time. We additionally determin the autoactivation behavior of wild-type trypsinogen and PRSS1-E79K with the addition of a lower amount of trypsin (1 nmol/L). We could confirm that autoactivation mechanism is influenced by the amount of added trypsin. The addition of a lower amount of trypsin led to a decreased autoactivation for wild-type trypsinogen and also for mutant PRSS1-E79K (Additional file 1: Figure S1). We can therefore conclude that the E79K mutation did not significantly contribute to the observed decreased autoactivation in PRSS1-sc under the conditions used in this study.

Activation with enteropeptidase

In its physiological setting, trypsinogen is activated by enteropeptidase. We therefore examined the activation behavior of the human wild-type and the surface-charged trypsinogen by human enteropeptidase to see if the surface-charged variant could be activated in a normal manner. The characterization of the trypsin activity after enteropeptidase activation showed that the surface-charged trypsinogen could be activated to normal wild-type levels after 100 to 120 min (Figure 6). In comparison, the wild-type trypsinogen was already completely active after 40 min. The activity of surface charged trypsinogen increases up to 3.5 U/mL compared to wild-type trypsinogen with 2.5 U/mL activity. These differences are based on different specific activities of the variants (PRSS1-sc 69 U/mg and PRSS1-WT 54.6 U/mg). The detailed kinetic parameters of wild-type and surface charged trypsinogen can be found in Table 1. The surface-charged and wild-type trypsinogen were incubated with enteropeptidase over 4 hours at 37°C without showing a decreasing activity, indicating that both trypsinogen mutants are in a stable conformation.

The activation rate constant for the activation of PRSS1-WT by enteropeptidase was 86.4 nmol/L/min, which is slightly higher (a factor of 6.5) compared to the activation rate constant for PRSS1-sc (13.3 nmol/L/min). This means that the activation rate of mutant PRSS1-sc is slowed down by a factor of more than 5000 when trypsin is used as the activating enzyme, whereas the same mutant can still be activated to normal levels when enteropeptidase is used as the activating enzyme, albeit at a 6.5-fold slower rate.

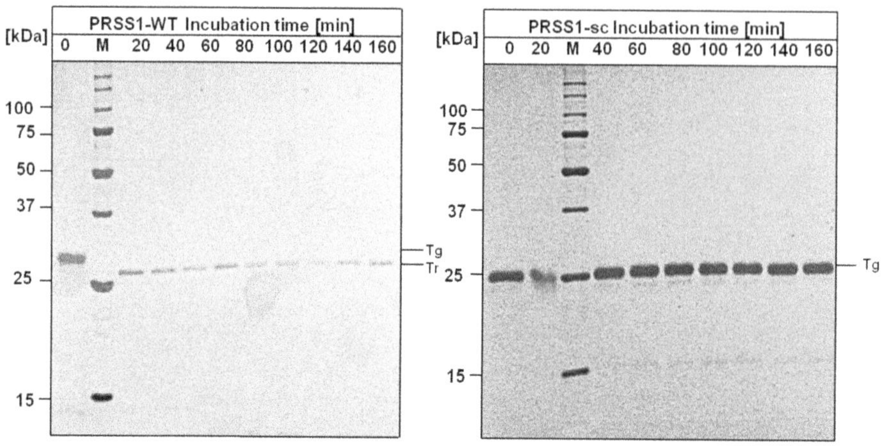

Figure 4 SDS-PAGE analysis of autoactivation of human trypsinogen. The samples were analyzed on a 15% reducing SDS-PAGE gel using silver staining. Reactions were terminated by adding SDS-PAGE sample buffer followed by an immediate denaturation step for 5 min at 95°C (Tg = trypsinogen; Tr = trypsin).

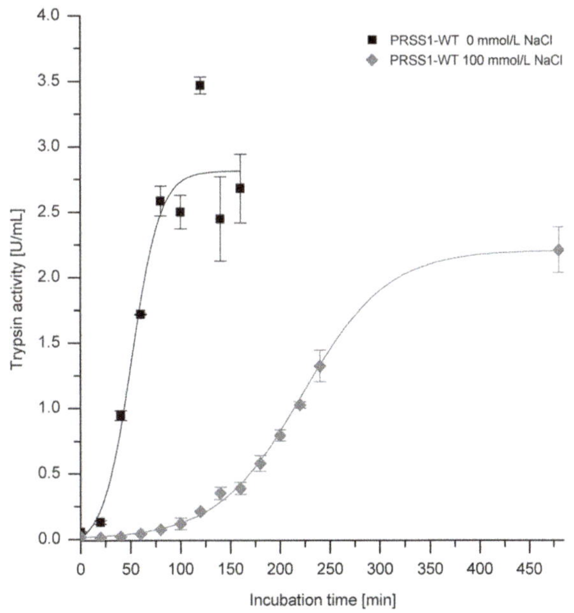

Figure 5 Autoactivation of trypsinogen at pH 8 and different salt concentrations. Trypsinogen autoactivation was measured in 100 mmol/L Tris-HCl pH 8, 1 mmol/L $CaCl_2$ and 100 mmol/L NaCl at 37°C with 2 μmol/L trypsinogen and 10 nmol/L trypsin as an initial starting concentration. Trypsin activity was determined with Cbz-GPR-pNA as substrate.

The course of this human enteropeptidase–mediated activation behavior of the surface-charged trypsinogen vs. wild-type trypsinogen was also examined via SDS-PAGE analysis (Figure 7). The results showed that it is possible to activate the surface-charged mutant to a full extent by using human enteropeptidase.

Two degradation bands are visible after activation. In previous studies, it was shown that these degradation bands are connected to the cleavage of the Arg122–Val123 peptide bond [21]. Arg122 is exchanged to a lysine residue in the surface-charged trypsinogen, which indicates that the Lys122–Val123 bond can also be cleaved by trypsin.

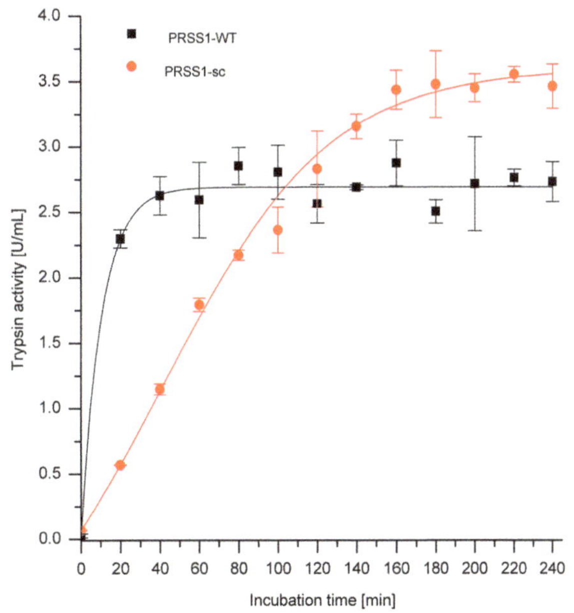

Figure 6 Activation of trypsinogen with human enteropeptidase. Enteropeptidase-mediated trypsinogen activation was measured in 0.1 mol/L Tris-HCl pH 8, 1 mmol/L $CaCl_2$ at 37°C with 2 μmol/L trypsinogen and hEPI-Sc-C112S (1 nmol/L final concentration).

Table 1 Kinetic parameters for trypsinogen with substrate Cbz-GPR-pNA

	K_m[µmol/L]	k_{cat}[1/s]	k_{cat}/K_m[L/µmol/s]
PRSS1-WT	78.0 ± 12.2	324.1 ± 13.8	4.7 ± 0.5
PRSS1-sc	129.8 ± 8.4	489.0 ± 18.9	3.7 ± 0.3
PRSS1-WT*	27 ± 3	99 ± 5	3.7

*Parameters as given by [31].

Kinetic parameters

To determine whether the mutations of surface-charged trypsinogen had an influence on the enzyme activity, Michaelis–Menten kinetic parameters were determined. Although the K_m value of the surface-charged trypsinogen was slightly increased compared to wild-type trypsinogen, no major differences for K_m, k_{cat} or k_{cat}/K_m values were observed (Table 1, Additional file 2 and Additional file 3). PRSS1-WT had a specific activity of 54.6 U/mg. In comparison, PRSS1-sc had a slightly increased specific activity of 69 U/mg. The catalytic efficiency (k_{cat}/K_m value) showed no difference for wild-type trypsinogen and surface-charged trypsinogen, indicating that the active center of the enzyme was not influenced.

To summarize these results, we found that the directed mutation of the trypsinogen surface charge led to an almost complete loss of trypsinogen autoactivation, while the enzyme could still be completely activated by its natural activation enzyme human enteropeptidase. These results were confirmed by activity measurements and by SDS-PAGE. Although the kinetic parameters of wild-type and surface-charged human trypsinogen for small peptide substrates were essentially the same, the activation of the surface-charged variant by human enteropeptidase occurred slightly slower than in the case of the wild-type enzyme.

Discussion

Enzyme autoactivation is a specific type of protein–protein interaction that is not only known to trypsinogen, but also for a number of other enzymes, including proteases such as pepsinogen and prothrombin [3,22-24]. Although different autoactivation mechanisms for the above mentioned enzymes exist, the intermolecular mechanism behind the trypsinogen autoactivation could potentially be a model for investigating the influence of protein surface charges on protein-protein interactions.

In this study, a human trypsinogen mutant was created with a distinct distribution of a surface charge pattern. This idea was based on our finding, that a trypsinogen variant from guinea pig which is defective in autoactivation shows a very specific distribution of surface charges. When we created a human variant in which a similar surface charge distribution was introduced, this led to a mutant of the human trypsinogen that showed a dramatically reduced autoactivation.

Lawrence et al. demonstrated that proteins can be highly supercharged [15] without abolishing their folding or function. Specifically, the net charge of GFP was increased by several amino acid exchanges that led to variants which remained entirely soluble upon thermal and chemical treatment. Supercharging altered the intermolecular properties of the proteins, providing aggregation resistance and the ability to associate in folded form with oppositely charged macromolecules [15]. The supercharging of GFP was quite intense, with an increase of the net charge from −7 to +36 or −30 [15,16]. In the structure-based design of supercharged, highly thermoresistant antibodies, the best mutant had an increased positive charge by eight [14]. In that study, fewer substitutions in the single-chain variable fragment antibody

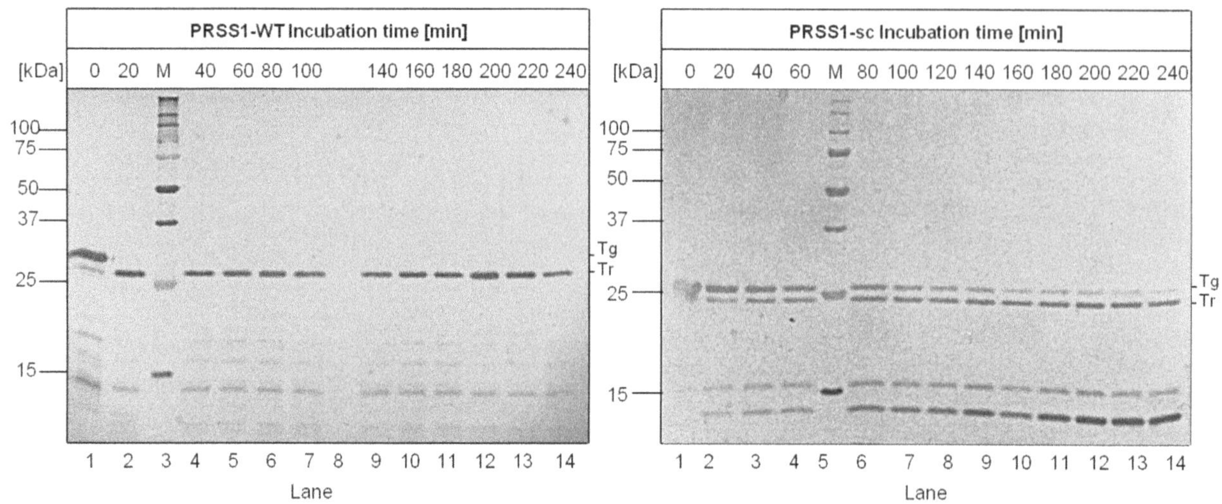

Figure 7 Activation of trypsinogen with human enteropeptidase. The samples were analyzed on a 15% reducing SDS-PAGE gel using silver staining. Reactions were terminated by adding SDS-PAGE sample buffer followed by an immediate denaturation step for 5 min at 95°C.

were required to achieve improvements in thermostability and affinity. The method of surface charging was also used by us in an earlier study to improve the solubility and thermostability of human enteropeptidase [17]. In this case, several mutations were introduced that led to six additional negative charges. The increased net charge led to a more than 100-fold higher solubility of the enzyme.

In this study, the surface charge of trypsinogen was changed by mutation of several amino acids, which resulted in a variant with practically no autoactivation.

Numerous amino acid substitutions have been discussed in the literature with respect to their effect on the stability of human trypsin. A mutation that was described in the literature and also examined in this study is located at position Arg122. This position is known to influence the autolytic – not the autoactivation – properties of the enzyme. The autolytic cleavage of the Arg122–Val123 peptide bond was suggested to trigger rapid trypsin degradation by increasing structural flexibility and exposing further tryptic sites [25]. Mutations at position 122, such as exchange to histidine or cysteine, inhibited the autolysis of trypsin [26-28] and enhanced the autoactivation of trypsinogen [26,28]. In this study, the mutation R122K did influence the self-cleavage of the enzyme, as shown by the SDS-PAGE gels. Furthermore, no decreasing activity was determined during activation with enteropeptideas and no increasing effect on the autoactivation was observed. We assume that the effect of a mutation at position 122 is too weak to cause a significant effect in the context of the surface charging we performed.

In this study, we could also show that the surface charge influences the autoactivation behavior of trypsinogen by investigating this property at different salt concentrations. Based on our results we could conclude that autoactivation is influenced by the surface charge of the protein because the autoactivation is decreased at higher ionic strength conditions.

The net charge of the modified protein was changed from a negative charge (−2) in the wild-type trypsinogen to a positive charge (+5) in the mutant. In total the human surface-charged trypsinogen has 16 negatively charged residues and 21 positively charged residues. More specifically, human wild-type trypsinogen shows a distinct negative surface potential on one side of the protein (Figure 1). Interestingly, these negative charges are mostly present close to the activation sequence of the protein. We therefore hypothesize that the negatively charged domains in the human wild-type trypsinogen allow an interaction with the positively charged surface areas of another trypsinogen molecule, and that this interaction leads to the autoactivation of the enzyme. While this study does not prove that the wild-type autoactivation mechanism is directly controlled by the surface charge of the protein and combinatorial effects may apply, it is nevertheless clear that a modified surface charge strongly affects the autoactivation behavior. At the same time, enteropeptidase can still activate the modified trypsinogen with an only moderately reduced efficiency, indicating that the interaction of these two proteins is determined by a different pattern of electrostatic interactions.

Additionally, it is also possible that reduced autoactivation of surface-charged trypsinogen is based on other effects like post-translational modifications, higher stability of the enzyme and/or conformational changes. As already stated, the advantage of a rational protein design is that the mutations on protein surface are less likely to significantly affect the core structure and therefore the function of the protein [13]. Trypsinogen mutant was designed by reducing possibility of post-translational modifications and we could also show that the catalytic efficiency has not changed. We also could show that autoactivation of wild-type trypsinogen is decreased at higher ionic strength (Figure 5). This proves that the modified surface charge of the protein can not only control protein–protein interactions in general, but also selectively, as in case of our model enzyme trypsinogen. This selective influence of a protein surface charge on its interaction with other proteins has not been demonstrated by us or, to the best of our knowledge, by anybody else until now.

Conclusion

In this study, we were able to reduce the autoactivation of human trypsinogen by the modification of its protein surface charge. The mutation of ten amino acids led to an increase in the net charge of the enzyme and resulted in a reduced autoactivation, which is probably due to specifically disturbed protein–protein interactions between two trypsin(ogen) molecules. The activation of trypsinogen by enteropeptidase was, in contrast, only influenced in a minor way. This gives a strong indication, that the protein surface charge of trypsinogen has a specific regulatory function on the protein-protein interaction pattern of the enzyme.

Methods

Materials

Ecotin was expressed in *E. coli* BL21 as described by Pál et al., 1996 [29,30] and purified via a trypsin affinity column. Purified ecotin was coupled to a pre-packed NHS-activated Sepharose column (GE Healthcare) according to the manufacturer's protocol.

Host strains *E. coli* DH5α, *E. coli* BL21 (DE3) and plasmid pET28a(+) were from Novagen (Darmstadt, Germany). *N*-Cbz-Gly-Pro-Arg-*p*-nitroaniline was purchased from Bachem AG (Bubendorf, Switzerland). Human enteropeptidase hEPL-Sc-C112S was prepared according to Simeonov et al., 2010 [17]. Chemicals were of the highest purity available.

Construct human trypsinogen mutant

Full-length cDNA of PRSS1 (human cationic trypsinogen) was from Deutsches Ressourcenzentrum für Genomforschung GmbH (Berlin, Germany). The plasmid (PRSS1 N-2X in vector full ORF shuttle clone IOH 43660) contained the full-length cDNA of PRSS1 [GenBank: NM_002769.4] and a stop codon.

The gene was amplified via polymerase chain reaction (PCR) using the forward primer 5'-ggagatataccatgggcttt gatgatgatgacaag-3' and the reverse primer 5'-cttgtcatcat catcaaagcccatggtatatctcc-3' resulting in the N-terminal end of the protein Met-Gly-Phe-Asp_4-Lys. It was shown that the modification of the N-terminus of trypsinogen enhances the expression level [31]. For the reason in the present study the N-terminus was also changed to Met-Gly-Phe-Asp_4-Lys.

The PCR was performed by heating for 2 min at 95°C, followed by 18 cycles of 95°C for 30 s, 54°C for 1 min, 68°C for 1 min and an additional extension time at 68°C for 2 min. The amplified PCR product was analyzed via a 1% agarose gel and purified via a gel extraction kit (Qiagen, Hilden, Germany). The purified DNA sequence was cloned into the pET28a vector via restriction sites *Nco*I and *Xho*I. *E. coli* BL21 (DE3) was transformed with construct pET28a-PRSS1dAP.

The DNA of the surface-charged trypsinogen was obtained via gene synthesis (Eurofins MWG Operon, Ebersberg, Germany). The N-terminal end was modified as described above for wild-type trypsinogen.

Expression and purification of human trypsinogen

Trypsinogen was expressed as inclusion bodies in *E. coli* cells. *E. coli* BL21 (DE3) cells harboring the expression vector with the PRSS1 gene were grown in TB medium with 30 µg/mL kanamycin at 37°C until an OD_{600} of 0.8 to 1 was reached. The expression was induced by adding 1 mmol/L IPTG and glucose to a final concentration of 1% (w/v). The cells were incubated for 18 h at 25°C and then harvested by centrifugation (20 min, 5000 *g*).

Inclusion bodies were purified by resuspending the cell pellet in 100 mmol/L Tris-HCl (pH 7.0) 1 mmol/L EDTA and 3 mmol/L $MgCl_2$ containing 0.1 mg/mL lysozyme and 10 µg/mL DNase I. The cells were disrupted with a high-pressure homogenizer (two passages of 1000 bar). After incubation for 30 min at 24°C, disrupted cells were centrifuged (20 000 *g*, 4°C, 30 min).

The inclusion bodies comprising the trypsinogen were prepared from the insoluble fraction after the cell disruption as follows: inclusion bodies were resuspended in IB washing buffer I (20 mmol/L EDTA pH 8.0, 500 mmol/L NaCl, 2% Triton X-100), stirred for 30 min at room temperature and centrifuged for 30 min at 20 000 *g*. The resulting pellet was washed again with IB washing buffer I and twice with IB washing buffer II (20 mmol/L EDTA, 100 mmol/L Tris-HCl, pH 7.0).

1 g of inclusion bodies were solubilized in 20 mL 4 mol/L Gnd-HCl, 0.1 mol/L Tris-HCl pH 8.0, 5 mmol/L EDTA, and then reduced with DTT (final concentration 30 mmol/L) for 20 min at 60°C. Afterwards, DTT was removed via a desalting column (GE Healthcare). The refolding was done via fast dilution of protein (final concentration 100 µg/mL) in 0.7 mol/L Arg-HCl (pH 8.6), 1 mmol/L EDTA, 2 mmol/L reduced glutathione and 2 mmol/L oxidized glutathione. The refolding mixture was incubated for 16 h at 4°C. The refolded trypsinogen was subsequently purified via an ecotin affinity column.

Protein concentrations of trypsinogen were determined spectrophotometrically at 280 nm using an extinction coefficient of 37525 L/mol/cm for PRSS1-WT [calculated by Protparam from http://www.expasy.org/].

Trypsin activity assay

Trypsin activity was determined using the synthetic substrate *N*-Cbz-Gly-Pro-Arg-*p*-nitroaniline (final concentration 200 µmol/L) with an absorption readout at 405 nm in 0.1 mol/L Tris-HCl (pH 8.0) 1 mmol/L $CaCl_2$ at 25°C.

Activation of trypsinogen by enteropeptidase

Trypsinogen (2 µmol/L final concentration) was activated by human enteropeptidase hEPl-Sc-C112S (1 nmol/L final concentration) in 0.1 mol/L Tris-HCl (pH 8.0) 1 mmol/L $CaCl_2$ at 37°C. At given times, aliquots for the trypsin activity assay and SDS-PAGE analysis were taken.

The activation rate constant was calculated for wild-type and surface charged trypsinogen based on the linear slope of the increasing trypsin activity. Together with the specific activity we then calculated how many trypsinogen molecules are activated by trypsin per minute.

Autoactivation of trypsinogen

Trypsinogen in 50 mmol/L HCl was diluted to a 2 µmol/L final concentration in 0.1 mol/L Tris-HCl pH 8.0 in the presence of 1 mmol/L $CaCl_2$. Autoactivation was initiated by addition of 10 nmol/L bovine trypsin and the reaction mixture was incubated at 37°C. At given times, aliquots for the trypsin activity assay and SDS-PAGE analysis were taken.

SDS-PAGE

Autoactivation of trypsinogen was also visualized by gel electrophoresis and silver staining. Samples were heat denatured at 95°C for 5 min and loaded onto 15% mini-gels (Miniprotean; Bio-Rad Laboratories, Hercules, CA, USA). Silver staining was performed according to Shevchenko et al., 1996 [32].

Modeling

Models for the structure of guinea pig trypsinogen and human trypsinogen were generated with the program MODELLER using the bovine trypsinogen structure [PDB: 1TGN] as template [33]. The atomic coordinates of the structures were obtained from the Protein Databank at the RCSB [http://www.rcsb.org/pdb/home/home.do]. The model with the lowest DOPE score [34] obtained from MODELLER was refined with fragment-guided MD simulation (FG-MD) [http://zhanglab.ccmb.med.umich.edu/FG-MD/] [35].

The program Coot was used for modeling of mutations in the human trypsin structure [36]. For visualization of the protein structure and the surface potential PyMol was used [37]. The surface potential was calculated by PDB2PQR [38,39].

Additional files

Additional file 1: Autoactivation of wild-type and E79K trypsinogen at pH 8. Trypsinogen autoactivation was measured in 100 mmol/L Tris-HCl pH 8, 1 mmol/L $CaCl_2$ at 37°C with 2 µmol/L trypsinogen and 1 nmol/L or 10 nmol/L trypsin as an initial starting concentration. Trypsin activity was determined with CBZ-GPR-pNA as substrate.

Additional file 2: Kinetic analysis of PRSS1-WT. Kinetic parameters are given in Table 1. Enzyme was activated by human enteropeptidase (hEPI-Sc-C112S) for 200 min at 37°C in 100 mmol/L Tris-HCl pH 8.0, 1 mmol/L $CaCl_2$. Assays were performed in 100 mmol/L Tris-HCl pH 8.0, 1 mmol/L $CaCl_2$ and 10 µmol/L to 1000 µmol/ L CBZ-GPR-pNA at room temperature. The reaction was started by adding PRSS1 variants (12.5 ng) and monitored continuously for 5 min by an absorbance measurement at 405 nm (extinction coefficient for p-nitroaniline $\varepsilon = 10092$ L/mol/cm).

Additional file 3: Kinetic analysis of PRSS1-sc. Kinetic parameters are given in Table 1. Enzyme was activated by human enteropeptidase (hEPI-Sc-C112S) for 200 min at 37°C in 100 mmol/L Tris-HCl pH 8.0, 1 mmol/L $CaCl_2$. Assays were performed in 100 mmol/L Tris-HCl pH 8.0, 1 mmol/L $CaCl_2$ and 10 µmol/L to 1000 µmol/L CBZ-GPR-pNA at room temperature. The reaction was started by adding PRSS1 variants (12.5 ng) and monitored continuously for 5 min by an absorbance measurement at 405 nm (extinction coefficient for p-nitroaniline $\varepsilon = 10092$ L/mol/cm).

Competing interests

A european patent application about the surface-charged human trypsinogen mutant was filed (EP13187194).

Authors' contributions

KB designed and performed the experiments and wrote the manuscript. TK helped with the discussion and wrote the manuscript. NS helped with protein modeling and wrote the manuscript. TZ helped with design of all experiments and wrote the manuscript. All authors corrected the manuscript. All authors read and approved the final manuscript.

Acknowledgements

We gratefully acknowledge Stefanie Langanke for the expert technical assistance, Renato Weiße for his support with modeling trypsinogen structures and Dr. Andrew Hagan for proofreading. This work was supported by the Federal Ministry of Education and Research (BMBF) Go-Bio project no. 0315988 to TZ.

References

1. Gitlin I, Carbeck JD, Whitesides GM: **Why are proteins charged? Networks of charge-charge interactions in proteins measured by charge ladders and capillary electrophoresis.** *Angew Chem Int Ed Engl* 2006, **45:**3022–3060.
2. Whitcomb D, Lowe M: **Human pancreatic digestive enzymes.** *Dig Dis Sci* 2007, **52:**1–17.
3. Kay J, Kassell B: **The autoactivation of trypsinogen.** *J Biol Chem* 1971, **246:**6661–6665.
4. Sahin-Tóth M: **Biochemical models of hereditary pancreatitis.** *Endocrinol Metab Clin North Am* 2006, **35:**303–312.
5. Teich N, Rosendahl J, Tóth M, Mössner J, Sahin-Tóth M: **Mutations of human cationic trypsinogen (PRSS1) and chronic pancreatitis.** *Hum Mutat* 2006, **27:**721–730.
6. Kiraly O, Guan L, Szepessy E, Toth M, Kukor Z, Sahin-Toth M: **Expression of human cationic trypsinogen with an authentic N terminus using intein-mediated splicing in aminopeptidase P deficient Escherichia coli.** *Protein Expres Purif* 2006, **48:**104–111.
7. Nemoda Z, Sahin-Tóth M: **The tetra-aspartate motif in the activation peptide of human cationic trypsinogen is essential for autoactivation control but not for enteropeptidase recognition.** *J Biol Chem* 2005, **280:**29645–29652.
8. Ózsvári B, Hegyi P, Sahin-Tóth M: The Guinea Pig Pancreas Secretes a Single Trypsinogen Isoform, Which is defective in autoactivation. *Pancreas* 2008, **37**(2):182–188.
9. Pasternak A, Liu X, Lin T, Hedstrom L: **Activating a zymogen without proteolytic processing: mutation of Lys15 and Asn194 activates trypsinogen†.** *Biochemistry* 1998, **37:**16201–16210.
10. Broderick JW, LARGMAN C, JOHNSON JH, GEOKAS MC: **Human cationic trypsinogen. Purification, characterization, and characteristics of autoactivation.** *J Biol Chem* 1978, **253:**2732–2736.
11. Teich N, Le Maréchal C, Kukor Z, Caca K, Witzigmann H, Chen J, Tóth M, Mössner J, Keim V, Férec C, Sahin-Tóth M: **Interaction between trypsinogen isoforms in genetically determined pancreatitis: mutation E79K in cationic trypsin (PRSS1) causes increased transactivation of anionic trypsinogen (PRSS2).** *Hum Mutat* 2004, **23:**22–31.
12. Strickler SS, Gribenko AV, Gribenko AV, Keiffer TR, Tomlinson J, Reihle T, Loladze VV, Makhatadze GI: **Protein stability and surface electrostatics: a charged relationship.** *Biochemistry* 2006, **45:**2761–2766.
13. Sanchez-Ruiz JM, Makhatadze GI: **To charge or not to charge?** *Trends Biotechnol* 2001, **19:**132–135.
14. Miklos AE, Kluwe C, Der BS, Pai S, Sircar A, Hughes RA, Berrondo M, Xu J, Codrea V, Buckley PE, Calm AM, Welsh HS, Warner CR, Zacharko MA, Carney JP, Gray JJ, Georgiou G, Kuhlman B, Ellington AD: **Structure-based design of supercharged, highly thermoresistant antibodies.** *Chem Biol* 2012, **19:**449–455.
15. Lawrence MS, Phillips KJ, Liu DR: **Supercharging proteins can impart unusual resilience.***J Am Chem Soc* 2007, **129:**10110–10112.
16. McNaughton BR, Cronican JJ, Thompson DB, Liu DR: **Mammalian cell penetration, siRNA transfection, and DNA transfection by supercharged proteins.** *Proc Natl Acad Sci U S A* 2009, **106:**6111–6116.
17. Simeonov P, Berger-Hoffmann R, Hoffmann R, Sträter N, Zuchner T: **Surface supercharged human enteropeptidase light chain shows improved solubility and refolding yield.** *Protein Eng Des Sel* 2011, **24:**261–268.
18. Gaboriaud C, Serre L, Guy-Crotte O, Forest E, Fontecilla-Camps JC: **Crystal structure of human trypsin 1: unexpected phosphorylation of Tyr151.** *J Mol Biol* 1996, **259:**995–1010.
19. Sahin-Tóth M, Kukor Z, Nemoda Z: **Human cationic trypsinogen is sulfated on Tyr154.** *FEBS J* 2006, **273:**5044–5050.
20. Sahin-Toth M, Gráf L, Tóth M: **Trypsinogen stabilization by mutation Arg117→His: a unifying pathomechanism for hereditary pancreatitis?** *Biochem Biophys Res Commun* 1999, **264:**505–508.
21. Kukor Z, Toth M, Pal G, Sahin-Toth M: **Human cationic trypsinogen - Arg (117) is the reactive site of an inhibitory surface loop that controls spontaneous zymogen activation.** *J Biol Chem* 2002, **277:**6111–6117.
22. Bustin M, Conway-Jacobs A: **Intramolecular activation of porcine pepsinogen.** *J Biol Chem* 1971, **246:**615–620.
23. al-Janabi J, Hartsuck JA, Tang J: **Kinetics and mechanism of pepsinogen activation.** *J Biol Chem* 1972, **247:**4628–4632.
24. Barranco-Medina S, Pozzi N, Vogt AD, Di Cera E: **Histone H4 promotes prothrombin autoactivation.** *J Biol Chem* 2013, **288:**35749–35757.

25. Várallyay É, Pál G, Patthy A, Szilágyi L, Gráf L: Two mutations in Rat trypsin confer resistance against autolysis. *Biochem Biophys Res Commun* 1998, **243**:56–60.
26. Sahin-Toth M, Toth M: Gain-of-function mutations associated with hereditary pancreatitis enhance autoactivation of human cationic trypsinogen. *Biochem Bioph Res Co* 2000, **278**:286–289.
27. Sahin-Toth M: The pathobiochemistry of hereditary pancreatitis: studies on recombinant human cationic trypsinogen. *Pancreatology* 2001, **1**:461–465.
28. Simon P, Weiss FU, Sahin-Toth M, Parry M, Nayler O, Lenfers B, Schnekenburger J, Mayerle J, Domschke W, Lerch MM: Hereditary pancreatitis caused by a novel PRSS1 mutation (Arg-122 - Cys) that alters autoactivation and autodegradation of cationic trypsinogen. *J Biol Chem* 2002, **277**:5404–5410.
29. Pál G, Sprengel G, Patthy A, Gráf L: Alteration of the specificity of ecotin, an E. coli serine proteinase inhibitor, by site directed mutagenesis. *FEBS Lett* 1994, **342**:57–60.
30. Pál G, Szilágyi L, Gráf L: Stable monomeric form of an originally dimeric serine proteinase inhibitor, ecotin, was constructed via site directed mutagenesis. *FEBS Lett* 1996, **385**:165–170.
31. Sahin-Toth M: Human cationic trypsinogen - Role of Asn-21 in zymogen activation and implications in hereditary pancreatitis. *J Biol Chem* 2000, **275**:22750–22755.
32. Shevchenko A, Wilm M, Vorm O, Mann M: Mass spectrometric sequencing of proteins silver-stained polyacrylamide gels. *Anal Chem* 1996, **68**:850–858.
33. Eswar N, Webb B, Marti-Renom MA, Madhusudhan MS, Eramian D, Shen M, Pieper U, Sali A: Comparative protein structure modeling using Modeller. *Curr Protoc Bioinformatics* 2006, **5**:Unit 5.6.
34. Shen M, Sali A: Statistical potential for assessment and prediction of protein structures.*Protein Sci* 2006, **15**:2507–2524.
35. Zhang J, Liang Y, Zhang Y: Atomic-level protein structure refinement using fragment-guided molecular dynamics conformation sampling. *Structure* 2011, **19**:1784–1795.
36. Emsley P, Lohkamp B, Scott WG, Cowtan K: Features and development of Coot. *Acta Crystallogr D Biol Crystallogr* 2010, **66**:486–501.
37. Schrödinger LL: *The PyMOL Molecular Graphics System, Version 1.3r1*. 2010.
38. Dolinsky TJ, Czodrowski P, Li H, Nielsen JE, Jensen JH, Klebe G, Baker NA: PDB2PQR: expanding and upgrading automated preparation of biomolecular structures for molecular simulations. *Nucleic Acids Res* 2007, **35**:W522–W525.
39. Dolinsky TJ, Nielsen JE, McCammon JA, Baker NA: PDB2PQR: an automated pipeline for the setup of Poisson-Boltzmann electrostatics calculations. *Nucleic Acids Res* 2004, **32**:W665–W667.

25

Efficient expression of codon-adapted affinity tagged super folder green fluorescent protein for synchronous protein localization and affinity purification studies in *Tetrahymena thermophila*

Gürkan Yilmaz[1†] and Muhittin Arslanyolu[2*†]

Abstract

Background: A superior Green Fluorescent Protein (GFP) mutant, known as superfolder GFP (sfGFP), is more soluble, faster folding, and is the brightest of the known GFP mutants. This study aimed to create a codon-adapted sfGFP tag (TtsfGFP) for simultaneous protein localization and affinity purification in *Tetrahymena thermophila*.

Results: In vivo fluorescence spectroscopic analyses of clones carrying a codon-adapted and 6 × His tagged TtsfGFP cassette showed approximately 2–4-fold increased fluorescence emission compared with the control groups at 3 h. Fluorescence microscopy also revealed that TtsfGFP reached its emission maxima at 100 min, which was much earlier than controls expressing EGFP and sfGFP (240 min). A *T. thermophila* ATP-dependent DNA ligase domain containing hypothetical gene (H) was cloned into the 3' end of 6 × His-TtsfGFP to assess the affinity/localization dual tag feature. Fluorescence microscopy of the 6 × His-TtsfGFP-H clone confirmed its localization in the macro- and micronucleus of vegetative *T. thermophila*. Simultaneous affinity purification of TtsfGFP and TtsfGFP-H with Ni-NTA beads was feasible, as shown by Ni-NTA purified proteins analysis by SDS-PAGE and western blotting.

Conclusions: We successfully codon adapted the N-terminal 6 × His-TtsfGFP tag and showed that it could be used for protein localization and affinity purification simultaneously in *T. thermophila*. We believe that this dual tag will advance *T. thermophila* studies by providing strong visual traceability of the target protein in vivo and in vitro during recombinant production of heterologous and homologous proteins.

Keywords: Codon adaptation, Super folder GFP, *Tetrahymena thermophila*, Affinity purification, Protein localization, Dual tag

Background

Homologous and heterologous expression of recombinant proteins in the unicellular ciliate *Tetrahymena thermophila* is frequently used in biological or biotechnological studies. The use of *T. thermophila* as an alternative eukaryotic host for the expression of recombinant proteins is based on some of its advantages, such as shorter cell division time, applicability of sterile cell culture techniques, maintenance of strains in liquid nitrogen, possibility of transformation using biolistic guns and electroporation, and the ability to introduce post-translational modifications, such as glycosylation [1,2]. However, *T. thermophila* has a disadvantage for heterologous protein expressions because of its use of an alternative codon dictionary with biased codon frequencies [3-5]. Additionally, the lengthy transformation protocol, which is associated with low transformation efficiency, needs to be improved to enable *T. thermophila* to become a more widely used eukaryotic host for recombinant protein production. The requirement for improved protein tags for protein localization in *T. thermophila* studies has triggered efforts to develop or adapt tags with different features. In *T. thermophila* studies, the most commonly used protein localization tag is the enhanced green fluorescent protein (EGFP), which is more stable and emits

* Correspondence: marslanyolu@anadolu.edu.tr
[†]Equal contributors
[2]Department of Biology, Faculty of Science, Anadolu University, Yunusemre Campus, Eskisehir, Turkey
Full list of author information is available at the end of the article

brighter fluorescence than wild-type GFP (WT-GFP). EGFP has been codon adapted for *T. thermophila* recently [6]. However, a novel, more advanced version of EGFP has been developed that needs to be adapted for *T. thermophila* studies.

The wild-type GFP (WT-GFP) is a ~26 kDa protein that emits green fluorescent light when exposed to the blue to ultraviolet spectral range [7,8]. The WT-GFP tag suffers from low fluorescence, comparative dimerization, aggregation, and high sensitivity to pH, making it inefficient. A variety of mutations have been introduced to make WT-GFP more soluble, stable, and have brighter emission characteristics [7,9]. GFP mutants have been widely used as visual marker proteins in developmental and cell biology studies [10,11]. Recently, a mutant, of sfGFP [GenBank accession no. HI069813.1 and Synthetic Sequence 3 from Patent EP1853717] was developed, which has superior features among GFP mutants, such as higher solubility, brighter fluorescence, faster folding, and higher resistance to denaturants such as urea and formamide [12]. These biochemical properties are attributed to the introduction of enhanced GFP mutations (F64L and S65T), cycle-3 GFP mutations (F99S, M153T, and V163A), and Super folder GFP mutations (S30R, Y39N, N105T, Y145F, I171V, and A206V) [12-14]. Hence, sfGFP also imparts solubility and enables proper folding of poorly folding fusion proteins [12,15]. Additionally, the introduction of a short affinity sequence, such as polyhistidine, to GFP resolves the lack of an affinity tag [16-20].

In some cases of heterologous protein expression, translation could be interrupted or terminated if the tag protein is used without codon adaptation. Therefore, tag and/or target gene sequences must be codon adapted to the host organism by the introduction of silent mutations [21-25]. The most commonly used protein localization tag in *T. thermophila* is a S65T mutant version of EGFP [3,6,10,11,26-32], which has recently been codon adapted for knockout studies in *T. thermophila* as a C-terminal protein localization tag [6].

To reduce the cost and experimental time of *T. thermophila* studies, this study aimed to develop an advanced dual fluorescence tag based on a *T. thermophila* codon-adapted sfGFP and an affinity tag such as 6 × His for simultaneous protein localization and affinity purification of the desired fusion protein. This type of dual tag will enable the tracking of protein production during in vivo and in vitro studies in *T. thermophila* studies.

Results

Tetrahymena codon adaptation of the Superfolder GFP gene

Analysis of the *Escherichia coli*-adapted synthetic sfGFP nucleotide sequence (encoding a 237 aa protein) showed that 57.72% of its codons (126 of 239 codons) or 21.2% of its sequence (153 of 714 bp) was not suitable for expression in *T. thermophila*. However, it was reported that the introduction of frequently used codons would help to improve gene expression [4,5]. Therefore, these non-adapted codons of sfGFP were re-assigned to *T. thermophila*'s frequently used codons as silent mutations, based on the codon frequencies, possible wobbling, and copy numbers of tRNA gene in the *T. thermophila* macronuclear genome, to avoid problems in protein expression [33]. The comparison of sfGFP and TtsfGFP sequences showed that more than 97% of the optimized codons (123 of 126 codons) had a change at the third position. However, only 20% of the changes (26 of 126 codons) were made at the first position, and changes in the middle base were limited to only two codons (Figure 1). The AT richness of TtsfGFP changed by only 6.1%, increasing from 57.8% to 63.9%.

Codon-adapted TtsfGFP shows earlier and higher emission

In this study, replacing the EGFP-Drp1 (Dynamin Related Protein 1) fusion gene in vector pVGF [10,34] with the 6 × His-TtsfGFP dual tag produced vector pVTtsfGFP (Figure 2). *T. thermophila* clones expressing TtsfGFP, sfGFP, and EGFP were grown and induced. Fluorescence microscopic analysis showed that TtsfGFP carrying clones began to emit fluorescence at 20 min, attaining maximum emission at 100 min (Figure 3). By contrast, sfGFP and EGFP carrying control clones began their emission at about 60 min, but did not reach the maximum level of emission comparable to TtsfGFP carrying clones until 240 min of incubation (Figure 3). The *T. thermophila* codon adaptation of sfGFP appeared to have increased the rate of protein synthesis rate and the overall expression level.

Fluorescent emission spectral counting analyses showed that *T. thermophila* clones carrying TtsfGFP had ~2.2 fold and ~4 fold higher emission compared with clones expressing sfGFP and EGFP, respectively, at 180 min (Figure 4). However, the TtsfGFP-carrying clone showed much earlier acceleration of emission (beginning from 30 min) than the others. The distinguishable difference in fluorescence emission among the clones began after 60 min and remained until 180 min.

Increased protein expression of *Tetrahymena thermophila* codon-adapted TtsfGFP tag

The total protein extracted from the induced *T. thermophila* clones at 180 min was analyzed by western blotting. Equal quantities of total protein (30 μg) from each clone were loaded onto SDS-PAGE. Western blotting was performed with a mouse monoclonal anti-GFP antibody as the primary antibody. TtsfGFP, sfGFP, and EGFP were observed as ~34 kDa proteins (theoretically molecular mass, 29 kDa, Figure 5, Additional file 1). The quantity of recombinant protein was the highest in the TtsfGFP clone compared with sfGFP and EGFP. These results demonstrated that using *T. thermophila* codon adaptation in the sfGFP gene improved translation efficiently.

```
      M  S  K  G  E  E  L  F  T  G  V  V  P  I  L  V  E  L  D  G  20
  1 atgtctaaaggtgaagaattattcactggtgttgttcctatttagttgaattagacggt
    atgagcaaaggagaagaacttttcactggagttgtcccaattcttgttgaattagatggt
      D  V  N  G  H  K  F  S  V  R  G  E  G  E  G  D  A  T  N  G  40
 61 gacgttaacggtcacaaattctctgttagaggtgaaggtgaaggtgacgctactaacggt
    gatgttaatgggcacaaattttctgtccgtggagagggtgaaggtgatgctacaaacgga
      K  L  T  L  K  F  I  C  T  T  G  K  L  P  V  P  W  P  T  L  60
121 aaattaactttaaaattcatttgcactactggtaaattacctgttccttggcctacttta
    aaactcacccttaaatttatttgcactactggaaaactacctgttccatggccaacactt
      V  T  T  L  T  Y  G  V  Q  C  F  S  R  Y  P  D  H  M  K  R  80
181 gttactacttaacttacggtgttcaatgcttctctagataccctgaccacatgaaaaga
    gtcactactctgacctatggtgttcaatgcttttcccgttatccggatcacatgaaacgg
      H  D  F  F  K  S  A  M  P  E  G  Y  V  Q  E  R  T  I  S  F  100
241 cacgacttcttcaaatctgctatgcctgaaggttacgttcaagaaagaactatttctttc
    catgacttttcaagagtgccatgcccgaaggttatgtacaggaacgcactatatctttc
      K  D  D  G  T  Y  K  T  R  A  E  V  K  F  E  G  D  T  L  V  120
301 aaagacgacggtacttacaaaactagagctgaagttaaattcgaaggtgacactttagtt
    aaagatgacgggacctacaagacgcgtgctgaagtcaagtttgaaggtgatacccttgtt
      N  R  I  E  L  K  G  I  D  F  K  E  D  G  N  I  L  G  H  K  140
361 aacagaattgaattaaaaggtattgacttcaaagaagacggtaacatttaggtcacaaa
    aatcgtatcgagttaaaaggtattgattttaaagaagatggaaacattc tcggacacaaa
      L  E  Y  N  F  N  S  H  N  V  Y  I  T  A  D  K  Q  K  N  G  160
421 ttagaatacaacttcaactctcacaacgttacattactgctgacaaacaaaaaaacggt
    ctcgagtacaactttaactcacacaatgtatacat cacggcagacaaacaaaagaatgga
      I  K  A  N  F  K  I  R  H  N  V  E  D  G  S  V  Q  L  A  D  180
481 attaaagctaacttcaaaattagacacaacgttgaagacggttctgttcaattagctgac
    atcaaagctaacttcaaaattcgccacaacgttgaagatggttccgttcaactagcagac
      H  Y  Q  Q  N  T  P  I  G  D  G  P  V  L  L  P  D  N  H  Y  200
541 cactaccaacaaaacactcctattggtgacggtcctgtttattacctgacaaccactac
    cattatcaacaaaatactccaattggcgatggccctgtccttttaccagacaaccattac
      L  S  T  Q  S  V  L  S  K  D  P  N  E  K  R  D  H  M  V  L  220
601 ttatctactcaatctgtttatctaaagaccctaacgaaaaaagagaccacatggtttta
    ctgtcgacacaatctgtcctttcgaaagatcccaacgaaaagcgtgaccacatggtcctt
      L  E  F  V  T  A  A  G  I  T  H  G  M  D  E  L  Y  K  *     239
661 ttagaattcgttactgctgctggtattactcacggtatggacgaattatacaaatga
    cttgagtttgtaactgctgctgggattacacatggcatggatgagctctacaaatga
```

Figure 1 Comparison of the DNA sequence of codon-optimized TtsfGFP with that of sfGFP. Top DNA sequence is the codon-adapted TtsfGFP and the bottom sequence is that of sfGFP. The 126 mutated bases are shown in bold letters. There was no change in the amino acid sequence of sfGFP.

Macronuclear and micronuclear localization of a *Tetrahymena thermophila* hypothetical protein as a fusion protein with the TtsfGFP tag in *Tetrahymena thermophila*

A hypothetical gene containing a DNA ligase domain (H) was cloned into the C-terminal end of the 6 × His-TtsfGFP tag in pVTtsfGFP. After the simultaneous induction of *T. thermophila* cells carrying pVTtsfGFP-H and control cells carrying pVTtsfGFP with 0.25 μg/mL of $CdCl_2$ for 1 h, microscopic analyses showed increasing GFP fluorescence in the Hoechst 33258 stained macronucleus and micronucleus of *T. thermophila* carrying pVTtsfGFP-H. However, in the control cells, a strong green fluorescence was emitted only from the cytoplasm, even after 1 h (Figure 6). Therefore, the nuclear

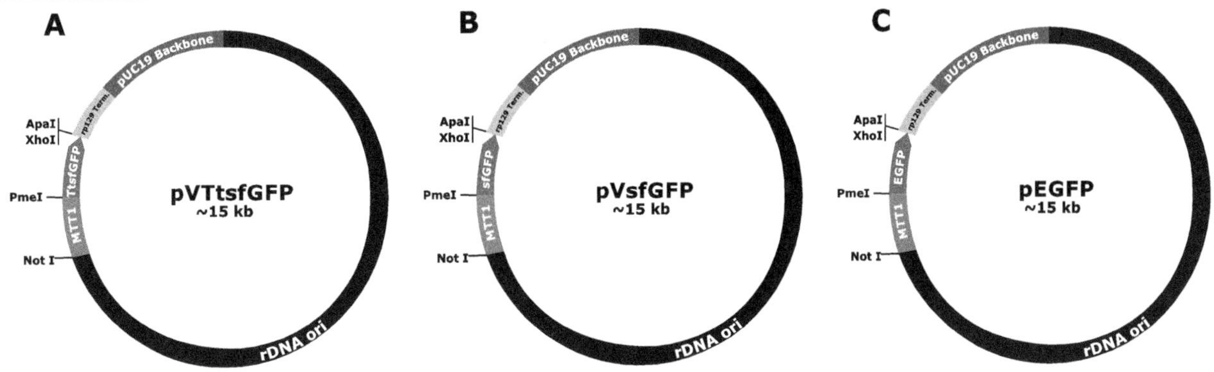

Figure 2 Constructed protein expression vectors of *T. thermophila*. A. Vector pVTtsfGFP includes the *T. thermophila* codon-adapted TtsfGFP expression cassette; **B.** Vector pVsfGFP carries a non-codon-adapted sfGFP expression cassette; **C.** Vector pEGFP includes the non-codon-adapted EGFP expression cassette. All vectors were derived from pVGF. MTT1 is a cadmium-inducible promotor. The transcription is terminated using an rpL29 termination sequence.

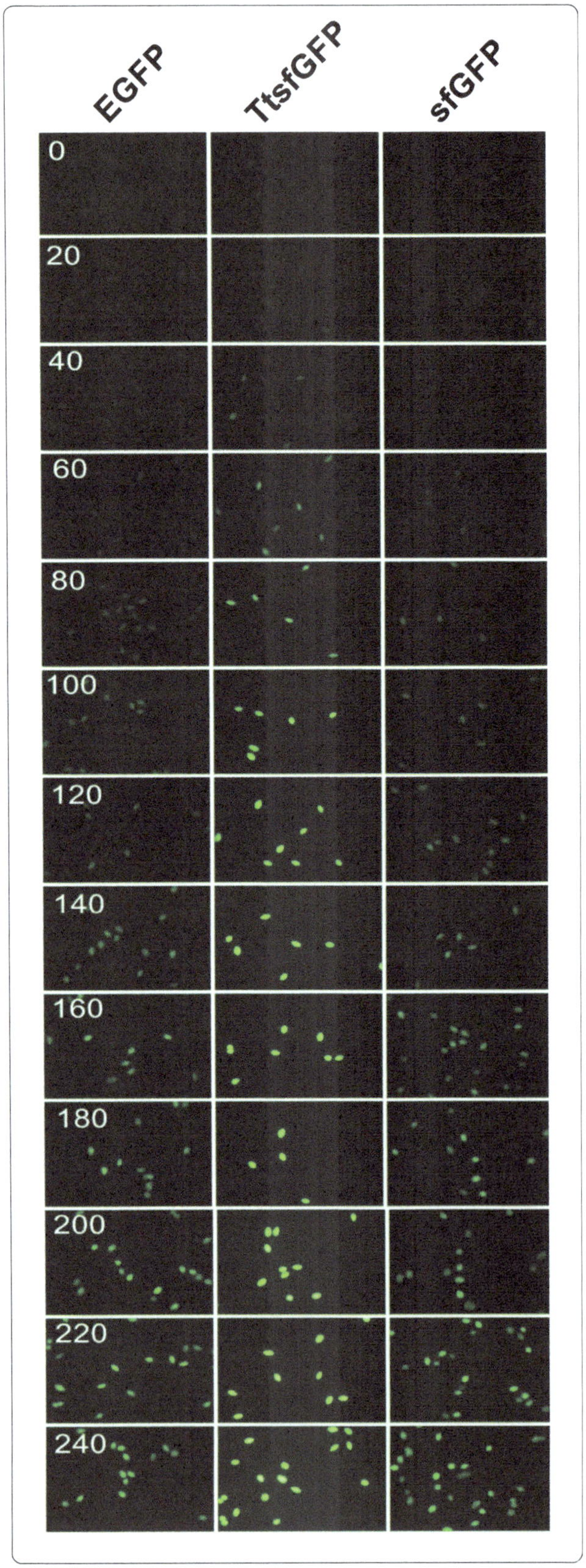

Figure 3 Comparative in vivo fluorescence microscope analyses of EGFP, TtsfGFP, sfGFP expression. The TtsfGFP-expressing cells emitted brighter green fluorescent light than the sfGFP and EGFP expressing clones at all period after induction with 2 μg/mL of $CdCl_2$. A Leica DM6000 B fluorescence microscope equipped with a 20× objective and a GFP filter was used. Photo revision using PhotoScape 3.6.3 software was performed without any misleading changes because of the low quality of the images.

localization signal sequence in the N-terminus of the hypothetical protein containing the DNA ligase domain (H), as predicted the CELLO program (Data not shown) [35], must be functional.

Use of the 6 × His-TtsfGFP tag for affinity purification

T. thermophila clones expressing the 6 × His-TtsfGFP tag were induced for 3 or 18 h with 2 μg/mL of $CdCl_2$. The total soluble protein was extracted and the 6 × His-TtsfGFP was purified using an Ni-NTA affinity column. The ~34 kDa TtsfGFP protein was detected by anti-GFP antibodies upon western blotting of the affinity purified samples (Figure 7). These results suggested that the TtsfGFP could be successfully purified using Ni-NTA beads, although in a reduced amount (Figure 8).

In the *T. thermophila* cell line expressing the hypothetical gene encoding the ATP-dependent DNA ligase domain as a 6 × His-TtsfGFP-H fusion protein, the fusion

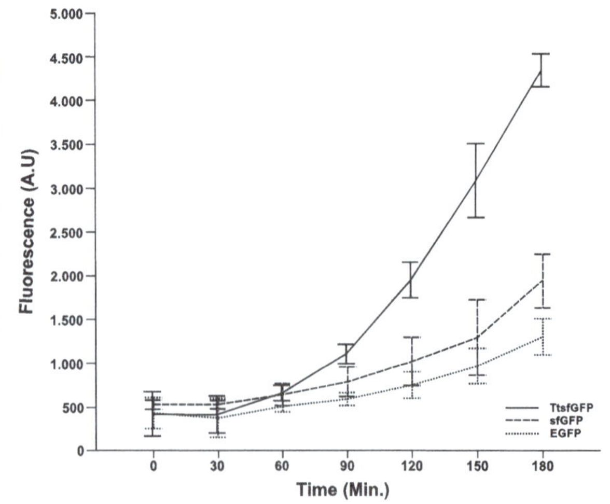

Figure 4 Comparative in vivo fluorescent spectrophotometric analyses of cells expressing EGFP, TtsfGFP, and sfGFP. Tetrahymena clones carrying TtsfGFP-, sfGFP-, and EGFP-expressing constructs were grown until the mid-logarithmic phase and cell density adjusted to 3 × 105 cells/mL. The cells were induced with 2 μg/mL of CdCl2. In vivo fluorescence spectrophotometric analyses were performed every 30 min at 488 nm excitation and 510 nm emission. The difference in emission between TtsfGFP, sfGFP, and EGFP began at 60 min and continued until 180 min. At the end of the time interval, TtsfGFP clones were found to emit ~2.2 fold and ~4 fold more fluorescence than sfGFP and EGFP, respectively.

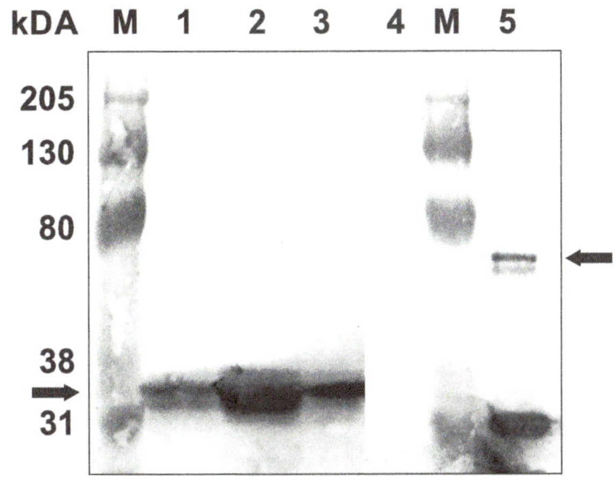

Figure 5 Western blot analysis of total proteins from *T. thermophila* cells expressing TtsfGFP, sfGFP, or EGFP. Equal amounts (30 μg) of EGFP (Lane 1), TtsfGFP (Lane 2), and sfGFP (Lane 3) were loaded. The left arrow shows a molecular mass of around 34 kDa corresponding to TtsfGFP. The quantity of TtsfGFP in the total protein extract appeared to be approximately 5–10-fold higher than sfGFP and EGFP. Total protein extracted from a *Tetrahymena* B2086 and CU428 cell mixture was used as a negative control (Lane 4). The positive control was TtsfGFP, which was constructed, expressed, and purified using Ni-NTA affinity purification from *E. coli*. The ~68 kDa band was predicted to be an sfGFP dimer (right arrow). Western blotting was performed with a monoclonal mouse anti-GFP antibody (1:1000). M: Bio-Rad Kaleidoscope western markers.

protein was localized to the micro- and macronucleus (Figure 6). In western blotting using the Ni-NTA affinity purified lysate, the 6 × His-TtsfGFP-H fusion protein appeared as the predicted ~95 kDa protein, but with many short truncated fusion protein fragments, not only in the total protein (Figure 8-A), but also in the affinity purified samples (Figure 8-B). However, western blotting indicated that some target protein was lost during the wash and flow-through steps of affinity purification. These analyses nevertheless indicated that the 6 × His-TtsfGFP tag is suitable for the affinity purification of a fusion protein, acting as a tracking marker in column purification after the determination of its protein localization.

Discussion

There have been attempts to develop better protein tags for recombinant protein production and functional studies in *T. thermophila*. For example, a *T. thermophila* codon-adapted version of EGFP was produced as a C-terminal epitope tag [28], but it was not evaluated for its translation and fluorescence efficiency [6]. Additionally, a potential dual affinity tag based on the *T. thermophila* glutathione S-transferase zeta and a 6 × His tag was reported [36]. In the present study, we developed an N-terminal affinity epitope tag using a more advanced GFP mutant, known as sfGFP, after codon adaption to *T. thermophila*. Cells carrying this new construct could be used not only for protein localization by fluorescent microscopy, but also for simultaneous affinity purification, and for protein function and protein-protein interaction studies. Indeed, we demonstrated that the *T. thermophila* codon preferences used in the 6 × His-TtsfGFP tag clearly improved its translational efficiency, resulting in an increase in the emitted fluorescence compared with the control groups (sfGFP and EGFP) (Figure 5). Additionally, TtsfGFP appears to possess superior properties compared with the controls, such as earlier and brighter fluorescence emission following induction with $CdCl_2$. Other researchers have reported similar findings; i.e., that codon adaptation is necessary for efficient translation of heterologous proteins in *T. thermophila*. For example, the codon adaptation of human DNase-I and neomycin resistance gene (Neo4) improved their translation in *T. thermophila* [37,38]. These findings support the conclusion that codon adaptation increases the protein synthesis rate and the overall expression level of recombinant proteins in *T. thermophila* [23,24,39].

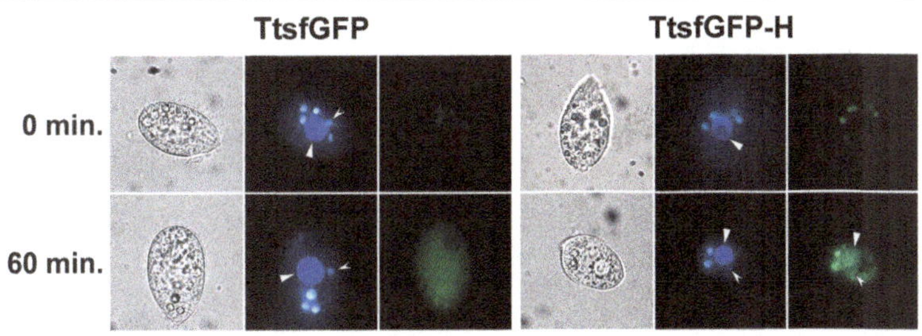

Figure 6 The macronuclear and micronuclear localization of TtsfGFP-H fusion protein in *T. thermophila*. pVTtsfGFP carrying *T. thermophila* cells as the control and *T. thermophila* carrying pVTtsfGFP-H cells were induced with 0.25 μg/ml of $CdCl_2$ for one hour. 6 × His-TtsfGFP-H was localized to the macro- (arrowed large structure) and micronucleus (arrowed small structure). However, there was no detectable localization of the 6 × His-TtsfGFP tag protein except in the cytoplasm. The images were taken by a Leica DM6000 B fluorescence microscope equipped with a 63× objective. GFP filter was used for TtsfGFP (last column) and A filter for DAPI/Hoechst 33258 staining (middle column). The first picture columns were taken with light microscopy. The 2–3 vesicles that are seen in the cytoplasm are unknown structures.

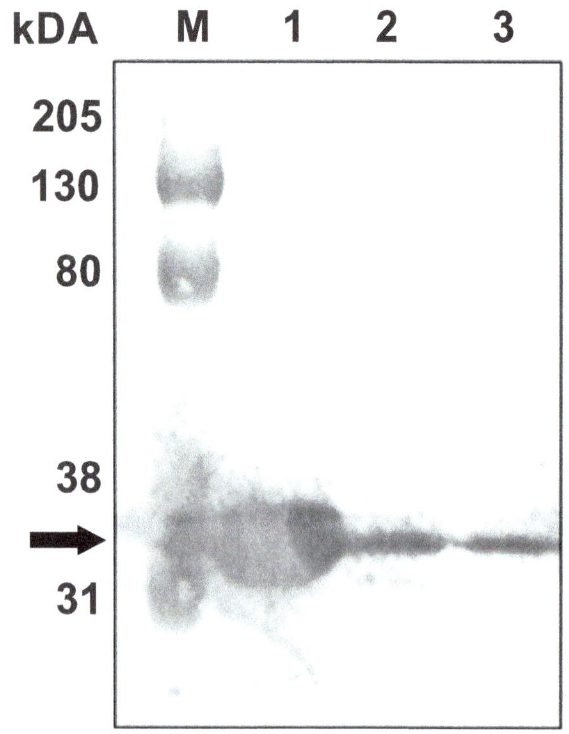

Figure 7 Western blot analysis of 6 × His-TtsfGFP purified with by a Ni-NTA column. The purified 6 × His-TtsfGFP isolated from cells with 2 μg/mL of $CdCl_2$ for 18 h. From the Ni-NTA affinity purification columns, the recombinant 6 × His-TtsfGFP was observed as ~34 kDa protein in the first elution (Lane 1, arrow), the flow-through (Lane 2), and the first wash (Lane 3) determined using a monoclonal mouse anti-GFP antibody as the primary antibody(1:1000). M: Bio-Rad Kaleidoscope western markers.

The utility of GFPs as protein localization tags may be severely compromised because of the absence of an affinity feature for protein purification, except for their immune precipitation with specific anti-GFP antibodies [28]. This deficiency can be overcome by the addition of a 6 × His-like affinity tags at appropriate positions in GFPs as N- or C-terminal poly-His affinity tag. Although this strategy has already been applied to GFPs used in *E. coli* and human studies [16-18,40], there has been no report on their use in *T. thermophila*. In this study, an N-terminal 6 × His tag, as a representative peptide affinity tag, was added to the codon-adapted TtsfGFP tag. This affinity feature of the 6 × His-TtsfGFP and 6 × His-TtsfGFP-H products was confirmed using Ni-NTA based affinity purification and western blotting. It is also possible that other peptide affinity tags, such as the Strep-tag II (WSHPQFEK) or the Calmodulin-tag (KRRWKKNFIA-VSAANRFKKISSSGAL) could replace the 6 × His tag to further improve the affinity purification of TtsfGFP [12]. Moreover, microscopic analyses showed that the localization of 6 × His-TtsfGFP is clearly intracellular, whereas the 6 × His-TtsfGFP-H fusion protein was specifically localized in the macronucleus and micronucleus in *T. thermophila* (Figure 6), although there are number of short fusion protein products (Figure 8). From these data, one could easily propose a hypothesis that the hypothetical gene containing DNA ligase domain (H) must have a nuclear localization signal sequence in its N-terminal region, as predicted by the CELLO program [35] when analyzing the N-terminal

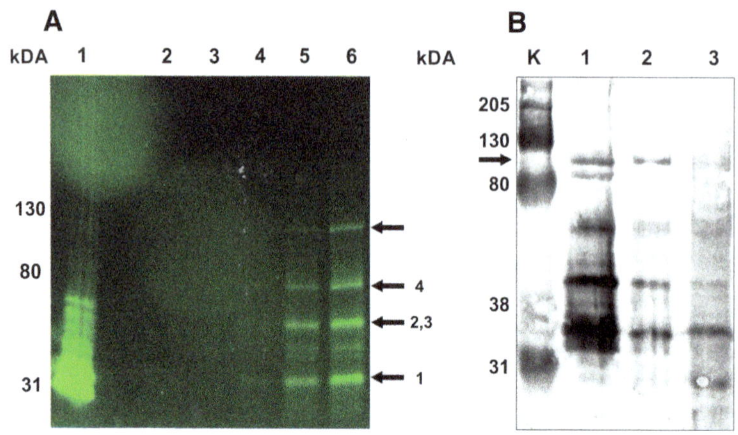

Figure 8 Fluorescent and western blot analysis of the hypothetical ATP dependent DNA ligase domain containing protein. A. SDS-PAGE was carried out in a discontinuous slab gel under semi-denaturing conditions by omitting mercaptoethanol from the sample buffer and without boiling. About 15 μg of total protein were loaded in each lane. Gels were visualized with Bio-Rad Gel Doc EZ using the blue sample tray for GFP. Lane 1: affinity purified 6 × His-TtsfGFP from *Escherícha coli*; lane 2: untransformed *T. thermophila* total cell protein (negative control); *T. thermophila* with pVTtsfGFP-H induced with 0.25 μg/mL of $CdCl_2$ for 3 h; lane 3: zero time, lane 4: 1 h, lane 5: 2 h, lane 6: 3 h. **B.** The 6 × His-TtsfGFP-H fusion protein purified from *T. thermophila* pVTtsfGFP-H clone was induced for 18 h and analyzed with western blotting by using monoclonal mouse anti-GFP antibody (1:1000). The 6 × His-TtsfGFP-H was approximately 95 kDa (Lane 1, black arrow), as expected. Many fragmented proteins were also visible. Moreover, some of the target and fragmented fusion proteins were lost during washing (Lane 2) and flow-through (Lane 3) steps of Ni-NTA affinity purification. The predicted size of the fragments based on the rare codons plus 6 × His-TtsfGFP would be approximately 36.7 kDa (Arrow 1), 48 kDa (Arrow 2) and 50 kDa (Arrow 3). The roughly 70 kDa band could be dimer of these broken fusion proteins caused by the dimerization of sfGFP (Arrow 4). M: Bio-Rad Kaleidoscope western marker.

10, 24 and 143 amino acids or the full length amino acid sequence of H protein (Data not shown). These data together proved that localization and affinity purification of a target protein could be performed using the same protein expression construct, which would reduce costs and save time.

The fusion protein fragments in Figure 8 most likely resulted from incomplete translation and/or degradation of the protein. However, degradation of proteins is unlikely during short (1–3 h) and long (18 h) induction time because the size of the fragments were unchanged in both conditions (Figure 8). In addition, these fragments could not be produced by protein degradation because of the very short duration time on ice during the protein isolation. Moreover, Wu et al. reported that the presence of sfGFP helped the poorly folding TEV protease to fold properly and increased the yield of the TEV protease to ~22% [15]. Therefore, the presence of sfGFP in fusion proteins should reduce the amount of broken fragments because of the reduced folding stress. Conversely, the relatively low yield of Ni-NTA affinity purified 6 × His-TtsfGFP-H might be explained in two ways. First, the partially translated or degraded recombinant protein fragments could bind competitively to the Ni-NTA beads, thereby reducing the binding efficiency of the full-length 6 × His-TtsfGFP-H fusion proteins. Consequently, most of the fusion protein could have been lost during the washing steps as flow through (Figure 8-B; lane 2 and 3). Second, the absence of codon optimization of the homologous H gene may cause multiple pauses in translation, producing broken fusion proteins because of the presence of rare codons (lower than 2–3% in all genes), producing a 2.7 kDa peptide from CTG_{24} for Leu, a 14 kDa peptide from GCG_{127} for Ala and a 16 kDa peptide from CTG_{144} for Leu [22,23,33]. The size of the predicted broken fusion proteins based on the rare codons plus 6 × His-TtsfGFP could be about 36.7 kDa, 48 kDa and 50 kDa, which correspond to the sizes observed in Figure 8-A and -B. The approximately 70-kDa band could be dimer of these broken fusion proteins caused by dimerization of sfGFP (Figure 5 and Figure 8). Strong transcription of 6 × His-TtsfGFP-H fusion gene under the MTT1 promoter with an endogenously expressed H gene may lead rapid depletion of CTG and GCG tRNAs, causing ribosome stalling and finally inhibition of translation to produce these fragments [22,23]. Therefore, the codon adaptation of homologous genes, in our case the H gene, could be required to increase the recovery of the intact 95 kDa fusion protein for this type of fusion protein production. However, if the problem of contaminating protein fragments persists after codon optimization of the H-protein, the positioning of the affinity tag at the C-terminus of TtsfGFP-6 × His should be considered. The advantages such as a faster-folding, brighter emission and facilitator of the fusion protein folding and the codon adaptation of the sfGFP tag make TtsfGFP an alternative tag for protein localization and affinity purification in *T. thermophila* studies.

Conclusions

In this study, we showed that the *T. thermophila* codon-adapted sfGFP mutant with an N-terminal 6 × His affinity tag was superior to EGFP and sfGFP, based on the improved translation efficiency, for simultaneous protein localization and affinity purification. Thus, we believe that this dual tag will help to advance *T. thermophila*-based studies by enabling target proteins to be visually traceable under in vivo and in vitro conditions. In addition, the 6 × His-TtsfGFP dual tag will enable proper folding and stability, and will extend the shelf life of target heterologous and homologous proteins expressed in *T. thermophila*.

Methods

Codon adaptation of Superfolder GFP and plasmid construction

Based on the codon use frequency, the absence and/or presence of tRNA gene(s), and gene copy numbers of *T. thermophila* [33], the sfGFP protein coding sequence was adapted by the introduction of silent mutations based on the *T. thermophila* codon dictionary (Figure 1).

Sequences encoding *T. thermophila* codon-adapted sfGFP (TtsfGFP) and non-codon-adapted control sfGFP were synthesized and cloned into a pUC57 vector with the addition of restriction and protease recognizing sites and 6 × Histidine affinity sequence by Shanghai Shine Gene Company (Molecular Bio-Technologies Inc., Shanghai, China).

The pVGF vector used in this study was a *T. thermophila* replicative protein expression vector carrying a paromomycin-resistant mutant rDNA origin and a Drp1 gene [10,34]. The gene was cloned under the control of the MTT1 promoter, which was inducible by $CdCl_2$ [41].

The 6 × His-TtsfGFP and 6 × His-sfGFP coding regions were digested by PmeI-ApaI restriction enzymes from the pUC57 vectors and cloned into the pVGF vector after releasing the EGFP-Drp1 fusion gene. The vectors were named pVTtsfGFP and pVsfGFP, respectively. *T. thermophila* pEGFP, for use as a control, was derived from pVGF by deleting the Drp1 gene using XhoI and ApaI enzymes and re-ligating it with a linker carrying three stop codons (constructed by Küçükoğlu, N. and Arslanyolu, M.). Cell lines expressing sfGFP and EGFP were used as control groups.

A cDNA fragment of a *T. thermophila* hypothetical gene carrying an ATP-dependent DNA ligase domain (named "H" in this study, 1667 bp) [GenBank: XM_001011861.1] was used to test the localization feature of the TtsfGFP tag. cDNA fragments of "H" were digested with XhoI-ApaI restriction enzymes and cloned into the 3′ end of the 6 × His-TtsfGFP gene in the pVTtsfGFP, which was named pVTtsfGFP-H. NEB-10-Beta competent *E. coli*

(C3019l, New England BioLabs Inc, Ipswich, MA, USA) were used in all cloning steps, and all vector constructs were confirmed by DNA sequencing (Quick Start Kit, 608120, Beckman Coulter CEQ8000,Brea, CA, USA). The Thermo GeneJet Plasmid MiniPrep kit (K0502,Thermo,-Waltham, MA, USA) and QIAquick Gel Extraction Kit (28706, Qiagene, Hilden, Germany) were used for plasmid DNA purification from *E. coli* and for extraction from agarose gels, respectively.

Electroporation of *T. thermophila*

Electroporation of conjugant *T. thermophila* strains-CU428 and B2086, (Tetrahymena Stock Center, Cornell University, Cornell, NY, USA) was performed as reported previously, with minor modifications [42,43]. During conjugation, pairing efficiency was monitored until it reached at least 80% (usually ~2–3 h). Cells were pelleted and washed with 10 mM HEPES (pH 7.5) and re-suspended in 10 mM HEPES, such that there were 1.7×10^7 cells/mL during the "macronuclear development stage 1" of conjugation (ca. 9–10 h after the start of conjugation). A mixture of 15–20 μg of plasmid DNA and 230 μL of cell suspension (in 20 mM HEPES pH 7.5) was placed in a Gene Pulser electroporation cuvette (0.4-cm gap), and pulsed with Bio-Rad Gene PulserXcell (440 V, 25-μF, 200 Ω; Hercules, CA, USA). Cells were incubated for 18–24 h for the recovery and execution of conjugation. After the addition of paromomycin at 100 μg/mL, the cells were incubated for 3 more days. This last step was repeated with gradually increasing concentrations of paromomycin (100–800 μg/mL) for 7–10 days. To assess the degree of transformation, 10 mL of cell suspension was induced with 2 μg/mL of $CdCl_2$ [44] and observed under a fluorescence microscope to detect the emission of fluorescence (Leica DM6000, GFP HP Filter Cube 11532366; Wetzlar, Germany). The transformed cells were maintained in liquid nitrogen at −80°C until further use.

Fluorescence microscopy and fluorescence analyses

To determine the comparative fluorescent emission times, TtsfGFP, sfGFP, and EGFP carrying transformed *T. thermophila* clones were grown until the mid-logarithmic phase in a PPY medium containing 100 μg/mL paromomycin at 30°C with 120 rpm agitation. After induction with 2 μg/mL of $CdCl_2$, the clones were incubated in an orbital shaker at room temperature with 50-rpm agitation. Samples were drawn every 20 min, starting at time zero, and fixed with 0.5 μL of 20% formamide per 500-μL cell sample. The fixed clones were immediately photographed under a Leica DM6000 fluorescence microscope (GFP HP Filter Cube 11532366, 20× objective).

For protein localization studies, *T. thermophila* clones carrying the 6 × His-TtsfGFP-H fusion protein plasmid were grown in PPY medium (containing 100 μg/mL paromomycin) until the mid-logarithmic phase at 30°C with 120 rpm agitation. The cells were induced with 0.25 μg/ml of $CdCl_2$ for low MTT1 promoter transcription [44]. The cells were fixed with 20% formamide for 1 h after induction and photographed using a Leica DM6000 fluorescence microscope (GFP HP Filter Cube 11532366, 40× objective). For fluorescence analyses, clones carrying plasmids expressing TtsfGFP, sfGFP, and EGFP were grown until the mid-logarithmic phase. The concentration of the cells was adjusted to 3×10^5 cells/mL before induction. All the clones were induced with 2 μg/mL of $CdCl_2$ and incubated at room temperature in an orbital shaker with 50-rpm agitation. After induction, 200 μL of samples were taken every 30 min, starting at time zero. Each sample was placed in a 96-well microtiter plate (Falcon 353915) and the fluorescence recorded at 488-nm excitation-510 nm emission in a Molecular Device Spektramax M2 [7,12]. The emission profiles of the *T. thermophila* clones were collected every 30 min for 180 min and the analysis was performed for three independent experiments.

Additional file

Additional file 1: Expression of Recombinant TtsfGFP in *Escherichia coli*. In this study, codon adaptation of TtsfGFP was performed not only for *T. thermophila*, but also for *E. coli*. Here, TtsfGFP and sfGFP (control) were cloned into the NdeI and BamHI sites of pET 16b. Recombinant 6 × His-TtsfGFP and 6 × His-sfGFP were expressed in *E. coli* BL21 (DE3). The **Figure S1.** shows Coomassie staining of the SDS-PAGE gel (Left gel): M. BioRad Standard protein marker, 1. Total protein of *E. coli* expressing 6 × His-TtsfGFP, 2. Affinity purified 6 × His-TtsfGFP, 3. Total protein of *E. coli* expressing 6 × His-sfGFP, 4. Affinity purified 6 × His-sfGFP. Ni-NTA affinity purification of 6 × His-TtsfGFP and 6 × His-sfGFP is shown in the right panel. 1E: first elution, 2E: second elution, and W: washing. These results showed that there is no difference in terms of recombinant protein production of TtsfGFP and sfGFP in *E. coli* BL21 (DE3).

Abbreviations

PVR: Polymerase chain reaction; Bp: Base pairs; GFP: Green fluorescent protein; sfGFP: Superfolder GFP; TtsfGFP: *T. thermophila* sfGFP; EGFP: Enhanced GFP; H: *T. thermophila* ATP-dependent DNA ligase domain containing hypothetical gene; WT-GFP: Wild-type GFP; Neo4: Neomycin resistance gene 4.

Competing interests

The authors declare that they have no competing interests.

Authors' contributions

GY participated in the experimental design, performed all the experiments, and drafted the manuscript. MA designed the experiment, evaluated the data, and revised the manuscript. Both authors read and approved the final manuscript.

Acknowledgments

We first thank Aaron Turkewitz for providing vector pVGF. We owe special thanks to Nurçin Küçükoğlu for her help in the localization experiments and Serkan Dereli for his help in the digital preparation of the figures. This study was supported by the Anadolu University Scientific Research Projects Commission (AUBAP) under Grant No. 1109F147 and TUBITAK (The Scientific and Research Council of Turkey) under Grant Nos. 113Z036 and 112T422 offered to Muhittin Arslanyolu.

Author details
[1]Graduate School of Science, Department of Advance Technologies, Biotechnology Program, Anadolu University, Yunusemre Campus, Eskisehir, Turkey. [2]Department of Biology, Faculty of Science, Anadolu University, Yunusemre Campus, Eskisehir, Turkey.

References

1. Orias E, Hamilton EP, Orias JD. *Tetrahymena* as a laboratory organism: useful strains, cell culture, and cell line maintenance. Methods Cell Biol. 2000;62:189–211.
2. Nanney DL, Simon EM. Laboratory and evolutionary history of *Tetrahymena thermophila*. Methods Cell Biol. 2000;62:3–25.
3. Heim R, Cubitt AB, Tsien RY. Improved green fluorescence. Nature. 1995;373:663–4.
4. Larsen LK, Andreasen PH, Dreisig H, Palm L, Nielsen H, Engberg J, et al. Cloning and characterization of the gene encoding the highly expressed ribosomal protein l3 of the ciliated protozoan *Tetrahymena thermophila*. Evidence for differential codon usage in highly expressed genes. Cell Biol Int. 1999;23:551–60.
5. Wuitschick JD, Karrer KK. Codon usage in *Tetrahymena thermophila*. Method Cell Biol. 2000;62:565–8.
6. Kataoka K, Schoeberl UE, Mochizuki K. Modules for C-terminal epitope tagging of *Tetrahymena* genes. J Microbiol Methods. 2010;82:342–6.
7. Tsien RY. The green fluorescent protein. Annu Rev Biochem. 1998;67:509–44.
8. Shimomura O. Discovery of green fluorescent protein. Methods Biochem Anal. 2006;47:1–13.
9. Zimmer M. Green Fluorescent Protein (GFP): applications, structure, and related photophysical behavior. Chem Rev. 2002;102:759–81.
10. Elde NC, Morgan G, Winey M, Sperling L, Turkewitz AP. Elucidation of clathrin-mediated endocytosis in *Tetrahymena* reveals an evolutionarily convergent recruitment of dynamin. PLoS Genet. 2005;1:e52.
11. Hosein RE, Williams SA, Haye K, Gavin RH. Expression of GFP-actin leads to failure of nuclear elongation and cytokinesis in *Tetrahymena thermophila*. J Eukaryot Microbiol. 2007;50:403–8.
12. Pédelacq J-D, Cabantous S, Tran T, Terwilliger TC, Waldo GS. Engineering and characterization of a superfolder green fluorescent protein. Nat Biotechnol. 2006;24:79–88.
13. Andrews BT, Schoenfish AR, Roy M, Waldo G, Jennings P. The rough energy landscape of superfolder GFP is linked to the chromophore. J Mol Biol. 2007;373:476–90.
14. Fisher AC, DeLisa MP. Laboratory evolution of fast-folding green fluorescent protein using secretory pathway quality control. PLoS One. 2008;3:e2351.
15. Wu X, Wu D, Lu Z, Chen W, Hu X, Ding Y. A novel method for high-level production of TEV protease by superfolder GFP tag. J Biomed Biotechnol. 2009;2009:591923.
16. Liu H, Naismith JH. A simple and efficient expression and purification system using two newly constructed vectors. Protein Expr Purif. 2009;63:102–11.
17. Al-homsi L, Assaad JM, Kweider M, Al-okla S, Abbady QA. Construction of pRSET-sfGFP plasmid for fusion-protein expression. Jordan J Biol Sci. 2012;5:279–88.
18. Lee C-H. Recombinant green fluorescent protein derivatives as a fusion tag for in vitro experiments. Interdiscip Bio Cent. 2009;1:1–15.
19. Drew D, Newstead S, Sonoda Y, Kim H, von Heijne G, Iwata S. GFP-based optimization scheme for the overexpression and purification of eukaryotic membrane proteins in Saccharomyces cerevisiae. Nat Protoc. 2008;3:784–98.
20. Cha HJ, Dalal NG, Pham MQ, Bentley WE. Purification of human interleukin-2 fusion protein produced in insect larvae is facilitated by fusion with green fluorescent protein and metal affinity ligand. Biotechnol Prog. 1999;15:283–6.
21. Scorer C a, Carrier MJ, Rosenberger RF. Amino acid misincorporation during high-level expression of mouse epidermal growth factor in Escherichia coli. Nucleic Acids Res. 1991;19:3511–6.
22. Zahn K. Overexpression of an mRNA dependent on rare codons inhibits protein synthesis and cell growth. J Bacteriol. 1996;178:2926–33.
23. Kane JF. Effects of rare codon clusters on high-level expression of heterologous proteins in Escherichia coli. Curr Opin Biotechnol. 1995;6:494–500.
24. Baca AM, Hol WG. Overcoming codon bias: a method for high-level overexpression of Plasmodium and other AT-rich parasite genes in Escherichia coli. Int J Parasitol. 2000;30:113–8.
25. Kim R, Sandler SJ, Goldman S, Yokota H, Clark J, Kim S. Overexpression of archaeal proteins in Escherichia coli. 1998, 20:207–210.
26. Malone CD, Falkowska K a, Li AY, Galanti SE, Kanuru RC, LaMont EG, et al. Nucleus-specific importin alpha proteins and nucleoporins regulate protein import and nuclear division in the binucleate *Tetrahymena thermophila*. Eukaryot Cell. 2008;7:1487–99.
27. Cui B, Liu Y, Gorovsky MA. Deposition and function of histone H3 variants in *Tetrahymena thermophila*. Mol Cell Biol. 2006;26:7719–30.
28. Howard-Till R, Lukaszewicz A, Novatchkova M, Loidl J. A single cohesin complex performs mitotic and meiotic functions in the protist Tetrahymena. PLoS Genet. 2013;9:e1003418.
29. Busch CJ-L, Vogt A, Mochizuki K. Establishment of a Cre/loxP recombination system for N-terminal epitope tagging of genes in Tetrahymena. BMC Microbiol. 2010;10:191.
30. Kilburn CL, Pearson CG, Romijn EP, Meehl JB, Giddings TH, Culver BP, et al. New *Tetrahymena* basal body protein components identify basal body domain structure. J Cell Biol. 2007;178:905–12.
31. Morise H, Shimomura O, Johnson FH, Winant J. Intermolecular energy transfer in the bioluminescent system of Aequorea. Biochemistry. 1974;13:2656–62.
32. Cormack BP, Valdivia RH, Falkow S. FACS-optimized mutants of the green fluorescent protein (GFP). Gene. 1996;173(1 Spec No):33–8.
33. Salim HMW, Ring KL, Cavalcanti ARO. Patterns of codon usage in two ciliates that reassign the genetic code: *Tetrahymena thermophila* and Paramecium tetraurelia. Protist. 2008;159:283–98.
34. Gaertig J, Gu L, Hai B, Gorovsky MA. High frequency vector-mediated transformation and gene replacement in Tetrahymena. Nucleic Acids Res. 1994;22:5391–8.
35. Yu CS, Chen YC, Lu CH, Hwang JK. Prediction of protein subcellular localization. Proteins Structure Function Bioinformatics. 2006;64:643–51.
36. Ozic C, Arslanyolu M. Characterization of affinity tag features of recombinant *Tetrahymena thermophila* glutathione-S-transferase zeta for *Tetrahymena* protein expression vectors. Turk J Biol. 2012;36:513–26.
37. Weide T, Herrmann L, Bockau U, Niebur N, Aldag I, Laroy W, et al. Secretion of functional human enzymes by *Tetrahymena thermophila*. BMC Biotechnol. 2006;6:19.
38. Mochizuki K. High efficiency transformation of *Tetrahymena* using a codon-optimized neomycin resistance gene. Gene. 2008;425:79–83.
39. Angov E, Hillier CJ, Kincaid RL, Lyon J. Heterologous protein expression is enhanced by harmonizing the codon usage frequencies of the target gene with those of the expression host. PLoS One. 2008;3:e2189.
40. Cabantous S, Terwilliger TC, Waldo GS. Protein tagging and detection with engineered self-assembling fragments of green fluorescent protein. Nat Biotechnol. 2005;23:102–7.
41. Boldrin F, Santovito G, Gaertig J, Wloga D, Cassidy-Hanley D, Clark TG, et al. Metallothionein gene from *Tetrahymena thermophila* with a copper-inducible-repressible promoter metallothionein gene from *Tetrahymena thermophila* with a copper-inducible-repressible promoter. Eukaryot Cell. 2006;5:422–5.
42. Gaertig J, Gorovsky MA. Efficient mass transformation of *Tetrahymena thermophila* by electroporation of conjugants. Proc Natl Acad Sci U S A. 1992;89:9196–200.
43. Barchetta S, La Terza A, Ballarini P, Pucciarelli S, Miceli C. Combination of two regulatory elements in the *Tetrahymena thermophila* HSP70-1 gene controls heat shock activation. Eukaryot Cell. 2008;7:379–86.
44. Shang Y, Song X, Bowen J, Corstanje R, Gao Y, Gaertig J, et al. A robust inducible-repressible promoter greatly facilitates gene knockouts, conditional expression, and overexpression of homologous and heterologous genes in *Tetrahymena thermophila*. Proc Natl Acad Sci U S A. 2002;99:3734–9.

Permissions

All chapters in this book were first published in BMCB, by BioMed Central; hereby published with permission under the Creative Commons Attribution License or equivalent. Every chapter published in this book has been scrutinized by our experts. Their significance has been extensively debated. The topics covered herein carry significant findings which will fuel the growth of the discipline. They may even be implemented as practical applications or may be referred to as a beginning point for another development.

The contributors of this book come from diverse backgrounds, making this book a truly international effort. This book will bring forth new frontiers with its revolutionizing research information and detailed analysis of the nascent developments around the world.

We would like to thank all the contributing authors for lending their expertise to make the book truly unique. They have played a crucial role in the development of this book. Without their invaluable contributions this book wouldn't have been possible. They have made vital efforts to compile up to date information on the varied aspects of this subject to make this book a valuable addition to the collection of many professionals and students.

This book was conceptualized with the vision of imparting up-to-date information and advanced data in this field. To ensure the same, a matchless editorial board was set up. Every individual on the board went through rigorous rounds of assessment to prove their worth. After which they invested a large part of their time researching and compiling the most relevant data for our readers.

The editorial board has been involved in producing this book since its inception. They have spent rigorous hours researching and exploring the diverse topics which have resulted in the successful publishing of this book. They have passed on their knowledge of decades through this book. To expedite this challenging task, the publisher supported the team at every step. A small team of assistant editors was also appointed to further simplify the editing procedure and attain best results for the readers.

Apart from the editorial board, the designing team has also invested a significant amount of their time in understanding the subject and creating the most relevant covers. They scrutinized every image to scout for the most suitable representation of the subject and create an appropriate cover for the book.

The publishing team has been an ardent support to the editorial, designing and production team. Their endless efforts to recruit the best for this project, has resulted in the accomplishment of this book. They are a veteran in the field of academics and their pool of knowledge is as vast as their experience in printing. Their expertise and guidance has proved useful at every step. Their uncompromising quality standards have made this book an exceptional effort. Their encouragement from time to time has been an inspiration for everyone.

The publisher and the editorial board hope that this book will prove to be a valuable piece of knowledge for researchers, students, practitioners and scholars across the globe.

List of Contributors

Pablo A Villegas-Silva
Centro de Investigación Científica de Yucatán AC, Calle 43 No. 130 Col. Chuburná de Hidalgo, Mérida, Yucatán 97200, Mexico

Tanit Toledano-Thompson
Centro de Investigación Científica de Yucatán AC, Calle 43 No. 130 Col. Chuburná de Hidalgo, Mérida, Yucatán 97200, Mexico

Blondy B Canto-Canché
Centro de Investigación Científica de Yucatán AC, Calle 43 No. 130 Col. Chuburná de Hidalgo, Mérida, Yucatán 97200, Mexico

Alfonso Larqué-Saavedra
Centro de Investigación Científica de Yucatán AC, Calle 43 No. 130 Col. Chuburná de Hidalgo, Mérida, Yucatán 97200, Mexico

Luis F Barahona-Pérez
Centro de Investigación Científica de Yucatán AC, Calle 43 No. 130 Col. Chuburná de Hidalgo, Mérida, Yucatán 97200, Mexico

Charles J Rosser
MD Anderson Cancer Center Orlando, Orlando, FL, USA
Nonagen BioScience Corp, Jacksonville, FL, USA
University of Hawaii Cancer Center, Honolulu, HI, USA

Yunfeng Dai
Department of Biostatistics, The University of Florida, Gainesville, FL, USA

Makito Miyake
MD Anderson Cancer Center Orlando, Orlando, FL, USA

Ge Zhang
MD Anderson Cancer Center Orlando, Orlando, FL, USA

Steve Goodison
MD Anderson Cancer Center Orlando, Orlando, FL, USA
Nonagen BioScience Corp, Jacksonville, FL, USA
Department of Health Sciences Research, Mayo Clinic, Jacksonville, FL, USA

Reda M El-Shishtawy
Chemistry Department, Faculty of Science, King Abdulaziz University, P.O. Box 80203, Jeddah 21589, Saudi Arabia

Saleh A Mohamed
Biochemistry Department, Faculty of Science, King Abdulaziz University, Jeddah, Kingdom of Saudi Arabia

Abdullah M Asiri
Chemistry Department, Faculty of Science, King Abdulaziz University, P.O. Box 80203, Jeddah 21589, Saudi Arabia
The Center of Excellence for Advanced Materials Research, King Abdulaziz University, Jeddah 21589, Saudi Arabia

Abu-bakr M Gomaa
Biology Department, Faculty of Science, King Abdulaziz University, Jeddah, Kingdom of Saudi Arabia

Ibrahim H Ibrahim
Biochemistry Department, Faculty of Science, King Abdulaziz University, Jeddah, Kingdom of Saudi Arabia

Hasan A Al-Talhi
Biochemistry Department, Faculty of Science, King Abdulaziz University, Jeddah, Kingdom of Saudi Arabia

Marta V Semkiv
Institute of Cell Biology, NAS of Ukraine, Drahomanov Street, 14/16, Lviv 79005, Ukraine

Kostyantyn V Dmytruk
Institute of Cell Biology, NAS of Ukraine, Drahomanov Street, 14/16, Lviv 79005, Ukraine

Charles A Abbas
Archer Daniels Midland Co Research Center, 1001 N Brush College Rd, Decatur, IL 62521, USA

Andriy A Sibirny
Institute of Cell Biology, NAS of Ukraine, Drahomanov Street, 14/16, Lviv 79005, Ukraine
University of Rzeszow, Zelwerowicza 4, Rzeszow 35-601, Poland

Monica Normark
Department of Chemistry, Umeå University, Umeå SE-901 87, Sweden

Sandra Winestrand
Department of Chemistry, Umeå University, Umeå SE-901 87, Sweden

Torbjörn A Lestander
Department of Forest Biomaterials and Technology, Swedish University of Agricultural Sciences, Umeå SE-901 83, Sweden

Leif J Jönsson
Department of Chemistry, Umeå University, Umeå SE-901 87, Sweden

Laura S Marder
Centro de Pesquisas em Biologia Molecular e Funcional (CPBMF), Instituto Nacional de Ciência e Tuberculose (INCT-TB), Pontifícia Universidade Católica do Rio Grande do Sul (PUCRS), Av. Ipiranga 6681, 90619-900 Porto Alegre, Brazil
Programa de Pós-Graduação em Biologia Celular e Molecular, PUCRS, Porto Alegre 90619-900, Brazil

Juleane Lunardi
Programa de Pós-Graduação em Biologia Celular e Molecular, PUCRS, Porto Alegre 90619-900, Brazil
Quatro G Pesquisa & Desenvolvimento, LTDA, Porto Alegre 90619-900, Brazil

Gaby Renard
Quatro G Pesquisa & Desenvolvimento, LTDA, Porto Alegre 90619-900, Brazil

Diana C Rostirolla
Centro de Pesquisas em Biologia Molecular e Funcional (CPBMF), Instituto Nacional de Ciência e Tuberculose (INCT-TB), Pontifícia Universidade Católica do Rio Grande do Sul (PUCRS), Av. Ipiranga 6681, 90619-900 Porto Alegre, Brazil
Programa de Pós-Graduação em Medicina e Ciências da Saúde, PUCRS, Porto Alegre 90619-900, Brazil

Guilherme O Petersen
Centro de Pesquisas em Biologia Molecular e Funcional (CPBMF), Instituto Nacional de Ciência e Tuberculose (INCT-TB), Pontifícia Universidade Católica do Rio Grande do Sul (PUCRS), Av. Ipiranga 6681, 90619-900 Porto Alegre, Brazil
Programa de Pós-Graduação em Biologia Celular e Molecular, PUCRS, Porto Alegre 90619-900, Brazil
Quatro G Pesquisa & Desenvolvimento, LTDA, Porto Alegre 90619-900, Brazil

José E S Nunes
Quatro G Pesquisa & Desenvolvimento, LTDA, Porto Alegre 90619-900, Brazil

Ana Paula D de Souza
Instituto de Pesquisas Biomédicas (IPB), Laboratório de Imunologia Molecular, PUCRS, Porto Alegre 90619-900, Brazil

Ana Christina de O Dias
Quatro G Pesquisa & Desenvolvimento, LTDA, Porto Alegre 90619-900, Brazil

Jocelei M Chies
Quatro G Pesquisa & Desenvolvimento, LTDA, Porto Alegre 90619-900, Brazil

Luiz A Basso
Centro de Pesquisas em Biologia Molecular e Funcional (CPBMF), Instituto Nacional de Ciência e Tuberculose (INCT-TB), Pontifícia Universidade Católica do Rio Grande do Sul (PUCRS), Av. Ipiranga 6681, 90619-900 Porto Alegre, Brazil
Programa de Pós-Graduação em Biologia Celular e Molecular, PUCRS, Porto Alegre 90619-900, Brazil
Programa de Pós-Graduação em Medicina e Ciências da Saúde, PUCRS, Porto Alegre 90619-900, Brazil

Diógenes S Santos
Centro de Pesquisas em Biologia Molecular e Funcional (CPBMF), Instituto Nacional de Ciência e Tuberculose (INCT-TB), Pontifícia Universidade Católica do Rio Grande do Sul (PUCRS), Av. Ipiranga 6681, 90619-900 Porto Alegre, Brazil
Programa de Pós-Graduação em Biologia Celular e Molecular, PUCRS, Porto Alegre 90619-900, Brazil
Programa de Pós-Graduação em Medicina e Ciências da Saúde, PUCRS, Porto Alegre 90619-900, Brazil

Cristiano V Bizarro
Centro de Pesquisas em Biologia Molecular e Funcional (CPBMF), Instituto Nacional de Ciência e Tuberculose (INCT-TB), Pontifícia Universidade Católica do Rio Grande do Sul (PUCRS), Av. Ipiranga 6681, 90619-900 Porto Alegre, Brazil
Programa de Pós-Graduação em Biologia Celular e Molecular, PUCRS, Porto Alegre 90619-900, Brazil

Raffaele Altara
Department of Pharmacology, Cardiovascular Research Institute Maastricht (CARIM), Maastricht University, Maastricht, The Netherlands

Marco Manca
Department of Pathology, Cardiovascular Research Institute Maastricht (CARIM), Maastricht University, Maastricht, The Netherlands

Marleen HM Hessel
Department of Pharmacology, Cardiovascular Research Institute Maastricht (CARIM), Maastricht University, Maastricht, The Netherlands

Ben J Janssen
Department of Pharmacology, Cardiovascular Research Institute Maastricht (CARIM), Maastricht University, Maastricht, The Netherlands

Harry H A Struijker-Boudier
Department of Pharmacology, Cardiovascular Research Institute Maastricht (CARIM), Maastricht University, Maastricht, The Netherlands

Rob JJ Hermans
Department of Pharmacology, Cardiovascular Research Institute Maastricht (CARIM), Maastricht University, Maastricht, The Netherlands

W Matthijs Blankesteijn
Department of Pharmacology, Cardiovascular Research Institute Maastricht (CARIM), Maastricht University, Maastricht, The Netherlands

Wesley Smith
VTT Technical Research Centre of Finland, Espoo FI-02044 VTT, Finland

Jussi Jäntti
VTT Technical Research Centre of Finland, Espoo FI-02044 VTT, Finland

Merja Oja
VTT Technical Research Centre of Finland, Espoo FI-02044 VTT, Finland

Markku Saloheimo
VTT Technical Research Centre of Finland, Espoo FI-02044 VTT, Finland

Tian Shen
Justus Liebig University Giessen, Institute of Food Chemistry and Food Biotechnology, Heinrich-Buff-Ring 58, Giessen 35392, Germany

Lena M Hof
Justus Liebig University Giessen, Institute of Food Chemistry and Food Biotechnology, Heinrich-Buff-Ring 58, Giessen 35392, Germany

Heike Hausmann
Justus Liebig University Giessen, Institute of Organic Chemistry, Heinrich-Buff-Ring 58, Giessen 35392, Germany

Marc Stadler
Department Microbial Drugs, Helmholtz Centre for Infection Research, Inhoffenstrasse 7, Braunschweig 38124, Germany

Holger Zorn
Justus Liebig University Giessen, Institute of Food Chemistry and Food Biotechnology, Heinrich-Buff-Ring 58, Giessen 35392, Germany

Diana Sanchez-Herrera
Renewable Energy Unit, Centro de Investigación Científica de Yucatán A.C (CICY), Calle 43 No. 130 Col. Chuburná de Hidalgo, C.P. 97200 Mérida, Yucatán, México

Daniella Pacheco-Catalan
Renewable Energy Unit, Centro de Investigación Científica de Yucatán A.C (CICY), Calle 43 No. 130 Col. Chuburná de Hidalgo, C.P. 97200 Mérida, Yucatán, México

Ruby Valdez-Ojeda
Renewable Energy Unit, Centro de Investigación Científica de Yucatán A.C (CICY), Calle 43 No. 130 Col. Chuburná de Hidalgo, C.P. 97200 Mérida, Yucatán, México

Blondy Canto-Canche
Biotechnology Unit, CICY A.C, Mérida, Yucatán, México

Xochitl Dominguez-Benetton
Separation and Conversion Technology, VITO–Flemish Institute for Technological Research, Boeretang 200, Mol 2400, Belgium

Jorge Domínguez-Maldonado
Renewable Energy Unit, Centro de Investigación Científica de Yucatán A.C (CICY), Calle 43 No. 130 Col. Chuburná de Hidalgo, C.P. 97200 Mérida, Yucatán, México

Liliana Alzate-Gaviria
Renewable Energy Unit, Centro de Investigación Científica de Yucatán A.C (CICY), Calle 43 No. 130 Col. Chuburná de Hidalgo, C.P. 97200 Mérida, Yucatán, México

Laura von Schantz
Department of Immunotechnology, Lund University, Lund, Sweden
Alligator Bioscience AB, Lund, Sweden

Herje Schagerlöf
Department of Biochemistry & Structural Biology, Lund University, Lund, Sweden
Camurus AB, Lund, Sweden

Eva Nordberg Karlsson
Department of Biotechnology, Lund University, Lund, Sweden

Mats Ohlin
Alligator Bioscience AB, Lund, Sweden

Joshua L Fuqua
Owensboro Cancer Research Program, 1020 Breckenridge St., Suite 201, Owensboro, KY 42303, USA
University of Louisville School of Medicine, James Graham Brown Cancer Center, 529 S Jackson Street, Louisville, KY 40202, USA

Valentine Wanga
Institute for Health Metrics and Evaluation, University of Washington, 2301 5th Ave, Suite 600, Seattle, WA, USA

Kenneth E Palmer
Owensboro Cancer Research Program, 1020 Breckenridge St., Suite 201, Owensboro, KY 42303, USA
University of Louisville School of Medicine, James Graham Brown Cancer Center, 529 S Jackson Street, Louisville, KY 40202, USA

Martin Ploss
Institute of Technical Biochemistry, University of Stuttgart, Allmandring 31, 70569 Stuttgart, Germany

Sandra J Facey
Institute of Technical Biochemistry, University of Stuttgart, Allmandring 31, 70569 Stuttgart, Germany

Carina Bruhn
Eduard-Zintl-Institute of Inorganic and Physical Chemistry, Technische Universität Darmstadt, Alarich-Weiss-Str. 12, 64287 Darmstadt, Germany

Limor Zemel
Center of Smart Interfaces, Physics of Surfaces and Institute of Materials Sciences, Technische Universität Darmstadt, Alarich-Weiss-Str. 10, 64287 Darmstadt, Germany

Kathrin Hofmann
Eduard-Zintl-Institute of Inorganic and Physical Chemistry, Technische Universität Darmstadt, Alarich-Weiss-Str. 12, 64287 Darmstadt, Germany

Robert W Stark
Center of Smart Interfaces, Physics of Surfaces and Institute of Materials Sciences, Technische Universität Darmstadt, Alarich-Weiss-Str. 10, 64287 Darmstadt, Germany

Barbara Albert
Eduard-Zintl-Institute of Inorganic and Physical Chemistry, Technische Universität Darmstadt, Alarich-Weiss-Str. 12, 64287 Darmstadt, Germany

Bernhard Hauer
Institute of Technical Biochemistry, University of Stuttgart, Allmandring 31, 70569 Stuttgart, Germany

Emma Johansson
Department of Chemical and Biological Engineering, Chalmers University of Technology, 412 96 Göteborg, Sweden
SP Processum AB, 891 22 Örnsköldsvik, Sweden

Charilaos Xiros
Department of Chemical and Biological Engineering, Chalmers University of Technology, 412 96 Göteborg, Sweden

Christer Larsson
Department of Chemical and Biological Engineering, Chalmers University of Technology, 412 96 Göteborg, Sweden

Wei Jiang
State Key Laboratory of Agricultural Microbiology, College of Life Science and Technology, Huazhong Agricultural University, Wuhan 430070, P. R. China

Lin Chen
State Key Laboratory of Agricultural Microbiology, College of Life Science and Technology, Huazhong Agricultural University, Wuhan 430070, P. R. China

Nan Hu
College of Biotechnology and Pharmaceutical Engineering, Nanjing Tech University, Nanjing 211800, P. R. China

Shaohui Yuan
State Key Laboratory of Agricultural Microbiology, College of Life Science and Technology, Huazhong Agricultural University, Wuhan 430070, P. R. China

Bin Li
State Key Laboratory of Agricultural Microbiology, College of Life Science and Technology, Huazhong Agricultural University, Wuhan 430070, P. R. China

Ziduo Liu
State Key Laboratory of Agricultural Microbiology, College of Life Science and Technology, Huazhong Agricultural University, Wuhan 430070, P. R. China

Agneta A Prasse
Institute of Bioanalytical Chemistry, Faculty of Chemistry and Mineralogy and Centre of Biotechnology and Biomedicine, Universität Leipzig, Deutscher Platz 5, 04103 Leipzig, Germany

Thomas Zauner
Institute of Bioanalytical Chemistry, Faculty of Chemistry and Mineralogy and Centre of Biotechnology and Biomedicine, Universität Leipzig, Deutscher Platz 5, 04103 Leipzig, Germany

Karin Büttner
Institute of Bioanalytical Chemistry, Faculty of Chemistry and Mineralogy and Centre of Biotechnology and Biomedicine, Universität Leipzig, Deutscher Platz 5, 04103 Leipzig, Germany

Ralf Hoffmann
Institute of Bioanalytical Chemistry, Faculty of Chemistry and Mineralogy and Centre of Biotechnology and Biomedicine, Universität Leipzig, Deutscher Platz 5, 04103 Leipzig, Germany

Thole Zuchner
Institute of Bioanalytical Chemistry, Faculty of Chemistry and Mineralogy and Centre of Biotechnology and Biomedicine, Universität Leipzig, Deutscher Platz 5, 04103 Leipzig, Germany
Octapharma Biopharmaceuticals GmbH, Im Neuenheimer Feld 590, 69120 Heidelberg, Germany

Albert J Paul
Institute of Applied Biotechnology (IAB), Biberach University of Applied Sciences, 88400 Biberach, Germany

Karen Schwab
Institute of Applied Biotechnology (IAB), Biberach University of Applied Sciences, 88400 Biberach, Germany

Friedemann Hesse
Institute of Applied Biotechnology (IAB), Biberach University of Applied Sciences, 88400 Biberach, Germany

Arlen Peña-Cardeña
Departamento de Ingeniería Celular y Biocatálisis, Instituto de Biotecnología, Universidad Nacional Autónoma de México, Cuernavaca, Morelos C.P. 62210, México

María Elena Rodríguez-Alegría
Departamento de Ingeniería Celular y Biocatálisis, Instituto de Biotecnología, Universidad Nacional Autónoma de México, Cuernavaca, Morelos C.P. 62210, México

Clarita Olvera
Departamento de Ingeniería Celular y Biocatálisis, Instituto de Biotecnología, Universidad Nacional Autónoma de México, Cuernavaca, Morelos C.P. 62210, México

Agustín López Munguía
Departamento de Ingeniería Celular y Biocatálisis, Instituto de Biotecnología, Universidad Nacional Autónoma de México, Cuernavaca, Morelos C.P. 62210, México

Christin Fischer
Anhalt University of Applied Sciences, Department of Applied Biosciences and Process Engineering, Bernburger Str. 55, 06366 Köthen, Germany

Annett Krause
Anhalt University of Applied Sciences, Department of Applied Biosciences and Process Engineering, Bernburger Str. 55, 06366 Köthen, Germany

Thomas Kleinschmidt
Anhalt University of Applied Sciences, Department of Applied Biosciences and Process Engineering, Bernburger Str. 55, 06366 Köthen, Germany

Silvia Cesarini
Department of Microbiology, University of Barcelona, Av. Diagonal 643, 08028 Barcelona, Spain

Belén Infanzón
Department of Microbiology, University of Barcelona, Av. Diagonal 643, 08028 Barcelona, Spain

F I Javier Pastor
Department of Microbiology, University of Barcelona, Av. Diagonal 643, 08028 Barcelona, Spain

Pilar Diaz
Department of Microbiology, University of Barcelona, Av. Diagonal 643, 08028 Barcelona, Spain

Soo Kyung Lyu
Department of Bioscience and Biotechnology and Protein Research Center for Bio-Industry, Hankuk University of Foreign Studies, Yongin 449-791, Republic of Korea

Hyockman Kwon
Department of Bioscience and Biotechnology and Protein Research Center for Bio-Industry, Hankuk University of Foreign Studies, Yongin 449-791, Republic of Korea

Stefan Bauer
Energy Biosciences Institute, University of California, Berkeley, CA 94720, USA

Ana B. Ibáñez
Energy Biosciences Institute, University of California, Berkeley, CA 94720, USA

Valeska Hauptmann
Leibniz Institute of Plant Genetics and Crop Plant Research, Corrensstrasse 3, 06466 Stadt Seeland, OT Gatersleben, Germany

Matthias Menzel
Fraunhofer Institute for Mechanics of Materials, Walter-Hülse-Strasse 1, 06120 Halle/Saale, Germany

Nicola Weichert
Leibniz Institute of Plant Genetics and Crop Plant Research, Corrensstrasse 3, 06466 Stadt Seeland, OT Gatersleben, Germany

Kerstin Reimers
Department of Plastic, Hand and Reconstructive Surgery, Hannover Medical School, Podbielskistr. 380, 30659 Hannover, Germany

Uwe Spohn
Fraunhofer Institute for Mechanics of Materials, Walter-Hülse-Strasse 1, 06120 Halle/Saale, Germany

Udo Conrad
Leibniz Institute of Plant Genetics and Crop Plant Research, Corrensstrasse 3, 06466 Stadt Seeland, OT Gatersleben, Germany

Karin Buettner
Institute of Bioanalytical Chemistry, Center for Biotechnology and Biomedicine, Universität Leipzig, 04103 Leipzig, Germany

Thomas Kreisig
Institute of Bioanalytical Chemistry, Center for Biotechnology and Biomedicine, Universität Leipzig, 04103 Leipzig, Germany

Norbert Sträter
Institute of Bioanalytical Chemistry, Center for Biotechnology and Biomedicine, Universität Leipzig, 04103 Leipzig, Germany

Thole Zuchner
Institute of Bioanalytical Chemistry, Center for Biotechnology and Biomedicine, Universität Leipzig, 04103 Leipzig, Germany
Octapharma Biopharmaceuticals GmbH, Im Neuenheimer Feld 590, Heidelberg 69120, Germany

Gürkan Yilmaz
Graduate School of Science, Department of Advance Technologies, Biotechnology Program, Anadolu University, Yunusemre Campus, Eskisehir, Turkey

Muhittin Arslanyolu
Department of Biology, Faculty of Science, Anadolu University, Yunusemre Campus, Eskisehir, Turkey